Citrus Genetics, Breeding and Biotechnology

Citrus Genetics, Breeding and Biotechnology

Edited by

Iqrar Ahmad Khan

www.cabi.org

CABI is a trading name of CAB International

CABI Head Office
Nosworthy Way
Wallingford
Oxfordshire OX10 8DE
UK

Tel: +44 (0)1491 832111
Fax: +44 (0)1491 833508
E-mail: cabi@cabi.org
Website: www.cabi.org

CABI North American Office
875 Massachusetts Avenue
7th Floor
Cambridge, MA 02139
USA

Tel: +1 617 395 4056
Fax: +1 617 354 6875
E-mail: cabi-nao@cabi.org

A catalogue record for this book is available from the British Library, London, UK.

A catalogue record for this book is available from the Library of Congress, Washington, DC.

ISBN-13: 978 0 85199 019 4

Typeset by MRM Graphics Ltd, Winslow, Bucks.
Printed and bound in the UK by Biddles Ltd, Kings Lynn.

Contents

Contributors

Castle W. S., University of Florida, Institute of Food and Agricultural Sciences Citrus Research and Education Center, Lake Alfred, FL 33850, USA

Cervera M., Departamento Protección Vegetal y Biotecnología, Instituto Valenciano de Investigaciones Agrarias (IVIA), Apartado Oficial, 46113 Moncada, Valencia, Spain

Chen C., University of Florida, Institute of Food and Agricultural Sciences Citrus Research and Education Center, Lake Alfred, FL 33850 USA

Deng X. X., Citrus Research Institute, Hauzhong Agricultural University, Wuhan 430070, PR China

Deng Z., University of Florida, Institute of Food and Agricultural Sciences, Gulf Coast Research and Education Center, Bradenton, FL 34203, USA

Domínguez A., Departamento Protección Vegetal y Biotecnología, Instituto Valenciano de Investigaciones Agrarias (IVIA), Apartado Oficial, 46113 Moncada, Valencia, Spain

Fagoaga C. J., Departamento Protección Vegetal y Biotecnología, Instituto Valenciano de Investigaciones Agrarias (IVIA), Apartado Oficial, 46113 Moncada, Valencia, Spain

Febres V. J., University of Florida, Horticultural Sciences Department, Plant Molecular and Cellular Biology Program, PO Box 110690, Gainesville, FL 32611, USA

Germanà M. A., Dipartimento SENFIMIZO Sezione di Frutticoltura Mediterranea, Tropicale e Sutropicale, Facolta di Agraria, Università degli Studi di Palermo, Viale delle Scienze, 11, 90128 Palermo, Italy

Ghorbel R., Departamento Protección Vegetal y Biotecnología, Instituto Valenciano de Investigaciones Agrarias (IVIA), Apartado Oficial, 46113 Moncada, Valencia, Spain

Gmitter F. G. Jr, University of Florida, Institute of Food and Agricultural Sciences Citrus Research and Education Center, Lake Alfred, FL 33850, USA

Goodrich R. M., University of Florida, IFAS, Citrus Research and Education Center, 700 Experiment Station Road, Lake Alfred, FL 33850, USA

Grosser J. W., University of Florida, Citrus Research and Education Center (CREC), 700 Experiment Station Road, Lake Alfred, FL 33850, USA

Guo W., National Key Laboratory of Crop Genetic Improvement, Hauzhong Agricultural University, Wuhan 430070, PR China

Jaskani M. J., Institute of Horticultural Sciences, University of Agriculture, Faisalabad 38040, Pakistan

Juárez J., Instituto Valenciano de Investigaciones Agrarias (IVIA), Apartado Oficial, 46113 Moncada, Valencia, Spain

Kender W. J., University of Florida, Citrus Research and Education Center (CREC), 700 Experiment Station Road, Lake Alfred, FL 33850, USA

Kepiro J. L., Department of Botany and Plant Sciences, University of California, Riverside, CA 92521, USA

Khan I. A., Institute of Horticultural Sciences, University of Agriculture, Faisalabad 38040, Pakistan *and* Nuclear Institute for Agriculture and Biology, Faisalabad, Pakistan *and* Sultan Qaboos University, Muscat, Oman

Krueger R. R., USDA-ARS National Clonal Germplasm Respository for Citrus and Dates, 1060 Martin Luther King Blvd, Riverside, CA 92507, USA

Lee R. F., USDA-ARS National Clonal Germplasm Respository for Citrus and Dates, 1060 Martin Luther King Blvd, Riverside, CA 92507, USA

Louzada E. S., Texas A&M University-Kingsville Citrus Center, 312 N, International Blvd, Weslaco, TX 78596, USA

Luro F., SRA, INRA/CIRAD, San Giuliano, 20230 San Nicolao, France

Moore G. A., University of Florida, Horticultural Sciences Department, Plant Molecular and Cellular Biology Program, PO Box 110690, Gainesville, FL 32611 USA

Navarro L., Instituto Valenciano de Investigaciones Agrarias (IVIA), Apartado Oficial, 46113 Moncada, Valencia, Spain

Nicolosi E., Dipartmento di OrtoFloroArboricoltura e Tecnolgie Agroalimentari, University of Catania, Italy

Ollitrault P., Center for International Cooperation in Agricultural Research for Development (CIRAD), Avenue Agropolis TA 71/09, 34398 Montpellier, Cedex 5, France

Omura M., National Agricultural Research Organization (NARO), Okitsu 458–6, Shimizu, Shizuoka, 424–0292, Japan

Peña L., Departamento Protección Vegetal y Biotecnología, Instituto Valenciano de Investigaciones Agrarias (IVIA), Apartado Oficial, 46113 Moncada, Valencia, Spain

Pina A., Departamento Protección Vegetal y Biotecnología, Instituto Valenciano de Investigaciones Agrarias (IVIA), Apartado Oficial, 46113 Moncada, Valencia, Spain

Roose M. L., Department of Botany and Plant Sciences, University of California, Riverside, CA 92521, USA

Williams T. E., Department of Botany and Plant Sciences, University of California, Riverside, CA 92521, USA

Yamamoto M., Faculty of Agriculture, Kagoshima University, 1-21-24, Korimoto, Kagoshima, 890-0032, Japan

Preface

Citrus fruits are regarded as major household items in more than 100 countries around the world. The world juice industry is also dominated by citrus juices. Citriculture is the foundation for the livelihood of millions of farm and industrial workers, entrepreneurs and businesses. There is a rich endowment of knowledge and technologies developed by dedicated researchers and academics around the world which addresses the issues faced by the citrus industry worldwide. Integration and application of knowledge from different disciplines has made its impact by guiding growers and investors to grow more and better citrus. Yet, we can see that the major citrus cultivars grown in different parts of the world are not the product of systematic breeding efforts. This book is intended to provide consolidated information on citrus breeding in the era of biotechnology which is likely to hasten the pace of variety development aimed at resolving the problems faced by grove owners growing currently available cultivars.

Citrus breeders and geneticists have pursued development of new cultivars for nearly a century. They do have clear goals to achieve. But, the traditional techniques have simply been insufficient to achieve the desired goals within reasonable time frames. The early researches focused on botanical and taxonomic subjects followed by interest in cytogenetics. The knowledge base has strengthened, but the outcome, in the form of commercially successful varietal releases, has remained slow, with a few success stories among rootstocks and fewer among scion cultivars. The recent advances in genetics, molecular biology and biotechnology have changed the pace of citrus breeding and genetics research. New approaches allow recombination within a broader and better understood germplasm pool, and more efficient selection methods have been developed for many traits. Progress has been also accelerated by the ease of communications via electronic media.

The scientific community is better organized than it was 50 years ago. The International Society of Citriculture has been holding its congresses at regular intervals since 1968. The amount of information presented on genetic improvement has grown manifold. There is hope that the flow of materials will follow the trends in the flow of information.

The idea of this book evolved from discussions held during citriculture congresses at Orlando, Florida (in 2000) and then at Agadir, Morocco (in 2004). Nearly all active citrus breeders of the world were willing to join hands to consolidate up-to-date information to accelerate citrus breeding. The subjects covered in the book are focused on citrus while providing information equally useful to the breeders of other tree crops. This will also help students of genetics and breeding identify appropriate applications of biotechnology in citrus breeding. While providing information on future avenues, the authors have also reviewed the past progress and achievements ensuring continuity of the subject. Several chapters include protocols for novel techniques that should facilitate their broader application by citrus breeders.

Many colleagues at CREC-Lake Alfred, where most of the editorial work was undertaken, helped in vetting the drafts. While the ideas and refereeing help from many is acknowledged, the editor takes the responsibility for errors and omissions. Patience of CABI management and staff has been a source of encouragement. Many thanks to them for doing an excellent job of compilation. This task could not have been accomplished without support from the Sultan Qaboos University, Muscat. It is hoped that the book will prove beneficial to a wide range of readership.

Iqrar Ahmad Khan

1 Citrus Breeding: Introduction and Objectives

Iqrar A. Khan and Walter J. Kender

Institute of Horticultural Sciences, University of Agriculture, Faisalabad 38040, Pakistan and *Nuclear Institute for Agriculture and Biology, Faisalabad, Pakistan*

As our early ancestors evolved from the process of nomadic food gathering to developing permanent food sources, crop breeding became an established practice. Over the centuries, humans selected, propagated and disseminated useful species that they chose to cultivate to feed the expanding populations. The hundreds of fruits, vegetables and grains that are now found on the supermarket shelves are each the result of the activities of plant breeders.

In 1865, the discovery of Gregor Mendel's laws of heredity became a turning point in the founding of the science of modern genetics, which transformed plant breeding into a modern scientific discipline. Mendel's critical and painstaking experiments allowed other plant breeders to conduct even more advanced studies on hybridization, which evolved into a remarkable success story. Consider the abundance of nutritious cultivars, the development of hybrid vigour in agronomic crops such as sorghum and maize, the development of short stature wheat and rice cultivars leading to the Green Revolution and elimination of extensive crop failures caused by southern corn leaf blight, or wheat rust or historic potato blight.

Until Mendel's paper on inheritance and its rediscovery at the turn of last century, plant breeding was based on traditional knowledge. Selection of desirable phenotypes was the primary means of perpetuating species. Plant breeding, now a scientific discipline, has resulted in major changes in how we feed the world's population through superior cultivars.

It can be said that the successful development of commercial crop industries around the world is directly related to the success of plant breeding programmes. The subject of citrus breeding and allied techniques has been reviewed previously (Cameron and Frost, 1968; Soost and Roose, 1996; Grosser and Gmitter, 2005). This book is a consolidation of the current status of science and technology relevant to citrus breeding.

Outlook

Tracking the ancestry of citrus is a complex process because of the great diversity and the distant centres of origin. Indeed, the evolution from wild citrus of Asia to modern day cultivars took hundreds of thousands of years. Now grown in more than 100 countries in tropical, subtropical and Mediterranean climates, citrus (including oranges, grapefruit, tangerines and mandarins, and lemons and limes) is the leading fruit crop grown in the world. In 2004, world production of citrus was 108,535,000 Mt (www.fao.org) which is dominated by oranges (Fig. 1.1). Brazil and the USA (Florida and California) were leading producers of sweet oranges. The USA is the primary producer of grapefruit. China, Spain and Japan produce 65% of the tangerines grown in the world. Lemons are produced primarily in Argentina, Spain and the USA, while Mexico is the largest producer of small fruited limes. Lime is also a traditional crop in South Asia and the Middle East. While consumers are generally familiar with the edible citrus types, there is an equally large component of non-edible citrus used as rootstocks for successful production of edible citrus. Additionally, many citrus species have industrial significance as a raw material for cosmetic and pharmaceutical products.

In general, the citrus production areas are located within 35° north and south of the equator. The main citrus regions, however, are in the subtropics, which are more than 20° north or south of the equator. As with most agricultural crops, many factors are known to limit the production and processing of citrus. Most are dependent on problems related to scion and rootstock deficiencies. Major constraints to citrus production involve management inefficiencies, susceptibility to pests and diseases, and environmental challenges. These lead to increasing production costs, declining labour supply in many parts of the world and urban encroachment, especially on the most productive farm lands. Thus, new and improved scion and rootstock cultivars aimed at controlling these production and marketing constraints have been the primary aim of citrus breeding efforts.

The development of new and improved citrus cultivars by conventional methods is a slow and costly process. It may take as long as 20–35 years or longer to release a new cultivar from the time of making the cross. Kinnow mandarin was bred and released at the University of California-Citrus Research Center, Riverside (Frost, 1935). The parental cross was made in 1915 and official release took 20 years. However, it took another period of more than 30 years

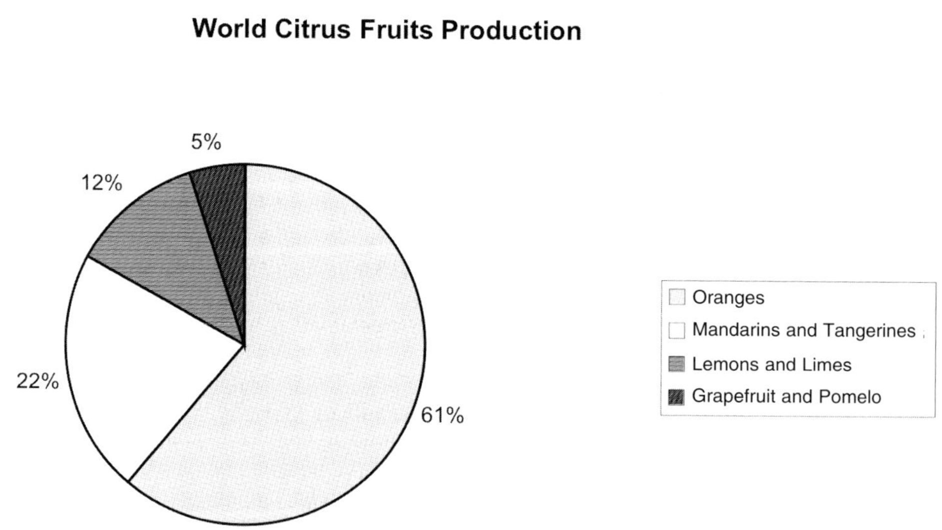

Fig. 1.1. Proportion of citrus fruits produced in the world.

before Kinnow became a successfully grown commercial cultivar in the Punjab region of Pakistan and India.

Several serious obstacles exist that hamper citrus hybridization. For example: (i) citrus is highly heterozygous; (ii) its unique reproductive biology such as apomixis and embryony; (iii) pollen and ovule sterility causing incompatibility; (iv) a long juvenile period taking as long as 5–10 years to express first flowering; and (v) adventitious embryos in the nucellar of developing ovules limit hybrid production. In developing seeds of polyembryonic cultivars, nucellar embryos compete with zygotic embryos. In addition to the long generation cycle, large seedling populations are needed (depending on species) that require extensive field space and labour.

The first formal citrus breeding programme was started by USDA in Florida in 1893 (Cooper *et al.*, 1962) which is still expanding. The University of California established the Citrus Research Center, Riverside in 1907. The University of Florida initiated its citrus breeding programme in 1924 which is now one of the largest breeding programmes, centred at the Citrus Research and Education Center, Lake Alfred (CREC). Today, there are numerous citrus breeding programmes spread in all major citrus-producing countries.

Most of the present day scion and rootstock cultivars of citrus are the progeny of chance seedlings or a mutant branch of a tree, called 'budsports'. The commercially successful cultivars now grown have resulted from the selection, propagation and advanced testing of thousands of such superior chance seedlings. This process of screening of potentially superior scion and rootstock candidates under commercial conditions involves a critical evaluation of tree performance and fruit quality traits. The above-mentioned breeding programmes played a major role in shaping today's citrus industry by providing information on the use of numerous scion and rootstock cultivars.

Breeders identify the most advanced candidates selected from initial screens and usually propagate them on certain rootstocks, establish them in replicated blocks in different regions and environmental conditions, and compare them with known cultivars. After such rigorous evaluation of horticultural performance, economic analysis and test marketing for fresh market and processing use, superior survivors may finally result in the introduction of a named variety.

In some citrus breeding programmes, new selections are patented for economic and scientific protection. After receiving a patent, the new cultivar may be licensed to nurseries to multiply the trees for commercial distribution. Certification programmes may be established to ensure that disease- and pest-free budwood is provided to commercial nurseries.

Achievements in Cultivar Improvement

Although citrus breeding programmes have existed in citrus-producing areas around the world for many years, most commercially important scion and rootstock cultivars were developed by means of selection of mutations or by chance seedlings. Relatively few have arisen from conventional citrus breeding. Examples of cultivars from breeding programmes in the USA are the mandarin hybrids such as tangelos (mandarin × grapefruit hybrids) and tangors (mandarin × sweet orange hybrids) – specifically Minneola, Orlando, Nova, Page, Robinson, Sunburst, Fall Glo and Ambersweet. Swingle citrumelo (grapefruit × trifoliate orange hybrid) and sweet orange × trifoliate orange hybrids Carrizo and Troyer citranges are examples of citrus rootstocks originating from breeding efforts. Three mandarin varieties were released from the Citrus Research Center, Riverside (Frost, 1935). Only one of the three, Kinnow mandarin (King × Willow Leaf), became a major success in the subcontinent of India and Pakistan. That reflects the value of international exchange of breeder's efforts. The process of release of new cultivars has now become more frequent, with expected impact on the citrus industry in the years to come.

Breeding Objectives

Depending on the needs of a specific region of the world, the objectives of citrus breeding programmes may vary considerably. For example, some of the following objectives may have high priority in some regions and low or no priority in others.

Scion breeding

See Table 1.1 for the objectives of scion breeding.

Rootstock breeding

Nearly all commercial citrus in the world is grown as grafted trees, with the scion cultivar budded on a selected rootstock cultivar. A good scion and rootstock combination supports development of trees that bear large quantities of high quality fruit. Such a stoinic combination can maintain health and productivity for 50 years or more with modest management. However, many available rootstocks are inadequate to meet the emerging needs and challenges. A large proportion of the problems faced by the citrus industry could be overcome by use of improved rootstocks (Wutscher and Hill, 1995; Bowman, 2000).

Benefits of the rootstock

EARLY BEARING. Budding a mature scion on to a rootstock plant takes advantage of the scion variety that has already passed juvenility which can start production within 1–2 years of planting in the grove, whereas citrus seedlings typically require many years to begin bearing fruit when grown from seed (3–15 years, depending on the species). In addition, citrus seedlings often maintain undesirable juvenile characteristics, such as excessively large thorns, for many years after fruiting begins.

TRUE TO TYPE. Propagation by budding/grafting on to a rootstock ensures that the trees will produce fruit that is identical to the source of budwood and thus allow plantings of a uniform type.

APOMIXIS. Most citrus rootstocks are apomictic, which can produce uniform plants from seed at a low cost. Seed propagation costs less and produces more vigorous and uniform nursery stock than by cuttings or tissue culture.

COMBINING TRAITS. Propagating citrus through budding on a rootstock is the benefit that probably has the most relevance in terms of rootstock breeding. It allows the graft creation of a tree that combines the best genetic fruit characteristics above ground with the strongest genetic root traits (adaptations to soil type, tolerance to salinity, resistance to diseases and nematodes, etc.) below ground as two separate units. Creating such a combination

Table 1.1. Objectives of scion breeding.

Tree performance	*Fruit characteristics*	*Postharvest of fruit*
Yield	Exterior appearance	Handling for fresh market
Cold hardiness	Size and shape	Economic and cultural importance
Adaptation to adverse climatic and soil conditions	Quality (ss/acid ratio)	Processing quality (processed into frozen concentrate or single strength juice products)
	Juice content, flavour/colour	
Adaptation to mechanical harvest	Ease of peeling	
Disease and pest resistance	Seedlessness	Storage life
	Season of ripening	Juice content and composition

of genetic traits in a single (self-rooted) genotype would be considerably more difficult.

Problems created by using a rootstock

EXPERTISE. Budding/grafting for propagation requires a significant expertise and adds cost when compared with seed propagation.

LACK OF APOMIXIS. A uniform nursery stand typically relies upon nucellar polyembryony (Frost and Soost, 1968) which results in the production of genetically uniform clonal seedlings. If a rootstock cannot be propagated uniformly by seed, clonal propagation must be accomplished by cuttings or tissue culture. The problem is further compounded when a rootstock strain produces a mixture of nucellar and zygotic seedlings which will require a rouging procedure (Khan and Roose, 1988).

GRAFT-TRANSMISSIBLE DISEASES. Vegetative propagation of the scion by budding can carry many virus and virus-like diseases and other pathogens. Hence, citrus has a long list of graft-transmissible pathogens that can debilitate trees and must be carefully excluded from propagation material (Roistacher, 1991).

GRAFT INCOMPATIBILITY. An important problem associated with using a rootstock is potential graft incompatibility between certain rootstocks and scions. Although generally not well understood, graft incompatibilities apparently arise when there is a conflict between the physiology of the rootstock and scion that causes the tree to grow weakly or die. A similar reaction can be induced by some virus diseases, such as incompatability which occurs when trees of sweet orange (*Citrus sinensis*) on sour orange (*C. aurantium*) are infected by isolates of citrus tristeza virus (Bowman and Garnsey, 2000).

ROOTSTOCK AFFECTS THE SCION. It must also be kept in mind that rootstock does affect many important traits of the scion grafted on to it, including tree size, productivity and fruit quality. Hence, a rootstock may degrade the tendency of a scion to be highly productive or to yield good quality fruit. One example of this is the rough lemon (*C. jambiri*) rootstock, which produces a vigorous and highly productive tree, but also induces the scion to yield low quality, low sugar fruit.

The Rootstock Development Process

Development of improved rootstocks is a long and many-faceted process. Typically, it takes 30–35 years from the time a new rootstock is created by cross-hybridization until it is released to the local citrus industry for commercial use. Commonly pursued objectives of rootstock breeding are listed in Table 1.2.

Molecular Genetics and Biotechnology

Citrus breeding and genetics has been fortunate to have an early and continuing history

Table 1.2. Commonly pursued objectives of rootstock breeding.

Apomixis
Improved productivity
Tree size reduction
Adaptation to soil conditions (salt, calcareous soils)
Improvement in poor nutrition
Resolution of bud union problems
Resistance to diseases (*Phytophthora*, citrus tristeza virus, citrus blight, etc.)
Resistance to soil borne pests (diaprepes, nematodes, etc.)
Improvement in fruit quality
High seed production

of meeting the needs of a changing industry through scientific research on genetic techniques. A new generation of scientific methodologies is now available for the study of genomics. It is possible to locate genes rapidly on chromosomes, to isolate genes from plants to study their function at the molecular level, to modify genes and to reintroduce them into living organisms. This is, indeed, a far cry from Gregor Mendel's 19th century tenant that plant characters are controlled by discrete hereditary units – which are now called genes.

Modern genetic engineering evolved from the discovery of DNA as the genetic material by Avery and co-workers followed by Watson and Crick's discovery in 1953 of the double helix structure of deoxyribonucleic acid (DNA). Since then, modern day geneticists have been able to use more precise and sophisticated technology to overcome the barriers and greatly to facilitate genetic procedures.

It is essential that future programmes for citrus cultivar improvement emphasize understanding the inheritance of fundamental qualitative and quantitative traits and also be comprehensive. For example, the study of genomics involves a molecular-based knowledge of the structure and function of genes. Such information will afford a precise use of such traits in the tree and fruit. Numerous examples can be cited where genomic technology has benefited plant improvement efforts. For example, in citrus, the long duration of field evaluation is a major obstacle for developing specific characters such as tree size, yield, disease and insect resistance. The use of marker-assisted selections linking molecular markers to genes, which control important traits, will significantly reduce the time required to screen mature tree characters. Other tools of biotechnology have been employed with equal ease to facilitate the progress of work in citrus breeding programmes. Shoot-tip grafting techniques have been used for sanitation of virus-infected germplasm, which has allowed wide dissemination and conservation of parental stocks.

Cytogenetics

Cytogenetic investigations in citrus have been restricted due to very small chromosome size and little implication for ploidy manipulations except for an interest in triploid breeding for seedlessness. In recent years, *Citrus* has been accepted as a model for the application of somatic hybridization and cybridization for crop improvement (Khan and Grosser, 2004; Grosser and Gmitter, 2005). Despite a huge variability in citrus, evolution has been restricted at the diploid level largely because of apomictic seed propagation. Somatic hybridization has offered a unique situation for citrus breeders to pursue citrus evolution at higher ploidy levels, particularly that based on allotetraploid combinations (Grosser *et al.*, 1996).

Research Diversity

To understand better the scope and depth of the citrus breeding/genetics research being conducted in the world today, the reader is encouraged to read the 2000 Proceedings of the International Society of Citriculture (ISC) held in Orlando, Florida (published in 2003) and ISC 2004 proceedings. Citrus breeders and geneticists presented a total of 114 papers in 2000 (71 oral and 43 posters). Of the oral presentations, topics included scion cultivar evolutions (seven), rootstock cultivar evolutions (eight), breeding (34), genetics and genes (eight) and biotechnology (14). These papers represented work done in over 20 different citrus-producing countries, including the USA, Brazil, Spain, France, China, Italy, Japan, Argentina, Cuba, South Africa, Australia, Oman, Israel, Morocco, Turkey, India, Pakistan, Egypt, Lebanon, Korea and New Zealand. In 2004, the number of genetics and breeding papers went up to 126 where a major shift was seen towards molecular genetics, genomics and transgenic citrus research.

The diversity and scope of global citrus genetic research projects are impressive.

Preservation of Citrus Germplasm

Citrus geneticists are concerned over the reported loss of citrus germplasm especially in the areas of origin. Such potential losses must be addressed on a global and cooperative basis before more valuable and irretrievable citrus materials are lost forever. Breeding of citrus requires a large diversity of citrus cultivars in a reliable and protected source. Since such citrus populations are rare, worldwide, habitat destruction threatens existing materials. The FAO sponsored Global Citrus Germplasm Network recommended in 1996 that (i) activities be initiated to identify the locations of gene banks of citrus and its relatives; (ii) it should be determined what needs to be preserved in collections and data banks; (iii) citrus germplasm should be characterized; (iv) procedures to conserve available and unique citrus germplasm should be developed; and (v) citrus germplasm should be evaluated and catalogued using molecular markers. Such activities, if properly conducted, would help ensure the availability of important germplasm for use by future citrus breeders and geneticists.

About this Book

Basic knowledge in plant genetics and biotechnology has opened the door to new concepts and methodologies, which have greatly advanced the scope and abilities of the citrus breeder. To define the progress in the field of citrus genetics, key scientists working on citrus cultivar improvement around the citrus-producing world have combined their talents to write chapters in areas of their expertise. Such information at this time of the genetic revolution will be of great value to the 'students of citrus'. These chapters characterizing the remarkable advances made in recent times will point out the success of past and current citrus genome research and the need for continued financial support for such research in the future.

The major topics covered in this book are:

- Origin and taxonomy
- Germplasm resources and shoot-tip grafting
- Cytogenetics
- Somoclonal variation
- Somatic hybridization and single chromosome transfer technology
- Mutation breeding
- Triploid hybrids (hybridization between diploids and tetraploids)
- Marker-assisted selection by linkage maps
- Gene cloning
- Genetic transformation.

References

Bowman, K.D. (2000) New hybrid citrus rootstocks developed by US Department of Agriculture. In: *Proceedings of the International Society for Citriculture, IX Congress*, p. 51.

Bowman, K.D. and Garnsey, S.M. (2000) Comparison of decline reaction from citrus tristeza virus infection in trees on sour orange hybrid rootstocks. In: *Proceedings of the International Society for Citriculture*, IX Congress, p. 996.

Cameron, J.W. and Frost, H.B. (1968) Genetics, breeding and nucellar embryony. In: Reuther, W., Batchelor, L.D. and Webber, H.J. (eds) *The Citrus Industry*, Vol. 2. University of California, Berkeley, pp. 325–370.

Cooper, W.C., Reece, P.C. and Furr, J.R. (1962) Citrus breeding in Florida – past, present and future. *Proceedings of the Florida State Horticultural Society* 75, 5–13.

Frost, H.B. (1935) Four new citrus varieties – the Kara, Kinnow and Wilking mandarins and the Trovita orange. *University of California Agricultural Experimental Station Bulletin* 597.

Frost, H.B. and Soost, R.K. (1968) Seed reproduction: development of gametes and embryos. In: Reuther, W., Batchelor, L.D. and Webber, H.J. (eds) *The Citrus Industry*, Vol. 2. University of California, Berkeley, pp. 290–324.

Grosser, J.W. and Gmitter, F.G., Jr (2005) Applications of somatic hybridization and cybridization in crop improvement, with citrus as a model. *In Vitro Cellular and Developmental Biology-Plant* 41, 220–225.

Grosser, J.W., Mourao-Fo, F.A.A., Gmitter, F.G., Jr, Louzada, E.S., Jiang, J., Baergen, K., Quiros, A., Cabasson, C., Schell, J.L. and Chandler, J.L. (1996) Allotetraploid hybrids between *Citrus* and seven related genera produced by somatic hybridization. *Theoretical and Applied Genetics* 92:577–582.

Khan, I.A. and Grosser, J.W. (2004) Regeneration and characterization of somatic hybrid plants of *Citrus sinensis* (sweet orange) and *Citrus micrantha*, a progenitor species of lime. *Euphytica* 137, 271–278.

Khan, I.A. and Roose, M.L. (1988) Frequency and characteristics of nucellar and zygotic seedlings in three cultivars of trifoliate orange. *Journal of the American Society for Horticultural Science* 113, 105–110.

Roistacher, C.N. (1991) *Graft-transmissible Diseases of Citrus*. FAO Publication, Rome.

Soost, R.K. and Roose, M.L. (1996) Citrus. In: Janick, J. and Moore, J.N. (eds). *Fruit Breeding, Vol. I: Trees and Tropical Fruits*. John Wiley and Sons, Inc., pp. 257–323.

Wutscher, H.K. and Hill, L.L. (1995) Performance of 'Hamlin' orange on 16 rootstocks in east-central Florida. *HortScience* 30, 41–43.

2 A Comprehensive Citrus Genetic Improvement Programme

Fred G. Gmitter Jr[1], Jude W. Grosser[1], William S. Castle[1] and Gloria A. Moore[2]

[1]University of Florida, Institute of Food and Agricultural Sciences, Citrus Research and Education Center, Lake Alfred, FL 33850 USA; [2]University of Florida, Institute of Food and Agricultural Sciences, Department of Horticultural Science, Gainesville, FL 32611 USA

Introduction

All the world's citrus industries will benefit from genetic improvements leading to the release of superior new rootstock and scion cultivars. Improvements in pest and disease resistance, tolerance of various environmental stress factors, horticultural performance and productivity, and fruit quality characteristics should result in greater economic returns to growers, processors and shipper/packers by reducing input unit costs and increasing product desirability and demand. Likewise, improvements in fruit quality, nutritive value and content of health-protective components should benefit consumers and increase consumption of this well recognized, nutrient-rich fruit. Until recent times, however, much of the process of genetic improvement of citrus was based on identification of chance mutations and seedlings displaying useful characteristics. Structured and targeted breeding programmes were generally inefficient and ineffective because of a lack of genetic knowledge of important traits, incomplete understanding of the significance of taxonomic distinctions and relationships, and the absence of breeding tools that could be employed to achieve necessary and desired goals. However, in the past two decades, there has been a tremendous explosion in the understanding of plant genetics and genomes, as well as associated technologies to enable application of new fundamental knowledge to genetic improvement of crop plants in general, and specifically of citrus; this volume itself is a testament to that technological revolution, specifically as it relates to citrus. As a consequence, a comprehensive approach to citrus genetic improvement and cultivar development, utilizing the most appropriate methods and technologies to achieve given improvement goals, makes the greatest sense.

The authors of this chapter are members of a team dedicated to developing genetically superior new citrus cultivars, in a comprehensive fashion. The individual members bring diverse expertise to the team, from fundamental genomics and

 Citrus Genetics, Breeding and Biotechnology (ed. I.A. Khan)

genetics, from tissue culture-based systems of somatic hybridization and genetic transformation, from traditional breeding techniques frequently assisted by tissue culture methods, to screening newly created materials for specific traits and characters in specialized challenges, and ultimately proving the value of new selections to growers and for commercial release following extensive replicated field trials. The basic elements of the University of Florida's citrus genetic improvement and cultivar development programme are outlined in Fig. 2.1. All of the research efforts of our team, whether they are traditional breeding for multitrait improvements or genetic experiments to clone genes that can precisely modify a specific trait, are directed ultimately toward the central and unifying goal of releasing new and improved rootstock and scion cultivars for the industry. All forms of genetic modification, listed in general terms in Fig. 2.1, yield new plants that represent a pool of genetic diversity upon which screening and selection are imposed to find and to verify truly superior performing individuals for release as new cultivars. Contributions from genomic research feed into the process as cloned genes that can be introduced to the pool via transformation; alternatively, genomic research also leads to the development of screening tools via

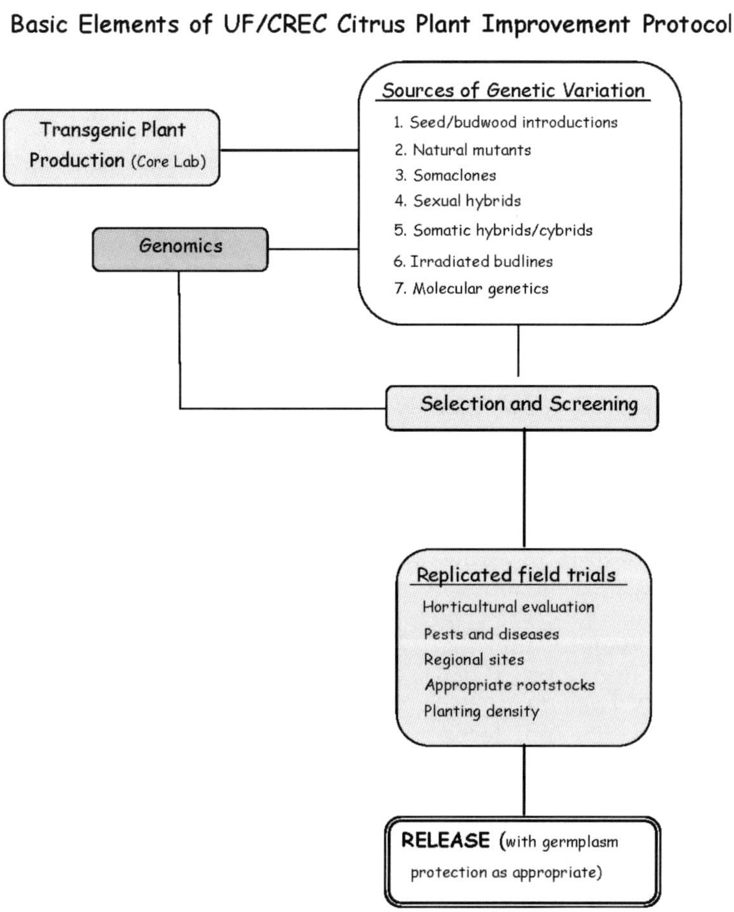

Fig. 2.1. Overview of the basic elements of the genetic improvement process for citrus.

marker-assisted selection (MAS), which enables much more efficient selection of superior recombinants improved for multiple traits from conventional breeding efforts. Once superior candidate individuals have been identified; their value and performance are assessed and documented in many locations and conditions to identify the most valued of the candidates for release.

The information and the figures contained within this chapter are extracted from a working document, titled 'A Comprehensive Citrus Genetic Improvement Program', which the team uses as a guide and a plan for the research efforts in which we are engaged; the document is reviewed frequently and updated periodically as needed. This document has been provided to various stakeholders in the process, including university administrators, citrus growers, funding agencies, researchers and any other interested individuals or groups, such that the comprehensive nature of citrus genetic improvement as we view it can be understood and supported. It represents an effort to explain in lay terms some of the complexities of genetic science from the perspective of outcome, and not necessarily from the perspective of the details of the science itself. As such, it enables administrators and funding agencies to approach questions relating to the needs of the programme, as well as its potential to deliver new improved products, from a holistic point of view.

Scion Protocol

Various sources of genetically variant plant materials are listed in Fig. 2.2. These include sources that have been used for many of the advances over centuries of citrus scion improvement: seed and budwood introductions from elsewhere, or naturally occurring mutations that may be found. Other types of mutations are also exploited, such as those that arise from tissue culture cycles (somaclones) or from induced mutations following irradiation. Given the taxonomic nature of many of the commercial citrus cultivar groups (sweet oranges, grapefruit, lemons and several types of mandarins) and their naturally occurring diversification through somatic mutations, these groups are not amenable to breeding strategies based on sexual hybridization. Therefore, mutational approaches to genetic improvement have been and will continue to be critical, along with the great potential now in hand by virtue of the ability to transform such cultivars with targeted genes. Also included in the list of sources of genetic variation are plants that arise from hybridization by sexual or somatic processes. Although the goal is to create new cultivars, the process likewise continues to identify new recombinants to be used as breeding parents in cases where breeding remains a viable improvement strategy. The protocol is robust; thus, plant material from any source or developed by any technique can be entered into the system at any point.

Seedlings, irradiated budlines or other scion candidate materials are propagated on to rootstocks or grown on their own roots, planted in a fruiting-out orchard, and potentially desirable or valuable individuals are selected according to their fruit and tree characteristics. Evaluations for specific characters that can be accomplished prior to first field planting (e.g. simplified disease resistance selection, the presence and/or performance of transgenes, etc.) take place when possible. Upon first fruiting, evaluations are made of both tree and fruit characteristics, with the objective of identifying those individuals that may potentially prove valuable as new cultivars. This objective is accomplished by a process of elimination of individuals with undesirable fruit characteristics, and by selection of individuals exhibiting positive characteristics (appropriate fruit size, colour, flavour, productivity, and so on). These observations are made for three years at least, to assess stability of trait expression over time and to avoid excluding potential candidates that may appear undesirable during their first

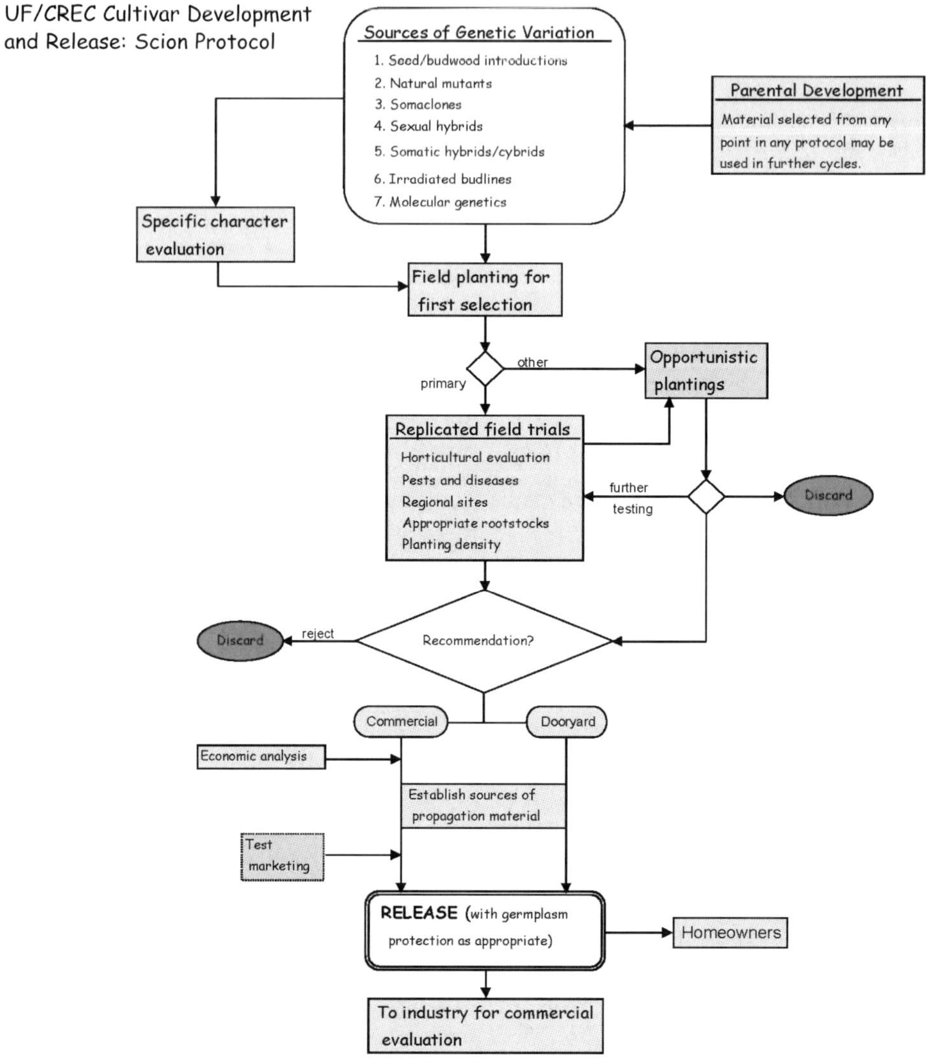

Fig. 2.2. Protocol for citrus scion improvement.

fruiting. The specific criteria for selection are determined by the team member producing the given materials, based on the type of fruit and the specific objectives of the genetic manipulation performed. The outcome of this phase of the protocol is a collection of candidates for advanced field evaluations, as well as the identification of individuals that can be used in further breeding cycles based on their expression of desirable combinations of genes and alleles.

Individuals selected from primary screening are propagated on to one or several rootstocks and planted in replicated field trials in multiple locations representing the environmental diversity in Florida. The objectives of this stage of the protocol are to identify truly superior individual selections and to acquire adequate information on their performance, in comparison with standard cultivars, to warrant further development leading to possible release of new cultivars. These objectives are accomplished by evaluating horticultural performance, susceptibility/resistance to pests and diseases, regional environmental effects on performance and quality, rootstock interactions and other relevant factors. Also important at this stage is the identification of special problems that may be encountered by growers attempting to produce fruit from these new cultivars. At times, opportunistic plantings may be made, for example, to test for resistance to a pathogen that may be widespread in a particular growing area; such plantings are not replicated field trials, but none the less they can provide valuable insight and information regarding the performance and potential of certain primary selections.

As data are collected from replicated field trials, financial analysis is used as a tool to determine the potential value of new processing cultivars compared with existing ones, based on yields and qualities of the fruit of the new cultivars. The performance and potential value of new scion cultivars intended for the fresh market can be better determined by test marketing. This phase of the protocol is applied only to the most promising, elite selections identified through previous stages of the protocol. Testing may be of the potential of processed products, as determined by sensory panel analysis, biochemical profiles as related to flavour and quality, and other approaches. Cultivars intended for the fresh fruit market can be test-marketed by cooperating agencies and organizations, on a small but commercial scale; fruit performance in important aspects of the market channels is carefully evaluated, including handling ability and consumer/retailer preferences and responses. Poor performance or results from economic analyses or from test marketing would probably preclude release of a given selection, in most cases.

As the latter stages of the protocol are entered, sources of budwood of the elite selections are established, to have sufficient propagation materials available for rapid multiplication and commercialization at the time of release. These selections must be certified through the Citrus Budwood Registration Bureau, Division of Plant Industry (DPI) of the Florida Department of Agriculture and Consumer Services (FDACS), to ensure the distribution of disease-free budwood to nurseries for propagation as required by state law. The programme supports an insect-proof repository facility in which pathogen-free budwood sources of advanced selections are held and increased as needed, for field trials or to provide budwood to the nursery industry if necessary. Throughout the entire process, measures are taken to provide appropriate protection for new cultivars and selections; these measures include the use of Material Transfer Agreements when materials are tested in the orchards of commercial growers cooperating in the evaluations, as well as patent protection prior to official release. When the protocol has been completed and the potential value of new scion selections has been demonstrated, then the cultivar(s) will be released upon acquisition of a plant patent and licensing of nurseries to produce trees for the industry and/or growers to produce the crop. Two types of releases are possible: first, the release of cultivars to the commercial industry; and, second, releases for selections that would be of interest to the home gardening clientele. The latter may be released following a less stringent approach to the protocol, compared with those to be released commercially.

Rootstock Protocol

Citrus rootstock requirements and characteristics evaluated are quite different in most cases from those of scions, so the sources of genetic variation exploited and the screening methods used are likewise different in nature and scope (Fig. 2.3). Hybridization, both sexual and somatic, is a primary tool for rootstock improvement. The introduction of rootstocks from other industries or breeding programmes is also an important part of the overall programme to bring our industry the best possible cultivars to use, regardless of their source of

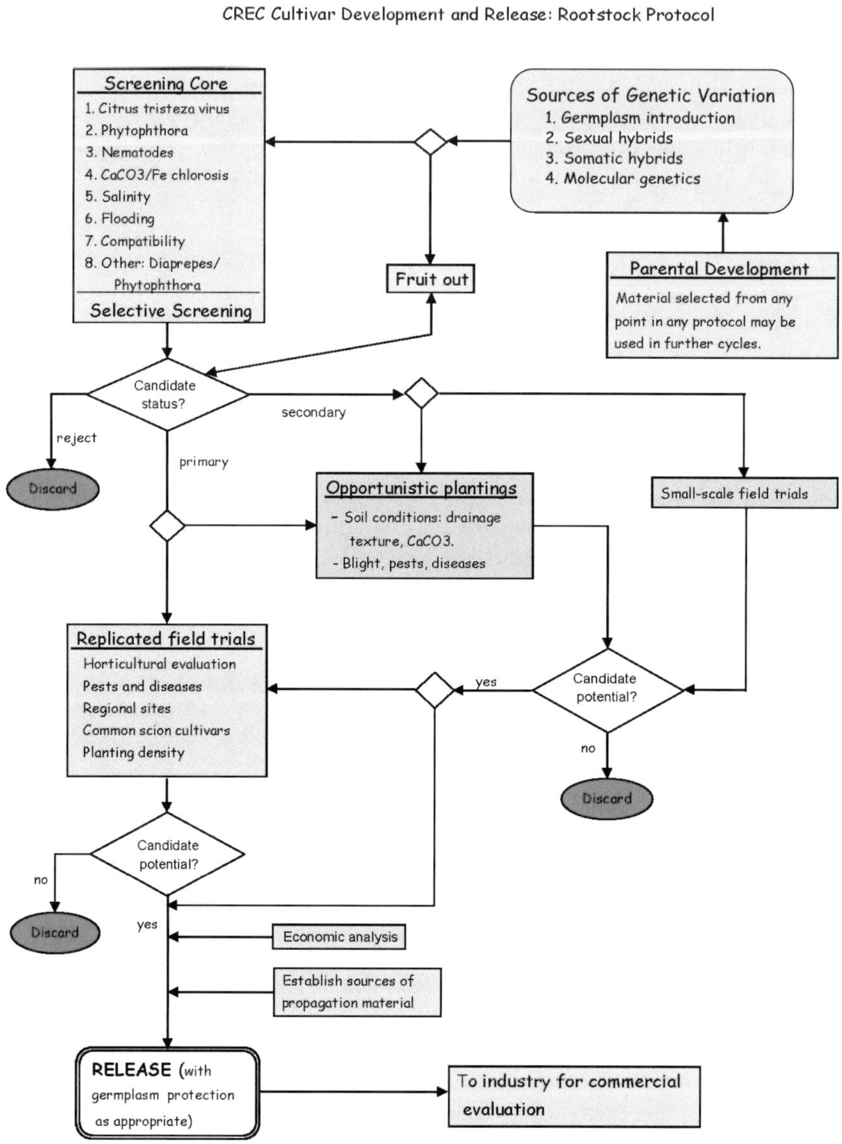

Fig. 2.3. Protocol for citrus rootstock improvement.

origin; such legally acquired selections or cultivars are frequently included in replicated field trials with our own advanced selections. As genetic engineering comes to play a greater role in the genetic improvement process over time, there will probably be greater exploitation of the technology to make important single trait improvements in rootstock cultivars or selections with already proven value. As with scion improvement, the rootstock protocol is robust and plant material from any source or developed by any technique can be entered into the system at basically any point.

Seedlings from crosses or introductions, or equivalent material such as cuttings of seedlings or somatic hybrids, receive their initial evaluations in the screening core facility or the fruiting-out orchard. The screening core facility is supported by the programme to provide a location where screening can be carried out on a repeatable and routine basis; implicit in this activity is the interaction with experts in the respective areas of plant pathology and physical stress physiology. The traits selected for screening are at the discretion of the team member responsible for introduction or production of the materials. Selection within the fruiting-out orchard is generally based on polyembryonic seed production and tree performance, two critical traits for rootstocks that are propagated by nucellar seeds. Primary candidates are those individuals that pass the screening core tests for citrus tristeza virus (CTV) resistance or tolerance, *Phytophthora* resistance and other tests as established by the investigator, such as tolerance of salinity or iron chlorosis due to high soil pH, abundant production of polyembryonic seeds, and so on. These candidates are entered into replicated trials designed primarily to conduct the first field evaluation involving soil, horticultural, pest and disease factors. The emphasis is on representative regional sites regarding soil conditions, scion cultivars and plant spacing. Following a first round of field trials, additional trials may be appropriate to confirm or otherwise expand the knowledge about a given rootstock candidate. Secondary candidates are those that do not meet all of the screening criteria for a primary candidate, but in the judgement of the investigator they may merit field evaluation none the less. These candidates are placed in smaller sized field trials or opportunistic plantings where they may be subjected generally to a lower level of evaluation. Opportunistic plantings are not formal field trials, but are plantings that may involve, for example, using new selections in reset situations, small groups of trees on one rootstock, placing trees in a particular field condition (heavy blight incidence; $CaCO_3$ soil) or any situations that are more commercial-like in size and planting arrangement; opportunistic plantings generally are used to assess trait performance for a limited number of site-specific problems for which the selections tested may prove useful.

When sufficient data are available from long-term field trials, financial analysis is used as a tool in rootstock evaluation, leading to cultivar release. When evaluation is nearing the point when a release is anticipated, it is critical to have seeds or vegetative material available for distribution to support commercialization of the new releases, and appropriate measures are taken to ensure adequate supplies for estimated needs. Measures are taken to provide protection, as appropriate, for new cultivars and selections; again this protection includes Material Transfer Agreements with cooperating growers participating in the commercial evaluation, and proper patent protection, as needed. Release will generally occur when the preceding protocol has been completed and a rootstock is judged as having sufficient potential, either as an alternative or as a replacement for current commercial rootstocks. Germplasm with potential use as breeding lines, even though they have not met all criteria for new cultivar status, may be released from any stage of the protocol.

Molecular Genetics Protocol

Molecular genetics is used as a term in the broad sense to include all activities related to citrus genetic improvement that are based on use of molecular techniques to manipulate genes at the level of nucleic acids (DNA or RNA). These activities also include fundamental genetic studies, as they provide a basis for molecular dissection of key traits based on plant performance (Fig. 2.4). Understanding the inheritance of important traits is fundamental to all genetic improvement strategies whether the traits sought are under qualitative or quantitative genetic control. Molecular genetics, likewise, includes all

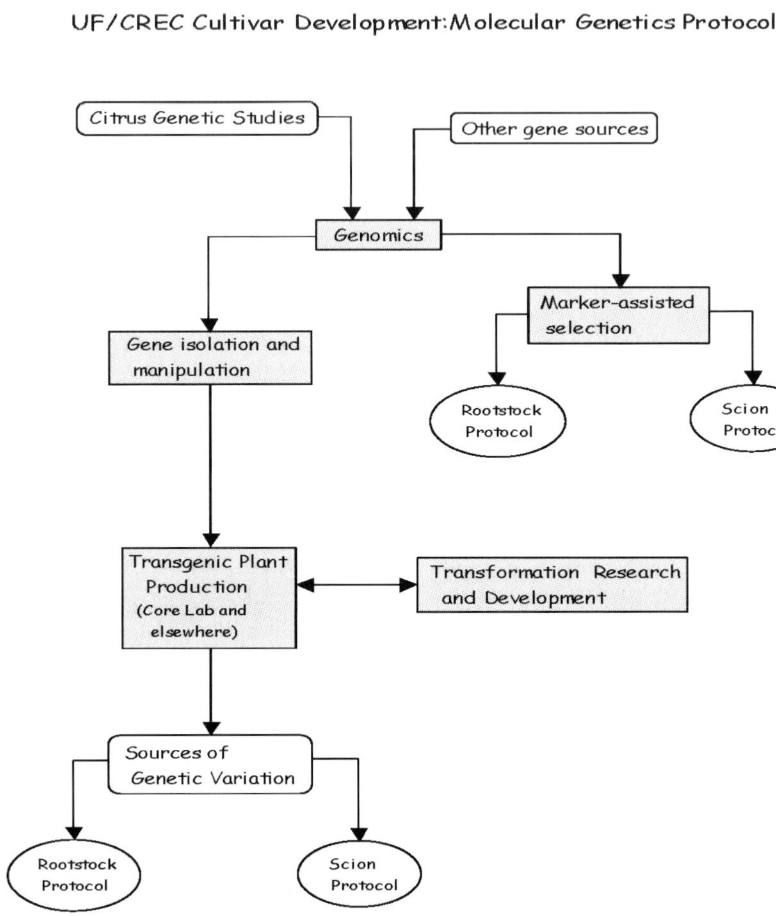

Fig. 2.4. Protocol for molecular genetics and its relationship with other protocols.

efforts linked to genomic sciences relating genetic information on important phenotypic traits to a molecular-based understanding of the structure and function of genes, to provide the basis for more precise and specific manipulation of tree and fruit characteristics. Genomic science also includes the development of DNA marker-based maps, as well as integrated physical and genetic maps, and ultimately whole-genome sequence information, for greater understanding and manipulation opportunities in the future. Genetic maps, in turn, by directly linking molecular markers to genes controlling economically important traits, make it possible for selection targeting certain goals of the cultivar development process to be accomplished more quickly and efficiently than by actual evaluation of plant traits. Because more efficient methods result in the capacity to evaluate many more plants, the likelihood of identifying plants recombinant for multiple valuable and desirable characteristics is greatly increased.

One outcome of genomic research is the ability to modify very specific individual characteristics in powerful ways: by isolating genes from other sources within the citrus gene pool for incorporation directly into commercial cultivars; by modification of existing genes to produce specific changes in plant characteristics; or by utilizing genes from any other organism that potentially could have beneficial phenotypic effects on citrus tree performance or fruit characteristics. Any genes that are isolated and/or modified can be incorporated into existing cultivars or breeding lines using genetic transformation technologies. The programme also includes a Citrus Core Transformation Facility, supported jointly by industry grants and university funds; this service is available to researchers within and outside of the citrus genetic improvement community, to increase the opportunities for testing potentially useful genes from outside the citrus gene pool. Ultimately, transformation of citrus germplasm with new and/or modified genes feeds into the scion and rootstock cultivar development protocol as another source of genetic variation for evaluation and selection. Research is conducted by some of the team members to improve the efficiency of the existing *Agrobacterium*-, protoplast- or biolistic-based transformation systems. A major goal is to develop an efficient system for all commercially important cultivars and elite breeding parents. Other objectives include minimizing the impact of citrus juvenility on the time from transformation to fruiting, minimizing technology fees associated with certain component elements of the transformation process, and developing more environment- and consumer-friendly transformation systems.

Conclusion

To summarize, our citrus cultivar development team has developed a collaborative and comprehensive approach to maximize efforts and opportunities for development of the genetically improved new rootstock and scion cultivars required for the future viability of our citrus industry, by maximizing the synergistic interactions of expertise and interests we possess as individual team members. Furthermore, by organizing activities into generalized and specific flow charts, with annotation to describe the essential components and outcomes from each step, we have assisted growers, funding agencies, research administrators and other interested parties in understanding how the various approaches to genetic improvement fit together. This, in turn, aids such parties in determining where additional investments into the research programme are needed and best placed for maximum benefit to the industry, through the outcome of newly released and superior cultivars.

This comprehensive programme is truly a partnership in both the narrow and the broad sense. It is a partnership first and foremost among the team participants, as well as the related experts who become a part of the process, through collaboration and sharing of expertise. On a broader level, it clearly represents a partnership of the research community with the industry. The industry supports the research programme not only by grants through the Florida Citrus Production Research Advisory Council (FCPRAC), but also through interaction and direct participation in the research efforts by providing land, grove care expenses, and labour for dozens of field trials, plantings and evaluations that take place. This involvement is a critical and essential downstream component of the comprehensive programme. All of the upstream research activities have no value whatsoever to the industry without this vital partnership in the final stages. The industry–research community interface provides a valuable feedback mechanism, serving to focus the upstream research efforts in appropriate directions, and bringing important new developments and opportunities arising from the genetic research to the attention of the industry's leaders and innovators, as well.

The team recognizes that this programme represents a complex and long-term process. It includes not only our experiments to create, select and validate genetically improved plant material but also the experiments that require industry cooperation and collaboration to assess the value and utility of the plant material in the orchard. However, the final stage of the process is seen once new cultivars are released, and they find their way into commercial production and the marketing of products to consumers. It is at that point that the benefit of many years of research, and the involvement of many different participants in the process, may be seen and judged.

3 Origin and Taxonomy

Elisabetta Nicolosi

Dipartimento di OrtoFloroArboricoltura e Tecnologie Agroalimentari, University of Catania, Italy

The History of Citrus

Different hypotheses have been formulated on the history and geographical origin of citrus. It would appear that all the species belonging to *Citrus* and its related genera originated in the tropical and subtropical regions of South-east Asia – north-eastern India, southern China, the Indo-Chinese peninsula – and the Malay Archipelago, and then spread to other continents (Webber, 1967; Chapot, 1975) (Fig. 3.1). Tolkowsky (1938) considers that the mountainous regions of southern China and north-east India are the centre of origin, 'where sheltered valleys and southern

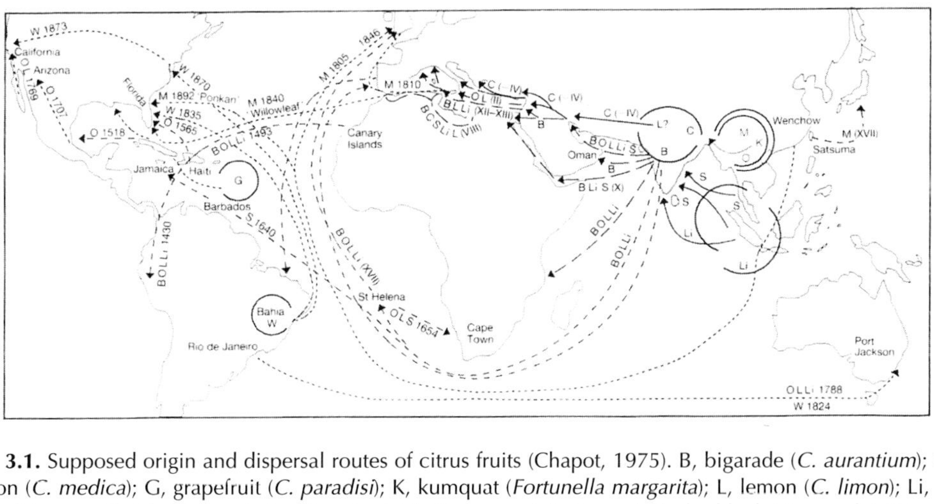

Fig. 3.1. Supposed origin and dispersal routes of citrus fruits (Chapot, 1975). B, bigarade (*C. aurantium*); C, citron (*C. medica*); G, grapefruit (*C. paradisi*); K, kumquat (*Fortunella margarita*); L, lemon (*C. limon*); Li, lime (*C. aurantifolia*); M, mandarin (*C. reticulata*); O, orange (*C. sinensis*); S, shaddock (*C. grandis*); W, 'Washington' navel; ——, BC; — —, AD 1–700; – – – , AD 700–1492 (711: Arab occupation of Spain); - - - , AD 1493–1700 (1493: second journey of Christopher Columbus); ......, after AD 1700 (first appearance of grapefruit). Centuries are given in Roman numerals (a minus sign indicates BC); years are given in arabic numerals. Source: Davies and Albrigo (1994).

slopes are sufficiently protected from the cold dry winds ...' while, according to Tanaka (1954), the citrus may have originated in north-eastern India and Burma; China should be considered only a secondary centre of distribution. He also maintains that it is likely that several species started in China and spread into Indo-China, Malaysia, north-eastern Asia and finally Japan. Tanaka (1954) proposed a theoretical dividing line running from the north-western border of India, above Burma, to the Yunnan province of China, to south of the island of Hainan (Fig. 3.2). Several citrus species such as the citron, the lemon, the lime, the pummelo and the sour and sweet orange originated south of this line, while mandarins and others originated north of the line. The wide diversity of citrus in Yunnan has been described by Gmitter and Hu (1990), who considered this province, through which the Tanaka line runs, to be the major centre of origin for citrus.

Calabrese (1998) indicated that the primordial genetic nucleus of citrus originated in China and the citrus slowly passed from its original location to other oriental regions and from there followed the paths of civilization. It is well known that the Chinese reached cultural goals before other people, and they have domesticated many fruit-bearing plants, including citrus, for production purposes and for aesthetic reasons since the second millennium BC.

The first indication regarding the presence of citrus in China is contained in the book 'Tribute of Yu': during the kingdom of Ta Yu (from 2205 to 2197 BC) *Citrus* names occur in the list of duty tribute sent to the imperial court at An-Yang, a territory near the big bend in the Yellow River. The term *Chu* found in this document probably refers to small sized mandarins and the kumquat (*Fortunella*), while the term *Yu* indicates the pummelo (*Citrus grandis*) and the 'Yuzu' (*Citrus junos*).

Fig. 3.2. Geographical map showing Tanaka's dividing line.

During the Chou dynasty (1027–256 BC), literature flourished and a lot of poems and books were written in this period. Among them, 'The five Canons' are the most important mythological, historical, philosophical and literary document of China for a period of about a millennium. In all this literature it is easy to find evidence regarding citrus; there are two types of citrus, the *Chu* and the *Yu*, which are mentioned most often.

The Ch'in and the Han dynasties followed the Chou, and during the Han dynasty (202 BC–220 AD) other citrus fruits were mentioned, *Cheng*, *Lu Chu* and *Huang Kan*. The most important information comes from Ssu Hsiang-ju (died in 118 BC), who wrote in his prose poem about *Cheng* (sour orange), *Lu Chu* (kumquat) and *Huang Kan* (yellow mandarin, probably including large mandarin-type fruits and oranges). After the Han dynasty, there was the 'Three Kingdoms' dynasty which came to an end in 265 AD. From 265 until 420 AD, China was governed by Chin (Tsin) and there are numerous historical records referring to the citrus in this period. In a work written in 304 AD by Chi Han, the first description of citron appears with the name *Kuo Han*. In Chinese culture, the citron, mainly the 'Fingered citron' (Fig. 3.3) also called 'The hand of Buddha', had become important; it was considered to be a lucky talisman and was exchanged for good luck. Citron is most probably native to India, and there the earliest references are to be found in the *Vajasaneyi Samhita*, a collection of sacred Brahma texts, called *Yajur-Veda*, written around the 8th century BC (about a millennium before the first description in China); *Jambila* or *Jambira* is the term which indicates the citron and the lemon (Tolkowsky, 1938).

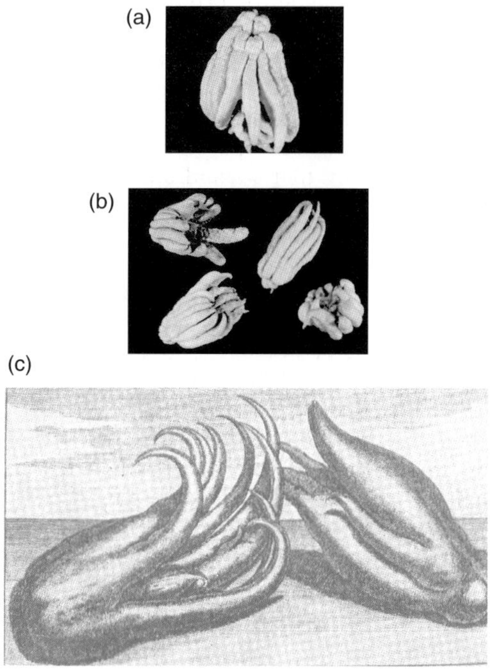

Fig. 3.3, a, b, c. Pictures showing different fingered citrons

The citron

The citron (*C. medica* L.) is probably native to India; it is well accepted by all authorities that it was the first citrus fruit known to Europeans, but there are different opinions about the exact period and the steps by which it was first brought from its native land. Gallesio (1811) accepts that the Jews brought the citron to Palestine upon their return from their Babilonian exile (~500 BC), while Tolkowsky assumes that the citron was introduced only following the return of Alexander the Great from India (~300 BC). According to Tolkowsky, the adoption of the citron by the Jews for worship during the feast of the Tabernacles (*Sukkoth*) took place only during the period of the Hasmonaite dynasty (~120 BC). The citron played a prominent part in Jewish religious rituals (Figs 3.4 a and b); it appeared on Jewish coins during 66–70 AD, and has been a favourite motif in Jewish art since then (Spiegel-Roy and Goldschmidt, 1996). The use of the citron during the rites of the Tabernacles at the beginning of the first millennium is also clearly evident from what Joseph Flavius, the most important writer of the Jewish tradition, writes in his *Judean Antiquity*, a volume which goes

Fig. 3.4a. A citron tree in a mosaic floor in Cesarea (Israel's coastal plain) from the 6th century.

through the steps of Jewish history from the Creation until the war against Rome.

There is an accurate description of the citron by Theophrastus in his *Historia Plantarum*, written around 313 BC. He says that there is a plant in Persia which produces a fruit called the 'Fruit of Persia' or 'Fruit of Media' or *Persian citron* or *Median Apple* (Book IV, Chapter IV). He describes the plant with thorns, similar to the pear but thinner, sharper and harder; the fruit is not edible. However, the fruit as well as the leaves are highly perfumed and are used to protect laundry from moths. Theophrastus goes on to say that the citron produces fruit continually and, while some fall off during maturation, others grow to maturity; the fruits come from specially formed flowers with a style, while the other flowers which are not formed in this way fall off and do not produce fruit. Anyway, there is ample evidence that the Greeks were familiar with

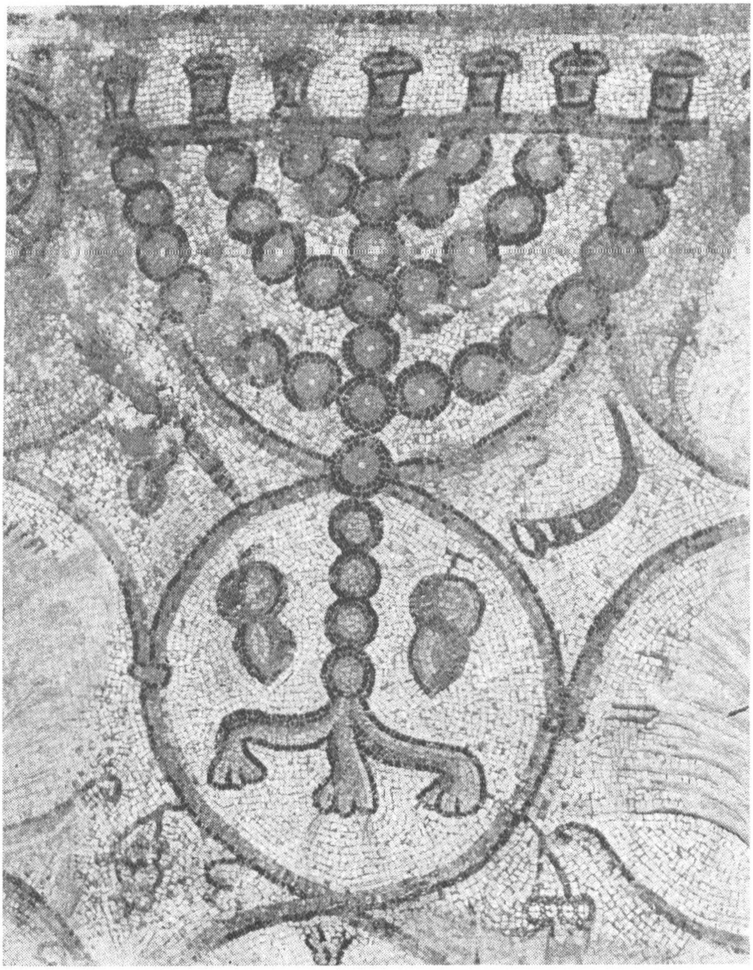

Fig. 3.4b. Two citron fruit on the sides of the 'Menorah' (the holy lamp of the temple) in the mosaic of synagogue at Nirim, southern coastal plain, Israel.

citron tree culture in that period; in fact, Theophrastus, who never left Greece, gave an excellent morphological description of the citron tree, describing how to produce and utilize its fruit (Fersini *et al.*, 1973). On the contrary, Virgil (70–19 BC), the first of the Latin writers, in his *Georgiche* (Book II-146) only briefly describes the citron, again using the name *Median apple*. He writes: 'in Media the apples of happiness (*Felix malum*) grow, whose juice has a persistently wretched taste, but is an excellent remedy against poisons'.

In the first century AD, the citron was known in Rome where the cultured Pliny (24–79 AD) in his *Historia Naturalis* gave the citron several names, e.g. *Malus medica* and *Malus assyria*, and also changed the Greek name *Kedros* into the Latin *Citrus*. Like Virgil, Pliny says that the citron, brought from Persia to Rome, was used as an antidote to poisons. There are other references to the citron written at the same time as Pliny and afterwards.

Certainly the most accurate technical description regarding the citron plant and its cultivation comes from Palladius, an agronomist who lived in the 3rd or 4th century AD. In his *Agriculturae Opus*, he dedicates a chapter to the citron, *De citreo*, where all the methods of propagation are well described, and he affirms that the citron needs enormous work and care in pruning.

Moreover, further proof of the knowledge that the Romans possessed about the citron comes from many sculptures, pictures and mosaics found in different places. Tolkowsky identified quinces and citrons on a tomb fragment of a 2nd century AD villa in Rome; moreover, he recognized leaves, flowers and citron fruits on the remains of a pictorial mural of Pompeii kept in the museum in Naples (Italy) and also on a mosaic belonging to a Roman Villa in Carthage. A mosaic in which the citrons are perfectly recognizable is in the Villa del Casale (probably built in the 2nd or 4th century AD) situated near Piazza Armerina (Sicily).

The lemon

The exact area of the origin of the lemon (*C. limon* (L.) Burm. *f.*) is still uncertain. It must have originated somewhere in southeastern Asia.

Tolkowsky and others suggest that the lemon is native to India: *Jambila* or *Jambira* is the term which indicates the citron and the lemon; both are part of the historical culture in certain Indian regions, and various forms of lemons grow wild (namely in the southern part of Himalaya and in Assam) and, because of this, many taxonomists believe India to be the exact location where the citron and lemon originated. It is possible that a population of citrus trees with characteristics similar to both the lemon and the citron existed there. Moreover, it is well accepted that lemons derive from citrons. There is evidence of a mixture of characteristics between lemons and citrons in the modern Indian *Citrus jambhiri*, which is analogous to the lemon in certain ways and to the citron in terms of the wrinkles of its skin.

Webber *et al.* (1967) consider southern China and probably Upper Burma to be the native home of the lemon. Gallesio, De Candolle and other early writers consider that the lemon was unknown to the Mediterranean people until the 10th century, when the Arabs brought the first lemons from India. According to Tolkowsky, these authors are mistaken: in fact, some branches with citron and lemon fruits can be seen in a mosaic in a Roman villa at Carthage (2nd century AD). Calabrese (1998) adds some more pictorial and musive depictions to those of Tolkowsky, such as the mosaics of the Villa del Casale di Piazza Armerina in Sicily and the Pompeian fresco of the Casa del Frutteto in Pompei and of the National Museum of Naples, and states that the lemon was well known by the Romans since the imperial period, although there are no literary references to this species.

To confirm the above-mentioned theory, other lemons are admirably represented in one of the best preserved Roman mosaics: the decoration of the vault of the Mausoleum of Saint Constance in Rome built by the Emperor Constant, son of the Emperor Constantine (306–337 AD).

Webber *et al.* (1967) say, with reasonable certainty, that the lemon, as well as the citron, the sour orange and the pummelo, had been introduced by the Arabs into Spain and the countries of North Africa by 1150 AD. The lemon is described in all the works of Arabian writers of the 12th century, and it is certain that the culture of the lemon was furthered by the Arabs in Palestine and Persia and commonly grown there. The first literary mention of the lemon, as well as the sour orange, in the Arabic language is in the 'Book of Nabatian Agriculture', a collection of ancient writings edited by the Iraqi agronomist Ibn el-Wahshya in 904 AD. The author calls the lemon *hasia* in the Nabatian language and affirms that in Persia it was known by the term *limun*, and it derives from the citron and produces a round, aromatic fruit. However, not all authors accept that this description corresponds to the lemon; it could be a citron or a lime and, to support this theory, some scholars speak of a *limun* or a *limu* with a non-acid pulp and so the fruit referred to could be the Mediterranean lime.

Following the activity of the Arabs in spreading citrus, the role of the Crusades in the West was very important. It was the first four crusades (1096–1204) which facilitated the spread of citrus fruit to France. During the 13th century, lemons were grown in Italy and in the South of France and Spain. Matteo Silvatico, a ligurian doctor at the beginning of the 14th century, claims in his *Opus Pandectarum Medicinae* that lemon juice is used as a medicine and as an appetite stimulant; moreover, the lemon was considered an effective remedy against worms and in curing pestilential fevers, and nausea in pregnant women.

Since the 15th century, the Spanish, the Portuguese and other navigators have been responsible for the spread of citrus in different subtropical areas of the world. It seems that the lemon, as well as other species, reached the USA thanks to Christopher Columbus who brought seeds and plants there from the island of Haiti during his voyage in 1493.

Sour and sweet oranges

The sour orange (*Citrus aurantium* L.) is believed to be native to South-east Asia, possibly India, while the sweet orange (*Citrus sinensis* (L.) Osbeck) originated in southern China and possibly as far south as Indonesia (Webber *et al.*, 1967).

The sour orange was the first species that spread from its native land to the West. Gallesio maintains that it was unknown to the Romans, while Tolkowsky, in total disagreement, asserts that the Romans were familiar with sour oranges. However, his theory loses credibility when he claims that the Romans were also familiar with sweet oranges. Webber *et al.* consider that sour oranges may have been brought to the Mediterranean basin after the Arab conquests; during the 10th century into Persia, Iraq, Syria, Palestine and Egypt, and later into North Africa, Sicily, Sardinia and Spain. It seems that the first description of the sour orange was given by Albertus Magnus (1193–1280) who called it *Arangus*.

According to Tolkowsky, the orange tree (sour orange) was known in Italy before Pompeii was destroyed (~70 AD) since there is a clear representation of an orange in a Pompeian mosaic.

Tolkowsky asserted that the sweet orange grew in Italy during the early Christian era, but no written evidence was found until the 15th century. By the beginning of the 16th century, evidence appears of the commercial importance of sweet oranges in Southern Europe. It is not so easy to trace how the sweet orange reached Europe for the first time since indications regarding its introduction into the Mediterranean area are often conflicting. Some authors support the view that the Portuguese brought it from India after they discovered the direct sea route around the Cape of Good Hope (1497) or after they reached China for the first time in 1518. Valmont de Bomare (1764) states that the first imported tree from which all the sweet orange trees of Europe are derived was imported from China and acclimatized in Lisbon at that time in the garden belonging to Count St Laurent. In conflict with this theory, Gallesio (1811) describes two documents found in the archives of Savona, one dated 1471 and the other a year later, 1472, which indicate sweet oranges; they are called *citruli* in the first document and *citranguli* in the second one. The hypothesis that the Portuguese brought oranges to Italy for the first time has little credibility; in fact, the culture of the sweet orange in Liguria was very extensive at the beginning of the 16th century but the Portuguese did not go there before about 1520. The question is, what is the origin of these oranges? Gallesio reported that they reached Europe from South Asia during the commercial trade route established and maintained by the Genoese during the Crusades. This hypothesis is based on the fact that several mentions of the sweet orange can be found in the literature of the 15th century. Tolkowsky also says that the sweet orange was known in Italy, Spain and Portugal before the voyage of Vasco de Gama (1498). In all three of these countries we can find specific literary references dating to the beginning of 1500, so it seems that the sweet orange reached Europe some time in the early part of that century, probably over the Genoese trade route.

The mandarin

China is one of the native homes of the mandarin (*Citrus reticulata* Blanco) and has a cultivation history of about 4000 years. Hirai *et al.* (1986) put forward the suggestion that cultivated mandarins have three origins: India, China and Japan. The mandarins in China were originally divided into two groups: *Kan* or *Gan* (*Macroacrumen*) and *Chu* or *Ju* (*Microacrumen*) on the basis of colour, shape and size of the fruit, surface, thickness and smell of the fruit peel, taste of the pulp and morphological tree characteristics (Li *et al.*, 1992).

The earliest record of *Ju* was in the book 'Tribute of Yu' during the kingdom of Ta Yu (2205–2197 BC).

The word *Kan* first appeared as *Huang Kan*, *Cheng* and *Lu Chu* during the Han dynasty (202 BC–220 AD). According to recorded history, although mandarins were cultivated in China as early as the time before the Xia dynasty (21st–16th century BC), the concept of fruit varieties was not developed until about the 3rd century during the Chin dynasty (265–420 AD). Before this period, mandarins had been called *Ju*, but this word was used for all citrus plants, such as the sour orange and the Yuzu (*C. junos* Tanaka), cultivated at that time (Li *et al.*, 1992).

The mandarin is the foremost citrus in Japan. Mandarin seeds were brought to Japan from China, probably to Kagoshima on Kyushu island; the first reference to the mandarin in Japanese literature was made by Kokwan (1278–1346 AD) (Spiegel-Roy and Goldschmidt, 1996).

Tolkowsky says that the introduction of the mandarin into Europe is fairly recent. The first European country to grow the small, loose-skinned mandarin orange was England; the first mandarin tree was brought there from China in 1805, and it was from England that the tree spread, first to Malta and then to Sicily and to continental Italy.

The pummelo and the grapefruit

The pummelo (*C. grandis* (L.) Osbeck) is of tropical origin; according to Hodgson (1967), it seems reasonably certain that the pummelo is indigenous to the Malayan and East Indian archipelagos. The number of varieties in the Malay Archipelago indicates an ancient cultivation.

It may have spread from the Malay and Indian Archipelagos to China and not vice versa (Spiegel-Roy and Goldschmidt, 1996); it later spread to Persia, Palestine and Europe. According to Tolkowsky (1938), it was mentioned in Palestine in 1187 AD and also in Spain, in the same period. In 1646, Ferrarius in his *Hesperides, sive de malorum aureorum cultura et usu* described and illustrated several kinds of pummelo growing in Italy (Fig. 3.5).

According to Webber (1943), the pummelo was described for the first time in Jamaica by Sloane (1696) who used the

Fig. 3.5. *Aurantium maximum* (pummelo) as represented by Ferrarius in *Hesperides* (1646) pp. 439 and 441.

name 'shaddock', and the same author added, later, in 1707, that the seed of the pummelo was brought to Barbados by Captain Shaddock, the Commander of an East Indian ship.

The grapefruit (*C. paradisi* Macf.) appears to have originated as a mutation or hybrid of the shaddock in the West Indies, perhaps Barbados. The fruit was mentioned for the first time by Griffith Hughes in his *The Natural History of Barbados* (1750); he writes of a '*forbidden fruit*', describing its pyriform shape and the absence of winged petioles. He describes the size of the fruit as '... between a large orange and the fruit of the smaller pummelo tree ... ' and deems it to be 'truly exquisite'.

The lime

Limes comprise a varied group of types – both acid and sweet varieties – which are so different from one another in tree and fruit characteristics that they have been given separate species status. There are two kinds of acid limes, the small-fruited Mexican (West Indian, Key) type, *C. aurantifolia* Swing., and the large-fruited Tahiti (Persian, Bearss) lime, *C. latifolia* Tan., which is triploid and therefore seedless.

Citrus aurantifolia is native to the Malaysian region of south-western Asia, while *C. latifolia* probably originated in the East and then spread to Persia, and then to Tahiti, possibly via Brazil and Australia, and finally to California.

Citrus limettioides Tan., the sweet lime, commonly referred to as the Indian or Palestine sweet lime, is native to north-eastern India where it is known as *Mitha nimboo*, while in Egypt it goes by the name of *Limun helou*.

Papedas

The subgenus *Papeda* (according to the Swingle taxonomy system) comprises six species (*C. ichangensis*, *C. latipes*, *C. micrantha*, *C. celebica*, *C. macroptera* and *C. hystrix*) which are native to different areas. *Citrus ichangensis* is native to south-western and west-central China where it grows in a truly wild state (Swingle, 1967); on the contrary, Scora (1975) considers that this species, along with *C. latipes*, originated in the sub-Himalayan area and then spread in south-western China, north-eastern India and northern Burma, far north of the range of the other Papedas. Swingle considers that *C. latipes* originated in a collection of plants in the Khasi Hill of north-eastern India. However, very similar types can be found in the mountains of northern Burma. *Citrus micrantha* seems to be native to the southern Philippine Islands, while *C. celebica* originated in the Celebes Islands in Indonesia. *Citrus macroptera* was discovered by Father Montrouzier on the Island of Art, situated a few miles to the north-west of the north end of New Caledonia (Swingle, 1967). Finally, the last species belonging to Papeda is *C. hystrix*, the best known and most widely distributed species of this subgenus in Sri Lanka, the Philippines and numerous Pacific Islands, but its exact origin is unknown.

Taxonomy

The majority of taxonomists consider that citrus species belong to the Geraniales order, the Rutaceae family and the subfamily Aurantioideae. Rutaceae is one of the 12 families in the Geraniineae suborder, and the Aurantioideae subfamily – one of the seven belonging to the Rutaceae (Engler, 1931) – is rather numerous and comprises the commercial citrus species and also several important related genera.

Aurantioideae, the 'Orange' subfamily, has been subdivided by Swingle into two tribes: Clauseneae with five genera and Citreae with 28 genera including *Citrus* and related genera, i.e. *Fortunella*, *Poncirus*, *Eremocitrus*, *Microcitrus* and *Clymenia*. The taxonomic situation of tribes, subtribes, genera and species within the Aurantioideae is controversial, complex and sometimes confusing. Citrus and many related genera hybridize readily and have done so in the wild for centuries. In Fig. 3.6, the tribes, subtribes, genera and species are listed after Swingle (1967).

Clauseneae comprise the more primitive genera: the fruit usually consists of small, semi-dry or juicy berries, except in *Merrillia* in which the fruit is of ovoid shape with a thick, leathery exocarp.

The tribe Citreae comprises three subtribes: Triphasiinae, Balsamocitrinae and Citrinae; the latter, with 13 genera, has been classified into three groups (Swingle, 1967). Group A 'the primitive citrus fruit trees' with five genera, *Severinia*, *Pleiospermium*, *Burkillanthus*, *Limnocitrus* and *Hesperethusa*, group B 'near citrus fruit trees' with only two genera, *Citropsis* and *Atalantia*, and group C 'true citrus fruit trees' includes six genera, *Fortunella*, *Eremocitrus*, *Poncirus*, *Clymenia*, *Microcitrus* and *Citrus*.

The first descriptions and classifications of citrus species and varieties date back to the 17th century, while the morphological description of the trees, the floral morphology and biology or the different uses of the fruit (i.e. citron fruit) date back to ancient times (Teophrastus 310 BC, Virgil 70–19 BC, Pliny 27–79 AD, etc.).

John Baptista Ferrarius, a Jesuit priest of Siena, made the first organic contribution

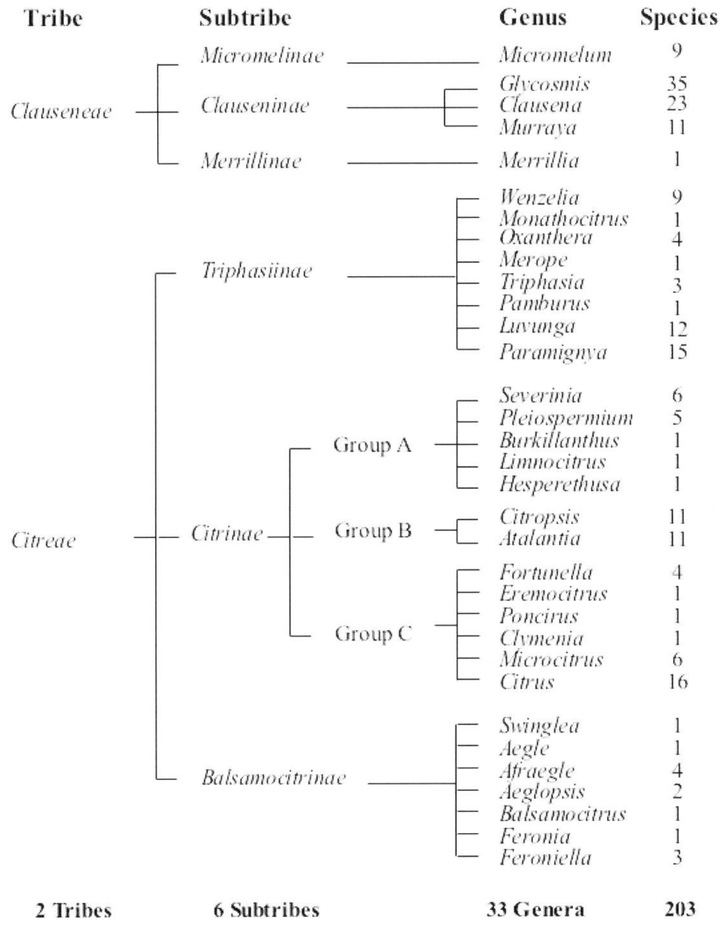

Fig. 3.6. *Aurantioideae* subfamily (after Swingle, 1967).

(a) (b)

Fig. 3.7. (a) *Limon citratus primae notae laevior,* Ferrarius (1646) p. 265. (b) *Malum citreum vulgare,* Ferrarius (1646) p. 61.

to citrus studies with his book *Hesperides, sive de malorum aureorum cultura et usu* (1646). The assistance of Cassiano dal Pozzo (1588–1657), who provided the original drawings and all the information to complete the book, was invaluable. The drawings, done in tempera, form part of Cassiano dal Pozzo's so-called 'paper museum', which is preserved today in the Royal Library of Windsor and in some private collections.

Ferrarius's work is divided into four books: in the first one, a table is inserted with the reproduction of various Greek and Roman coins, while the other three books are dedicated to the *Hesperides*: *Aegle*, *Aretusa* and *Hesperetusa* (the nymphs who guarded the garden of golden apples at the western extremity of the earth). The citron, the lemon and the orange are well described in them.

In an attempt to contain all the citrus in only three taxonomic groups, Ferrarius also places the chinotto orange (*Aurantium sinense*) and the pummelo (*Aurantium maximum*) with oranges; he puts with oranges and lemons some phenotypes (e.g. *Limon citratus* and *Aurantium citratum*) whose names and fruit characteristics indicate that they could be interspecific hybrids.

The Ferrarius iconography was certainly created for a taxonomic purpose, but it represents a happy meeting point between figurative culture and scientific literature. The fruits are drawn to their real size and are often cut into two (Fig. 3.7a and b). Even though the drawings that illustrate the work are not coloured, the fineness of the features is such that they succeed in rendering the three-dimensionality of the fruits and their morphological characteristics with great effectiveness and provide excellent visual documentation (Baldini, 1997).

A few years later, other authors described citrus fruits, although in less detail (Steerbeck, 1682; Hermann, 1687; Tournefort, 1700).

The first volume of the *Nürnbergische Hesperides* by Johann Christoph Volkamer was published in 1708, followed 6 years later by another volume complementary to the first one.

Even though this monumental work was written in German and printed in Nuremberg, it describes the citrus fruit of the north of Italy and particularly those cultivated on the shores of Lake Garda, on the coast of the Brenta and on the Ligurian coast, and can be therefore considered an Italian citrology. The scientific approach traces that of Ferrarius: the first parts of both volumes, dedicated to the nymph *Aegle*, deal with citrons; the second one, dedicated to *Aretusa*, concerns lemons (Fig. 3.8), and the third one, dedicated to *Hesperetusa*, explains the orange and pummelo that are distinguished from one another and marked with a specific terminology. Even though the number of genotypes represented in Volkamer's book is greater than those described by Ferrarius, it remains uncertain if each one of them corresponds to a different genotype.

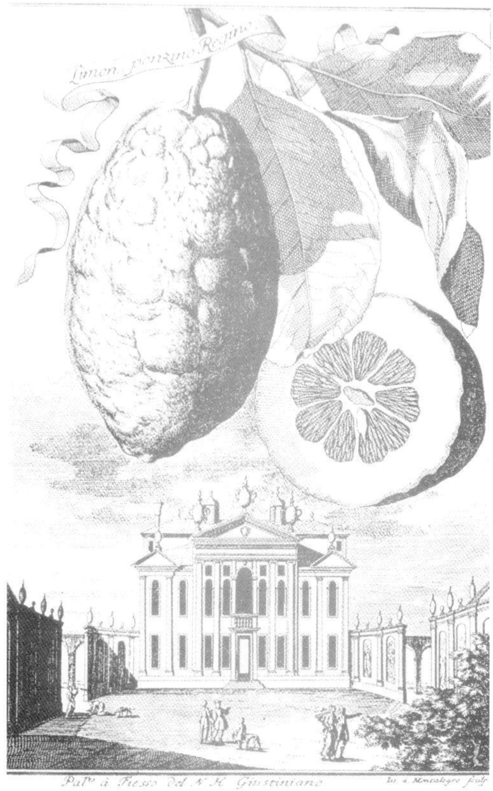

Fig. 3.8. A lemon, *Limon ponzino Regino*, as represented by Volkamer (1708) in *Nürnbergische Hesperides*.

A complete change of the classification system of citrus is given by Linneus (1737) who, in his work *Genera plantarum*, created the genus *Citrus*, attributing three main species to it: *Citrus medica* (citrons and lemons), *Citrus aurantium* (sweet and sour oranges and the pummelo) and *Citrus trifoliata*.

In 1767, Linneus added in *Systema naturae* the species *Citrus decumana*, thus separating the pummelo from the complex species *Citrus aurantium*.

Linneus collaborated with Osbeck, and together they formulated the binomial names of three species: *Citrus grandis*, *Citrus limonia* and *Citrus sinensis*.

Proceeding with this short review on the classification of the species, we get to 1768, when Burmann raised the lemon to the rank of the species, giving it the name of *Citrus limon*.

In 1790, while he was studying the citrus of Indochina and Canton, De Loureiro discovered some species such as *Citrus nobilis* (King mandarin), *Citrus madurensis* (calamondin) and *Citrus margarita* (later *Fortunella margarita* Lour. (Swing.)).

In 1813, De Candolle studied and classified *Citrus hystrix* that was the first species of the *Papeda* subgenus according to the classification later done by Swingle.

In 1837, Blanco named a new species, *Citrus reticulata*, a type of mandarin of the Philippines.

In the first years of the 19th century, a

lot of authors published papers on the citrus. The first one must have been Giorgio Gallesio who, in 1811, published the *Traité du Citrus* in Paris. It is a modest work from a publishing aspect, but it stands out for its innovative contribution to citrus taxonomy. The work constitutes the first organic attempt at the systematic botany of the *Citrus* genus, of its species and of its cultivated varieties. Citrons, lemons, sour oranges, sweet oranges and their hybrids are well described in the volume.

In the sole illustration of the work, a synoptic table of *Citrus*, citrus are distributed on the ramifications of four leading branches of a symbolic genealogical tree, distributed in four species: *Citrus medica cedra* (citron), *Citrus medica limon* (lemon), *Citrus aurantium indicum* (sour orange) and *Citrus aurantium sinense* (sweet orange) (Fig. 3.9).

Gallesio's work should have been accompanied by a citrografic atlas containing colour tables of the main varieties of citrus. However, the atlas was not actually published and the drawings mentioned in the preface were unknown for almost two centuries; they were found only recently among the papers and the books that Gallesio had left to his Genoan heirs (Baldini, 1996, 1997).

Risso's work followed in 1813. New classifications were elaborated on the diagram proposed by Linneus. In particular, Risso distinguished sour oranges with the binomial terminology of *Citrus vulgaris* and later of *Citrus bigaradia* (this synonym was used in Mediterranean countries for a long time); Risso changed the *Citrus limon* of Burmann to *Citrus limonium*, and included all the sweet orange varieties under the name *Citrus aurantium*. Risso also created a new species, *Citrus limetta*, distinguishing it from citrons and lemons. Subsequently he added some new species: *Citrus aurata*, *Citrus peretta*, *Citrus mellarosa* and *Citrus rissoi*; all these species refer to intermediate types of citrus between limes and pummelo.

Finally, the work Risso published in collaboration with Poiteau in 1818 is bound to be of considerable importance: *Histoire naturelle des orangers* in which the bergamot and the 'lumia' that were called *Citrus bergamia* and *Citrus lumia* were separated, as different species.

The following taxonomists were interested in the systematics of the Aurantioideae at the same time as the scholars already cited and afterwards: Blume (1823), Blanco (1837), Macfadyen (1837), Tenore (1840), Fortune (1848), Oliver (1861) and Pasquale (1867).

In 1875, the work of Hooker, *Flora of British India*, was published in which the citrus and related genera found a botanical place in 13 genera and 43 species (Hooker attributed only four species to the *Citrus* genus).

In 1888, Bonavia published a voluminous treaty on the oranges, lemons and other citrus of India and Ceylon and, subsequently, in 1890, he added the publication of an atlas. Bonavia's theories on the morphology and evolution of citrus were extremely clever and original, but they diverged widely from the standard of other taxonomists (Swingle, 1967).

In 1896, Adolph Engler published a classification of the Aurantioideae that was subsequently examined and fixed in 1931. In this second edition of the *Die natürlichen Pflanzenfamilien*, Engler, in an effort to describe the Rutaceae family, divided the subfamily of the Aurantioideae into 29 genera including 180 species, 11 of which belong to *Citrus*.

Around the same time as Engler's work, other studies were carried out on the Aurantioideae by several taxonomists, such as Bailey (1895, 1903) or Guillaumin (1911). The latter, for example, worked on the citrus of Indochina and he re-examined the previous citrus taxonomic classification and divided the Aurantioideae into 11 genera and 44 species, six of which belong to *Citrus*.

The Swingle and Tanaka systems

Two different classification systems are commonly accepted for the citrus taxon-

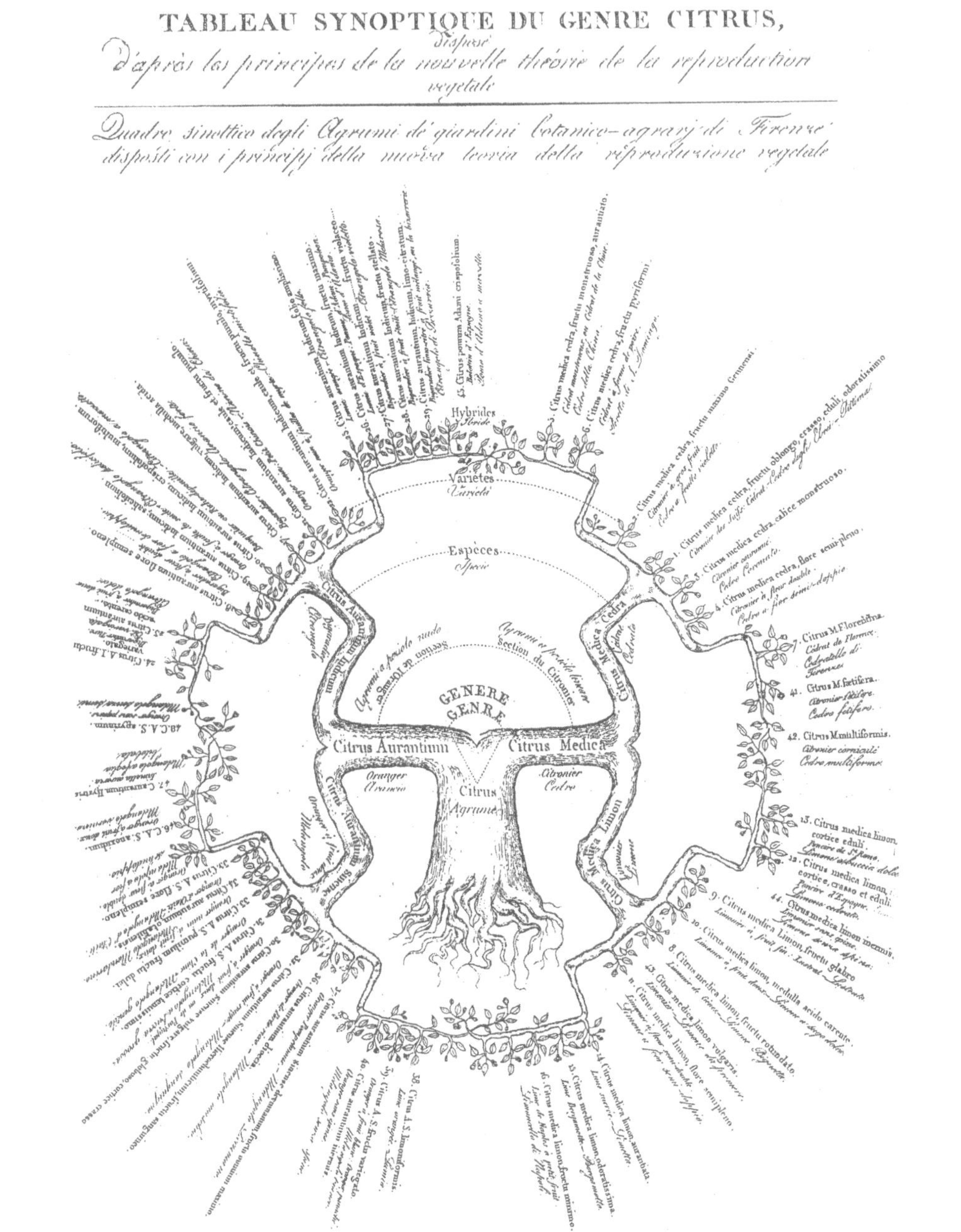

Fig. 3.9. The synoptic table of *Citrus* of Gallesio.

omy: the system of Swingle (1943, 1967) and that of Tanaka (1954, 1961).

Despite their collaboration, between 1915 and 1930, the differences in their ideas are obvious; their classifications represent two different taxonomic conceptions and their contrast is clearly emphasized just by observing the subdivision within the genus *Citrus*.

Between 1912 and 1926, W. T. Swingle,

of the United States Department of Agriculture, proposed a new system for the classification of the Aurantioideae, with a series of publications which were definitively coordinated in 'The botany of *Citrus* and its wild relatives', which appeared in the first volume of the *The Citrus Industry* edited by Reuther *et al.* (1967). Swingle accepted the Engler classification and divided the genus *Citrus* into two subgenera: *Citrus* (in 1943, 1st edition of the *The Citrus Industry*, *Eucitrus*) and *Papeda*, which included, respectively, ten and six species.

The two subgenera were separated according to their morphological characteristics and to the chemical components of flowers, leaves and fruits. To be more precise, the species of the first subgenus have flowers with grouped and perfumed stamens, and a small petioles no larger than three-quarters of the whole leaf; the fruit has edible flesh with little or no bitter oil. On the contrary, the *Papeda* have small flowers with free separate stamens, large winged petioles, that can be very big, reaching even larger dimensions than those of the same lamina; the fruit is inedible due to the high content of bitter oil in its flesh.

In 1926, almost at the same time as Swingle's studies, Marcovitch published his classification which divided the *Citrus* into three groups, *Aurantium*, *Intermedium* and *Medica*, and 20 species.

From 1915 to 1961, Tyozaburo Tanaka supplied a considerable contribution to the classification of the Aurantioideae and published a series of information-packed papers on the taxonomy entitled '*Revisio aurantiacearum*'.

In 1954, in *Species problems in Citrus*, he published a classification of the genus *Citrus* in which two subgenera, *Archicitrus* and *Metacitrus*, eight sections, 13 subsections, eight groups, two subgroups, two microgroups and 145 species were distinguished. Seven years later, in 1961, he added two new subsections, another group and 12 new species to his system, taking the total to 157 species.

He hypothesized that the citrus originated in Asia about 30 million years ago from *C. hystrix*, *C. latipes*, *C. macroptera* and *C. combara*.

Tanaka's classification is more complex compared with that of Swingle because of the far greater number of species included in each subgenus; on the contrary, Swingle's classification appears easier to understand although it does not provide an exhaustive description of citrus systematics. Even so, the Swingle system is the most used, but there are some cases in which it is amplified with the inclusion of some species belonging to Tanaka's system.

There is a big difference between the two systems regarding the taxonomic classification of mandarins. This is particulary relevant because it concerns a group that includes many genotypes that are widely cultivated and of great economic importance. So, Swingle includes in the species *Citrus reticulata* all the mandarins except *C. tachibana*, a wild species from Japan, and *C. indica*, a wild species from India, whereas Tanaka separates mandarins into 36 species. Swingle affirms that in some cases the sole distinguishing characteristic is the size of the leaves or of the fruits, and affirms that such differences are easily influenced by different factors: the rootstock, the humidity and the soil fertility.

Tanaka gave special consideration to the cultivated species (not wild) and, in 1928 at the International Congress of Horticulture of Vienna, he proposed the adoption of the term *Hortulanorum* next to the name of the taxonomist (e.g. *Citrus clementina* Hort., ex Tanaka).

In order to heal the rift between the American school of Swingle and the Japanese school of Tanaka, Hodgson proposed a new classification in 1961. He increased the number of the species from 16 to 36, dividing them into four groups: 'acid fruits', 'orange group', 'mandarins group' and 'other'.

Modern taxonomic systems

All the citrus taxonomic classifications elaborated in the past were based only on the morphological and anatomical differences and on the geographical area of origin.

Subsequently, from the beginning of the last century, chemical characters have also been used for this purpose. Swingle (1943) first mentioned the possibility of using glycosides as a taxonomic marker, in addition to classical morphological characters. After Swingle, several researchers employed various techniques in order to analyse the biochemical composition of different parts of the citrus plant i.e. leaves, flowers and fruits; in particular, these include long chain hydrocarbon profiles (Nagy and Nordby, 1972), flavonoids (Tatum *et al.*, 1974), leaf and rind oils (Malik *et al.*, 1974), polyphenol oxidase-catalysed browning of young shoots (Esen and Soost, 1978) and seed teguments (Gorgocena and Ortiz, 1988), root peroxidase isoenzymes (Button *et al.*, 1976), leaf isozymes (Soost and Torres, 1982; Hirai *et al.*, 1986), and fraction I protein in leaves (Handa *et al.*, 1986).

Taxonomic studies received a boost from the work of Barrett and Rhodes in 1976: 'A numerical taxonomic study of affinity relationships in cultivated *Citrus* and its close relatives'. Barrett and Rhodes performed a comprehensive phylogenetic study that evaluated 146 morphological and biochemical tree, leaf, flower and fruit characteristics. This study suggested that only three citrus types, namely the citron, *Citrus medica*, the mandarin, *Citrus reticulata*, and the pummelo, *Citrus grandis* (now called *C. maxima* Burm. Merrill), constituted true or valid species. Similar results have been communicated by Scora in 1975 and, in 1988, he added another true species, the *Citrus halimii*, a species previously described by Stone *et al.* in 1973 as a new species from Malaya and peninsular Thailand.

Barrett and Rhodes postulated relationships among citrus, indicating a probable origin of cultivated citrus and some *Citrus* and related genera. In Fig. 3.10, some of the affinity relationships according to Swingle

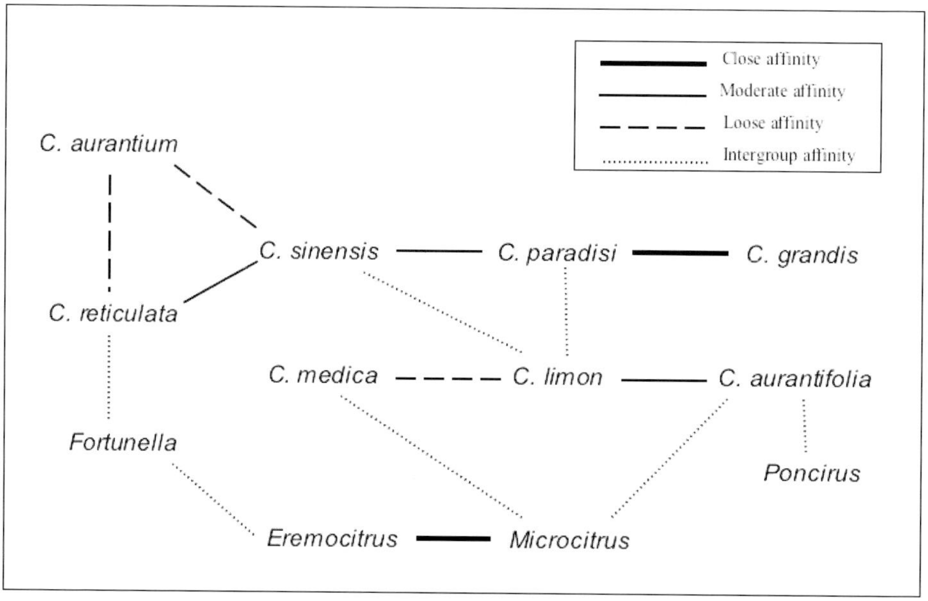

Fig. 3.10. Affinity relationships between some *Citrus* species and relatives after Swingle (1943) and Barrett and Rhodes (1976).

(1943) and Barrett and Rhodes (1976) are listed.

Moreover, modern techniques have been instrumental in deciphering the taxonomic situation in *Citrus*; the development of various biochemical and molecular markers has provided some answers regarding the relationships of various citrus types. In particular, DNA markers, with their phenotypic neutrality, abundance and imperviousness to environmental conditions, have been most useful.

The concept of the true valid species and the other genotypes derived from hybridization between them has gained further support from various studies using biochemical and molecular markers, including isozymes (Torres *et al.*, 1978; Fang *et al.*, 1993; Herrero *et al.*, 1996), organelle genome analysis (Green *et al.*, 1986; Yamamoto *et al.*, 1993; Luro *et al.*, 1995; Nicolosi *et al.*, 2000), microsatellites (Fang and Roose, 1997; Fang *et al.*, 1998), RFLP (restriction fragment length polymorphism), RAPD (random amplified polymorphic DNA) and SCAR (sequence characterized amplified region) analyses (Luro *et al.*, 1992; Federici *et al.*, 1998; Nicolosi *et al.*, 2000).

Even though different authors have studied the phylogenetic relationships in *Citrus*, the work carried out by the group of M. L. Roose, University of California appears very interesting because of the large number of genotypes investigated and the results obtained (Federici *et al.*, 1998). They analysed the relationships among 88 accessions of *Citrus* and related genera using RFLP and RAPD analysis. The main results obtained suggest that *C. maxima* has some affiliation with some papedas; *C. medica* clustered with *C. indica*; *Fortunella* is not separate from the genus *Citrus*; and mandarin species do not cluster in the groups Tanaka used. Another important result is that of *C. ichangensis* (a *Papeda* species in the Swingle classification, but in the *Metacitrus* subgenera of Tanaka) that is assumed to be a distinct species, very different from most other *Citrus* species, loosely aligned with *C. hystrix* and *C. micrantha*, but not easily placed in relation to other species. Moreover, the researchers arrived at the conclusion that it is not appropriate to place *C. ichangensis* into the subgenus *Metacitrus* of Tanaka with all the mandarins.

Many authors have investigated the relationships in the mandarin group; among these, the studies carried out by Li *et al.* (1987, 1988, 1992) on the basis of isozyme data are particularly interesting. The authors suggest that it would be better to divide mandarins into 'three primary species', the wild species *C. mangshanensis*, *C. daoxianensis* and *C. chuana*. The mangshan wild mandarin, *C. mangshanensis*, should be a transitional wild species of mandarin from one type of ichang papeda, *C. ichangensis*.

As also appears from the results of the literature cited above, the taxonomy of mandarins is very complex and is in continual development, so that new species are still emerging among the wild species placed in the primary centre of origin.

To clarify phylogenetic relationships among *Citrus* and its relatives, Nicolosi *et al.* (2000) employed different molecular markers such as RAPDs, SCARs and chloroplast DNA analysis. The analysis based on the RAPD and SCAR data of the 40 citrus genotypes, including four related genera, allowed the elaboration of different dendrograms which all indicated that the genus *Citrus* is quite distant from the related genera *Poncirus*, *Microcitrus* and *Eremocitrus*, but not from *Fortunella* (Figs 3.11 and 3.12 were obtained, respectively, by PAUP and polymorphism parsimony). Moreover, a clear separation between the two subgenera indicated by Swingle, *Citrus* and *Papeda*, was observed, with the exception of *C. indica* and *C. celebica*.

The three valid species belonging to *Citrus*, citron, mandarin and pummelo were separated into three distinct clusters, and each one also comprises some other hybrid genotypes.

In Fig. 3.11, the first cluster, the Citron cluster, comprises, of course, *C. medica* and also *C. aurantifolia*, *C. macrophylla*, *C. limon*, *C. bergamia*, *C. limettioides*, *C.*

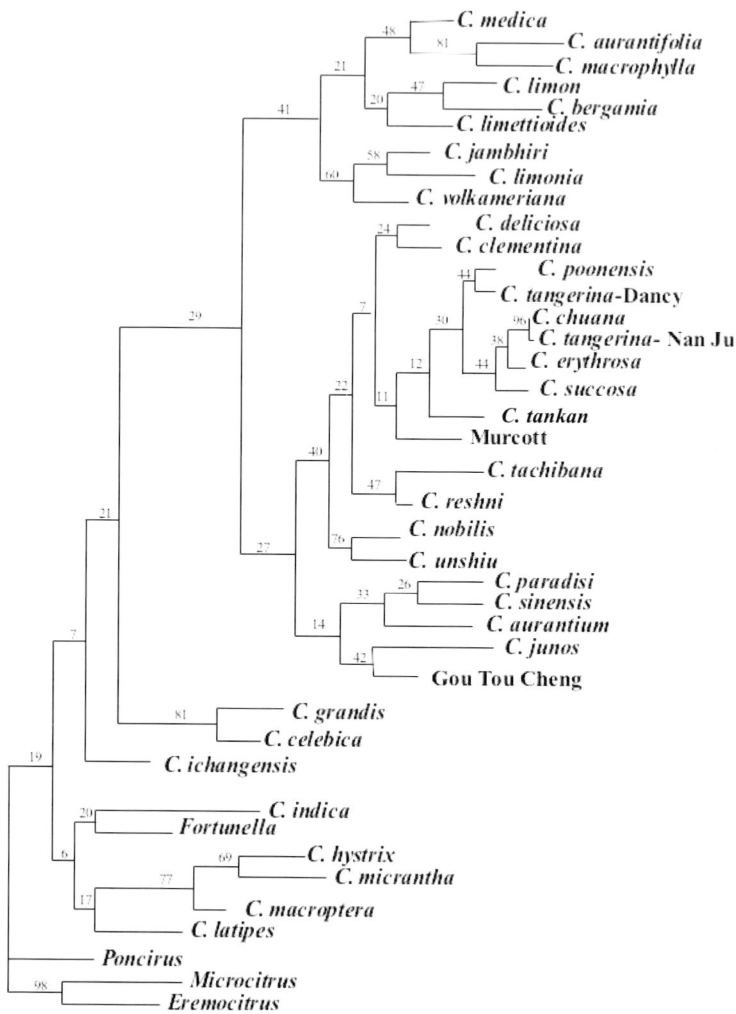

Fig. 3.11. A 50% majority rule consensus tree (PAUP tree) for 40 genotypes of *Citrus* and related genera derived from bootstrap analysis (500 replications) of RAPD and SCAR data with confidence levels for arms (Nicolosi *et al.*, 2000).

jambhiri, *C. limonia* and *C. volkameriana*. The second, called the Mandarin cluster, consists of all the mandarin and mandarin-like accessions as well as *C. tachibana* and *C. paradisi*, *C. aurantium*, *C. sinensis* and *C. junos*. The third is the Pummelo cluster with *C. grandis* and *C. celebica*, a papeda species. Moreover, other clusters were separated as reported in Fig. 3.11. Analysing the second dendrogram (Fig. 3.12), that was elaborated using another analysis system (polymorphism parsimony), the position of both *C. aurantifolia* and *C. macrophylla* is different: they moved from the Citron cluster to the Micrantha cluster; *C. micrantha* is considered as another species which contributed to the origin of the cultivated citrus. The clustering differences between the two phylogenetic trees compared with several genotypes may be explained by their hybrid origin. Besides the position of *C. aurantifolia* and *C. macrophylla*, *C.*

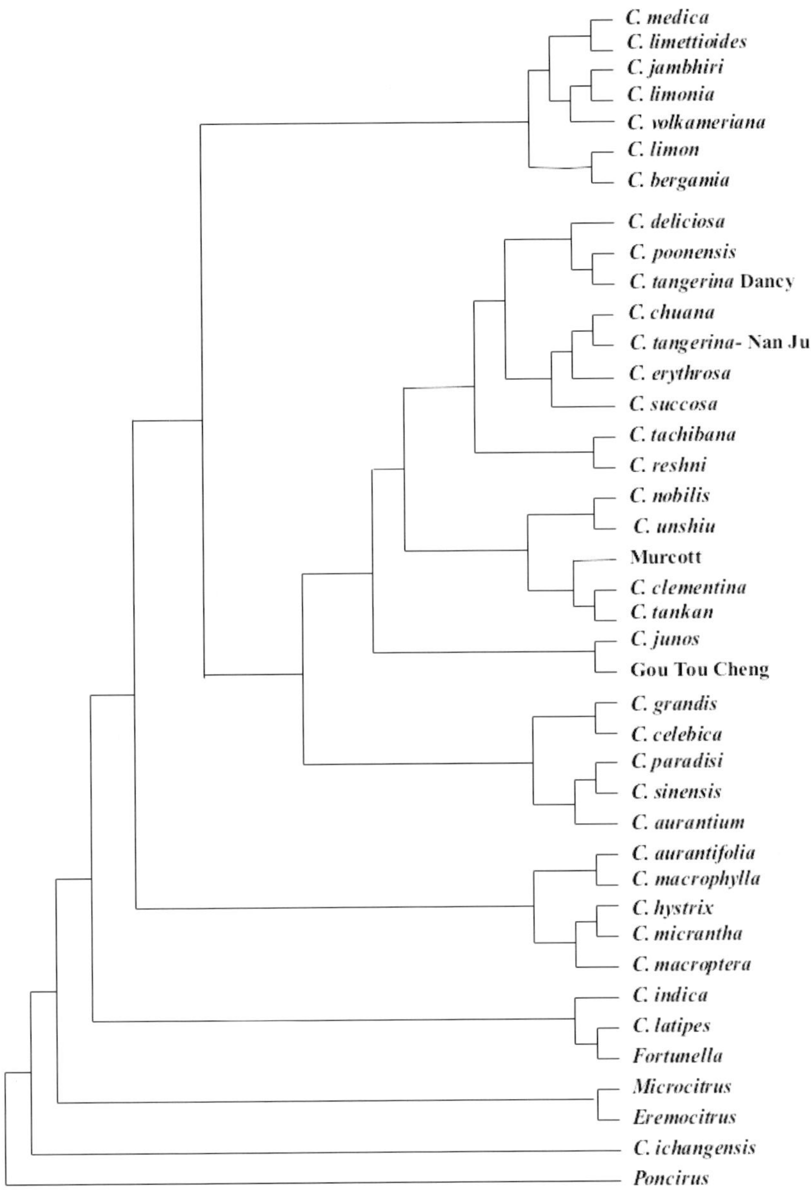

Fig. 3.12. Dendrogram (PP tree) for 40 genotypes of *Citrus* and related genera obtained by Polymorphism Parsimony analysis of RAPD and SCAR data (Nicolosi *et al.*, 2000).

sinensis, *C. aurantium* and *C. paradisi* also stay in different clusters and form a sub-cluster of the Mandarin cluster in the first dendrogram while they stay with Pummelo in the second one.

The chloroplast DNA (cpDNA) data resulted in a different phylogenetic tree which can be explained by the different nature of the two genomes: nuclear (total) DNA and cpDNA. The latter is conservative and maternally inherited and its analysis permits us to trace backwards to the original type. Each cluster located in Fig. 3.13, a phylogenetic tree elaborated by PAUP, includes accessions ranging from cultivated types to ancient ones. By analysing the dendrogram, it is possible to observe that the four related genera form a single cluster each and are distant from each other and from *Citrus*, as observed in 1986 by Green *et al.*

In contrast to total DNA data, in the analysis of the dendrogram obtained after the cpDNA data, the *Citrus* are not separated in the two subgenera suggested by Swingle but most of the *Citrus* genotypes analysed fall into two subgenera described by Tanaka, *Archicitrus* and *Metacitrus*, except for the Citron cluster, including *C. medica* and *C. indica*, which belong to two separate subgenera, i.e. *C. medica* to *Archicitrus* and *C. indica* to *Metacitrus* in the section *Acrumen*.

As reported previously, Tanaka sug-

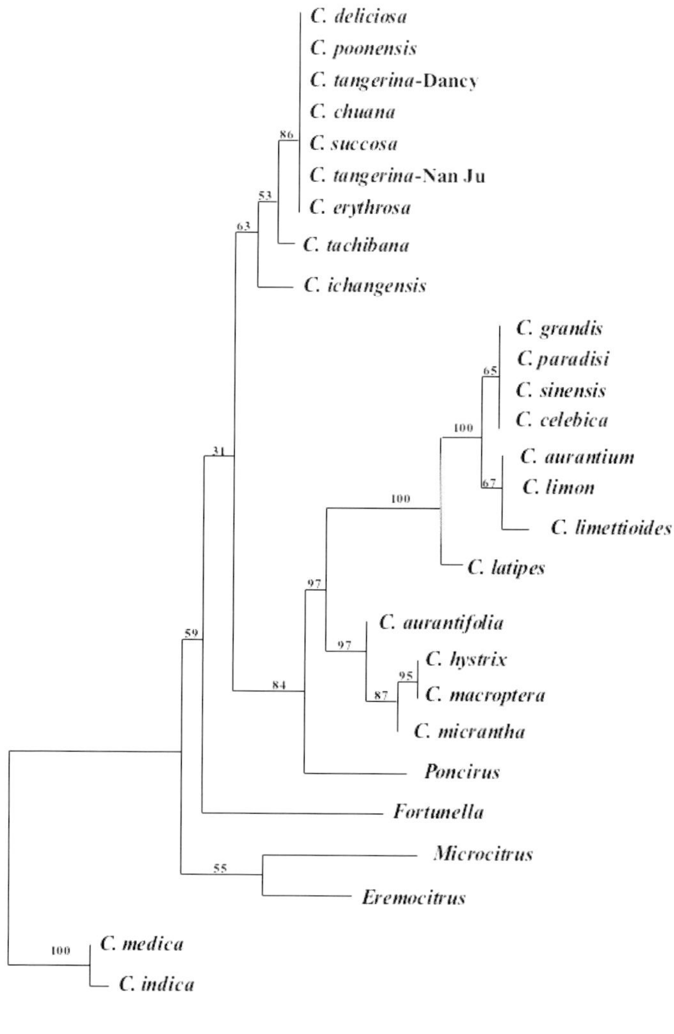

Fig. 3.13. A 50% majority rule consensus tree for 27 genotypes of *Citrus* and related genera derived from bootstrap analysis (500 replications) of cpDNA data with confidence levels for arms (Nicolosi *et al.*, 2000).

gested that citrus originated from *C. hystrix*, *C. latipes*, *C. macroptera* and *C. combara*. The presence in the dendrogram (Fig. 3.13) of *C. latipes*, *C. hystrix* and *C. macroptera*, respectively the first species in the Pummelo cluster and the other two in the Micrantha cluster, might indicate that the ancient maternal relationship is in the cluster. Unfortunately, *C. combara* was not included in this analysis. *Citrus ichangensis*, that is considered by Li *et al.* (1992) to be an ancestor of the mandarin through an intermediate genotype, in the result with cpDNA analysis, is clustered with all the mandarin accessions and *C. tachibana* that is considered to be a kind of wild mandarin species of Japan.

Origin

Which genotypes are the parents of the main citrus species? This question has interested many researchers, and a lot of hypotheses have been formulated.

It is bound to be a difficult task to go back to the relationships of different citrus species, considering the large number of genotypes that can be found in different citrus world areas.

A lot of research has already been carried out using both classical and innovative methods. The citrus germplasm is immense, and it is important to study the main centre of origin and spread. Naturally, in order to get a better understanding of the genetic origin, it is necessary to know about those ancestral species.

At this point some results obtained by different methods with regard to the origin of some citrus species will be reported. The species analysed are those considered to be of primary importance due to their economical and territorial spread.

The hypotheses of Barrett and Rhodes, and Scora regarding the three valid species found support in several other pieces of research; the pummelo, the citron and the mandarin are considered to be the ancestors of most of all the other citrus genotypes.

The group of sweet oranges (*C. sinensis* Osbeck), also called by different names such as Portugal oranges or Malta oranges, were already known in the past to Ferrarius (1646) as *Aurantium vulgare medulla dulci* and to Volkamer (1708) as *Aurantium fructu dulci* (both authors distinguished them from sour oranges). Linneus (1753) placed the sweet orange as a variety of sour orange, naming it *Citrus aurantium* var. *sinensis*. Today, the separate origin of sweet and sour oranges is well accepted. As regards the sweet orange, the most widely distributed citrus species, full agreement on its hybrid origin has been reached among different researchers. Although the presence of a lot of varieties originated by mutations, sweet oranges are thought to be hybrids between a mandarin genotype with pummelo (Barrett and Rhodes, 1976; Torres *et al.*, 1978; Scora, 1988; Fang and Roose, 1997; Nicolosi *et al.*, 2000).

The sour orange (*C. aurantium* L.) has a separate, parallel origin; as is well known, sour oranges are easily distinguished from sweet varieties by many morphological and taste characteristics, such as their broadly winged petioles, the distinctly different odour of the essential oil, the taste of the fruit, and so on. Also, different names have been used for sour oranges: *Citrus vulgaris* Risso (1813), *Citrus bigaradia* Risso and Poiteau (1818) and others; *Citrus aurantium* L. is the scientific binomial adopted after the description provided by Linnaeus in 1753. A hybrid origin can certainly be attributed to this species: it is believed by Barrett and Rhodes to be of predominantly *C. reticulata* genotype introgressed with genes from *C. grandis*, and this idea is further supported by different molecular analyses. Moreover, cpDNA analyses have revealed that, as expected, the pummelo is the maternal parent due to its monoembryonic nature (Nicolosi *et al.*, 2000).

In spite of its uncertain geographical origins, it seems certain that the grapefruit (*C. paradisi* Macf.) is very closely related to the pummelo; in fact, the grapefruit derives from a backcross between it, or an ancestral type, and a sweet orange (Barrett and

Rhodes, 1976; Torres *et al.*, 1978; Scora, 1988; Nicolosi *et al.*, 2000).

It is more difficult to go back to the exact genetic origin of the lemon. There are divergent hypotheses regarding it: both Swingle (1943) and Malik *et al.* (1974) considered the lemon to be a hybrid of citron and lime. Later, Barrett and Rhodes speculated that the lemon is a complex trihybrid of citron, pummelo and a species of *Microcitrus*, but one carrying a greater proportion of citron genes acquired by further introgression from the citron. Torres *et al.* (1978) excluded the hypotheses of Swingle and Malik *et al.* because, using isozyme analyses, they found that the W allele in the isozyme phosphoglucose isomerase was absent both in citron and lime, so they indicated the origin of lemon to be sour orange × lime. The sour orange was also suggested as a candidate parent by Hirai and Kozaki (1981), and also molecular marker data indicate that the lemon originated from citron and sour orange, the latter being the maternal parent (Nicolosi *et al.*, 2000). Using other molecular markers, inter-sequence simple repeat (ISSR) markers, Fang *et al.* (1998) suggested a possible polyphyletic origin for the lemon.

Different considerations were also put forward for the acid Mexican lime (*C. aurantifolia* Christm.): a trihybrid intergeneric cross involving the citron, the pummelo and a *Microcitrus* species was considered by Barrett and Rhodes (1976), while Torres *et al.* (1978) indicated that this lime may derive from a cross between citron and a papeda species. Using RAPD and SCAR markers, Nicolosi *et al.* (2000) found that the citron and *C. micrantha* (a papeda species that produces inedible fruit) should be the parents; also RFLP data support the notion that the citron is a parent (Federici *et al.*, 1998).

Similar results were obtained for *C. macrophylla*: Nicolosi *et al.* (2000) assumed that *C. micrantha* and *C. medica* were involved in its origin, while Swingle was of a different opinion. He considered *C. macrophylla* to be a hybrid of *C. celebica*, or some different species of the subgenus *Papeda*, with *C. maxima*. Also, the results obtained by Federici *et al.* (1998) indicate that *C. macrophylla* is closely related to *C. micrantha*.

The sweet lime *C. limettioides* (Palestine sweet lime) was considered by Webber (1943) to be a hybrid of the Mexican lime with a sweet lemon or a sweet citron. Barrett and Rhodes agreed on the involvement of the Mexican lime as a parent, but they thought that a sweet orange rather than a lemon or citron was involved. A different hypothesis emerges from molecular marker analysis (Nicolosi *et al.*, 2000) which indicates that the Palestine sweet lime might be a backcross hybrid between sweet orange and citron.

Moreover, the citron is involved in the origin of other important citrus hybrids such as the bergamot (*C. bergamia* Risso and Poiteau), the Volkamer lemon (*C. volkameriana* Ten. and Pasq.), the rough lemon (*C. jambhiri* Lush.) and the Rangpur lime (*C. limonia* Osbeck). The bergamot orange is a hybrid of *C. aurantium* × *C. medica* (Scora, 1988; Deng *et al.*, 1996), and the Volkamer lemon has a similar parallel origin. However, as reported by Nicolosi *et al.* (2000), in this genotype a greater proportion of mandarin genes was observed, probably due to backcrossing with mandarin. The citron and the mandarin are involved in the origin of the rough lemon and the Rangpur lime.

The highly varied group of mandarin includes numerous species, most of which derive from intergeneric and interspecific crosses, while others, which are surely commercially important, derive from man-made crosses.

The clementine is of great importance in the Mediterranean citrus industry. Despite fairly divergent theories about its origin, the consensus is that it is a tangor, a hybrid between orange and mandarin. Trabut (1902) referred to the discovery of an interesting new specimen among the seedlings bred by Father Clèment Rodier at the orphanage of the Father of the Holy Spirit in Misserghin, near Oran, in Algeria. The Algiers Horticultural Society called it clementine, the name suggested by Trabut, who also put forward the mandarin and the 'Granito' sour orange as the parents of the clementine. Webber (1943) and Chapot (1963) disagreed with this theory. Webber held the clementine to be a variant within the mandarin-like group or, if it were an interspecific cross, a mandarin and sweet orange hybrid. Chapot, on the other hand, found some remarkable similarities to the Canton mandarin. However, it seems strange that the Canton mandarin should have escaped the notice of such an expert on the far-eastern citrus industry as Tanaka, who claimed he had only seen the clementine under cultivation and coined the term *Citrus clementina* Hortulanorum for it (1954). Webber's hypothesis on the interspecific cross between mandarin and sweet orange is supported by the results obtained using molecular markers (Deng *et al.*, 1996; Nicolosi *et al.*, 2000). A similar origin is considered by these authors for other genotypes such as tankan and satsuma, while they have some doubts as to the origin of Murcott and King, as to whether it is the sour or sweet orange that is involved.

References

Bailey, F.M. (1895) *Citrus inodora. Queensland Acclimatisation Society Transactions* 1, 194.

Bailey, F.M. (1903) *Contributions to the Flora of British New Guinea.* Governative Printer, Brisbane.

Baldini, E. (1996) *L'Atlante Citrografico di Giorgio Gallesio.* Accademia dei Georgofili, Firenze.

Baldini, E. (1997) Scienza e arte nella citrologia italiana. *Memorie e Rendiconti dell'Accademia degli Zelanti e dei Dafnici* Serie IV Vol. VII, 37–87.

Barrett, H.C. and Rhodes, A.M. (1976) A numerical taxonomic study of affinity relationships in cultivated Citrus and its close relatives. *Systematic Botany* 1, 105–136.

Blanco, M. (1837) *Flora de Filipinas.* Lopez, Manila.

Blume, K. (1823) *Catalogus van eenige der merkwardigste 200 in-alsa uit-hee msche gewassen, te vinden in Lands Plantentuin te Buitenzorg.* Batavia Lands-Drukkerij.

Bonavia, E. (1888) *The Cultivated Oranges and Lemons, etc of Indie and Ceylon.* W.H. Allen and Co., London

Burmann, N.L. (1768) *Flora indica.* Lugd-Bat., Leiden.

Button, J., Vardi, A. and Spiegel-Roy, P. (1976) Root peroxidase isozymes as an aid in *Citrus* breeding and taxonomy. *Theoretical and Applied Genetics* 47, 119–123.

Calabrese, F. (1998) *La Favolosa Storia degli Agrumi.* L'Epos Società Editrice, Palermo.

Chapot, H. (1963) La Clèmentine. *Al Awamia* 7, 1–34.

Chapot, H. (1975) The citrus plant. In: *Citrus Technical Monograph, no. 4.* Ciba-Geigy Agrochemicals, pp. 6–13.

Davies, F.S. and Albrigo, L.G. (1994) *Citrus.* CAB International, Wallingford, UK.

De Candolle, A.P. (1813) *Catalogus Horti Botanici Monspeliensis.* Typ Martel, Monspelii.

De Loureiro, J. (1790) *Flora Cochinchinensis.* Typ. Academiae, Ulyssipone, Lisbon.

Deng, Z.N., Gentile, A., Nicolosi, E., Continella, G. and Tribulato, E. (1996) Parentage determination of some citrus hybrids by molecular markers. *Proceedings of the International Society of Citriculture* 2, 849–854.

Engler, A. (1896) Rutaceae. In: Engler, A. and Prantl, K. (eds) *Die natürlichen Pflanzenfamilien,* Theil 3 Abth. 4, Engelmann, Leipzig.

Engler, A. (1931) Rutaceae. In: Engler, A. and Prantl, K. (eds) *Die natürlichen Pflanzenfamilien,* 2nd edn. Engelmann, Leipzig, pp. 19, 187–359.

Esen, A. and Soost, R.K. (1978) Separation of nucellar and zygotic citrus seedlings by use of polyphenol oxidase-catalyzed browning. *Proceedings of the International Society of Citriculture* 2, 616–618.

Fang, D.Q. and Roose, M.L. (1997) Identification of closely related citrus cultivars with inter-simple sequence repeat markers. *Theoretical and Applied Genetics* 95, 408–417.

Fang, D.Q., Zhang, W.C. and Xiao, S.Y. (1993) Studies on taxonomy and evolution of *Citrus* and its related genera by isozyme analysis. *Acta Phytotaxon Sinica* 31, 329–352. (in Chinese with English abstract.)

Fang, D.Q., Krueger, R.R. and Roose, M.L. (1998) Phylogenetic relationships among selected *Citrus* germplasm accessions revealed by inter-simple sequence repeat (ISSR) markers. *Journal of the American Society of Horticulture Science* 123, 612–617.

Federici, C.T., Fang, D.Q., Scora, R.W. and Roose, M.L (1998) Phylogenetic relationships within the genus *Citrus* (*Rutaceae*) and related genera as revealed by RFLP and RAPD analysis. *Theoretical and Applied Genetics* 96, 812–822.

Ferrarius, G.B. (1646) *Hesperides sive de malorum aureorum cultura et usu libri quatuor.* Sumptibus Hermanni Scheus, Roma.

Fersini, A., Giorlando, A. and Tuoto, G. (1973) *Il cedro (Citrus medica, L.).* Patron Editore, Bologna.

Fortune, R. (1848) New plants from the Society's garden. *Citrus japonica. Journal of the Royal Horticultural Society London.*

Gallesio, G. (1811) *Traité du Citrus.* Libr. L. Fantin, Paris.

Gmitter, F.G. and Hu, X. (1990) The possible role of Yunnan, China, in the origin of contemporany citrus species (Rutaceae*). Economic Botany* 44, 267–277.

Gorgocena, Y. and Ortiz, J.M. (1988) Seed teguments as an aid in citrus chemotaxonomy. *Journal of Horticulture Science* 63, 687–694.

Green, R.M., Vardi, A. and Galun, E. (1986) The plastome of *Citrus*. Physical map, variation among *Citrus* cultivars and species, and comparison with related genera. *Theoretical and Applied Genetics* 72, 170–177.

Guillaumin, A. (1911) Rutaceae. In: Lecomte, H. (ed.) *Flore Gènèrale de l'Indo-Chine.*

Handa, T., Ishizawa, Y. and Ogaki, C. (1986) Phylogenetic study of fraction I protein in the genus Citrus and its close related genera. *Japanese Journal of Genetics* 61, 15–24.

Hermann, P. (1687) *Horti Academici Lugduno-Batavi Catalogus.* Boutesteyn, Leiden.

Herrero, R., Asìns, M.J., Carbonell, E.A. and Navarro, L. (1996) Genetic diversity in the orange subfamily Aurantioideae. I. Intraspecies and intragenus genetic variability. *Theoretical and Applied Genetics* 92, 599–609.

Hirai, M. and Kozaki, I. (1981) Isozymes of citrus leaves. *Proceedings of the International Society of Citriculture* 1, 10–13.

Hirai, M., Kozaki, I. and Kajiura, I. (1986) Isozyme analysis and phylogenic relationship of citrus. *Japan Journal of Breeding* 36, 377–389.

Hodgson, R.W. (1961) Classification and nomenclature of Citrus fruit. *California Citrograph* p. 48.

Hodgson, R.W. (1967) Horticultural varieties of Citrus. In: Reuther, W., Batchelor, L.D and Webber, H.D. (eds) *The Citrus Industry.* University of California Press, California, pp. 431–591.

Hooker, J.D. (1875) *The Flora of British India.* 7 Volumes, Reeve & Co., London.

Hughes, G. (1750) *The Natural History of Barbados.* Printed for the author, London.

Li, W.B., He, S.W. and Liu, G.F. (1987) A study of citrus in Hunan province by analysis of leaf peroxidase isozymes. *Acta Horticulturae Sinica* 14, 153–160.

Li, W.B., Liu, G.F. and He, S.W. (1988) A study of citrus natural resources in Hunan by analysis of leaf isozymes. *Proceedings of the International Symposium on Horticultural Germplasm, Wild and Cultivated.* Part I. Fruit Trees, pp. 252–257.

Li, W.B., Liu, G.F. and He, S.W. (1992) Leaf isozymes of mandarin. *Proceedings of the International Society of Citriculture* I, 217–220.

Linneus, C. (1737) *Genera Plantarum.* Holmiae, Stokholm.

Linneus, C. (1753) *Species Plantarum.* Holmiae, Stokholm.

Linneus, C. (1767) *Systema Naturae.* Holmiae, Stokholm.

Luro, F., Laigret, F. and Bové, J.M. (1992) Application of random amplified polymorphic DNA (R.A.P.D.) to Citrus genetics and taxonomy. *Proceedings of the International Society of Citriculture* I, 225–228.

Luro, F., Laigret, F., Bovè, J.M. and Ollitrault, P. (1995) DNA amplified fingerprinting, a useful tool for determination of genetic origin and diversity analysis in *Citrus. Hort Science* 30 (5): 1063-1067.

Macfadyen, J. (1837) *The Flora of Jamaica; a Description of the Plants of the Island.* Longmans, London.

Malik, M.N., Scora, R.W. and Soost, R.K. (1974) Studies on the origin of the lemon. *Hilgardia* 42, 361–382.

Marcovitch, V.V. (1926) Indeeling van het geslacht *Citrus. Landbouw* 2 (4).

Nagy, S. and Nordby, H.E. (1972) Saturated and mono-unsaturated long-chain hydrocarbon profiles of lipids from orange, grapefruit, mandarin and lemon juice sacs. *Lipids* 7, 660–670.

Nicolosi, E., Deng, Z.N., Gentile, A., La Malfa, S., Continella, G. and Tribulato, E. (2000) Citrus phylogeny and genetic origin of important species as investigated by molecular markers. *Theoretical and Applied Genetics* 100, 1155–1166.

Oliver, D. (1861) The natural order *Aurantiaceae,* with a synopsis of the Indian species. *Linnean Society Botanical Journal* 5.

Pasquale, G.A. (1867) *Catalogo del Real Orto Botanico di Napoli.* Napoli.

Reuther, W., Batchelor, L.D. and Webber, H.J. (eds) (1967) *The Citrus Industry,* 2nd edn. University of California Press, California, pp. 1–39.

Risso, J.A. (1813) *Essai sur l'Historie Naturelle des Orangers, Bigaradies, Limettiers, Cedratiers, Limoniers ou Citronniers, Cultivès dans le Dèpartment des Alpes Maritimes.* Dufour, Paris.

Risso, J.A. and Poiteau, A. (1818) *Histoire Naturelle des Orangers.* Herissant de Doux, Paris.

Scora, R.W. (1975) On the history and origin of Citrus. *Bulletin of the Torrey Botanical Club* 102, 369–375.

Scora, R.W. (1988) Biochemistry, taxonomy and evolution of modern cultivated citrus. *Proceedings of the International Society of Citriculture* 1, 277–289.

Soost, R.K. and Torres, A.M. (1982) Leaf isozymes as genetic markers in *Citrus. Proceedings of the International Society of Citriculture* 1, 7–10.

Spiegel-Roy, P. and Goldschmidt, E. (1996) Biology of Citrus. *Cambridge University Press.*

Steerbeck, F. (1682) *Citricultura, oftregeringhe der uythemsche boomen te weten oranien.* J. Jakobs, Antwerp.

Stone, B.C., Lowry, J.B., Scora, R.W. and Kwiton, J. (1973) *Citrus halimii*: a new species from Malaya and Peninsular Thailand. *Biotropica* 5. 102–110.

Swingle, W.T. (1943) The botany of citrus and its wild relatives of the orange subfamily. In: Webber, H.J. and Batchelor, L.D. (eds) *The Citrus Industry.* University of California Press, California, pp. 129–474.

Swingle, W.T. (1967) The botany of citrus and its wild relatives. In: Reuther, W., Batchelor, L.D. and Webber, H.J. (eds) *The Citrus Industry.* University of California Press, California, pp. 190–430.

Tanaka, T. (1954) Species problem in *Citrus* (Revisio aurantiacearum, IX). *Japan Society Prom. Sci.,* Veno, Tokyo.

Tanaka, T. (1961) Contribution to the knowledge of *Citrus* classification. *Reports Citrologia* 107–114.

Tatum, J.H., Berry, R.E. and Hearn, C.J. (1974) Characterization of citrus cultivars and separation of nucellar and zygotic seedlings by thin-layer chromatography. *Proceedings of the Florida State Horticultural Society* 87, 75–81.

Tenore, M. (1840) *Indices Seminum in Horto Botanico Neapolitano 1840 Collectorum.* Volume 9, Napoli.

Tolkowsky, S. (1938) *Hesperides: a History of the Culture and Use of Citrus Fruits.* John Bale, Sons and Curnow Ltd, London.

Torres, A.M., Soost, R.K. and Diedenhofen, U. (1978) Leaf isozymes as genetic markers in citrus. *American Journal of Botany* 65, 869–881.

Tournefort, J.P. (1700) *Institutiones Rei Herbariae Editio Altera.* Typ. Reg. Paris.

Trabut, L. (1902) L'hybridation des *Citrus*: un nouvelle tangèrine 'la Clèmentine'. *Revue Horticole* 74, 232–234.

Volkamer, J.C. (1708) *Nürnbergische Hesperides, oder gründliche Beschreibung der Edlen Citronat, Citronen, und Pomerantzen-Früchte.* 2 volumes. Bei Johan Andrea Endters feel. Sonn und Erben, Nürnberg.

Webber, H.J. (1943) Cultivated varietes of citrus. In: Webber, H.J. and Batchelor, L.D. (eds) *The Citrus Industry.* University of California Press, California, pp. 475–668.

Webber, H.J. (1967) History and development of the Citrus industry. In: Reuther, W., Batchelor, L.D. and Webber, H.J. (eds) *The Citrus Industry,* 2nd edn. University of California Press, California, pp. 1–39.

Yamamoto, M., Kobayashi, S., Nakamura, Y. and Yamada, Y. (1993) Phylogenic relationships of *Citrus* revealed by diversity of cytoplasmic genomes. In: Hayashi T, Omura, M. and Scott, N.S. (eds) *Techniques on Gene Diagnosis and Breeding in Fruit Trees.* Fruit Trees Research Station, Okitsu, Japan, pp. 39–46.

4 Citrus Germplasm Resources

Robert R. Krueger[1] and Luis Navarro[2]

[1]USDA-ARS National Clonal Germplasm Repository for Citrus & Dates, 1060 Martin Luther King Blvd, Riverside, CA 92507-5437, USA; [2]Instituto Valenciano de Investigaciones Agrarias, Apartado Oficial, 46113 Moncada, Valencia, Spain

Introduction

Plant 'germplasm' is living tissue from which new plants can be grown, containing the complete and unique genetic information that gives plants their individual characteristics and links generations of living plants to one another. In a broader sense, 'germplasm' can refer to the potential hereditary material within a species or within related taxa. Plant germplasm thus constitutes the raw materials for the creation of new plant types by conventional breeding or biotechnological techniques.

Citrus germplasm is represented by seeds, pollen, budwood and in some cases leaves and other vegetative organs of *Citrus* and some related genera. The genera of interest have traditionally been those in the subfamily Aurantioideae (also called Citroideae) of the family Rutaceae. This chapter summarizes the current state of knowledge of citrus germplasm resources. The status of current efforts in the conservation of these resources is reviewed, and the management and use of *ex situ* collections is briefly discussed.

General Concepts in Plant Germplasm Conservation

Plant germplasm is in reality biological information that is passed down through generations in an unbroken chain (Wilkes, 1988). Once this chain is broken, for instance by the destruction of a particular habitat and the consequent loss of certain gene pools, that unique information is lost forever. This has led to the necessity of protecting and preserving plant genetic diversity for current and future use.

Habitat or ecosystem destruction or alteration throughout the world has resulted in the loss of increasingly large numbers of plant species, reducing genetic diversity and threatening the availability of plant germplasm needed for future generations (World Wide Fund for Nature and International Union for Conservation of Nature and Natural Resources, 1994–1997). Consequently, preservation of the genetic diversity represented in all the plant ecosystems throughout the world has become a major issue of international concern (Holden and Williams, 1984; Brown *et al.*, 1989; Holden *et al.*, 1993; National

Research Council, 1993; Food and Agriculture Organization of the United Nations, 1996). Although in some cases habitat loss may be due to natural events or 'catastrophes', in most cases habitat loss is due to human activities. In historical perspective, McNeely *et al.* (1995) have summarized these losses as paralleling the human hunter-gatherer, farming, urban and modern high-energy phases. These phases of course greatly simplify the actual causes of (among many others) population growth, domestication of fire, pollution, exploitation of natural resources, conversion of natural habitat to agricultural or urban uses, deforestation, etc. The reader is referred to McNeely *et al.* (1995) for an overview, and to the World Wide Fund for Nature and International Union for Conservation of Nature and Natural Resources volumes (1994–1997) for more specific information on particular localities.

Ideally, genetic resources should be conserved *in situ*. However, the factors mentioned above make maintenance of genetic resources *in situ* somewhat precarious. This is especially true today when technology makes possible the loss or alteration of large areas of habitat in very short time frames. *Ex situ* conservation is therefore often necessary to save or conserve genetic resources. Genetic materials may be lost through disease, weather, etc., and so *ex situ* collections should be maintained in many cases even when there is not an immediate threat of habitat loss. *Ex situ* collections are also more accessible for researchers for characterization, evaluation and utilization. Maintenance of germplasm in a disease-free state is also desirable, and this is often possible only in *ex situ* collections. Unfortunately, losses can also occur in *ex situ* plant collections because of inadequate maintenance, natural disasters, etc. For more detailed information on general aspects of plant genetic conservation, the reader is referred to Given (1994).

Plant germplasm conservation by humans can be considered to have started in some ways with early human beings, who saved seeds of favoured varieties for food, feed, fibre, and medicinal and religious purposes, developed grafting (which allowed perpetuation of specific clones) and managed natural resources. Although with a small and technologically limited population, early human beings had a limited impact in these areas, they still must have exerted some pressure upon the particular varieties or types of plants conserved. Plants without obviously utilizable properties would probably not have received recognition as types being worth conserving. As humans increased in population and technology, this viewpoint led to the problems noted above.

Involvement of governmental bodies in germplasm conservation might be considered to have started long ago with royal gardens, woods, etc. However, as with the early humans and agriculturally oriented societies in general, plants were selected for inclusion in these areas based upon obvious traits, such as beauty, hunting habitat, etc. With the development of scientifically based agriculture, some agricultural types of plants were conserved almost by accident as types for use in breeding programmes and other investigations. This did lead in some instances to substantial amounts of crop plant genetic resources being conserved, but it was not until the latter half of the 20th century that governments began making concerted efforts to conserve crop plant germplasm. Perhaps it was Vavilov, the seminal figure in plant genetic resource conservation, who first recognized the importance of making a concentrated effort to conserve not only crop species but also their wild relatives in *ex situ* gene banks. Programmes or systems with this goal were developed in different countries and in some cases by intergovernmental entities such as the centres of the Consultative Group on International Agricultural Research (CGIAR) and some establishments of the Food and Agriculture Organization of the United Nations (FAO).

An example of government-sponsored conservation of plant genetic resources is the United States Department of Agriculture-Agricultural Research Service (USDA-ARS) National Plant Germplasm System (NPGS) in the USA, which had its origins in the 1970s (Shands *et al.*, 1988; White *et al.*, 1989; National Research Council, 1991; Shands, 1995). Efforts towards conservation of the so-called 'clonal' crops (such as citrus) began later, in the mid- to late 1980s (Brooks and Barton, 1977; Westwood, 1989). Although the USA has one of the largest and most advanced systems for plant germplasm conservation, it is ironic that, with a few notable exceptions such as maize, North America is not home to important crop plant species. Consequently, most of the materials maintained by the NPGS are acquired from other countries by exchange, plant exploration, etc.

In recent years, plant germplasm conservation and utilization has become complicated by such factors as the Convention on Biological Diversity, Globalization, Intellectual Property Rights, Indigenous Rights, Farmer's Rights, etc. Discussion of these controversial and political areas is beyond the scope of this review, and there is no current one-volume overview of these intertwined and complex competing rights. Some very general information on these areas as well as on plant germplasm conservation in general can be gleaned from popular writings such as Fowler and Mooney (1990), Busch *et al.* (1995) and Raeburn (1995). Perhaps the best and most complete information was available in the now-defunct magazine *Diversity*. Other sources are various publications and other disseminations from the International Plant Genetic Resources Institute (IPGRI) <http://www.ipgri.org/> (formerly the International Board for Plant Genetic Resources, IBPGR) and the FAO <http://www.fao.org>. Since most of the international political negotiations regarding plant genetic resources take place under the auspices of these entities, these webpages are perhaps the best place to start for up-to-date basic information.

Citrus Germplasm

The genus *Citrus* is one of 33 genera in the subfamily *Aurantioideae* (or *Citroideae*) of the family Rutaceae (Table 4.1). Most Aurantioideae genera are native to and have their centre of diversity in north-eastern India, southern China, the Indochinese peninsula and nearby archipelagos, although some related Aurantioideae genera are native to Asia, Africa and Australia.

Much of the descriptive information on these genera, as well as the establishment of one of the most widely accepted taxonomic systems used therefore, was published in various places by Walter T. Swingle of the United States Department of Agriculture (USDA) in the early decades of the 20th century. These publications, as well as earlier accounts from various flora, plant exploration reports, etc. were summarized in Swingle (1943) and its minor revision as Swingle and Reece (1967). This is still the most complete single source of information in this area and will form the basis for much of the following information. Except in the rare cases where new information is added in Swingle and Reece (1967), reference will be made to Swingle (1943) with the understanding that the same material is available in Swingle and Reece (1967).

Although some recent information has been generated on the Aurantioideae, much of the information on these genera is quite old and relies heavily on the work of Swingle, which in turn often refers to even older materials. In some cases, a species description by Swingle (1943) is based upon a single herbarium specimen or previous description. It is quite probable that not all of the species described by Swingle (1943) constitute valid taxa, due both to assumptions made about the materials observed by the authors and possible loss of species or gene pools since the original descriptions were published or herbarium specimens collected. On the other hand, as will be seen below, some genera that have had attention paid to them recently have actually had increases in the number of species catalogued.

Table 4.1. The Aurantioideae (orange subfamily) of the plant family Rutaceae.

Tribe	Subtribe	Subtribal group	Genus	Species	Origin
Clauseneae	Micromelinae	(none)	*Micromelum*	9	S.E. Asia, Oceania
	Clauseninae	(none)	*Glycosmis*	35	S.E. Asia, Oceania
			Clausena	23	S Asia, Oceania
			Murraya	11	S. and S.E. Asia, Oceania
	Merrilliinae	(none)	*Merrillia*	1	S.E. Asia
Citreae	Triphasiinae	Wenzelia group	*Wenzelia*	9	Oceania
			Monanthocitrus	1	Oceania
			Oxanthera	4	Oceania
			Merope	1	S.E. Asia, Oceania
		Triphasia group	*Triphasia*	3	S.E. Asia, Oceania
			Pamburus	1	S. and S.E. Asia, Oceania
		Luvugna group	*Luvugna*	12	S. and S.E. Asia, Oceania
			Paramignya	15	S. and S.E. Asia
	Citrinae	Primitive citrus fruit trees	*Severinia*	6	S. China, S.E. Asia
			Pleiospermium	5	S. Asia, Oceania
			Burkillanthus	1	S.E. Asia, Oceania
			Limnocitrus	1	S.E. Asia
			Hesperethusa	1	S. and S.E. Asia
		Near citrus fruit trees	*Citropsis*	11	Central Africa
			Atalantia	11	S. and S.E. Asia
		True citrus fruit trees	*Fortunella*	5	S. China
			Eremocitrus	1	Australia
			Poncirus	2	Central and N. China
			Clymenia	1	Oceania
			Microcitrus	7	Australia
			Citrus	16	S. and S.E. Asia, S. China
	Balsamocitrinae	Tabog group	*Swinglea*	1	Philippines
		Bael-fruit group	*Aegle*	1	India
			Afraegle	4	West Africa
			Aeglopsis	2	W. Africa
			Balsamocitrus	1	Uganda
		Wood-apple group	*Limonia*	1	S. and S.E. Asia
			Feroniella	3	S.E. Asia

Review of the Aurantioideae

The Rutaceae are a family of approximately 160 genera and 1650 species of mostly trees and shrubs. The leaves are compound, with glands and without stipules, often thorny; the flowers are 4- or 5-merous, regular and perfect, usually with a superior ovary and 3–5 locules; fruit varies, usually a capsule or berry. The subfamily Aurantioideae is further characterized by leaves and bark with oil glands, and most characteristically a fruit which is a hesperidium, i.e. a berry whose fleshy parts are divided into segments surrounded by a rind or hard shell, the seeds of which are without endosperm and sometimes with two or more nucellar asexual embryos.

The Aurantioideae is divided into tribes. Although Engler (1931) recognized only a single tribe and Tanaka (1932) eight, the most commonly used division is that of Swingle (1938), which recognizes two tribes (the Clauseneae and Citreae) as shown in Table 4.1 (see also Swingle, 1943). These two tribes are further divided into six subtribes, 33 genera, and 203 species as per Swingle (1943). For a complete but not up-to-date enumeration of published species, the reader is referred to Carpenter and Reece (1969).

The Clauseneae are considered to be the most primitive members of the Aurantioideae. This tribe is distinguished from the Citreae by never having axillary spines, having leaves alternately attached to a non-articulated rachis, and having a rachis that does not break into segments when the leaves fall. The Citreae are the opposite of the Clauseneae: axils have single or paired spines, leaves are oppositely attached to an articulated rachis, and the rachis breaks into segments when the leaves fall. This is the larger and more economically important of the two tribes, and includes *Citrus* and its closes relatives. The Clauseneae and Citreae are further divided into subtribes, and the subtribes of the Citreae (but not the Clauseneae) are further divided into subtribal groups (Table 4.1). The reader is referred to Swingle (1943) for more information on the distinguishing characteristics of these subdivisions.

Recent molecular phylogeny studies have suggested slightly different relationships within the Aurantioideae and Rutaceae. Herrero *et al.* (1996b) suggest that *Swinglea* is more closely related to *Murraya* than to *Afraegle* and that *Aeglopsis* is more closely related to *Fortunella*, *Microcitrus* and *Citrus* than to the other hard shell species, while Chase *et al.* (1999) suggest that *Luvunga* is more closely related to other subfamilies (Rutoideae and Toddaliodae) than to other members of the Aurantioideae. Recent work in molecular systematics (Samuel *et al.*, 2001) and karysystematics (Guerra *et al.*, 2000) found that the traditional division into tribes and subtribes did not reflect phylogenetic relationships. Because these arrangements are less intuitively obvious than the classical taxonomy of Swingle (1943), the latter will be utilized for this brief discussion of Aurantioideae genera. This is not a taxonomic treatment (see Nicolosi, Chapter 3, for that) but it is convenient for presentation and discussion.

For the current chapter, the genera of the Aurantioideae will be briefly discussed in a somewhat reverse order, starting with the most important genus, *Citrus*, and proceeding to the most distantly related genera at the end.

Citrus

The taxonomy of the genus *Citrus* is not precisely established. There have traditionally been two major systems of *Citrus* taxonomy utilized: Swingle and Tanaka. The Swingle system (Swingle, 1943) recognizes 16 species in two subgenera (*Citrus* and *Papeda*). Modifications of Swingle recognize 17 species (Bhattacharya and Dutta, 1956; Stone, 1994a), 36 species (Hodgson, 1961) or 31 species (Singh and Nath, 1969).

The other widely utilized taxonomic system for *Citrus* and related genera is that of Tyôzaburô Tanaka of the University of Osaka. The Tanaka version of *Citrus* taxonomy was developed more or less concurrently with the Swingle system (Tanaka, 1954, 1966, 1969a, b, 1977). The Tanaka taxonomy recognizes up to 162 species within 13 primary elements in its most highly developed form (Tanaka, 1977). This lack of agreement with Swingle reflects differences of opinion as to what degree of difference justifies species status and whether or not supposed hybrids among naturally occurring forms should be assigned species status. Although Tanaka's differences with Swingle primarily involved the genus *Citrus* itself, Tanaka also catalogued and described many related Aurantioideae genera.

There is no definitive work on *Citrus* taxonomy, and the two major systems are both currently in use. In practice, some germplasm banks (such as France, Spain and the USA) use a sort of hybrid system that is in many ways closer to Tanaka than to Swingle. In addition, international scientific citriculture societies, such as the International Society of Citriculture (ISC) and the International Organization of Citrus Virologists (IOCV), utilize both systems.

Recently it has been suggested that only citron (*C. medica*), mandarin (*C. reticulata*) and pummelo (*C. maxima*) constitute valid species (within the subgenus *Citrus*) and that other important types (orange, grapefruit, lemon and lime) originated from one or more generations of hybridization between these ancestral genera (Scora, 1975, 1989; Barrett and Rhodes, 1976; Mabberley, 1997). Interestingly, the earliest workers also believed that there were only three or four valid species of citrus (Linnaeus, 1753; Hooker, 1875), and recent molecular studies have supported this concept (Asíns *et al.*, 1996; Federici *et al.*, 1998; Herrero *et al.*, 1996b; Nicolosi *et al.*, 2000; Gulsen and Roose, 2001; Moore, 2001).

None of the systems is ideal, but the Tanaka system is better adapted to the horticultural traits of different groups. The Swingle system is poorly adapted to these traits, particularly in regard to the mandarin group. A system with only three species may be scientifically more rigorous, but is even less cognizant of horticultural traits than is that of Swingle. The Swingle system will be mainly followed here in order to discuss the various types of *Citrus*. A technical discussion of *Citrus* taxonomy is found in Nicolosi (Chapter 3).

The three ancestral species only reproduce sexually since they are not apomictic. Consequently, some mandarins, pummelos and citrons have higher levels of genetic diversity since many of the cultivars have arisen through sexual hybridization. On the other hand, most of the cultivars of orange, grapefruit, lemon and some mandarin groups such as satsumas and clementines originated from nucellar seedlings or budsports. Consequently, the amount of genetic diversity within these groups is relatively low, in spite of there being many named varieties with important differences in horticultural traits (Herrero *et al.*, 1996a). Table 4.2 summarizes the current understanding of the origin, mode of reproduction and level of genetic diversity within certain commercially important species of the genus *Citrus*.

Citrus originated and has as its centre of diversity in the south-eastern region of Asia, including India, southern China, the Indochinese peninsula and nearby archipelagos. Tanaka (1954) proposed a theoretical dividing line (the Tanaka line) that runs south-eastwardly from the north-west border of India, above Burma, through the

Table 4.2. The genus *Citrus*: a summary.

Species	Common name	Known age (years)	Year named	Probable origin	Probable native habitat	Seed reproduction	Genetic diversity
C. medica	Citron	2300	1753	True species	India	Sexual	Moderate
C aurantium	Sour orange	900	1753	Hybrid	China	Nucellar	Low
C. sinensis	Sweet orange	500	1757	Hybrid	China	Nucellar	Low
C. maxima	Pummelo	2000 (?)	1765	True species	China	Sexual	High
C. limon	Lemon	800	1766	Hybrid	India	Partly sexual	Moderate
C. reticulata	Mandarin	2000 (?)	1837	True species	China	Variable	High
C. aurantifolia	Lime	700	1913	Hybrid	Malaya	Partly sexual	Moderate
C. paradisi	Grapefruit	200	1930	Hybrid	Barbados	Nucellar	Low
C. tachibana	Tachibana	2000 (?)	1924	Unknown	Japan	Sexual	Moderate (?)
C. indica	Indian wild org	2000 (?)	1931	Unknown	India	Sexual	Moderate (?)
C. hystrix	Mauritius papeda	2000 (?)	1813	Unknown	S.E. Asia	Sexual	Moderate (?)
C. macroptera	Malesian papeda	2000 (?)	1860	Unknown	S.E. Asia	Sexual	Moderate (?)
C. celebica	Celebes papeda	2000 (?)	1898	Unknown	Celebes	Sexual	Moderate (?)
C. ichangensis	Ichang papeda	2000 (?)	1913	Unknown	China	Sexual	Moderate (?)
C. micrantha	Papeda	2000 (?)	1915	Unknown	Philippines	Sexual	Moderate (?)
C. latipes	Khasi papeda	2000 (?)	1928	Unknown	Assam	Sexual	Moderate (?)

Yunnan province of China, to south of the island of Hainan. Citron, lemon, lime, sweet and sour oranges, and pummelo originated south of this line, while mandarins, kumquats and trifoliates originated north of the line. The mandarins apparently developed along a line north-east of the Tanaka line, along the east China coast, through Formosa and to Japan, while the trifoliates and kumquats are found in a line crossing south-central China in an east–west direction. Yunnan, China, through which the Tanaka line runs, has recently been proposed as a major centre of origin for citrus (Gmitter and Hu, 1989, 1990).

The genus *Citrus* is by far the most economically important in the Aurantioideae and probably the entire Rutaceae. It includes the cultivated sweet and sour oranges, mandarins, lemons, limes, pummelos, grapefruits, etc., as well as other types not commonly consumed. It also includes the papedas (subgenus *Papeda*). The fruits of the latter subgenus are not edible due to their pulp vesicles having dense aggregations of acrid oil that give the juice a bitter, unpleasant flavor. The flowers are smaller than those of the subgenus *Citrus*, and the leaves have elongated petioles with broad wings.

Since most readers will be well acquainted with the cultivated types of *Citrus* and they are covered in more detail in a previous chapter, only a few comments will be made concerning them. For descriptions of some important commercial cultivars, the best concise sources are Hodgson (1967) and Saunt (2000). The use of *Citrus* spp. as germplasm resources is, of course, concerned with more than fruit quality and characteristics. Such traits as disease resistance and tolerance, adaptation to different soil and environmental conditions, resistance to insects, tolerance of cold conditions, etc. are important characteristics of genotypes maintained in germplasm collections and utilized by scientists. Table 4.3 presents examples of some characteristics

Table 4.3. Some reported attributes of Aurantiodeae genera.

Genus	Attribute	References
Aegle	Anti-insect properties	Roy, 1998
Aegle	Food use	Allen, 1967; Burkill, 1935; Corner, 1988; Parmar and Kaushal, 1982; Roy, 1998; Singh and Roy, 1984; Swingle, 1943; Swingle and Reece, 1967
Aegle	Medicinal properties	Allen, 1967; Burkill, 1935; Corner, 1988; Gupta and Banerjee, 1972; Jain, 1965; Parmar and Kaushal, 1982; Roy, 1998; Singh and Roy, 1984
Aegle	Reaction to canker	Lee, 1918; Peltier, 1918; Peltier and Frederich, 1920, 1924
Aegle	Reaction to citrus tristeza virus	Knorr, 1956; Müller and Garnsey, 1984; Yoshida, 1996
Aegle	Reaction to miscellaneous fungal diseases	Winston *et al.*, 1927
Aegle	Reaction to miscellaneous virus or viroid diseases	Iwanami *et al.*, 1993
Aeglopsis	Reaction to canker	Peltier, 1918; Peltier and Frederich, 1920, 1924
Aeglopsis	Reaction to citrus tristeza virus	Knorr, 1956; McClean, 1961; Müller and Garnsey, 1984; Yoshida, 1996
Aeglopsis	Reaction to miscellaneous fungal diseases	Winston *et al.*, 1927
Aeglopsis	Reaction to nematodes	Baines *et al.*, 1960
Afraegle	Anti-insect properties	Bowman *et al.*, 2001

Table 4.3. Continued

Genus	Attribute	References
Afraegle	Leaf characteristics	Hirano, 1931
Afraegle	Reaction to citrus tristeza virus	Knorr, 1956; McClean, 1961; Müller and Garnsey, 1984
Afraegle	Reaction to nematodes	Baines *et al.*, 1960; Ford and Feder, 1958
Atalantia	Anti-insect properties	Luthria *et al.*, 1989
Atalantia	As a preserve	Swingle, 1943; Swingle and Reece, 1967
Atalantia	Dye	Burkill, 1935
Atalantia	Food use	Awasthi, 1991
Atalantia	Leaf characteristics	Hirano, 1931
Atalantia	Medicinal use	Burkill, 1935
Atalantia	Reaction to canker	Lee, 1918; Peltier, 1918; Peltier and Frederich, 1920, 1924*Atalantia* Reaction to citrus tristeza virus Knorr, 1956; Müller and Garnsey, 1984; Yoshida, 1996
Atalantia	Reaction to miscellaneous fungal diseases	Winston *et al.*, 1927
Atalantia	Reaction to miscellaneous virus or viroid diseases	Iwanami *et al.*, 1993
Atalantia	Reaction to nematodes	Baines *et al.*, 1960
Atalantia	Salt tolerance	Traub and Robinson, 1937
Balsamocitrus	Anti-insect properties	Bowman *et al.*, 2001
Balsamocitrus	Leaf characteristics	Hirano, 1931
Balsamocitrus	Reaction to canker	Lee, 1918; Peltier, 1918; Peltier and Frederich, 1920, 1924
Balsamocitrus	Reaction to citrus tristeza virus	Müller and Garnsey, 1984
Balsamocitrus	Reaction to miscellaneous fungal diseases	Winston *et al.*, 1927
Balsamocitrus	Reaction to nematodes	Baines *et al.*, 1960
Citropsis	Adaptation to soil conditions	Swingle, 1914a
Citropsis	Anti-insect properties	Bowman *et al.*, 2001
Citropsis	Anti-insect properties	Swingle, 1940b
Citropsis	Disease resistance	Swingle, 1940b; Traub and Robinson, 1937
Citropsis	Leaf characteristics	Hirano, 1931
Citropsis	Photosynthesis	Khairi and Hall, 1976
Citropsis	Reaction to canker	Lee, 1918; Peltier, 1918; Peltier and Frederich, 1920, 1924
Citropsis	Reaction to citrus tristeza virus	Müller and Garnsey, 1980; Yoshida, 1996
Citropsis	Reaction to miscellaneous fungal diseases	Winston *et al.*, 1927
Citropsis	Reaction to nematodes	Ford and Feder, 1962
Citropsis	Reaction to nematodes	Ford and Feder, 1960
Citropsis	Rootstock use	Swingle, 1914a, 1940b
Citrus and hybrids	Essential oils	Verzera *et al.*, 2001
Citrus and hybrids	Salt tolerance	Caro *et al.*, 1973; Cooper, 1961; Cooper and Gorton, 1951, 1952; Cooper and Peynado, 1954; Cooper and Shull, 1953; Cooper *et al.*, 1958; Douglas and Walker, 1983; Embleton *et al.*, 1973b; Kirkpatrick and Bitters, 1969; Levy *et al.*, 1999; Maas, 1993; Ream and Furr, 1976; Storey and Walker, 1999; Wutscher *et al.*, 1973

Table 4.3. Continued

Genus	Attribute	References
Citrus and hybrids	Adaptation to soil conditions	Campbell and Goldweber, 1979; Cooper and Olson, 1951; Cooper and Peynado, 1954; El-Otmani, 1996; Ford, 1964; Hilgeman *et al.*, 1966; Rouse and Wutscher, 1985; Sagee *et al.*, 1992; Shaked *et al.*, 1988; Wutscher *et al.*, 1970
Citrus and hybrids	Allelopathy	Al-Saadawi and Al-Rubeaa, 1985a, b; Al-Saadawi *et al.*, 1985; Sinha-Roy and Chakraborty, 1976
Citrus and hybrids	Anti-fungal properties	Asthana *et al.*, 1992; Stange *et al.*, 1993
Citrus and hybrids	Anti-insect properties	Back and Pemberton, 1915; Beavers and Hutchison, 1985; Bhumannavar *et al.*, 1988, 1989; Bowman *et al.*, 2001; Cameron *et al.*, 1969; Chowdhury and Das, 1979; Howard, 1979a, b; Jacobson, 1989; Mendel *et al.*, 1991; Nguyen and Fraser, 1989; Pollard *et al.*, 1983; Sadana and Joshi, 1979; Serit *et al.*, 1992; Shapiro and Gottwald, 1995; Shapiro *et al.*, 1997; Spitler *et al.*, 1984; Yang and Tang, 1988
Citrus and hybrids	Cold hardiness	Cooper, 1952; Furr and Armstrong, 1959; Furr *et al.*, 1966; Gardner and Horanic, 1963; Hearn *et al.*, 1963; Yelenosky, 1985; Yelenosky and Hearn, 1976; Yelenosky and Young, 1977; Young, 1963a, b, c, 1966,1977; Young and Hearn, 1972; Young and Olson, 1963a, b; Young *et al.*, 1960
Citrus and hybrids	Fruit composition or development	Bitters and Batchelor, 1951; Kefford and Chandler, 1961; Roose *et al.*, 1985; Sinclair and Bartholomew, 1944; Woodruff and Olson, 1960; Wutscher and Bistline, 1988; Wutscher and Shull, 1975
Citrus and hybrids	Growth habit	Bowman, 1994
Citrus and hybrids	Growth rate	Maggs and Alexander, 1969
Citrus and hybrids	Hair and scalp preparations	Roia, 1966
Citrus and hybrids	Insect attractants or stimulants	Honda, 1990
Citrus and hybrids	Leaf characteristics	Azab and Hegazy, 1995; Halma, 1929; Hirano, 1931; Turrell, 1947, 1961
Citrus and hybrids	Level of apomixis	Frost and Soost, 1968; Minessy and Higazy, 1957; Moreira *et al.*, 1947: Ueno *et al.*, 1967
Citrus and hybrids	Medicinal use	Attaway, 1994; Awasthi, 1991; Baker, 1994; Benevente-García *et al.*, 1997; Bracke *et al.*, 1994; Calomme *et al.*, 1996; Dagar and Dagar, 1991; Eldridge, 1975; Ghazanfar and Al-Sabahi, 1993; Jain, 1965; Kalt, 2001; Kandaswami *et al.*, 1991; Lal and Lata, 1980; Lal and Yadav, 1983; Lam *et al.*, 1994; Manners and Hasegawa, 1999; Manthey *et al.*,

Table 4.3. Continued

Genus	Attribute	References
Citrus and hybrids	Medicinal use	1999; Middleton *et al.*, 1994; Miller *et al.*, 1992, 1994; Montanari *et al.*, 1997; Paul and Cox, 1995; Rouseff and Nagy, 1994; San Martín A., 1983; Shah and Joshi, 1971
Citrus and hybrids	Metaxenia	Burger, 1985; Hearn *et al.*, 1969; Wallace and Lee, 1999
Citrus and hybrids	Mineral nutrition	Castle and Krezdorn, 1975; Cooper and Peynado, 1955; Cooper *et al.*, 1952, 1955; El-Shazly *et al.*, 1992; Embleton *et al.*, 1962, 1973b; Fallahi, 1992; Fallahi and Rodney, 1992; Haas, 1945a, b; Hodgson and Eggers, 1938; Intrigliolo and Starrantino, 1988; Labanauskas and Bitters, 1974; Lin and Myhre, 1991a, b; Rasmussen and Smith, 1958; Shannon and Zaphrir, 1958; Sharples and Hilgeman, 1972; Smith, 1975; Smith *et al.*, 1949; Sudahono *et al.*, 1994; Wallace and Mueller, 1972; Wallace *et al.*, 1952, 1953; Wutscher, 1986, 1989; Wutscher and Dube, 1977; Wutscher and Shull, 1972, 1975, 1976a, b; Wutscher *et al.*, 1970
Citrus and hybrids	Photosynthesis	Iwasaki and Oogaki, 1985; Khairi and Hall, 1976; Morinaga and Ikeda, 1990; Syvertsen and Graham, 1985
Citrus and hybrids	Reaction to *Alternaria*	Gardner *et al.*, 1986; Kohmoto *et al.*, 1991; Solel and Kimchi, 1997
Citrus and hybrids	Reaction to canker	Gottwald *et al.*, 1993; Koizumi, 1981; Leite and Mohan, 1984; Peltier, 1918; Peltier and Frederich, 1920, 1924; Zubrzycki and Diamante de Zubrzycki, 1980
Citrus and hybrids	Reaction to citrus tristeza virus	Bitters, 1959, 1972; Bitters *et al.*, 1973b; Carpenter *et al.*, 1982; Costa *et al.*, 1949; Dornelles, 1976; Garnsey, 1992; Grant and Costa, 1948; Grant *et al.*, 1949, 1951; Knorr, 1956; Müller and Garnsey, 1984; Olson, 1960; Salibe, 1977; Stubbs, 1963; Xueyuan *et al.*, 1993
Citrus and hybrids	Reaction to diseases of recalcitrant aetiology	Beretta *et al.*, 1992; Donadio and Banzato, 1988; Marais and Lee, 1991; Marais *et al.*, 1993
Citrus and hybrids	Reaction to Mal secco	Chapot, 1963; Crescimanno *et al.*, 1973; De Cicco *et al.*, 1984; Russo, 1976/77; Solel and Spiegel-Roy, 1978; Thanassoulopoulos, 1991
Citrus and hybrids	Reaction to miscellaneous bacterial diseases	Laranjeira *et al.*, 1998; Li *et al.*, 1996a, b, c, d, 2000

Table 4.3. Continued

Genus	Attribute	References
Citrus and hybrids	Reaction to miscellaneous fungal diseases	Bender and Menge, 1986; Ieki, 1981; Inoue, 1984; Klotz and Fawcett, 1930; Olson and Godfrey, 1953; Peltier and Frederich, 1923; Winston, 1923; Winston *et al.*, 1927
Citrus and hybrids	Reaction to miscellaneous mycoplasmas	Bové *et al.*, 1996a; Fraser and Singh, 1969; Garnier *et al.*, 1991; McClean and Schwarz, 1970
Citrus and hybrids	Reaction to miscellaneous virus or viroid diseases	Bitters, 1950; Calavan and Christiansen, 1965; Carpenter and Furr, 1967; Childs, 1951; Cohen, 1974; Iwanami *et al.*, 1996; Olson *et al.*, 1962; Rossetti *et al.*, 1965; Salibe and Moreira, 1965; Vogel and Bové, 1971; Yamada and Sawamura, 1952
Citrus and hybrids	Reaction to nematodes	Baines *et al.*, 1960, 1967; Cameron *et al.*, 1969; Davis, 1984; Feldmesser and Hannon, 1969; Ferguson *et al.*, 1992; Ford and Feder, 1958, 1962, 1964, 1969; Ford and Hutchins, 1967; Hutchison and O'Bannon, 1972; Kaplan, 1981; McCarty *et al.*, 1979; Niles *et al.*, 1995; O'Bannon and Ford, 1977; Van Gundy and Kirkpatrick, 1964
Citrus and hybrids	Reaction to *Phytophthora*	Baines *et al.*, 1967; Bitters *et al.*, 1973b; Cameron *et al.*, 1972; Carpenter and Furr, 1962; Carpenter *et al.*, 1975, 1981; Davenport and Rouse, 1992; Furr and Carpenter, 1961; Gondell, 1946; Grimm and Hutchison, 1977; Hutchison and Grimm, 1972; Ippoloti *et al.*, 1997; Klotz and Fawcett, 1930; Klotz *et al.*, 1967, 1968; Lee, 1925; Smith *et al.*, 1987; Tuzcu *et al.*, 1984; Vanderweyen, 1980
Citrus and hybrids	Root characteristics	Bevington and Castle, 1982; Castle and Krezdorn, 1977; Castle and Youtsey, 1983; Cossman, 1939; Ford, 1952, 1954; Luxmore *et al.*, 1971; Savage *et al.*, 1945; Smith and Wallace, 1954; Syvertsen, 1981; Syvertsen and Graham, 1985
Citrus and hybrids	Storability of pollen	Ganeshan and Sulladmath, 1983
Citrus and hybrids	Tree size control	Castle, 1980; Fallahi, 1992; Fallahi and Rodney, 1992; Phillips and Castle, 1977; Roose *et al.*, 1985; Shannon and Zaphrir, 1958; Wallace *et al.*, 1953; Wutscher and Shull, 1975
Clausena	Anti-insect properties	Yang and Tang, 1988; Jones, 1995
Clausena	Essential oils	Brown, 1954; Swingle, 1943; Swingle and Reece, 1967; Molino, 1993

Table 4.3. Continued

Genus	Attribute	References
Clausena	Food use	Brown, 1954; Burkill, 1935; Swingle, 1943; Swingle and Reece, 1967
Clausena	Medicinal uses	Anderson, 1986a, b; Brown, 1954; Burkill, 1935; Jones, 1995
Clausena	Quality of wood	Burkill, 1935; Jones, 1995
Clausena	Reaction to canker	Gottwald *et al.*, 1993; Lee, 1918; Peltier, 1918; Peltier and Frederich, 1920, 1924
Clausena	Reaction to citrus tristeza virus	Müller and Garnsey, 1984; Yoshida, 1996
Clausena	Reaction to miscellaneous fungal diseases	Winston *et al.*, 1927
Clausena	Reaction to nematode	Baines *et al.*, 1960
Clausena	Rootstock use	Swingle, 1943; Swingle and Reece, 1967; Campbell, 1974
Clymenia	Anti-insect properties	Bowman *et al.*, 2001
Clymenia	Food use	Swingle, 1943; Swingle and Reece, 1967
Clymenia	Reaction to citrus tristeza virus	Yoshida, 1996
Clymenia	Reaction to miscellaneous virus or viroid diseases	Iwanami *et al.*, 1993
Eremocitrus	Anti-insect properties	Bowman *et al.*, 2001
Eremocitrus	Cold hardiness	Chapman and Brown, 1993; Yelenosky *et al.*, 1978
Eremocitrus	Drought tolerance	Swingle, 1914b
Eremocitrus	Leaf characteristics	Hirano, 1931
Eremocitrus	Reaction to canker	Gottwald *et al.*, 1993; Lee, 1918; Peltier, 1918; Peltier and Frederich, 1920, 1924
Eremocitrus	Reaction to citrus tristeza virus	Yoshida, 1996
Eremocitrus	Reaction to Mal secco	Russo, 1976/77
Eremocitrus	Reaction to miscellaneous virus or viroid diseases	Iwanami *et al.*, 1993
Eremocitrus	Reaction to *Phytophthora*	Carpenter and Furr, 1962; Grimm and Hutchison, 1972
Feroniella	Condiment	Traub and Robinson, 1937
Feroniella	Reaction to canker	Lee, 1918; Peltier, 1918; Peltier and Frederich, 1920, 1924
Feroniella	Reaction to citrus tristeza virus	Yoshida, 1996
Feroniella	Reaction to miscellaneous fungal diseases	Winston *et al.*, 1927
Feroniella	Resistance, tolerance, susceptibility to miscellaneous virus or viroid diseases	Iwanami *et al.*, 1993
Fortunella	Anti-insect properties	Bowman *et al.*, 2001
Fortunella	As a preserve	Swingle, 1943; Swingle and Reece, 1967
Fortunella	Cold hardiness	Hume, 1902; Young and Hearn, 1972
Fortunella	Level of apomixis	Ueno *et al.*, 1967
Fortunella	Reaction to *Alternaria*	Kohmoto *et al.*, 1991
Fortunella	Reaction to canker	Gottwald *et al.*, 1993; Lee, 1918; Peltier, 1918; Peltier and Frederich, 1920, 1924
Fortunella	Reaction to citrus tristeza virus	Grant and Costa, 1948; Yoshida, 1996
Fortunella	Reaction to Mal secco	Russo, 1976/77
Fortunella	Reaction to miscellaneous fungal diseases	Fulton, 1925; Winston, 1923; Winston *et al.*, 1927

Table 4.3. Continued

Genus	Attribute	References
Fortunella	Reaction to miscellaneous mycoplasmas	Moll and van Vuuren, 1982
Fortunella	Reaction to miscellaneous virus or viroid diseases	Childs, 1951; Iwanami *et al.*, 1993; Yamada and Sawamura, 1952
Fortunella	Reaction to *Phytophthora*	Carpenter and Furr, 1962; Grimm and Hutchison, 1972
Glycosmis	Anti-insect properties	Bandara *et al.*, 1990; Bowman *et al.*, 2001; Shapiro *et al.*, 1997, 2000
Glycosmis	Medicinal use	Burkill, 1935; Dagar and Dagar, 1991
Glycosmis	Poison	Burkill, 1935
Glycosmis	Quality of wood	Burkill, 1935
Glycosmis	Reaction to canker	Peltier, 1918; Peltier and Frederich, 1920, 1924
Glycosmis	Reaction to citrus tristeza virus	Knorr, 1956; Müller and Garnsey, 1984; Yoshida, 1996
Glycosmis	Reaction to miscellaneous fungal diseases	Winston *et al.*, 1927
Hesperethusa	Condiment	Swingle, 1943; Swingle and Reece, 1967; Traub and Robinson, 1937
Hesperethusa	Reaction to canker	Lee, 1918; Peltier, 1918; Peltier and Frederich, 1920, 1924
Hesperethusa	Reaction to citrus tristeza virus	Müller and Garnsey, 1984; Yoshida, 1996
Hesperethusa	Reaction to miscellaneous fungal diseases	Winston *et al.*, 1927
Limnocitrus	Salt tolerance	Swingle, 1940a
Limonia	Anti-fungal activity	Adikaram *et al.*, 1989
Limonia	Food use	Allen, 1967; Burkill, 1935; Corner, 1988; Swingle, 1943; Swingle and Reece, 1967
Limonia	Ink	Burkill, 1935
Limonia	Medicinal use	Allen, 1967; Burkill, 1935
Limonia	Reaction to canker	Lee, 1918; Peltier, 1918; Peltier and Frederich, 1920, 1924
Limonia	Reaction to citrus tristeza virus	Yoshida, 1996
Limonia	Reaction to miscellaneous fungal diseases	Winston *et al.*, 1927
Luvugna	Medicinal use	Burkill, 1935
Merope	Medicinal use	Jones, 1982
Merope	Rootstock use	Jones, 1982
Merope	Salt tolerance	Swingle, 1915a
Merrillia	Reaction to citrus tristeza virus	Yoshida, 1996
Merrillia	Rootstock use	Swingle, 1918
Merrillia	Wood quality	Burkill, 1935; Jones, 1995; Swingle, 1918
Microcitrus	Anti-insect properties	Bowman *et al.*, 2001
Microcitrus	As a preserve	Swingle, 1943; Swingle and Reece, 1967
Microcitrus	Drought tolerance	Traub and Robinson, 1937
Microcitrus	Essential oil composition	Brophy *et al.*, 2001
Microcitrus	Leaf characteristics	Hirano, 1931
Microcitrus	Reaction to canker	Lee, 1918; Peltier, 1918; Peltier and Frederich, 1920, 1924

Table 4.3. Continued

Genus	Attribute	References
Microcitrus	Reaction to citrus tristeza virus	Müller and Garnsey, 1984; Yoshida, 1996
Microcitrus	Reaction to Mal secco	Russo, 1976/77
Microcitrus	Reaction to miscellaneous fungal diseases	Fulton, 1925; Winston *et al.*, 1927
Microcitrus	Reaction to miscellaneous virus or viroid diseases	Iwanami *et al.*, 1993
Microcitrus	Reaction to nematodes	Baines *et al.*, 1960
Microcitrus	Reaction to *Phytophthora*	Carpenter and Furr, 1962
Micromelum	Medicinal use	Anderson, 1985b; Brown, 1954; Burkill, 1935; Cassady, 1979; Jain and Borthakur, 1980; Jones, 1995; Uhe, 1974
Murraya	Anti-insect properties	Bowman *et al.*, 2001; Dowell, 1989; Jones, 1995
Murraya	Condiment	Burkill, 1935; Joseph and Peter, 1985; Morton, 1984; Swingle, 1943; Swingle and Reece, 1967
Murraya	Dentifrice	Brown, 1954; Parmar and Kaushal, 1982
Murraya	Essential oils	Burkill, 1935; Joseph and Peter, 1985
Murraya	Leaf characteristics	Halma,1929; Hirano, 1931
Murraya	Medicinal use	Burkill, 1935; Iyer and Mani, 1990; Jain and Borthakur, 1980; Jones, 1995; Joseph and Peter, 1985; Manandhar, 1995; Morton, 1984; Parmar and Kaushal, 1982
Murraya	Quality of wood	Jones, 1995
Murraya	Reaction to canker	Gottwald *et al.*, 1993; Lee, 1918; Peltier, 1918; Peltier and Frederich, 1920, 1924
Murraya	Reaction to citrus tristeza virus	Knorr, 1956; Müller and Garnsey, 1984; Yoshida, 1996
Murraya	Reaction to miscellaneous fungal diseases	Fulton, 1925; Winston *et al.*, 1927
Murraya	Reaction to nematodes	Baines *et al.*, 1960
Murraya	Reaction to Huanglongbing	Aubert, 1987
Pamburus	Reaction to citrus tristeza virus	Knorr, 1956; Müller and Garnsey, 1984
Pamburus	Rootstock use	Swingle, 1916
Paramignya	Medicinal use	Burkill, 1935
Paramignya	Reaction to canker	Lee, 1918; Peltier, 1918; Peltier and Frederich, 1920, 1924
Paramignya	Reaction to miscellaneous fungal diseases	Winston *et al.*, 1927
Pleiospermium	Reaction to citrus tristeza virus	Yoshida, 1996
Poncirus	Adaptation to soil conditions	Shaked *et al.*, 1988; Shannon *et al.*, 1960a, b; Yelenosky *et al.*, 1973
Poncirus	Cold hardiness	Roose and Kupper, 1992; Traub and Robinson, 1937; Yelenosky *et al.*, 1973; Young, 1963b, c, 1977; Young and Hearn, 1972
Poncirus	Fruit composition or development	Bitters and Batchelor, 1951; Bitters *et al.*, 1973a; Kefford and Chandler, 1961; Roose *et al.*, 1985; Sinclair and Bartholomew, 1944; Woodruff and Olson, 1960; Wutscher and Bistline, 1988; Wutscher and Shull, 1975

Table 4.3. Continued

Genus	Attribute	References
Poncirus	Growth rate	Yelenosky *et al.*, 1968; Maggs and Alexander, 1969
Poncirus	Level of apomixis	Frost and Soost, 1968; Khan and Roose, 1988; Pio *et al.*, 1984; Ueno *et al.*, 1967
Poncirus	Mineral nutrition	Castle and Krezdorn, 1975; Embleton *et al.*, 1973b; Hodgson and Eggers, 1938
Poncirus	Photosynthesis	Morinaga and Ikeda, 1990; Syvertsen and Graham, 1985
Poncirus	Reaction to canker	Gottwald *et al.*, 1993; Peltier, 1918; Peltier and Frederich, 1920, 1924
Poncirus	Reaction to citrus tristeza virus	Bitters, 1959, 1972; Costa *et al.*, 1949; Grant and Costa, 1948; Grant *et al.*, 1949, 1951; Salibe, 1977
Poncirus	Reaction to Mal secco	Chapot, 1963; Russo, 1976/77
Poncirus	Reaction to miscellaneous fungal diseases	Fulton, 1925; Olson and Godfrey, 1953; Peltier and Frederich, 1923
Poncirus	Reaction to miscellaneous mycoplasmas	McClean and Schwarz, 1970
Poncirus	Reaction to miscellaneous virus or viroid diseases	Childs, 1951; Vogel and Bové, 1971; Yamada and Sawamura, 1952
Poncirus	Reaction to nematodes	Baines *et al.*, 1960; Cameron *et al.*, 1977; Feder, 1968; Ferguson *et al.*, 1992; Ford, 1969; Hutchison and O'Bannon, 1972; Kaplan and O'Bannon, 1981; McCarty *et al.*, 1979; Niles *et al.*, 1995
Poncirus	Reaction to *Phytophthora*	Carpenter and Furr, 1962; Davenport and Rouse, 1992; Furr and Carpenter, 1961; Grimm and Hutchison, 1972; Laville and Blondel, 1979; Raghavendra Rao and Prasad,, 1983
Poncirus	Root characteristics	Castle and Youtsey, 1983; Luxmore *et al.*, 1971; Smith and Wallace, 1954; Syvertsen and Graham, 1985
Poncirus	Salt tolerance	Embleton *et al.*, 1973b; Kirkpatrick and Bitters, 1969
Poncirus	Tree size control	Bitters, 1950; Castle and Phillips, 1977; Hutton *et al.*, 2000; Phillips, 1969; Phillips and Castle, 1977; Roose *et al.*, 1985
Severinia	Adaptation to soil conditions	Wutscher *et al.*, 1970
Severinia	Cold hardiness	Young, 1963c
Severinia	Fruit composition or development	Wutscher and Bistline, 1988
Severinia	Leaf characteristics	Hirano, 1931
Severinia	Mineral nutrition	Embleton *et al.*, 1973b; Wutscher *et al.*, 1970
Severinia	Nematode resistance or tolerance	Baines *et al.*, 1960; Hutchison and O'Bannon, 1972; Kaplan, 1981
Severinia	Reaction to canker	Traub and Robinson, 1937
Severinia	Reaction to canker	Gottwald *et al.*, 1993; Lee, 1918; Peltier, 1918; Peltier and Frederich, 1920, 1924

Table 4.3. Continued

Genus	Attribute	References
Severinia	Reaction to citrus tristeza virus	Grant and Costa, 1949; Knorr, 1956; Yoshida, 1996
Severinia	Reaction to Mal secco	Russo, 1976/77
Severinia	Reaction to miscellaneous fungal diseases	Fulton, 1925; Olson and Godfrey, 1953; Winston *et al.*, 1927
Severinia	Reaction to miscellaneous mycoplasmas	Hung *et al.*, 2001
Severinia	Reaction to miscellaneous virus or viroid diseases	Iwanami *et al.*, 1993
Severinia	Reaction to *Phytophthora*	Carpenter and Furr, 1962; Grimm and Hutchison, 1972
Severinia	Salt tolerance	Embleton *et al.*, 1973b; Traub and Robinson, 1937
Severinia	Tree size control	Bitters, 1950; Bitters *et al.*, 1977; Castle and Phillips, 1977; Phillips, 1969; Phillips and Castle, 1977; Traub and Robinson, 1937; Xuannan and Dehui, 1989
Swinglea	Hair and scalp preparations	Brown, 1954
Swinglea	Leaf characteristics	Hirano, 1931
Swinglea	Reaction to canker	Lee, 1918; Peltier, 1918; Peltier and Frederich, 1920, 1924
Swinglea	Reaction to citrus tristeza virus	Knorr, 1956; Müller and Garnsey, 1984; Yoshida, 1996
Swinglea	Reaction to miscellaneous fungal diseases	Fulton, 1925; Winston *et al.*, 1927
Swinglea	Reaction to miscellaneous virus or viroid diseases	Iwanami *et al.*, 1993
Swinglea	Reaction to *Phytophthora*	Ríos-Castaño and Tafut, 1972
Swinglea	Medicinal use	Brown, 1954
Triphasia	Anti-insect properties	Bowman *et al.*, 2001
Triphasia	Food use	Burkill, 1935
Triphasia	Gum	Burkill, 1935
Triphasia	Medicinal use	Burkill, 1935
Triphasia	Reaction to canker	Lee, 1918; Peltier, 1918; Peltier and Frederich, 1920, 1924
Triphasia	Reaction to citrus tristeza virus	Knorr, 1956; Müller and Garnsey, 1984; Yoshida, 1996
Triphasia	Reaction to miscellaneous fungal diseases	Winston *et al.*, 1927
Triphasia	Wood quality	Burkill, 1935

of *Citrus* and other members of the Aurantioideae that have been reported. It is by no means complete or exhaustive, but gives a general overview of some of the characteristics of the Aurantioideae that have been reported in the literature.

The citrons, *C. medica*, are one of the primal three species of citrus mentioned above. They are monoembryonic, and so all seedlings are zygotic. This group represents one of the greatest pools of genetic diversity within the genus *Citrus*. The actual place of origin of the citrons is not definitively known. It is assumed to be northern India or southern China, but Swingle (1943) speculates that citrons may have arisen in southern Arabia. In any case, the citron arrived in the Middle East in ancient times and became important to the ancient Hebrews; it is still used in the Jewish Feast of Tabernacles (Goor and Nurock, 1968). The citron is also believed to be the first type of citrus to arrive in Europe. Citrons are not a large crop in most areas, but are locally important in the Mediterranean area. The most important food use is of the candied peel in confections. It is also used as for pickle, and essential oils are sometimes distilled. The citron is an attractive and fragrant dooryard tree; for this use, the famous 'Buddha's Hand' (*C. medica* var. *sarcodactylis*) must be mentioned. There are also traditional medicinal uses of citron, which suggest that it may have medical benefits. Citrons are sensitive to cold and tend to make small, shrubby trees, especially when grown from cuttings.

The pummelos, *C. maxima*, are also a primal, monoembryonic species and reservoir of genetic diversity. They are native to South-east Asia and the nearby archipelagos, and it is in this area where they have their greatest use. Pummelo fruit are very variable in size, shape, and internal and external colour. They have in common a thick peel with non-adherent larger, pulp vesicles than other *Citrus* species.

The mandarins, *C. reticulata*, comprise both monoembryonic and polyembryonic (apomictic) types. This is a variable group and is sometimes divided into subgroups based upon secondary compounds (Scora, 1989). As used by Swingle, this name is applied to wild or primitive fruits as well as highly selected or hybridized fruits. The mandarins include sour types (*C. reticulata* var. *austera*), large-fruited elite selections and small-fruited types used as rootstocks. There are distinctive subtypes such as the satsumas of Japan (usually referred to as *C. unshiu*) and the clementines of the Mediterranean areas (usually referred to as *C. clementina*). The mandarins are the most cold-hardy species of the subgenus *Citrus* but are variable in other characteristics. Some rootstock types are citrus tristeza virus (CTV)-tolerant but these vary in their tolerance to salinity, pH and other soil conditions, as well as other graft-transmissible diseases (Wutscher, 1979; Castle, 1987). Mandarin trees in general are relatively small and spindly compared with many other types of citrus. Many natural and man-made hybrids of mandarins exist, the most notable being the tangelos (*C. reticulata* × *C. paradisi*, sometimes referred to as *C.* × *tangelo*) and the tangors (*C. reticulata* × *C. sinensis*, sometimes referred to as *C. nobilis*).

The relatively recently described monoembryonic *C. halimii* (Stone *et al.*, 1973), which has characteristics intermediate between *Citrus* and *Fortunella*, is sometimes considered a valid species (Scora and Kumamoto, 1983), but there is some evidence that this is not actually a *Citrus* species at all (Berhow *et al.*, 2000). Recent molecular studies have supported *C. halimii* as a distinct species (Luro *et al.*, 1992; Herrero *et al.*, 1996b; Fang *et al.*, 1998b), closely related to *Fortunella* (Herrero *et al.*, 1996b). Swingle (1943) also accepts the mandarin-like *C. indica* and *C. tachibana* as valid species. Molecular evidence also supports species status for *C. indica* (Fang *et al.*, 1998a; Federici *et al.*, 1998) and *C. tachibana* (Hirai *et al.*, 1990; Herrero *et al.*, 1996b). These three species are somewhat obscure and do not often figure in taxonomic schemes.

The other species of the subgenus *Citrus* are probably hybrids or successive hybrids of these primal types. The grapefruit, *C. paradisi*, should probably not be considered a distinct species but is accorded species status by Swingle (1943). Its probably hybrid origin is recognized in the designation *C.* × *paradisi* (Bailey Hortatorium, 1976). It probably arose relatively recently as a spontaneous hybrid of pummelo and sweet orange (Scora *et al.*, 1982). The origin of the grapefruit and its relationship to the 'forbidden fruit' of the Caribbean region is an interesting story (Gmitter, 1995). The history of the various grapefruit cultivars is well documented compared with most other types. All current cultivars arose from the original grapefruit type, being selected for seedlessness or pigmented flesh (Gmitter, 1995).

The lemon (*C. limon*) probably arose as a hybrid of citron (paternal) and sour orange (maternal) (Nicolosi, *et al.*, 2000;

Gulsen and Roose, 2001), sour orange probably being in turn a hybrid of mandarin and pummelo (Scora, 1975, 1989; Berhow *et al.*, 2000; Gulsen and Roose, 2001; Moore, 2001). Different varieties of mandarin and pummelo probably gave rise to the sweet orange (Barrett and Rhodes, 1976; Berhow *et al.*, 2000; Nicolosi *et al.*, 2000). Limes probably originated as hybrids between citron and either a *Microcitrus* species (Barrett and Rhodes, 1976; Scora, 1989; Federici *et al.*, 1998) or a papeda (Nicolosi *et al.*, 2000). Probably there is little justification for according separate status to such groups as sweet lemon, sweet lime and limetta (Federici *et al.*, 1998). The situation with other groups (rough lemons, rangpur limes) is less clear, but probably there are contributions to both groups from mandarin and citron (Scora, 1989; Federici *et al.*, 1998).

The papedas constitute the subgenus *Papeda*. The characteristics of this distinct group are stated above. Swingle (1943) recognized *C. ichangensis*, *C. latipes*, *C. micrantha*, *C. celebica*, *C. macroptera* and *C. hystrix*, and various subspecies and hybrids. More recently, a new putative species, *C. hongheensis*, has been described (Yinmin *et al.*, 1976). Although the original description of *C. ichangensis* by Swingle (1913a) was based on only a few specimens, there are local variants including variously petioled (Kiang and Zu-Zhao, 1984) and long- and short-fruited types (Ding *et al.*, 1990; Gmitter and Hu, 1989, 1990).

Fortunella

The genus *Fortunella* contains the kumquats. *Fortunella* spp. closely resemble *Citrus* spp. and were included in *Citrus* until Swingle (1915c) separated them into a separate genus based on: (i) differences in the ovary, ovules and stigma of the flower; (ii) abaxial leaf glands; (iii) small more or less angular flower buds; and (iv) a sweet, edible, more or less pulpy skin. It is probably the last characteristic that most people consider the most distinctive of the kumquats. Yamamoto *et al.* (1993) could not distinguish between *Fortunella* and *Citrus* using restriction fragment length polymorphism (RFLP), and Mabberley (1998) included *Fortunella* in his revision of *Citrus*. However, at this time, *Fortunella* is generally considered a separate genus.

Swingle (1915c) originally described four species of *Fortunella*: *F. margarita*, *F. japonica*, *F. crassifolia* and *F. hindsii*. In Swingle (1943), the 'Meiwa' and 'Changshou' kumquats were considered to be hybrids and were described but not accorded species status, removing *F. crassifolia* ('Meiwa') as a species. Swingle (1943) also rejected Tanaka's (1933) *F. obovata* ('Changshou') but accepted as a 'species of doubtful validity' the large-fruited *F. polyandra* from Malaysia (which is speculated to be a limequat). An updated nomenclature that recognizes the hybrid nature of *F. crassifolia* and *F. obovata* is presented in Fantz (1988).

One unique facet of *Fortunella* is the existence of a wild tetraploid form of *F. hindsii* in addition to the diploid forms of this species. *Fortunella hindsii* was formerly considered to be in the genus *Atalantia*, but was placed in *Fortunella* in 1915 (Swingle, 1915c). Swingle (1940b) later added the var. *chintou*, which is a diploid version of the 'normally' (?) tetraploid species. The tetraploid and diploid forms have the differences expected in identical varieties with different ploidy levels. The existence of both diploid and tetraploid forms in *ex situ* collections has been confirmed by root squash chromosome counts and flow cytometry (J. Juárez and L. Navarro, unpublished).

Fortunella trees are evergreen, small, and slow growing. Most of the intrageneric differences are in the fruit size and shape. In China, kumquats have been cultivated and eaten since ancient times. Here, different selections within these cultivars are recognized (Yin-min, 1985; Dashen and Fangcong, 1989). Kumquats are currently considered to exist only in cultivation, with the possible exception of *F. hindsii*. Their area of origin in northern China makes them one of the most cold tolerant of the Aurantioideae, partially due to their characteristic winter dormancy. *Fortunella* spp. are also slightly precocious and resistant to canker and *Phytophthora* (Table 4.3). The kumquats are attractive plants and recently there have been introductions of variegated and seedless types suitable for backyard use.

Kumquats, as with most Aurantioideae, hybridize readily with their closely related genera. This has led to their use in breeding programmes, particularly in programmes for breeding cold-tolerant citrus. This resulted in the creation of many different '-quats' in the early part of the 20th century (see Swingle, 1943 for additional references). These were bred with the objective of creating a 'citrus-like' fruit that was cold hardier than citrus. The '-quats' have not taken this role in commercial production but now represent a separate group in many *ex situ* collections. The procimequat ((*C. aurantifolia* × *F. japonica*) × *F. hindsii*), which is found in *ex situ* collections, is a triploid resulting from the hybridization of the tetraploid *F. hindsii* and a diploid limequat.

Fortunella has some interesting relationships with *Citrus* as regards rootstock/scion compatibilities. These two genera are closely related, but their graft compatibilities are somewhat less congenial than are those of *Citrus* and some of the other closely related Aurantioideae genera. *Fortunella* is best propagated on trifoliate rootstocks, and often develops incompatibilities with *Citrus* and even with citranges. A study in Florida indicated that after 4.5 years, 'Calamondin' 'Meiwa', and 'Nagami' on 'Benton', 'Cleopatra', 'Flying Dragon', 'Sun Chu Sha' and 'X-639' were free of bud union problems, whereas these same varieties showed various bud union problems on 'Carrizo', 80-18 citrumelo, 'Swingle' citrumelo, sweet orange and sour orange (Youtsey, 1997). This seems to be truer of 'Meiwa', which is sometimes grown with a 'Nagami' interstock. *Fortunella hindsii* is sometimes used as a dwarfing rootstock in China. The recently described citrus leaf blotch virus (Galipienso *et al.*, 2000, 2001; Vives *et al.*, 2001), originally found in 'Nagami' (Navarro *et al.*, 1984b) and later in other kumquat sources, produces incompatibility on trifoliate rootstocks and may be the cause of some incompatibilities of kumquats described in the past.

The calamondins represent a distinct group and are sometimes accorded species status. However, Swingle (1943) does not accept this as a valid species, stating that it is an apparent hybrid between a sour mandarin and a kumquat [*C. reticulata* var. *austera*? × *Fortunella* spp.?]. When the calamondins are accorded species status, they are usually referred to as *C. mitis* or *C. microcarpa*. Calamondins are sometimes referred to as *C. madurensis*, but this is incorrect as *C. madurensis* actually refers to *F. japonica* (Swingle, 1943; Wijnands, 1984). Ingram and Moore (1975) and Wijnands (1984) recognize the hybrid nature of this taxon with the coinages ×*Citrofortunella mitis* and ×*Citrofortunella microcarpa*, respectively. Calamondin is extensively reviewed by Mabesa (1990), with an emphasis on fruit quality and processing properties. Although the calamondin is used as a fresh fruit and a condiment and has some known medicinal uses, it is not as well characterized as many other types of Aurantioideae germplasm. It is graft compatible with *Citrus* and has been used as a rootstock, as has its apparent hybrid the 'Philippine calamandarin' (which does not appear to be a calamondin × mandarin hybrid despite its name).

Microcitrus

The genus *Microcitrus* is very closely related to *Citrus* and was originally included in that genus. Swingle (1915b) separated *Microcitrus* from *Citrus* and recognized four species: *M. australis*, *M. australasica*, *M. inodora* and *M. garrowayii*. Swingle (1943) added *M. maideniana* and *M. warburgiana*. *Microcitrus papuana* was published by Winters (1976). *Microcitrus warburgiana* and *M. papuana* are native to Papua New Guinea, while the remaining species are native to Australia. *Fortunella warburgiana* may be similar to the ancestral type from which both *Microcitrus* and *Citrus* evolved.

Recently, an additional species has been published: '*Citrus gracilis*' (Mabberley, 1998). Mabberley attempted a major taxonomic revision of *Citrus* (see above) and some closely related taxa in which *Microcitrus* as well as *Eremocitrus* are re-inserted into the genus *Citrus*. This system has not at this point been widely accepted and the species status of '*C. gracilis*' may be somewhat in doubt.

Microcitrus differs from *Citrus* in its dimorphic foliage, very small juvenile leaves, the distinct shape and venation of adult leaves, very small flowers having free stamens and a very short pistil, a few celled ovary with numerous ovules in each cell, and the subglobose stalked pulp vesicles. *Microcitrus* spp. are shrubs or small trees. The fruit of most species except *M. australasica* are round, with green rinds. In the right environmental conditions, *Microcitrus* trees are quite vigorous. This genus has been reported to be a source of precocity, drought tolerance, nematode resistance, *Phytophthora* resistance and adaptations to low fertility situations (Table 4.3). *Microcitrus* is graft compatible with *Citrus*. There has recently been some interest in the use of *Microcitrus* fruits in Australia, where they are among the 'bush foods' (Sykes, 2001), and in the USA, where they are sometimes used as a garnish.

Microcitrus hybridizes readily with *Citrus*, *Fortunella* and other closely related genera. Swingle (1943) describes trigeneric hybrids, such as the Faustrimedin (*Microcitrus australasica* × (*Fortunella* sp. × *Citrus* sp.)). Also of note are the 'Sydney Hybrid' (*M. australis* × *M. australasica*) and the red-fruited *sanguinea* variant of *M. australasica*, both of which are often found in *ex situ* collections.

Eremocitrus

Eremocitrus is a monospecific genus (*E. glauca*) that is very similar to *Microcitrus* (Swingle, 1914b). However, the flowers are smaller and the fruit has 3–5 locules with two ovules in each. It is native to central and northern New South Wales and south-eastern Queensland in Australia.

As a native of desert habitats, *E. glauca* is xerophytic and cold tolerant. Its xerophytic qualities include small, thick, greyish leaves with thick cuticles and small substomatal chambers and an extensive root system. As with some other species adapted to hot, arid climates, the leaves may abscise during severe drought conditions. There is some confusion as to the cold hardiness of *E. glauca* due to some temperature conversion errors in Swingle (1943) and Swingle and Reece (1967). However, *E. glauca* can probably tolerate temperatures of about –5.5°C (approximately +20°F) or less, consistent with the original paper (Swingle, 1914b) and more recent observations (Yelenosky *et al.*, 1978).

Eremocitrus is graft compatible with *Citrus* and hybridizes readily with that and other closely related genera. *Eremocitrus* is reported to be a source of tolerance to heat, boron, salt and *Phytophthora* root rot (Table 4.3). When grafted to *Citrus* as a rootstock, it has some unusual reactions. Two *E. glauca* trees were planted in the University of California Citrus Variety Collection in 1983. One, grafted on 'Carrizo' citrange, is a normal sized tree, while one grafted on *C. macrophylla*, not normally a dwarfing rootstock, was only about 1 m tall in 2003.

Clymenia

The genus *Clymenia* is another that is closely related to *Citrus* and was originally included in that genus. Swingle (1939) separated *Clymenia* from *Citrus* based upon the structure of the pulp vesicles, which are short, plump, blunt, oval or subglobose, sessile or very short stalked, and attached to the side walls of the 14–16 locules; this is not like any species in the genus *Citrus*. Furthermore, the leaves of *Clymenia* are unlike those of any *Citrus* species, and the flowers have enlarged disks with 10–20 times as many stamens as petals. Recent chemotaxonomic work (Berhow *et al.*, 2000) suggests that *Clymenia* is closely allied with *Fortunella* and may be a hybrid between *Fortunella* and *Citrus*. *Clymenia* is considered a primitive genus in the 'true citrus fruit trees group' and may be a link between that group and the 'near citrus fruit trees' group.

For many years, *Clymenia* was a monotypic genus (*C. polyandra*). However, Stone (1985a) described a 'problematical new species', *C. platypoda*, which differed from *C. polyandra* in its leaf form; it was speculated that *P. platypoda* may actually represent a hybrid between *Clymenia* and *Citrus*.

Clymenia is not a well characterized genus. The trees are described as being small and are of small size when cultivated in *ex situ* collections. The fruits resemble sweet limes and are eaten by the natives (who give it the name *a-mulis*) of its place of origin in the Bismarck Archipelago north-east of Papua New Guinea. *Clymenia* is considered rare in the wild (Jones, 1990). It is graft compatible and hybridizable with *Citrus* but has been little studied as far as its disease reaction and other adaptations.

Poncirus

Swingle (1943) described the genus *Poncirus* as 'remarkable' and it is, indeed, the most remarkable genus of at least the 'true citrus fruit trees' and probably the entire Aurantioideae. In fact, it is difficult to know how properly to approach *Poncirus*.

Poncirus, like the other genera of the 'true citrus fruit trees', was for many years included in *Citrus* despite its 'many strikingly aberrant characters' (Swingle, 1943). These characteristics were well known at least by the early part of the 20th century (Swingle, 1909), but *Poncirus* was apparently not separated from *Citrus* and the name *Poncirus* reapplied until 1916 (Swingle, 1914–1917). Its early inclusion in *Citrus* led to the still used common name (or misnomer) of 'trifoliate orange'.

Swingle (1909, 1914–1917) noted the differences between *Poncirus* and *Citrus*. The leaves are trifoliolate and deciduous, the tree being winter dormant. The flower buds are formed in the early summer, and overwinter protected by bud scales. In the spring, the flowers are borne on old wood. They are nearly sessile, with the petals opening flat, the stamens entirely free, and the ovary having 6–8 locules. The fruits are densely and finely pubescent. The pulp vesicles carry scattered hair-like organs that bear at their expanded tips thick-walled, fissured cells secreting a viscous fluid allowing the pulp vesicles to slip past one another. Transverse plates composed of thick-walled cells are found in the pith of the stem. *Poncirus* trees are relatively small and shrub-like. They are very thorny, with a relatively thick trunk and limbs.

These pronounced differences make it difficult to place *Poncirus* in an orderly course of evolutionary progress from some remote common ancestor of *Citrus* and its related genera, suggesting that there are many gaps therein and that *Poncirus* is the most 'isolated and aberrant of the "true citrus fruit trees"' (Swingle, 1943). Swingle's (1943) view is that the putative remote ancestor of the 'true citrus fruit trees' originated in a tropical or subtropical climate. While the other genera of the 'true citrus fruit trees' remained in those climates, *Poncirus* 'migrated' to the temperate climate of north-eastern Asia. In doing so, it developed the adaptations to winter cold noted in the previous paragraph. These adaptations permit *Poncirus* to survive in temperate climates that would kill all other Aurantioideae.

The trifoliate has apparently been cultivated since ancient times in China and was introduced into Japan several centuries before the end of the first millennium AD. It may be the oldest rootstock to be used in citriculture. It is also used as an ornamental or hedge in that area. The introduction of the trifoliate to the Western world was apparently later than the introduction of edible types of citrus. In the USA, despite several suggestions that it was introduced during Colonial times and cultivated at Monticello and other historic sites, its actual date of introduction was most probably about 1869 (Swingle, 1909).

The useful characteristics as well as the limitations of the trifoliate became evident in the last part of the 19th century. These are summarized in Table 4.3, and include resistance to *Phytophthora* root rot and CTV; cold tolerance; adaptation to heavy soils; small size of budded trees; lack of suitability to calcareous soils; etc. *Poncirus* is graft compatible with and hybridizes with *Citrus*. While the use of *Poncirus* as a rootstock in Asia is very old and it is also used in current commercial production in many locales, perhaps its major importance in this area has been as a parent in intergeneric hybridization with *Citrus*. This has produced the very important citranges (*P. trifoliata* × *C. sinensis*) and citrumelos (*P. trifoliata* × *C. paradisi*), as well as citrandarins (*P. trifoliata* × *C. reticulata*), citremons (*P. trifoliata* × *C. limon*) and citradias (*P. trifoliata* × *C. aurantium*). These have been coined ×*Citroncirus* species by Ingram and Moore (1975). While normally slow growing, when used as a rootstock *Poncirus* is stimulated to add girth more quickly than the scion variety, causing the characteristic 'bell' at the bud union.

The early observations of *Poncirus* by Swingle (1909, 1914–1917, 1943) indicate that the trifoliates were not considered a very variable group. This is somewhat misleading. Swingle (1909) indicated that most trifoliates in the USA had small flowers due to a stunting of the petals near the base. By 1916 Swingle (1914–1917) had recognized the small flowered forms as being distinct from the 'normal' large flowers. By the 1930s, the 'normal small-flowered form and a large-flowered form of seemingly greater vigor' had been recognized (Traub and Robinson, 1937). These large- and small-flowered forms were different from the 'normal' large-flowered and imperfectly developed small-flowered forms described earlier, however, being based upon the size and structure of the petals (Shannon *et al.*, 1960a, b).

Shannon *et al.* (1960a, b) collected material from suckers of trifoliate rootstocks used in commercial production in several areas of California and divided them into small- and large-flowered varieties. These selections, along with others from Florida and other areas of the southern USA, are still the basis for most collections of trifoliates in the USA and, to a lesser extent, in other countries. Seedlings of large- and small-flowered varieties produced similar biomass, but large-flowered varieties had an upright habit of growth with a single trunk whereas small-flowered varieties produced multiple trunks resulting in a bushy growth habit. Suckering of the multiple shoots of the small-flowered types reduced the weight of the small-flowered types as compared with the large-flowered types. Fruits from the small-flowered varieties matured earlier than those from the large-flowered varieties.

Shannon *et al.* (1960a, b) reported that 'Washington' navel trees on the large-flowered 'Christian' trifoliate were larger than those on the small-flowered 'Rubidoux' trifoliate. Similarly, Bitters *et al.* (1973a) compared 24 selections of trifoliates and found that trees were larger on almost all large-flowered selections than on small-flowered selections. However, yields and fruit quality were similar from trees on both large- and small-flowered varieties. The small-flowered types produced slightly larger fruit and were more efficient on a canopy volume basis.

This system of dividing the trifoliates into small- and large-flowered varieties is used in the USA and many other countries. Japan, which has a far longer history of cultivation of *P. trifoliata*, has a somewhat different system (Okamoto, 1935; Iwasaki, 1943; Iwasa and Shiraishi, 1957; Iwasaki and Nishiura, 1963; Tanaka, 1969). In the Japanese system, four different strains are recognized based upon leaf size: diploid large leaf; tetraploid large leaf; small leaf; and normal (medium leaf). Iwasa and Shiraishi (1957) report other strains in addition to these, including Hiryû (see below), 'multi-leaf strain', '*P. trifoliata* var. *microcarpa*' (which is described as a 'useless fruit – details not known'), a 'spotted' strain (i.e. variegated), a purple-flowered strain and a bent-thorn strain.

Iwasaki and Nishiura (1963) note that the common (Chûba-kei) strain is superior to either the large-leaf (Ôba-kei) or small-leaf (Koba) strains as a nursery tree and rootstock. Chûba-kei developed a larger root system with abundant fibrous roots, whereas Ôba-kei had a root system with few branches and thick roots, and Koba-kei had well-branched, slender roots. Chûba-kei produced a buddable diameter trunk sooner than the other strains. Tree growth was greatest and most uniform on Chûba-kei. Ôba-kei and Koba were considered variants not suitable for use as rootstocks.

It may be tempting to correlate the large-flower and large-leaf strains, and the small-flower and small-leaf strains. Iwasaki (1943) does correlate flower size and leaf size in the Japanese strains studied, and notes differences in growth habit similar to those reported by Shannon *et al.* (1960a, b). However, it is not entirely certain how these Japanese strains relate to the US strains studied by Shannon *et al.* (1960a, b). There is not enough information on gradations between small- and large-flowered types in the USA to determine if there could be an intermediate group. If this could be done, it might be possible to designate a third, medium-flowered group and see where its horticultural characteristics fit in with the small- and large-flowered groups. However, the fact that the most common strain in Japan is medium leafed and that the strains with small and large leaves are not suitable for rootstock use speaks against this. Although there were differences in performance as rootstocks of the varieties tested by Shannon *et al.* (1960a, b) and Bitters *et al.* (1973a), the differences were small and none of the varieties was apparently unfit for use as rootstocks. This might be related to disease status of the scion varieties. The early Japanese work may well have utilized scion varieties that were infected with Exocortis or other viroids (to which trifoliates are susceptible), which were common in Japan in those times. The scion varieties utilized by Shannon *et al.* (1960a, b) and Bitters *et al.* (1973a) were not likely to have been infected with viroids. These differences in pathogen status could have influenced horticultural performance and account for some of the differences noted.

More probably, the differences between the USA and Japanese classifications can be traced to the fact that there were few introductions of trifoliates into the USA, with a resultant lack of genetic diversity. Shannon *et al.* (1960a, b) took cuttings from rootstock suckers and developed these into selections. The original providence of the rootstocks from which the buds were taken is unknown, but probably represented very few introductions of *P. trifoliata*. This is supported by recent molecular work (Fang *et al.*, 1997). Inter-simple sequence repeat (ISSR) markers revealed that trifoliates in the UCR/USDA Citrus Variety Collection clustered into four groups. All small-flowered accessions clustered into group 3 and all large-flowered accessions clustered into group 4. Group 2 had only two accessions, which were apparently derived from zygotic seedlings of individuals similar to the common genotype. On the other hand, three trifoliate accessions introduced from China more recently had relatively different fingerprints and probably represented a more divergent genetic make-up. This would appear to support the idea that the genetic base of the US-derived accessions is relatively narrow. That there is more genetic variability in *Poncirus* than previously thought is demonstrated by the range of adaptations exhibited by *Poncirus* germplasm recently introduced into Australia from China (Broadbent *et al.*, 2003).

One variety of *P. trifoliata* of particular interest is 'Flying Dragon' (*P. trifoliata* var. *monstrosa*), also known as 'Hiryû' (Swingle, 1943). This is a dwarf variety, with very slender leaves and crooked, tortuously curved branches and spines. Although all trifoliates produce a small tree when used as a rootstock, 'Flying Dragon' is the only variety that is a true dwarfing rootstock. Trees budded on 'Flying Dragon' seldom exceed 2.5 m in height. Swingle (1943) stated that he had introduced it to the USA in 1915 but that it had never flowered or fruited and that if it could be induced to do so it would be interesting to 'study the chromosomes ... to see if perhaps they are abnormal'. More recently, Cheng and Roose (1995) demonstrated that the dwarfing characteristics were inherited as if they were a single dominant gene for which 'Flying Dragon' is heterozygous. The curved thorns and trunk appeared to be pleiotropic effects of the dwarfing gene. The data also suggested that 'Flying Dragon' originated as a mutant of a non-dwarfing genotype and has not undergone sexual recombination since. The phenotypic effects may be tied to plant growth hormone metabolism, since 'Flying Dragon' had significantly higher abscisic acid (ABA) in leaves and roots and significantly lower indoleacetic acid (IAA) in shoots than normal types of trifoliates (Xiao *et al.*, 2001).

The above statements were written based upon what is normally called 'Flying Dragon' in the USA. This may actually be different from the 'Hiryû' of Japan. According to some workers, Swingle's (1943) description of 'Flying Dragon' may be closer to the Japanese variety 'Unryû' (letter from M. Iwamassa to W. P. Bitters, ~1970). On the other hand, Iwasa and Shiraishi (1957) describe Hiryû as a 'decorative bonsai plant which has flexible stems', a description which seems consistent with that of Swingle (1943). The accessions present in germplasm collections in the USA (and others which obtained their material from the USA, particularly from Riverside) are morphologically similar to one another and to Swingle's (1943) description. That does not mean that the name was not misapplied. A separate accession arriving in the USA as 'Unryû' did not establish so it is not possible to compare it with the existing 'Hiryû' accessions. However, the imported 'Unryû' was from the same source as one 'Hiryû' accession, so several types of confusion are possible.

Poncirus was for many decades a monotypic genus. Recently, a putative new species, *P. polyandra*, was published (Ding *et al.*, 1984). This was discovered in 1978 in Yunnan Province (Duan, 1990). This differs from *P. trifoliata* in having larger leaves, some floral differences and, most strikingly, in being evergreen (Duan, 1990). Perhaps this last characteristic is related to its providence in Yunnan, the southernmost Province of China. *Poncirrus polyandra* is little known in the west (and perhaps in China). Duan (1990) suggests that it is dwarfing as a rootstock. Its other characteristics are not known to the writer other than possibly being susceptible or tolerant (rather than resistant) to CTV. Fang (1993) considered it as deserving of species status based on isozyme data, while others see it as a possible hybrid (F. Gmitter, personal communication, 1998). Recent ISSR data suggest that *P. polyandra* is genetically distinct from the other groups of *Poncirus*, falling outside the four groups of Fang *et al.* (1997) (Krueger and Roose, 2003).

Atalantia

The genus *Atalantia*, along with *Citropsis*, constitute the 'near citrus fruit trees' of Swingle (1943). As with many genera of the Aurantioideae, the taxonomic history of *Atalantia* is somewhat confusing; it was formerly included in *Limonia*. The background for the current status of *Atalantia* can be found in Swingle (1943). Swingle (1914–1917) enumerated ten species of *Atalantia*; this was expanded in 1940 (Swingle, 1940a, b). Some species of *Atalantia* are considered rare (Jones, 1990).

Members of *Atalantia* are small, attractive trees with fragrant white flowers and small, greenish-yellow fruits. The juice vesicles differ from those of *Citrus* in being sessile rather than stalked. These juice vesicles are conical and are arranged radially in the locules, attaching to the dorsal wall of the locules and imbedded in the inner layer of the rind. *Atalantia ceylanica*, *A. rotundifolia*, *A. guillaumini* and *A. hainanensis* have few or no vesicles, and large seeds that nearly fill the locules. These species constitute the subgenus *Rissoa*. The leaves of *Atalantia* are simple or unifoliolate like those of *Citrus*, but differ in having more prominent and numerous lateral veins.

Atalantia is graft compatible with *Citrus*. It has been reported to be resistant to the burrowing nematode (Table 4.3) but is not as well characterized as some other genera. Since the trees are attractive, there has been minor use of *Atalantia* in ornamental plantings. The wood is also very attractive.

Citropsis

Citropsis was formerly included as a section of *Limonia* until separated by Swingle (1914a). Additional species were added several decades later (Swingle, 1940b). Swingle (1943) recognizes 11 species and one variety. The members of this genus are native to the tropical forests of Africa rather than to South-east Asia.

Citropsis has fruits similar to those of *Atalantia* but has pinnate, trifoliolate or five-foliolate leaves. *Citropsis* is graft compatible with *Citrus* and has been experimented with as a rootstock. It has been tested for nematode and disease reaction, as well as for several other properties (Table 4.3). Although its fruits resemble those of *Atalantia*, *Citropsis* is not as attractive as a tree. The leaves of most species are fairly large and coarse, and the trunk and branches are quite spiny. The fruits, while reasonably attractive, have a strong, fetid odour.

Hesperethusa

Hesperethusa, *Limnocitrus*, *Burkillanthus*, *Pleiospermium* and *Severinia* have primitive pulp vesicles, a fact that was not recognized until the middle part of the 20th century. Swingle (1938, 1943) used this, at that time new, fact to separate these five genera into the new group, the 'primitive citrus fruit trees'.

Hesperethusa is a monotypic genus (*H. crenulata*) native to the Indian subcontinent and the Indochinese peninsula. There is some controversy as to whether *Hesperethusa* or *Naringi* is the correct genus name. *Naringi* has been put forth by Panigrahi (1975) and is accepted by Wiersma and León (1999).

Hesperethusa has leaves that are similar to those of *Citropsis* in having odd pinnate leaves with broad petioles. Unlike those of *Citropsis*, the leaves of *Hesperethusa* have prominent, slightly raised glands. The small black fruits have a few primitive pulp vesicles in each locule which are attached to the dorsal wall and also to the base of the locule. The trees are medium sized and attractive. *Hesperethusa* is described as 'subdeciduous' (Swingle. 1943) and, indeed, those in *ex situ* collections sometimes lose some or all of their foliage, apparently with no obvious environmental cue.

Hesperethusa is graft compatible with *Citrus*. Often an interstock improves compatibility. It is used in its native area as a condiment, but the fruits are not eaten as they are 'bitter (not acid)' (Swingle, 1943).

Pleiospermium

Pleiospermium was separated from *Limonia* by Swingle (1916b). The existence of the primitive vesicles mentioned above was not known in some of the other genera of the 'primitive citrus fruit trees' at that time, and consequently the relationships between *Pleiospermium* and other Aurantioideae genera stated by Swingle (1916b) were not correct. It was not until the 1930s that the relationship of *Pleiospermium* to the other genera in this group was published (Swingle, 1938, 1943).

The original paper by Swingle (1916b) published the trifoliate *P. alatum* and the polymorphic (1- to 3-foliate) *P. dubium*. Swingle (1939) later added three unifoliate species: *P. longisepalum*, *P. sumatranum* and *P. latialatum*. The leaves of all five species have long, articulate, winged petioles. The small fruits have small, ovoid or cylindrical pulp vesicles with bluntly rounded or acute tips arising from the inner wall of the ovary. A more or less well-differentiated outer cortical layer surrounds a central portion that disintegrates as the fruits mature into an oily mass. There are various anatomical differences in the fruits between the different species (Swingle, 1943).

Swingle (1943) considers *Pleiospermium* to be a significant genus from the evolutionary standpoint, with primitive pulp vesicles linking the 'near citrus fruit trees' and 'true citrus fruit trees' of the Citrinae with the genera from the other, more primitive tribes of Triphasiinae and Balsamocitrinae. The fact that some genera exhibit the more primitive trifoliolate leaves and others the more evolved unifoliolate leaves is cited to support this contention.

Pleiospermium is graft compatible with *Citrus*. It has been shown to be susceptible to CTV but is otherwise not well characterized (Table 4.3).

Burkillanthus

Burkillanthus, represented by only one species (*B. malaccensis*), is similar to *Pleiospermium* in its leaf characteristics but is quite different in its fruit. The pulp vesicles of *Burkillanthus* somewhat resemble those of *Pleiospermium*, but are much larger and less obvious corticate, and the central core does not disintegrate. The ovary of *Burkillanthus* has 22–26 ovules in each of the five locules; this is the largest number of ovaries of any plant in the Aurantioideae. The fruits are much larger than those of *Pleiospermium* and have a thick, leathery rind. This species is not well represented in *ex situ* collections and there has apparently been no characterization work on it. It is considered rare and threatened by Jones (1990).

Limnocitrus

The monotypic genus *Limnocitrus* (*L. littoralis*) was formerly included in several different genera until separated out by Swingle (1940a). It is similar to *Pleiospermium* but has simple leaves that are thick and sparingly veiny, with slightly crenulate margins and straight rigid thorns in the axils. *Limnocitrus* also has slender, fusiform pulp vesicles with acute tips and slightly narrowed bases. These are different from those of *Pleiospermium* and *Atalantia*. The fruits resemble small oranges externally.

Limnocitrus is native to the tidal marshes of the South-east Asian archipelagos and mainland. Because of this habitat, it probably possesses a high tolerance to soil moisture and salinity. Although it was said by Swingle (1943) to be widely distributed, it is not well represented in *ex situ* collections. It is graft compatible with *Citrus*. More information on the properties of this species is needed.

Severinia

Severinia is probably the best known of the 'primitive citrus fruit trees' but in actuality it is only the type species, *S. buxifolia*, that is well known. *Severinia* was generally included in *Atalantia*, which it superficially resembles, until Swingle (1916c) reapplied the 19th century generic name *Severinia*. *Severinia* differs from *Atalantia* in having a dark, usually nearly blackberry-like fruit at maturity. The pulp vesicles are quite primitive, in contrast to the relatively well developed ones found in *Atalantia*, and two or three oil glands are found in the mesocarp of the young ovary. The vesicles do not break down into a gelatinous mass as in *Pleiospermium*. The leaves of *Severinia* are somewhat similar to those of *Atalantia*, but have non-articulated petioles.

Severinia buxifolia is a relatively common plant in the southern part of China and adjacent areas. It is a handsome shrub, and is sometimes used for hedges. The leaves are (or were) used in China for making yeast cakes. According to Swingle (1943), there are different horticultural varieties in Asia that differ in height, growth habit, etc., but have not been recognized as subspecies or varieties. Even in *ex situ* collections there are variant forms and multiple accessions. A common variant is the 'brachytic form' with shortened internodes and dense foliage.

Severinia buxifolia has received more horticultural attention than the majority of citrus relatives (Table 4.3). It is graft compatible with *Citrus* and the resultant plants are generally long lived. The dwarfing effect of *S. buxifolia* as a rootstock is well known. It has been tested for reaction to diseases and nematodes more than most other citrus relatives. It is known to be tolerant of cold, boron and salinity. Probably the main reason for its more extensive characterization than other related genera is simply the fact that it is commonly found in *ex situ* collections and is easy to grow.

The other species of *Severinia* include *S. disticha*, *S. linearis*, *S. paniculata*, *S. retusa*, *S. trimera* and *S. lauterbachii*. These were named or re-named by Swingle (1938) from other species (mostly *Atalantia*). Of these, *S. disticha* is the one most similar to *S. buxifolia* and the other most common *Severinia* accession in *ex situ* collections. Some species of *Severinia* are considered rare by Jones (1990).

Luvunga

Luvunga, *Paramignya*, *Pamburus*, *Triphasia*, *Merope*, *Monanthocitrus* and *Wenzelia* are the 'minor citroid fruit trees'. These genera do not have pulp vesicles, but the fruit often contains mucilaginous gum.

Luvunga, along with the very similar *Paramignya*, comprises the *Luvunga* group of the 'minor citroid fruit trees'. *Luvunga* is a genus of 12 rather non-diverse species, which are not well known or characterized. Unlike many other genera within the Aurantioideae, it has retained the same generic name since the 1830s (Swingle, 1943). *Luvunga* is known mostly from the work of Swingle (1943), which relies mainly on old descriptions from herbaria specimens. A new species, *L. minutiflora*, was proposed by Stone (1985c). *Luvunga* is undoubtedly one taxon that, when definitively studied, will end up being different from the present description.

All species of *Luvunga* are woody vines of the south and South-east Asian continent and archipelagos. These vines cling to the forest trees by recurved thorns borne in the leaf axils. The leaves are usually trifoliolate with long, wingless clusters. Occasionally *Luvunga* bears unifoliolate leaves, which have shorter petioles than usual and thus resemble normal *Paramignya* leaves. The smallish fruits are yellowish and filled with mucilage. The rough skin has small oil glands on the peel and there are larger ones between the peel and the locules.

Paramignya

Paramignya is closely related to *Luvunga* and resembles it in general appearance. Surprisingly, most species were formerly included in *Atalantia* despite such pronounced differences as the lack of pulp vesicles in *Paramignya*. It is native to the same geographical area as *Luvunga* and shares its vinous habit. The main vegetative difference is that *Paramignya* has unifoliolate leaves borne on short petioles. The leaves have a well-defined pulvinus so that they can maximize interception of sunlight. The fruits of *Paramignya* and *Luvunga* are very similar.

Also like *Luvunga*, *Paramignya* is not a well characterized genus. The 15 species are apparently not very diverse and most are known mostly from herbarium specimens. Several species were added to *Paramignya* by Swingle (1938, 1940a). Like *Luvunga*, *Paramignya* will undoubtedly be very different from what is currently thought when it is definitively studied.

Triphasia

Triphasia and *Pamburus* comprise the *Triphasia* group of the 'minor citroid fruit trees'. In the work of Swingle (1943), this group differs from the *Luvunga* group primarily in leaf and spine characteristics. The *Triphasia* group are small trees or shrubs rather than vines, and do not have recurved thorns. The petioles are not articulated and there are no pulvini. They are native to the south-eastern Asian archipelagos.

Triphasia has small reddish or pink fruits with 3–5 locules of one or two ovules apiece in the ovaries. They are sometimes used for marmalade. The best-known species, *T. trifolia*, was for many years the only species in the genus. It is an attractive small tree or shrub with trifoliate leaves, and is grown as an ornamental in some tropical and subtropical regions. It is somewhat sensitive to cold, but has been said to show some salt tolerance. It is also the only species usually found in *ex situ* collections. Despite this relative commonality, *Triphasia* is not well characterized (Table 4.3). It is graft compatible with *Citrus*. Small (1933) and Brizicky (1962a) state that it has become naturalized in some areas of south Florida; however, current Floridians are not acquainted with this (K. D. Bowman, personal communication, ~2001). Observations in Riverside suggest that *Triphasia* is fairly tender. A 4*n* form, *T. trifolia* forma *tetraploidea*, was described by Swingle (1940c).

Other species are less well known. They include *T. grandifolia*, which apparently differ mainly in having large, simple leaves. *Triphasia brassii* was renamed by Swingle (1938) from the *Paramignya brassii* of White (1926).

Pamburus

Pamburus is a monotypic genus (*P. missionis*). It differs from *Paramignya* in the leaf characteristics and growth habit cited above, and from *Triphasia* in having single, rather than paired, spines; longer anthers; and lacking an inflation at the base of the filaments. *Pamburus missionis* is a robust, thorny tree with dark green foliage. It is native to southern India and Sri Lanka, where its wood is sometimes used for furniture. Swingle (1943) stated that it was almost as cold tolerant as commonly cultivated citrus, but it sometimes has been difficult to establish in the field in subtropical climates. Although it is not uncommon in *ex situ* collections, little more has been published on *Pamburus* since the old Swingle treatments (1916a, 1943).

Wenzelia

The *Wenzelia* group of the Triphasiinae consists of the genera *Wenzelia*, *Monanthocitrus*, *Oxanthera* and *Merope*. They have 4–8 ovules (most commonly six) in each locule, and simple or unifoliolate leaves. All four genera are native to the South-east Asian archipelagos; only *Merope* has been reported from the mainland (Myanmar/Burma). These genera constitute some of the least known of the Aurantioideae. They are not well represented in *ex situ* collections and little is known of their horticultural properties.

Wenzelia has large, simple leaves, unique in the 'minor citroid fruit trees' except for *Monanthocitrus*. The fruits are ellipsoid with 6–8 seeds per locule immersed in a mucilaginous substance. All species of *Wenzelia* were described and published in the 20th century, but sometimes from little material. Swingle (1943) recognized nine species divided into two subgenera (*Wenzelia* and *Papualimo*), which are perhaps somewhat dubious as valid subspecies (Stone, 1985b). Stone (1985b) reviewed the paucity of material with which Swingle was able to work and indicated a need for a modern revision based upon additional specimens as well as field collections. Unfortunately, Stone died before being able to accomplish this, so the genus remains somewhat confused. The 'clearly distinct' (Swingle, 1943) *W. palu-*

dosa was transferred to *Monanthocitrus paludosa* by Stone (1985b).

Wenzelia species are shrubs. As stated, little is known of their horticultural properties. However, their graft compatibility with *Citrus* is somewhat problematic.

Monanthocitrus

Monanthocitrus is very similar to *Wenzelia*, with the main difference being the spotted seeds of the former (Swingle, 1943). This seems a somewhat questionable characteristic upon which to separate genera, particularly since Swingle had little material to work with and not all specimens examined had seeds. Stone (1985b) states that 'The distinctness of *Wenzelia* and *Monanthocitrus* remains somewhat problematical' and 'needs reconsideration'. *Monanthocitrus* was construed as a monotypic genus (*M. corniculata*) by Swingle (1943). *Monanthocitrus grandiflora* was transferred to *Wenzelia* by Swingle (1938, 1943). Stone (1985b) believed that it belonged better to *Monanthocitrus* but did not publish that combination. Swingle also assigned *W. grandiflora* and *W. dolichophylla* to *Wenzelia* rather than *Monanthocitrus* even though seeds were not available for inspection. More recently, two new species have been published: *M. bispinosa* (Stone, 1985b) and *M. oblanceolata* (Stone and Jones, 1988). This genus is little known nor collected.

Oxanthera

Oxanthera is a distinctive genus of 4–5 species. The leaves of these shrubs or small trees are glabrous and glaucous, usually rather blunt at the tip and cuneate at the base, and borne on thornless branches. The leaves of three of the four species are thick and coriaceous, but *O. undulata* has thin leaves. The flowers are large and fragrant. The fruits are distinctive in appearance, being elongate and longitudinally ribbed. In at least one species (*O. neo-caledonica*), the fruit rind is fragrant. There are usually six (sometimes 4–8) ovules per locule. A new putative species, *O. brevipes*, was recently published (Stone, 1985b). Although apparently not extremely rare in its native southeast Asian archipelagos, this genus is not often found in *ex situ* collections and is consequently not well characterized horticulturally. Swingle (1943) speculated that it is xerophytic based upon its leaf morphology.

Merope

Merope is a monotypic genus (*M. angulata*) distinguished from *Wenzelia* and *Monanthocitrus* by having four ovaries per locule and large seeds in the triangular (in cross-section) fruit. *Merope angulata* is a spiny, salt-loving shrub with thick, leathery leaves that grows in tidal marshes in Java, Burma and the East Indian Archipelago. This species is considered salt tolerant due to its habit; this was confirmed in greenhouse tests by Swingle (1915a). This genus is not well represented in *ex situ* collections but probably has useful properties. Jones (1982) reported briefly on habitat loss affecting this species, as well as certain medicinal uses. *Merope* is considered threatened by Jones (1990).

Swinglea

Swinglea, *Aegle*, *Afraegle*, *Aeglopsis*, *Balsamocitrus*, *Limonia* and *Feroniella* constitute the Balsamocitrinae, or 'hard-shelled citroid fruit trees'. This curious branch of the Aurantioideae has large, heavy fruits with a hard, mostly wood-like exocarp. The locules are filled with resinous gum and the seeds are sometimes covered with fine hairs.

Swinglea comprises the Tabog group of this subtribe. 'Tabog' is a name used for this species in the Philippines. Swingle (1913b) published his description of this monotypic species with the name *Chaetospermum glutinosa*. It was renamed *Swinglea gluti-*

nosa by Merrill (1927), who pointed out that *Chaetospermum* had already been used as a name for a genus of fungus.

Swinglea glutinosa is distinct within the 'hard-shelled citroid fruit trees' in having a tough, leathery shelled fruit rather than a hard, woody shelled fruit. The cells of the fruits are lined with spongy tissue and the walls of the locules are lined with large mucilage glands, features not found in other members of the Balsamocitrinae. The seeds are hairy. *Swinglea* and *Burkillanthus* are somewhat similar in some external attributes, but have different fruit anatomies.

Swinglea is a small, thorny tree with trifoliolate leaves that is native to Luzon Island in the Philippines. It has been spread throughout many tropical areas, and is not uncommon in *ex situ* collections. Swingle (1943) states that it is cold-hardy to 35°F (1.67°C), but it has been difficult to establish in the field in Riverside, where winter temperatures are rarely lower than that. *Swinglea glutinosa* is graft compatible with *Citrus*, and there has been some horticultural characterization over the years (Table 4.3).

Aegle

Aegle, *Aeglopsis*, *Afraegle* and *Balsamocitrus* comprise the *Aegle* group of the 'hard-shelled citroid fruit trees'. They all have hard, woody fruits without longitudinal ribs. The locule walls do not have large, evident mucilage glands, but mucilage glands are present. Leaves are simple, unifoliolate or trifoliolate. The ancestral type was presumably from south Asia, but three of the genera (*Aeglopsis*, *Afraegle* and *Balsamocitrus*) 'migrated' to Africa, while none 'migrated' eastward to the South-east Asian archipelagos.

Aegle is distinct from the other members of the group in having very numerous (>30) stamens, a large number (up to nearly 20) of locules in the ovary, and wooly seeds. The leaves are trifoliolate but variable in size and other anatomical features. A monotypic genus (*A. marmelos*), it is deciduous when cold temperatures occur during the winter. It is native to India and has spread from there to adjacent areas. *Aegle marmelos* can grow to 15 m or more in height in the proper conditions.

Aegle marmelos is the bael fruit of India and adjacent areas. It has a long history of cultivation and use. It is a sacred tree to Shiva and its leaves are used as offerings. In addition to being edible, various medicinal uses have been ascribed to the bael (Table 4.3). Swingle (1943) describes large-fruited, cultivated varieties and small-fruited, wild varieties, and speculated that there are different strains of varieties. In fact, in India, there are named varieties; the names are derived from the localities in which it is grown (Singh and Roy, 1984; Roy, 1998). Differences in levels of sugar, mucilage, phenolics and tannins are documented, as well as physical differences such as shell thickness/hardness, fruit size, seediness, etc. Positive qualities are associated with larger fruit, some of which have shells that can be broken with only finger pressure.

Although there is apparently genetic diversity present in India, the genetic diversity of *A. marmelos* in *ex situ* collections is much lower. There have been few introductions to the USA and other Western countries. Most *A. marmelos* in these collections represents very hard-shelled, medium-fruited examples. *Aegle* is graft compatible with *Citrus* and its horticultural properties have been documented (Table 4.3).

Aeglopsis

Aeglopsis differs from *Aegle* in having fewer locules in the ovary, fewer stamens, and simple leaves. The fruits are smaller (and the shell less hard?) than those of *Aegle* and are pyriform. The two species, *A. chevalieri* and *A. eggelingii*, are both native to Africa. *Aeglopsis* is graft compatible with *Citrus* but has otherwise had only limited horticultural evaluation (Table 4.3).

Afraegle

Afraegle is also native to Africa, but is more confined to the western portion of the continent than is *Aeglopsis*. *Afraegle* is similar to *Aeglopsis*, but has (mostly) trifoliolate rather than simple leaves, a larger number of locules per ovary and slightly less woody shells. *Afraegle* trees are vigorous. *Afraegle* is graft compatible with *Citrus*. There has been little horticultural evaluation done (Table 4.3).

Balsamocitrus

The monotypic genus *Balsamocitrus* (*B. dawei*) is native to central Africa. It is similar to *Aeglopsis*, but has trifoliolate leaves and larger fruit. Both *Aeglopsis* and *Afraegle* were formerly included in *Balsamocitrus* until split off at different times (Swingle, 1912). Although its geographic distribution in Africa is apparently fairly limited, *B. dawei* appears to grow well in a fairly wide range of climates when maintained in *ex situ* collections. It is vigorous and exhibits more vertical growth, as compared with the more spreading growth of *Afraegle*. However, it has not had much horticultural evaluation. This is possibly due to the somewhat erratic fruit production (at least in Riverside). It is graft compatible with other hard-shelled citrus relatives and is *probably* graft-compatible with *Citrus*.

Limonia

Limonia and *Feroniella* constitute the distinctive wood-apple group of the 'hard-shelled citroid fruit trees'. They are distinguished by the fact that the locules fuse during development into a single cavity, which becomes lined with parietal placentae bearing many ovules. The leaves are odd pinnate with paired opposite leaflets.

Although confusion regarding generic names is by no means rare in the Aurantioideae, the case with *Limonia* is perhaps one of the least clear. This is a monotypic genus. Swingle (1914c, 1943) argued for *Feronia limonia*, stating that *Limonia* was ambiguous. Airy-Shaw (1939) supported the name *Limonia acidissima*. Since Swingle wrote the book (Swingle, 1943), *Feronia limonia* has generally been used. More recently, Panigrahi (1977) accepted *Limonia* as the genus name but rejected *L. acidissima*, instead substituting the older species name *elephantum*, resulting in *L. elephantum*. Stone and Nicolson (1978) rejected this and supported *L. acidissima*. *Feronia limonia*, with *F. elephantum* as a synonym, is supported by *Hortus Third* (Bailey Hortatorium, 1976), while Wiersma and León (1999) support *L. acidissima*.

In any case, these thorny trees are native to India, where they grow in the hills but not to the same elevation as *Aegle*. The growth habit is curious and somewhat variable. The branches often are clustered towards the top of the tree, with drooping branchlets. Other trees have the branches and branchlets spaced at regular intervals up the trunk. The fruits are edible and also have medicinal uses. *Limonia* is deciduous. Swingle (1943) states that the leaves are dropped in cold weather. However, in *ex situ* collections the leaves have been observed to drop during the warm periods of the year and even in greenhouses, so perhaps it is more of a generalized stress reaction. Although *Limonia* is present in many *ex situ* collections and is graft compatible with *Citrus*, it is not well characterized horticulturally (Table 4.3).

Feroniella

Feroniella is very similar to *Limonia* in its leaf and fruit characteristics. It differs in the form of its stamen and in having glabrous rather than pilose seeds. The stamens are unique among the Aurantioideae in having pilose, partly free, appendages at the base of the filaments, which are much longer than the anthers. This difference is not readily apparent and it is difficult to distinguish between *Limonia* and *Feroniella*.

The four species of *Feroniella* are native to the South-east Asian peninsula and archipelagos. It is rare (Jones, 1990). Although Swingle (1943) does not state that it is deciduous, in *ex situ* collections it appears similar in this regard to *Limonia*. Like *Limonia*, *Feroniella* is graft compatible but is not well characterized (Table 4.3).

Micromelum

The tribe Clauseneae (Very Remote and Remote Citroid Fruit Trees) contains 5 genera: *Micromelum*, *Glycosmis*, *Clausena*, *Murraya* and *Merrillia*. The first and last constitute the subtribes Micromelinae and Merrilliinae, respectively; the remaining genera, the subtribe Clauseninae. Swingle (1943) considered the Clauseneae to be primitive members of the Aurantioideae, primitiveness being characterized as: small, semi-dry or juicy berries; odd-pinnate leaves with alternately attached leaflets; and a lack of spines in the leaf axils. Chemotaxonomy (Kong *et al.*, 1988) suggest that *Micromelum* is a primitive species that gave rise to the two branches of *Murraya* and subsequently to other taxa (*Glycosmis* and *Clausena* from the *Bergera* section and *Merrillia* from the *Murraya* section).

Micromelum is a genus of about nine species distributed from Pakistan through south and South-east Asia to the archipelagos. They are shrubs or trees, with compound leaves of up to 23 leaflets. They are distinguished from other members of the 'very remote and remote citroid fruit trees' by having cotyledons that are flat and folded rather than thick and plano-convex. The ovaries have a twisting of the radial follicle walls. Both of these characteristics are unique in the Aurantioideae.

Swingle (1943) spelled out the species problem in *Micromelum* as being the variability of characteristics in the described species. In spite of Swingle's (1943) assertion that 'new characters are being found that promise to make possible the satisfactory definition and the easy recognition of the species', little more has been written or published in the last 60 years, and the genus remains obscure. Although *Micromelum* species are reported in various flora, there is little information about their distribution and a lack of references to taxonomic advances. This genus is not well represented in *ex situ* collections and needs horticultural characterization.

Glycosmis

Glycosmis, *Clausena* and *Murraya* are the Clauseninae, or 'remote citroid fruit trees'. These three genera have very simple and primitive flowers and fruits, which are usually borne in large terminal panicles. This subtribe is 'typical' of the Clauseneae, with the Micromelinae and Merrillinae being aberrant.

Glycosmis differs from the *Clausena* and *Murraya* in having a persistent rather than deciduous style, in usually having only one rather than two ovules per locule and in having rust-coloured, ferrugineous puberulence on new vegetative growth, inflorescence axes and buds. In addition, *Glycosmis* has equilateral leaflets rather than the non-equilateral leaflets of most *Clausena* and *Murraya* species.

In contrast to his treatment of most genera, Swingle (1943) presents a truncated view of *Glycosmis*. Rather than his usual thorough treatment of even questionable species, he presents only a 'tentative list of species of *Glycosmis*'. This list is simply an enumeration of published names and reported locations. Swingle (1943) admits in his treatment 'The species of *Glycosmis* are still very inadequately described and need study more urgently than those of any other genus of the orange subfamily ... the taxonomy of *Glycosmis* is in such a state of confusion that it is not yet possible to key out all the species or even to tell with certainty how many should be recognized as valid.'

Fortunately, relatively recently there has been an extensive revision and updating of *Glycosmis* by Stone (1978a, 1985b,

1994b). The first paper (Stone, 1978a) gives a review of the 95 published species of *Glycosmis* and the confusion surrounding this genus over several centuries. Much of the confusion revolves around obscure rules for taxonomic nomenclature and the ambiguity in the identity of *G. pentaphylla*.

Stone (1978a) reviews earlier work on *Glycosmis* by Tanaka (1937), Narayanaswamy (1941), Brizicky (1962b) and Mitra and Subramaniam (1969), after which it is concluded that some older materials and descriptions of *G. pentaphylla*, the type species, were based instead upon *G. mauritiana*. This was due to the incorrect application of *G. pentaphylla* (and its older identification as *Limonia pentaphylla*) by Roxburgh. The correct name of *G. pentaphylla* in the sense of Roxburgh is thus *G. mauritiana*. The *G. pentaphylla* (and its older identification as *L. pentaphylla*) of Retzius is considered to be the 'real' *G. pentaphylla*. Stone (1978a) is supported in this identification by Panigrahi (1985), who also argued that *G. pentaphylla* and *G. arborea* are synonyms, and by Huang (1987). The reader is referred to these works for the complete (and complicated) rationale behind this.

The result of the ambiguity in the identity of *G. pentaphylla* is that, according to Stone (1978a), many specimens identified as *G. pentaphylla* will have to be re-identified. It is also noted by Stone (1978a) that due to the ambiguity in the identity and description of *G. pentaphylla*, the *concept* (his italics) of this species had expanded over time and it became a sort of warehouse for any unidentified *Glycosmis* collected. Stone (1978a) identifies at least four species that he believed were mistakenly identified as *G. pentaphylla*: *G. mauritiana*, *G. citrifolia*, *G. parviflora* and *G. cochinchinensis*. In addition to casting doubt on the identity of herbarium specimens of *G. pentaphylla*, this information also calls into question the identity of some *G. pentaphylla* (and perhaps other *Glycosmis* species) found in *ex situ* collections. The fact that Swingle (1943) correctly identified *G. pentaphylla* as being the *G. pentaphylla* of Retzius *might* mean that *G. pentaphylla* as introduced as by the USDA is correctly identified. The identification of *G. pentaphylla* in *ex situ* collections needs to be verified.

In the second part of his revision, Stone (1985b) presents a systematic treatment of *Glycosmis*. In this treatment, he recognized 43 species, of which seven were new (some of these were published previously by Stone). A number of new varieties were also published. Stone (1985b) excludes *G. africana*, *G. bonii*, *G. cambodiana*, *G. crenulata*, *G. harmandiana*, *G. spinosa* and *G. subvelutina*. Stone (1994b) discusses some further aspects of his conspectus and also analysed three additional *Glycosmis* species published by Huang (1987). The conspectus (Stone, 1985b) includes a key to *Glycosmis* species, which Stone admitted is only 'fairly close to being a usable key'. The key divides *Glycosmis* into five groups based upon leaf and floral characteristics.

The monumental tripartite work of Stone (1978a, 1985b, 1994b) remains the most thorough treatment, with little or nothing being published after that. The death of Stone and other factors make it unlikely that much more will be written definitively about this genus in the near future.

Glycosmis is distributed throughout south and South-east Asia, from India through the archipelagos. The species are shrubs and have varying habitats, including limestone soils and riparian areas. Observed *Glycosmis* species are generally attractive plants, with shapely leaves and small, pinkish-red berries. Most of the published information on this species seems to involve the taxonomic issues. *Glycosmis* is represented in many *ex situ* collections, but species maintained are often limited to *G. pentaphylla* (but which one?) and in some cases *G. parviflora*. There has been little horticultural characterization of this genus. Small (1933) and Brizicky (1962a) report *Glycosmis* (but which species?) as being naturalized in south Florida, but this is somewhat doubtful (K. D. Bowman, personal communication, 1999).

Clausena

Clausena and *Murraya* differ from *Glycosmis* in the characteristics noted in the previous section. These species are very similar, but differ from one another in certain ways. *Clausena* has flower buds that are globose or broad-ovoid, never long and slender, while *Murraya* has cylindrical or long-ovoid flower buds. Similarly, the style of *Clausena* is short and thick, whereas that of *Murraya* is long and slender. A distinctive feature of *Clausena* is the large, hourglass-shaped gynophore. This, however, is a somewhat variable characteristic (Swingle, 1943). Molino (1994) bases the key on the same stylar characteristics, the structure of the gynophore and the structure of the stamen.

Swingle (1943) notes some of the same challenges in *Clausena* taxonomy and nomenclature as in that of *Glycosmis*: a lack of specimens for examination and variability within taxa. He was able to borrow and examine nearly a thousand herbarium specimens in his study of *Clausena*. Based upon this study, Swingle (1943) recognizes 23 genera, some of which he had published within a few years of his monograph (Swingle, 1940b, 1942). His key is divided into four parts based upon geographic distribution (basically the Indian subcontinent; the Indochinese peninsula; the archipelagos; and Africa). To this, a new species (*C. calciphila*) was added by Stone (1978b).

Molino (1994) based a revision of *Clausena* on new collections and a greater emphasis on floral characteristics, resulting in a collapsing of the genus to 15 species divided into four sections based upon floral structure. A new species was described (*C. poilanei*) and several new names and combinations were proposed (for instance, the *C. calciphilla* of Stone, 1978a, became *C. sanki* var. *calciphila*). Most notable was the establishment of synonymy for some of the species. This was especially pronounced in the case of *C. anisata*, which absorbed four species and seven varieties of Swingle's (1943) treatment. *Clausena anisata* became the only *Clausena* species to be found in Africa and was considered to have 'migrated' there relatively recently. Additional species noted by Swingle (1943) were also pulled into other species.

Clausena is widely distributed geographically, from the South-east Asian archipelagos, through the South-eastern and south Asian regions, and north to the subtropical parts of China. There are two 'satellite' areas: Africa, where *C. anisata* (in the sense of Molino, 1994) is widespread, and north-eastern Australia, where *C. brevistyla* and the recently published *C. smyrelliana* (Forster, 2000) are found. Some species are reported to grow at elevations of up to 3000 m, while others are decidedly tropical.

Clausena species range from small shrubs to 20 m tall trees. Molino (1994) states that in areas that are climatically marginal, particularly the two satellite areas, the leaves may be deciduous; this is not noted by Swingle (1943). Observations in Riverside, California, would seem to indicate that there is at least some deciduous aspect to the leaves. However, as with the other deciduous types maintained there, the trigger for leaf drop is not apparent. Likewise, flowering at Riverside is erratic. It is possible that this is due to climatic factors, but Molino (1994) states that flowering occurs throughout the year in many areas, but with a peak in the dry season. The flowers are borne on bunches and produce small fleshy berries surrounded by mucilage.

Several species of *Clausena* deserve particular mention. Although several *Clausena* species produce edible fruits, the only one commercially cultivated is *C. lansium*, the Wampee. *Clausena lansium* is actually quite distinctive among *Clausena* species, having a star-shaped bud and a five-angled ovary that has hairs on the lateral walls of the locules. In addition, there is a central gum canal in the twigs. For additional distinctive characteristics, see Swingle (1943). The Wampee is native to southern China, and that is its main area of cultivation. Cultivation has spread to a very limited extent to the countries around China. The golden-yellow berries, which are borne on large panicles, are the largest of any *Clausena* species and are eaten fresh as well as being used for preserves. The Wampee is relatively common in *ex situ* collections, but it is sometimes difficult to obtain seeds since the fruits are so irresistible to birds. Most *ex situ* collections have only a small representation of Wampee, but in China there are many local varieties. A few found in Kwantung Province are listed in Swingle (1943), while different named varieties are found in Fujian and Guangxi Provinces (Shu Yuan, personal communication, 1984); there are doubtless others. Wampee is graft compatible with *Citrus* but there are apparently some incompatibility problems since the resulting trees are small and weak (Swingle, 1943; Campbell, 1974). Campbell (1974) stated that cold winters can kill Wampees in southern Florida, but limited experience in the harsher climate of southern California indicates that they can survive temperatures near 0°C. In fact, *C. lansium* seems to be more cold tolerant than other *Clausena* accessions maintained in Riverside (*C. anisata* and *C. excavata*). The *Clausena* accessions seem more sensitive than the closely related *Murraya* accessions maintained. It is possible that it is some environmental factor other than cold that makes establishment in the field difficult.

A few other species will be mentioned in passing. Several other species produce berries that are harvested from wild trees and consumed locally. These include *C. dentata* var. *dulcis* of Madras, India; *C. indica* of India and Ceylon; and *C. dentata* var. *henryi* of south-western China. In the treatment of Molino (1994), *C. indica* was absorbed into *C. sanki* and both varieties of *C. dentata* were absorbed into *C. anisata*. *Clausena anisum-olens* (incorporated into *C. sanki* by Molino, 1994) is native to the Philippines, where it grows up to elevations of 1500 m. It is used in the preparation of *anisado*, an alcoholic beverage, and has a very high concentration of the essential oil *anethole*, the true anis oil. See Molino (1993) for more information on various specimens of this throughout the world.

Occasionally *Clausena* species other than *C. lansium* are found in *ex situ* collections. However, *C. lansium* is by far the most common. Consequently, although there has been some horticultural characterization of this genus, most of that work has been done on *C. lansium*, and on a rather limited genetic range of material at that.

Murraya

Murraya is closely related to *Clausena* with the differences noted above. Swingle (1943) describes 11 mostly closely-related species. Huang (1978) added *M. kwangsiensis*. More recently, data of various types has supported the division of *Murraya* into two species, *Murraya* and *Bergera*. Chemotaxonomic studies with alkaloids (Kong *et al.*, 1986) and terpenes (Li *et al.*, 1988) divided *Murraya* into two groups. The first group (*Murraya*) possesses yellowish stems and roots; flowers with larger petals (1–2 cm long); and red, ovoid to ellipsoid fruits. The other group (*Bergera*) possesses brown stems and roots; flowers with smaller petals (4–7 mm long); and purplish-black, globose to ellipsoid fruit (Jones, 1995). Kong *et al.* (1986) state that this division supports that of Swingle (1943); however, no division of *Murraya* is noted in either Swingle (1943) or its revision (Swingle and Reece, 1967). This is confirmed by Jones (1995), who then gives the group names cited above, even though they do not appear in Swingle (1943), Swingle and Reece (1967), nor Kong *et al.* (1986). Recently, this division was reinforced with molecular (Samuel *et al.*, 2001) and karyotypic (Guerra *et al.*, 2000) analysis. *M. koenigii* and *M. siamensis* were separated into the genus *Bergera*, with the remaining *Murraya* species being retained in *Murraya*. This division of *Murraya* seems valid and will probably become increasingly accepted.

The type species, *M. paniculata*, is the well known 'Orange jessamine' or 'Hawaiian mock orange' (among other names) that is sometimes cultivated as an ornamental in locales with a suitable climate. The flowers are large and fragrant, the fruits small and attractive, and the overall appearance pleasing. The distinctive variety *M. paniculata* var. *ovatifoliolata* is sometimes cultivated as well, but is decidedly less common. There are various horticultural selections of *M. paniculata*, and there are obviously wild ones as well.

There is some confusion with the status of *M. exotica* vis-à-vis *M. paniculata*. Swingle (1943) lists *M. exotica* as a synonym of *M. paniculata*. Huang (1959) considers this a variety (*M. paniculata* var. *exotica*). Stone (1985c) and Jones (1995) list *M. exotica* as a separate species, although Stone (1985c) states 'There is merit in Huang's proposal for subspecific status but at least equal merit in regarding them as species.' Huang later elevated them to species status (see discussion and citation in Kong *et al.*, 1986), so perhaps lacking a definitive treatment *M. exotica* should be considered a provisional species.

The main differences are the larger, more ovate and acuminate leaves and larger flowers of *M. paniculata* (Stone, 1985c). Thus, *M. exotica* would apparently become synonymous with Swingle's (1943) *M. paniculata*, while *M. paniculata* in the sense of Stone (1985c) and Jones (1995) would apparently become synonymous with the *M. paniculata* var. *ovatifoliolata* of Swingle (1943) based upon the descriptions. However, neither author establishes the synonymy. Swingle limited *M. paniculata* var. *ovatifoliolata* to Australia, while Stone (1985c) and Jones (1995) divide the geographic area of Swingle's (1943) *M. paniculata* between the two species. Mabberley (1998) considers '*M. exotica* var. *ovatifoliolata*' to be the 'wild plant ... distinguished from the cultivated forms in its straggling habit, rather hairy shoots and broadly oval or ovate leaflets'. Molecular studies by Ranade *et al.* (2006) found *M. paniculata* to be less heterogeneous than *M. koenigii* and that *M. exotica* did not represent a distinct species from *M. paniculata* but rather morphologically distinct forms of *M. paniculata*. Thus, this particular species complex (?) remains somewhat in doubt.

More distinct is *M. koenigii*, 'Curry Leaf'. As its name implies, the aromatic leaves of this species are used as condiments in the preparation of curries and other Indian/South Asian foods. It is found both wild and in cultivation from the Indochinese peninsula through the Indian subcontinent. It has been taken and established as a cultivated type by diasporaed Indians in areas with suitable climates, such as the south-east Asian archipelagos and east Africa. In addition to the leaves, there are various medicinal uses of *M. koenigii*, and the small black berries are edible. Essential oils are also sometimes distilled. *Murraya koenigii* is a small spreading shrub up to 2.5 m high. In south Asia, it is found as an understorey plant up to 1500 m elevation. In *ex situ* collections, it seems fairly hardy and can be established in the field in Florida and southern California. For more information on *M. koenigii*, the reader is referred to Parmar and Kaushal (1982), Morton (1984) and Joseph and Peter (1985).

Murraya is commonly found in *ex situ* collections. However, commonly only a few accessions of *M. koenigii* or *M. paniculata*/*M. exotica* are maintained. Because of the importance of *M. koenigii*, there has been more horticultural characterization of this genus (or at least of *M. koenigii*) than of most other Aurantioideae. *Murraya* is reported to be graft compatible with *Citrus*, but its reaction in this area is not as well established as with some other genera.

Merrillia

The monotypic genus *Merrillia* (*M. caloxylon*) has some unique characteristics compared to other genera in the Clauseneae. Swingle (1918), in his original description and naming, calls *M. caloxylon* 'the most aberrant of the citrus fruits'. Recent molecular phylogeny work (Samuel *et al.*, 2001) suggests that *Merrillia* should be included in *Murraya* s.s. rather than being considered a separate species.

It is similar to *Murraya* but with some unique flower and fruit characteristics. The pale yellow-green flowers are the largest in the Aurantioideae; however, they do not open completely and thus become trumpet-shaped. The oblong, yellow-green fruits are also very large, with a thick, leathery, lacunose epicarp and narrow, flattened, hairy seeds. *Merrillia caloxylon* is an attractive plant in its floral characteristics and wood; but the fruit is best described as 'interesting'. Tree heights of over 30 m were reported by Swingle (1943); this is much larger than most other specimens. The usual height is apparently half or less of the 30 m figure.

Merrillia caloxylon is native to Malaysia, where it is called 'ketenggah'. It is considered rare or at least endangered (Jones, 1987, 1990; Stone and Jones, 1988). In 1988, it had only been collected three times in the wild since 1969 (Stone and Jones, 1988). Because it is rare, *Merrillia* is not encountered very often in *ex situ* collections and has not been well characterized horticulturally (Table 4.3).

Conservation Status of Citrus Genetic Resources

Citrus is one of the most ancient crop species domesticated by humans. The long history of selection and vegetative propagation has led to the perpetuation of 'elite' germplasm lines. This has often led to the neglect and disappearance of progenitor wild types. In addition, *Citrus* hybridizes readily, and in many cases possesses nucellar embryony. These factors can also reduce genetic diversity. Consequently, 'wild' citrus are relatively rare, mostly existing as scattered trees in remote areas rather than as pure stands. Where 'natural' populations are located, it is often difficult to determine whether they represent wild ancestors or are derived from naturalized forms of introduced or selected varieties.

Agricultural utilization of citrus has involved a narrow range of genetic material, making this a genetically vulnerable crop. Genetic diversity in the centres of origin is also threatened or severely threatened by habitat loss caused by deforestation, population pressure, fire, hydroelectric development, clearance for agriculture or other development, tourism, etc. (World Wide Fund for Nature and International Union for Conservation of Nature and Natural Resources, 1994–1997). These factors may be especially important in countries such as India and China, which have rapidly expanding populations coupled with rapid economic/industrial development. This can result in very rapid losses of habitat and genetic diversity. This situation makes *ex situ* conservation of genetic resources of citrus and related genera imperative. This statement is not meant to diminish the importance of *in situ* conservation and habitat preservation, but to put into perspective the very real potential for loss of citrus genetic resources *in situ*.

Aurantioideae genera related to *Citrus* are utilized much less frequently and therefore exist most often as 'wild' unselected types. These 32 genera are mostly tropical and of limited commercial importance. Therefore, there has been less attention focused upon them except by local inhabitants. These areas are also often in danger of habitat destruction, and therefore the threat of losing genetic diversity is also present for these genera, particularly since their representation on *ex situ* collections is very limited.

Assessment of the genetic vulnerability of any crop requires knowledge of the extent and distribution of genetic diversity. This is acquired by systematic sampling and mapping of the flora of the geographical areas in which the species in question are found, as well as an assessment of *ex situ* collections. Unfortunately, information on natural and semi-natural *Citrus* and other Aurantioideae germplasm is limited on the international level. This is due to the remoteness of some of the material, a lack of resources devoted to assessing these areas, and in a few instances political unrest or other unsettled conditions. In some cases, information may be available at the local or national level, but not to the international genetic resource conservation community.

The information that is available is often simply a catalogue (sometimes quite old) of plants present in an area, with little more than names and phenotypic descriptions. Often even information on the frequency of occurrence is lacking. More detailed characterization and evaluation data are needed to assess adequately the actual amount of genetic diversity present. These data should include both descriptive phenotypic data and molecular genetic analysis of germplasm existing both *in situ* and *ex situ* (Albrigo *et al.*, 1997; Gmitter *et al.*, 1999).

The status of citrus genetic resources and their conservation has been reviewed by Reuther (1977), the International Board for Plant Genetic Resources (1982), Albrigo (1997, 2001) and Broadbent *et al.* (1999). A limited amount of information is found in Food and Agriculture Organization of the United Nations (1996). Rouse (1988) and Bettencourt *et al.* (1992) have summarized the world citrus collection situation, identifying major and minor citrus collections. Outside of the centres of origin/diversity, collections consist mostly of advanced lines and commercial varieties. Large *ex situ* citrus collections of this sort are found in Argentina, Australia, Brazil, Corsica (France), Morocco, New Zealand, South Africa, Spain, Turkey and the USA. Some of the larger collections contain many selections of the same variety, and so the genetic diversity is less than might be expected from the number of accessions.

Specific information on the status of citrus genetic resources in many countries

Table 4.4. Information sources for citrus germplasm status in various countries and areas.

Country or geographic area	References
South East Asia	Mehra and Sastrapodja, 1988; Jones, 1990; Verheij and Coronel, 1991; Coronel, 1995; Osman *et al.*, 1995; Saamin and Ko, 1997a, b; Hor *et al.*, 1999
Thailand	Anupunt, 1999
Philippines	Garces, 1999
Malaysia	Santiago and Sarkawi, 1962; Allen, 1967; Jones, 1985, 1989, 1991; Jones and Ghani, 1987; Saamin and Ko, 1997a, b; Ko, 1999
Vietnam	Ca, 1999; Le *et al.*, 1999
China	Yin-min, 1985; Hu, 1989; Gmitter and Hu, 1989, 1990; Zhaomin, 1989; Zhang *et al.*, 1992; Xueqin, 1995; Zhusheng, 1997, 1999; Zhusheng *et al.*, 1996; Deng *et al.*, 1997a, b; Weidong *et al.*, 2000
India	Bhattacharya and Dutta, 1956; Dutta, 1958; Singh, 1981; Singh, 1985; Dass, 1990; Singh and Chadha, 1993; Chadha, 1995; Singh and Uma, 1995; Chadha and Singh, 1996; Rai *et al.*, 1997a, b; Ghosh, 1999
Nepal	Chaudhary, 1999
Japan	Nishida *et al.*, 1981; Iwamasa, 1988; Omura, 1996, 1997; Nito *et al.*, 1999
Australia	Armstrong, 1975; Forsyth, 1988; Sykes, 1993, 1997, 1999; Mabberley, 1998
Spain	Ortiz *et al.*, 1988
Turkey	Tuzcu, 2001
Cyprus	Georgiou, 2001
Morocco	El-Otmani *et al.*, 1990
Latin America	Anderson, 2001
Brazil	Machado, 1997
USA	Cameron, 1974; Reuther, 1988

and geographical areas is available from various sources, as referenced in Table 4.4. There is undoubtedly more information available in governmental or non-governmental organization (NGO) offices in some countries, but this is not generally available to the public. Table 4.4 includes information on countries that have only *ex situ* collections, as well as countries having *in situ* resources. These *ex situ* collections may maintain elite lines, undeveloped material or both. In addition, they may contain breeding lines, synthetic hybrids and related genera. All these are sources of genetic material for the breeder or biotechnologist. The following brief comments summarize some of the information from the sources in Table 4.4, with an emphasis on areas near the centre of origin.

Southern PR China is one of the centres of diversity for *Citrus* and related genera and a wide range of genetic diversity is apparently still present *in situ*. However, some (though not all) areas are threatened with habitat degradation or lack of proper management that could result in decreases in genetic diversity. In PR China, exploration and collection of indigenous citrus genetic resources began in the 1950s and 1960s, but was interrupted by the Cultural Revolution of 1967–1972. Governmental surveys resumed during the 1970s and 1980s and uncovered a number of new putative species, including *Citrus hongheensis*, *C. mangshanensis*, *C. daoxianensis* and *Poncirus polyandra*. These putative new species are mostly incompletely known outside of PR China. Areas that have been explored include Guangxi district, Guangzi province; Shennon Jia, Hubei province; Sichuan, Gansu and Shanxi provinces; Hainan Island and Tibet. There are also a number of indigenous Aurantioideae in southern China. There is exploitation (use) of indigenous germplasm and some attempts at *in situ* preservation have been made. However, conservation of citrus genetic resources in PR China is mostly *ex situ* at present. Beginning in the early 1960s, a National Citrus Germplasm Repository was established at Beibei, Chongqing, Sichuan province. This is a large collection with over 1000 accessions. Regional citrus germplasm repositories were also established in Huangyan, Zhejiang province; Guiling, Guangxi province; Zhangsa, Hunan province; Guangzhou, Guangdong province; Jiangjin, Sichuan province; Wuzhung province and Hubei province. The exact composition of these collections is mostly unknown outside PR China, but a high percentage is indigenous germplasm and undoubtedly represents a substantial amount of diversity, although some of the germplasm, indigenous and otherwise, consists of advanced lines or selections. The accessions at the repositories have had a limited amount of characterization and evaluation done on them. There have apparently been some difficulties in maintenance owing to such factors as lack of funds, disease and weather (freezes and floods). Recently the Chinese government has provided increased resources for the support of the Beibei collection and it is again in a stable condition. The Beibei collection currently has slightly over 1000 accessions, including 296 local selections and 116 wild selections. Huazhong Agricultural University maintains collections of 280 accessions *in vivo* and 110 accessions *in vitro* (X.X. Deng, personal communication, 2007). Regional Citrus Research Institutes in Shantou, Guangdong; Ichang, Hubei; Thouyang, Hunan; Ganzhou, Jianxi; Yuchi, Yunnan and Wu, Jiangsu maintain small amounts of citrus germplasm, as do botanic gardens such as Xithanbanna and Guanzhon.

In India, the north-east region is the centre of origin/diversity. Unfortunately, this region has sometimes experienced civil unrest, making evaluation of genetic diversity and plant exploration difficult. There are apparently a few stands of 'wild' citrus in these areas, but many of the 'wild' populations apparently consist of dooryard plantings. A long history of cultivation and selection has produced many genotypes/landraces, which are difficult to separate from 'wild' citrus. Still, a wide range of genetic diversity undoubtedly exists in these areas. There is an *in situ* gene sanctuary for citrus in the Garo Hills of the north-east, which is a field gene bank reported to have over 600 accessions. Other regions of diversity include the central and northwest Himalayas, Maharashtra and the southern peninsula. *Ex situ* conservation of citrus germplasm began in the 1950s in India, and there are now collections at eight sites (Chetalli, Bangalore, Rahuri, Tirupati, Abohar, Bhatinda, Yercaud and New Delhi), with smaller collections at 14 additional sites (Akola, Barapani, Birouli, Hessaraghatta, Katol, Ludhiana, Nurpur, Parbhani, Pantnagar, Pedong, Periyakulam, Sirmour, Srirampur and Tinsukia). Total accessions are probably in the neighbourhood of 600 accessions. The *ex situ* collections in India are mostly of rootstock varieties and a few local cultivars, representing less genetic diversity than might be expected. Many of the indigenous types described in historical accounts such as Bonavia (1890) and later works such as those of Bhattacharya and Dutta (1956) and Dutta (1958) are apparently not in any of the collections. As in China, there are apparently problems with the maintenance of these collections. The intention of the Indian government is to concentrate accessions from the various collections at the National Research Centre for Citrus in Nagpur and/or at regional research centres at Bangalore Tirupati (south), Ludhiana/Abohar (north), Rahuri (central), and Shillong and Assam (north-east).

South-east Asia (including Malaysia) is rich in indigenous germplasm and, with chance seedlings, semi-wild and wild types. Most indigenous types of citrus are grown in the hot lowlands. One species (*C. halimii*) is still found wild in the highlands, while the majority of the others are cultivated. Some introduced Aurantioideae species (e.g. *Aegle marmelos* and *Limonia acidissima*) have become naturalized. This genetic diversity is threatened by deforestation, development and disease. In 1983–1988, IBPGR coordinated four collecting missions to Thailand, Malaysia, Indonesia and Brunei. The materials are maintained in Japan, the organizer of the missions. There are four collections in Malaysia, the main one being the University of Malaya (Rimba Ilmu) Botanical Garden. This was established at the request of IBPGR in 1986. Other Malaysian collections at Jerangau Station, Trengganu, Kuala Kangsar and Cameron Highlands are maintained by the Malaysian Agricultural Research and Development Institute. There are also some *in situ* conservation efforts, such as at the Taman Negara National Park in Pahang and the Danum Valley in Sabah. There are apparently three collections in Thailand with over 500 total accessions. The most important are Phichit Horticultural Research Center, which has a collection of mostly native pummelos; and Nan Horticultural Research Station, which has approximately 70 accessions of mandarins, sweet oranges and citrus relatives, of which approximately 25% are native. In the Philippines, the main collection of citrus genetic resources is maintained by the national Plant Genetic Resources Laboratory of the Institute for Plant Breeding in Los Baños, and consists mostly of commercial and imported varieties; there are said to be two other collections with slightly over 100 accessions. There are also three collections in Indonesia (~500 accessions) and at least two in Vietnam (National Institute of Agricultural Science and Technology and Phu Ho Fruit Research Center), which contain materials collected by an IBPGR-sponsored programme in 1992.

Asia's largest collections, outside of the centres of origin discussed above, are in Japan. Citrus entered Japan in ancient times, compared with its appearance in countries farther away from the centres of origin, and some types became semi-naturalized. There is a limited amount of *in situ* preservation of these naturalized types but, as in other areas, development is a threat. The Fruit Tree Research Stations in Tsukuba, Okitsu and Kuchinotsu have large collections that have a number of citrus relatives. Total accessions were said to be over 1200 in 1996. Of interest are the large numbers of mandarin types, especially satsumas. There are also three other collections of citrus germplasm in Nagasaki, Kagoshima and Okinawa. Japan has been active in collecting in South-east Asia (see above), Nepal (1983–1985) and Vietnam (1996) in IBPGR-coordinated cooperative programmes. Accessions collected from these ventures are maintained in Japan.

Australia has several *ex situ* collections maintained by State Government Departments of Primary Industries and the Commonwealth Scientific and Industrial Research Organization (CSIRO) that consist primarily of cultivated types. However, this island country is the centre of origin for several related genera (most notably *Eremocitrus* and *Microcitrus*), that are included in the collections, as well as in certain botanic gardens (e.g. Royal Botanic Garden, Sydney, and Brisbane Botanic Gardens) and arboreta (e.g. Waite Research Institute, University of Adelaide Arboretum). Also of interest are hybrids of these native types.

The situation with the related Aurantioideae genera is less well known, particularly from outside the southern/south-east Asian region. Although these genera are sometimes represented in collections, there is little information available about their status *in situ*. However, as many of them originated in countries which are currently rapidly developing, experiencing population growth and pressure, or being bothered with civil unrest, it is probable that at least some native populations exist in habitats which may be threatened. These factors also make assessment of the situation difficult.

Overall, the genetic diversity of *Citrus* and related Aurantioideae is vulnerable. Habitat loss is common in areas in which these plants are endemic, and eco-geographic assessments of these areas are often lacking. Although some efforts are being made in the areas of *in situ* and *ex situ* conservation, it is probable that there has been considerable genetic erosion for these species. Due to the lack of eco-geographic information, as well as characterization and evaluation data from the *ex situ* collections, it is impossible to say to what extent this erosion has occurred. It is imperative that more resources be devoted to these areas in the future.

Due to these factors, it has recently become evident that more intensive interactions and coordination between the various entities dealing with citrus germplasm conservation is necessary (Albrigo, 1999; Ramanatha Rao and Arora, 1999). A proposal to establish a global network on citrus genetic resources conservation and utilization was recommended during the meeting of the FAO Intergovernmental Group on Citrus in April 1996. Accordingly, this proposal was followed up and further elaborated during the Symposium on the Conservation of Genetic Resources of Citrus and its Relatives, held in South Africa in May 1996, where the major technical issues to be addressed by a global cooperative programme were analysed (Albrigo, 1997).

The Global Citrus Germplasm Network (GCGN) was formally constituted under the aegis of the FAO. The GCGN will function on a voluntary basis and will involve national institutions as well as existing regional and inter-regional networks dealing with citrus genetic resources conservation and utilization (Global Citrus Germplasm Network, 1998). It will serve to link different initiatives in different parts of the world dealing with citrus genetic resources exploration, conservation and utilization. The GCGN will also play a role in harmonizing and strengthening ongoing networking initiatives that are dealing with citrus germplasm conservation and utilization, and in promoting new undertakings in different regions of the world. The existing regional and inter-regional citrus networks (IACNET (Americas), MECINET (Mediterranean region)) and those under constitution (Asia-Pacific and sub-Saharan Africa) will participate in the GCGN. The Global Network will be guided by a Coordinating Board chaired by the General Coordinator of the Network, and will include the coordinators of the technical working groups and representatives of the different regional and inter-regional citrus networks. More information is available in Broadbent *et al.* (1999), Global Citrus Germplasm Network (1998), Albrigo (2001), and on the Internet (<http://www.lal.ufl.edu/CONGRESS/Gcgn rept.html>). It is hoped that this sort of international cooperation will increase the efficiency of citrus genetic resource conservation efforts.

Conservation of Citrus Germplasm

Although conservation of citrus genetic resources *in situ* is valuable and extremely important, conservation *ex situ* is vital for the utilization of these resources by breeders, horticulturists, phytopathologists, etc. The establishment of *ex situ* collections of citrus germplasm permits the efficient maintenance of appropriate amounts of specific true-to-type, (potentially) pathogen-tested varieties.

The composition of *ex situ* collections of citrus genetic resources varies with the mission of the collection. Some collections consist solely of commercial varieties, which provide propagative material to growers. Varieties maintained in these collections may be pathogen tested and utilized in a certification programme, or they may simply be trees of local varieties. These collections cannot be strictly considered as germplasm banks since they have very little variability and their objective is not the conservation of genetic resources. Other collections have the specific mission of conserving genetic resources. These collections also include non-commercial types of *Citrus* for utilization in agricultural research, as well as other genera or species within the Rutaceae (generally within the Aurantioideae).

This section reviews some concepts in the management of citrus *ex situ* collections. Some general information regarding conservation of the so-called 'clonal' crops (at least as practised in the USA) was reviewed some time ago by Westwood (1989). However, as with many other aspects of citriculture, management of citrus genetic resources differs from that of most other crops, including other clonals. The intent of this section is to provide an overview with general information on management of citrus germplasm resources. An in-depth discussion of the many subjects involved in citrus germplasm resource management is beyond the scope of this chapter. Due to the focus of this book, more information will be provided in the areas utilizing biotechnological techniques. Other areas will be discussed and some general references provided.

Much of this section is based upon the collections with which the authors are most familiar, the USDA-ARS National Clonal Germplasm Repository for Citrus and Dates (NCGRCD)/UCR Citrus Variety Collection at the University of California, Riverside, and the Citrus Germplasm Bank of the Instituto Valenciano de Investigaciones Agrarias (IVIA) in Moncada, Valencia, Spain. Some aspects of the Repository's functioning have been described by Krueger (1997a, b, 1999) and Williams (1991, 1992a, b); more information about the IVIA germplasm bank can be found in Navarro (1976), Navarro *et al.* (1980, 1981, 1988, 2002), and at the webpage <http://ivia.es/deps/biot/germop.htm >.

Overview of citrus genetic resource conservation

Conservation of citrus genetic resources may be thought of as a series of six main steps, not all of which are mutually exclusive or in strict sequence: acquisition; introduction; maintenance; characterization and evaluation; documentation and databases and utilization. The basic relationships between these various steps are shown in Fig. 4.1.

In order to establish a collection of *ex situ* citrus genetic resources, *acquisition* of germplasm resources is necessary. This can occur by *plant exploration* in an area of genetic diversity; by selection of cultivated

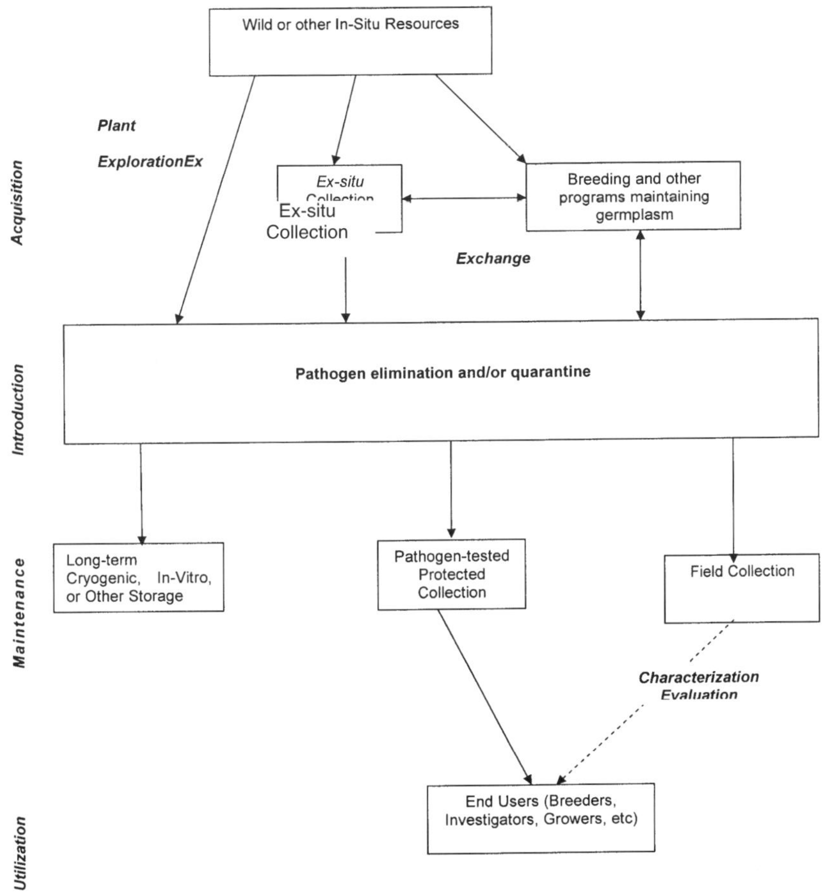

Fig. 4.1. Overview of the management of citrus genetic resources.

or new genotypes originating in an area; or by *exchange* of germplasm between cooperating persons, programmes or agencies. *Introduction* of germplasm into a germplasm bank has the attendant risk of also introducing exotic pests or pathogens for the area where the germplasm bank is located. If germplasm is introduced from another country, the importation may be regulated by a phytosanitary legislation that could include a mandated *quarantine* system to protect the phytosanitary health of a country, crop or industry. Even when the germplasm is acquired from within the country where the bank is located, it is advisable to apply therapy methods in order to introduce only healthy plant material, thus avoiding the presence of pathogens that may cause the loss of genotypes and/or interfere with their evaluation and utilization.

After introduction, citrus germplasm is *maintained* in various types of *collections* in order to conserve it and to make available propagative or other type of material for distribution and utilization. Efficient *utilization* is based upon thorough *characterization* and *evaluation* of pertinent traits. Adequate *documentation* (nowadays usually in a *database*) of the germplasm, its provenance and its characteristics is necessary for these activities. Because citrus germplasm is generally maintained as living trees, it is appropriate to have *back-up collections* or plantings. Ideally, long-term *preservation* of a *base collection* of citrus germplasm as a back-up to the *active* or *working collection* is desirable.

Acquisition of citrus germplasm

The first step in the acquisition of citrus germplasm is the identification of the source of the desired genotypes, either domestically or in a foreign country. In this last case, it will be necessary to satisfy legal and political requirements for its introduction, including possible property rights according to international agreements on germplasm exchange. In addition, the phytosanitary and/or quarantine regulations must be met during the following introduction phase.

Identification of citrus germplasm for introduction as new accessions may occur by several means. When a particular type of citrus germplasm is maintained by governmental or academic researchers or institutions, its existence is often public knowledge. This does not necessarily imply that its existence is widely known, but rather that it is knowable. That statement is meant to contrast the existence of citrus genetic resources in collections (both *ex situ* and *in situ*) with those that might be found in the wild or under local cultivation. In the latter circumstances, the exact nature of citrus genetic resources may be known only in very general terms or potentially not at all. This is particularly true for areas in which there may be civil unrest or political instability. New and interesting citrus germplasm may also originate by spontaneous budsport mutations in the field.

Citrus germplasm in governmentally or academically maintained collections should be well documented as to its background, availability, disease status and other pertinent facts. Unfortunately, this is not the case in many instances, particularly when a collection has its beginnings in the remote past. The germplasm and its characteristics, use, etc. may be discussed at scientific meetings, during exchange programmes, in correspondence and other similar professional activities. The same is true of commercial plantings. The public nature of information concerning these types of citrus germplasm makes identification of useful new accessions a matter of the citriculturist keeping up with events in their field. Obviously, a single individual cannot know of the existence of all types of germplasm. However, an informed researcher will make an attempt to know in general terms of the existence of these resources and will utilize their network of professional contacts to broaden the range of materials with which they are acquainted.

Acquisition of genetic resources from established persons, organizations or entities is termed *exchange*. There is an implication here of *free* exchange; however, this is becoming less clear-cut in recent years due to the political factors discussed previously. There have always been legal requirements in germplasm exchange as far as obtaining the necessary permissions, appropriate permits or licences, payment to donors or governmental entities, etc. However, the political factors noted have complicated matters further, and exchange of genetic materials is becoming less free in general than it has been in the past. In many instances, special permission, a Material Transfer Agreement spelling out conditions of use, or other documents must be signed and adhered to where in the past materials were sent freely and could be used without restriction. These requirements are sometimes complicated and may interfere with the acquisition of genetic resources. However, the laws of the countries 'owning' the genetic resources must be adhered to. An in-depth discussion of these ever-changing issues is beyond the scope of this review, but up-to-date information is available at the IPGRI website (<http://www.ipgri.cgiar.org/>).

Acquisition of citrus genetic resources in a foreign country from wild populations or local plantings owned by indigenous groups is more complicated. As stated, it is often difficult or impossible to determine the nature of the genetic resources present in an area. Sometimes information is found in governmental surveys, flora and other writings. However, these are often not available to researchers from outside the country or, in some cases, outside the immediate area. In this case, it is valuable to have cooperators in the 'exporting' country or region who can help identify the potential resources. In some cases, the existence of wild populations is not known to the government or to scientists but only to local inhabitants, and identification of genetic resources in an area becomes almost a matter of serendipity. This is the case in countries such as China and India, where *ex situ* collections are maintained but remote sources of uncollected germplasm may still exist.

When genetic resources are to be acquired by plant exploration in a foreign country, it is perhaps even more important to adhere rigidly to whatever guidelines are applicable. It is advisable to secure permission to collect plants from the local authorities as well as from established governments. Guidance on some of these issues, along with information on the extensive planning necessary for a plant collecting expedition, are available in Guarino *et al.* (1995). The remoteness of some desirable types of germplasm also complicates the physical act of acquiring new materials. This is especially true of germplasm that is often best obtained as perishable vegetative material (i.e. budwood), as is often desirable with *Citrus* and related taxa.

Form of propagative material

Citrus germplasm has usually been acquired as either budwood or seed. There are advantages and disadvantages to both forms, and there are often different phytosanitary legal requirements for the introduction from other countries of seed as compared with budwood. Recently it has been possible in special cases to exchange citrus germplasm as nucellar embryogenic callus cultured *in vitro*.

Budwood is in many cases the preferred form for exchange. Budwood will be true to type and will not have juvenile characteristics associated with it. However, citrus budwood potentially harbours pests and graft-transmissible pathogens. Therefore, the risk of introducing exotic pests and diseases is much greater with budwood than with seeds. Consequently, introduction of budwood is highly regulated, and often an extensive, expensive and time consuming post-entry quarantine process is legally required (Frison and Taher, 1991). This may limit the introduction of new germplasm. This is particularly true in the case of monoembryonic types, which will not come true to type from seed, and for elite varieties or types with unique characteristics, which are best acquired as budwood.

Introduction of new citrus germplasm by seeds is simpler from the phytosanitary standpoint, since systemic seed-borne diseases of citrus are very rare (Garnsey, 1999; Timmer *et al.*, 2000). (In addition to the old reports on psorosis transmission through seeds (Childs and Johnson, 1966), there have been recent reports of seed transmission of citrus leaf blotch virus (Guerri *et al.*, 2004) and citrus variegated chlorosis (Li *et al.*, 2003).) However, the extended juvenility period of seed-introduced material delays and complicates the use of accessions introduced in this form. In addition, monoembryonic types, including the basic species of citron and pummelo, several mandarin types and most species of related genera, will not come true to type. However, if the goal of an introduction is to increase the amount of genetic diversity present in a collection rather than a specific type, this may be acceptable. Even polyembryonic types may have off-types present that must be carefully rouged. Recently, it has been possible to identify nucellar seedlings using molecular markers (Ruiz *et al.*, 2000; Krueger and Roose, 2003; Krueger *et al.*, 2003). However, this is somewhat complicated and may be beyond the resources available to some programmes.

Citrus germplasm can also be exchanged in the form of embryogenic nucellar callus that can be recovered by ovule culture *in vitro* (Button and Bornman, 1971; Kobayashi *et al.*, 1983; Grosser and

Gmitter, 1990; Pérez *et al.*, 1998, 1999). This type of material has the advantage of being free of pests and diseases. Another advantage is that it can be used directly for breeding through protoplast fusion (Grosser *et al.*, 2000; Olivares-Fuster *et al.*, 2000) or genetic transformation (Fleming *et al.*, 2000). The disadvantage is that recovered plants have juvenile characters and that tissue culture laboratory facilities are needed for exchange.

Minimizing phytosanitary risks in the collection and shipping of citrus germplasm

All possible measures should be taken to avoid collecting propagative material that is infested with pests or pathogens, either if selected in the country or imported from other areas. This is not always possible, particularly when collecting material from wild populations or when relying on a collaborator who may or may not be able to comply with this concept. However, as far as possible, this principle should be adhered to.

If possible, germplasm should be obtained from a 'healthy collection'. If this is not possible, material should be obtained from the area with the lowest phytosanitary risk. When possible, source trees should be identified to allow future observation. If maintained in a formal collection, any information on the disease status or pathogen testing of the trees should be documented, as well as the prevalence of pests and diseases. Germplasm should not be collected from high risk areas: areas with diseases for which no indexing procedures are available; areas where background information on donor trees is lacking or very limited; areas with diseases of high economic threat; and areas with diseases that are known or suspected to be vector transmitted.

When collecting citrus seeds, sound, healthy appearing, true-to-type fruit should be collected from parts of the tree more than 1 m above the ground. The seeds should be extracted from the fruit and treated if necessary to remove the pulp from the seed, and then surface sterilized. When collecting citrus budwood, collecting tools should be sterilized by dipping in a dilute solution of sodium hypochlorite between each tree from which budwood is collected. Budwood should be labelled, packed in plastic bags and maintained at 25°C or lower until it can be further processed. The budwood should be rinsed, surface sterilized, re-rinsed, and air- or towel-dried completely. If possible, the budwood should be dipped in a solution containing a fungicide, insecticide and miticide. The ends of the budsticks or the entire stick can be dipped in melted paraffin wax to help reduce desiccation. It is important that both budwood and seeds be completely dry before packaging or storing.

Movement of citrus germplasm

Acquisition of citrus germplasm involves movement from one location to another. In a few instances, the distances involved may be short, but often the movements are international in nature. Germplasm may be shipped by several different means, or hand carried if possible. It is important to take all possible measures to prevent deterioration or loss of propagative materials during transit. This is particularly true of budwood, but seeds are also vulnerable to deterioration. Some of the phytosanitary considerations in this area are discussed in a later section.

Citrus budwood deteriorates rapidly at even moderately warm temperatures. Therefore, after collection (whether from an *ex situ* collection or from the wild), citrus germplasm should be stored if possible at refrigerator temperature (3–5°C) prior to packaging and shipment. If this temperature is for some reason unobtainable, the germplasm should be kept at as low a temperature as possible under the circumstances. Contamination from pathogenic organisms can also result in the deterioration of budwood, and for this reason it is important to maintain as high a state of cleanliness as possible. Generally it is desirable to pack budwood in sealed bags within a sturdier container, such as cardboard boxes. The budwood should be kept as cool as possible by using insulated containers, 'blue ice' or other appropriate means.

Citrus seeds are less prone to deterioration than budwood. Their lower sensitivity to temperature makes shipping seeds less complicated than shipping budwood. Seeds should be stored at 3–5°C after surface sterilization. They can then be taken from storage and shipped immediately or after a short period of drying (to remove surface moisture from condensation during storage). Citrus seeds can generally be successfully shipped in an envelope or padded envelope at ambient temperature. When transit times are long, it may be a good idea to use an insulated container. Maintaining a low temperature whenever possible is advisable but is not as critical as it is for budwood.

Citrus germplasm should be packaged with permanent identification labels or tags, the package should be marked to indicate that plant material is enclosed, and required documents should be included in the package (see below). Citrus germplasm should be shipped as soon as possible after packaging, using the most rapid and reliable form of transport available. Clear instructions should be given to the carrier to avoid exposure to extremes of temperature. The receiver (both the consignee and any relevant inspection stations) should be notified in advance of all shipping details.

In some cases, such as collecting germplasm in the wild, it is not possible to adhere to all appropriate packaging and shipping concepts. However, the closer that it is possible to adhere to these concepts, the greater the chance that the citrus propagative materials will arrive at their destination in a viable condition.

All required documentation should be included in the shipping container. In some cases, it is required that permits and other documents be affixed to the outside of the container. Required documentation will vary with the regulations of the importing country or political subdivision thereof. In some cases, no legal documents may be required. However, movement of citrus between countries (and sometimes between political subdivisions within a country) is, in general, highly regulated. In fact, citrus is one of the most highly regulated crops in regards to restrictions on its free movement. Usually, at a minimum, a phytosanitary certificate is required. In many cases, an import permit issued by the importing country or a political subdivision is required. In some cases, these imports spell out specific requirements that must be satisfied before the citrus germplasm is allowed in. These requirements may be either pre-entry or post-entry. Pre-entry requirements may include cutting material from pathogen-tested stock, area freedom from a particular pest or disease, specific testing results, pesticide treatment, etc. Post-entry requirements may include a specific quarantine protocol or test, establishment in an isolated area for observation, etc. In addition to the required legal documents, germplasm movements should include identification, passport data and other information required or useful to the consignee or end-user.

Introduction of citrus germplasm

Phytosanitary considerations play a very important role in the conservation of citrus genetic resources since in the majority of cases they determine how new accessions can be introduced in a germplasm bank and also have a strong influence in their management and utilization. Of all the tree and fruit crops of the world, citrus is affected by the greatest number of graft-transmissible pathogens in addition to pathogenic fungi and bacteria and pests. These graft-transmissible pathogens can debilitate trees and devastate industries, due to their virulent nature and transmission by use of infected budwood for propagation, mechanical transmission and insect vectors.

In the 1930s, the viral nature of psorosis was demonstrated by Dr Howard S. Fawcett, the father of citrus plant pathology, and soon thereafter many other important diseases of citrus were shown to be caused by viruses. Since those early years, many if not the most important graft-transmissible diseases of citrus have been established as being caused by viruses, viroids, mycoplasmas or bacteria. However, the causal agents of some diseases of quarantine significance of citrus have not yet been established. New diseases of citrus caused by graft-transmissible agents are occasionally reported, as are the causal agents of conditions previously thought to be genetic or physiological.

Citrus is native to the southern/Southeast Asia region of India, China and surrounding areas. It has spread outward from this centre of origin for many centuries as seed, budwood or small plants. For most of this period, there were no restrictions upon movement of propagative materials, and, along with the citrus material, pathogens and some of their vectors were undoubtedly moved to different parts of the world in which they had not been previously established. In addition, new severe diseases have appeared in several countries that are not present in the areas of origin.

Graft-transmissible pathogens and certain other plant pathogens of quarantine significance are unable to move on their own, and are not spread by environmental factors. Their dispersal is primarily by humans or insects. Humans may knowingly or unknowingly spread these pathogens by movement of plant parts. The pathogens may be moved inadvertently in the course of normal trade, travel or scientific exchange. This may occur either directly (i.e. in plant parts) or indirectly (i.e. in an insect contaminating a shipment of commodity). Once moved to a new area, these plant pathogens may infect previously uninfected plants through human activities (e.g. propagation with infected materials) or insects. Under appropriate conditions, these pathogens can become established as endemic diseases and in some cases may cause epidemics of plant diseases.

Accidental introduction of pathogen-infected citrus budwood can have catastrophic consequences. For instance, prior to the 1920s, the citrus industries of Argentina and Brazil flourished despite the use of tristeza-susceptible rootstocks. However, 20 years after tristeza-infected nursery stock was imported from South Africa and Australia, 20 million trees had perished. Another example is the case of Spain, where tristeza was probably introduced in the 1930s with nursery trees of navel oranges imported from California; up to now, the disease has killed over 40 million trees grafted on sour orange (Cambra *et al.*, 2000). These and similar catastrophes led to restrictions on the movement of citrus propagative materials and the establishment of quarantines and other regulations controlling the movement and propagation of citrus.

The risk of introductions of previously unestablished pests and pathogens is minimized by completely prohibiting introduction of the pest or pathogen either directly or indirectly. However, when regulatory agencies have considered the benefits of accepting new material into a country, they may allow the entrance of materials under an approved series of safeguards designed to minimize the risk of introducing exotic pests and pathogens. This generally applies to an introduction of limited amounts of material, initially for research purposes and only later released to the general public. The risks associated with large-scale commercial importations are unacceptably high in most instances.

In general terms, the steps usually taken to prevent introduction of exotic pests and pathogens include regulation of importations which may result in the introduction of an exotic pest or pathogen; foreign certification of materials before entry; post-entry inspection by domestic plant health authorities; and in some cases post-entry quarantine. Some general information on issues pertinent to quarantine has been reviewed recently by Foster and Hadidi (1998), Frison and Diekmann (1998) and Kahn (1999a, b).

As a consequence of these safeguards, introduction of new accessions from other countries to a germplasm bank often has to be done through post-entry quarantine stations, which are usually operated by agencies different from those operating the gene banks. This may pose important limitations to the germplasm bank's ability to acquire new germplasm. There are only a few countries in which the quarantine facilities are located in the germplasm banks. In these cases, it is much easier to schedule the importation of new germplasm from other countries. A few countries have post-entry quarantine stations for citrus that do not have any direct relationship with the germplasm banks, e.g. Australia, Brazil, Japan and New Zealand. These programmes are generally concerned more with importation of commercial varieties than with germplasm per se.

Citrus propagative material is highly regulated as compared with propagative material of most other crops. Generally, citrus and other commodities carrying similar risks require import permits that spell out the requirements for accepting new material and for releasing it. Phytosanitary certifications are usually required or preferred. Post-entry inspection is generally required. Some general aspects for the safe introduction of citrus propagative materials from other countries are found in Broadbent (1999), Frison and Diekmann (1999), Frison and Taher (1989), Knorr (1977), Lee *et al.* (1999), Roistacher (1977), Roistacher *et al.* (1977), Lee *et al.* (1999), Navarro (1992, 1993), and Navarro *et al.* (1984a, 1991).

In some countries without post-entry quarantine stations, budwood not known to be virus free is completely prohibited from entering a country. Therefore, the maintenance of pathogen-free materials is critical not only for phytosanitary health, but also to make possible or facilitate exchange of citrus germplasm. In any case, pathogen-free budwood is highly desirable for exchange as compared with budwood of unknown disease status. This situation complicates the exchange of germplasm and has led to the establishment of different collections around the world, since needed germplasm would not be readily available from a single source. In some instances, importing countries allow the importation of citrus germplasm from programmes which are recognized by the importing country as maintaining high-quality, pathogen-tested, usually protected collections. In these cases, there is usually a periodic review of the programme and re-certification (or decertification).

A brief comment should be made regarding scientific exchange of plant material or germplasm. Although it may seem that this sort of exchange would be less likely to result in the introduction of exotic pathogens than would commercial importations, this is not necessarily the case. It is unlikely that an exotic virus of citrus would be introduced on a shipment of fruit (although pests and other types of pathogens might be). However, most citrus pathogens of quarantine significance are graft transmissible in budwood, which is the preferred medium of exchange for research purposes in most instances. In the USA, introductions of budwood through approved channels for scientific research were routinely contaminated. During the period 1954–1966, 70% of budwood entering California was infected (Roistacher and Nauer, 1968), and from 1968 to 1978, 100% of the budwood entering the USA was infected (Kahn *et al.*, 1979); overall from 1958 through 1993, 65% of the budwood entering the USA was infected (Waterworth, 1993). Although the percentage of introduced budwood that is infected has decreased somewhat as knowledge of these diseases has increased and more clean stock programmes have been established, the threat of introduced budwood harbouring pathogens is still very real.

Scientists shipping plant germplasm to other scientists may be expected to be more aware of potential phytosanitary issues than commercial interests, but there are times when these issues may not be completely known. There is also a risk when exchanging cultures of pathogens, and in exchanging insects that may harbour viruses. Although generally there are regulations regarding handling of these types of materials both pre- and post-entry, no system is perfect. For scientists, there is a higher standard for careful and safe introductions of plant material than there is for the general public.

Introduction of accessions from other countries: the quarantine process

There are two different concepts for citrus quarantine procedures. The classical approach involves propagation of imported material, followed by observations and indexing for the presence of pests and pathogens; infected material can be destroyed or subject to a therapy procedure. The biotechnology-based approach (Navarro *et al.*, 1984a, 1991) involves culturing *in vitro* the imported budwood, the recovery of plants by shoot-tip grafting *in vitro*, and observations and indexing for the presence of pests and pathogens. This technique is covered in detail in Chapter 17. The material is released from quarantine only when no pests and pathogens are present. This sometimes involves repeated cycles of testing, therapy and re-testing. It is important to note that the concept of 'releasing' also includes the planting of the imported material in the field collection of the germplasm bank. Citrus seeds are not quarantine items in the USA and Spain. Although usually not strictly required from a regulatory standpoint, accessions received as seed are in some cases also routinely pathogen tested before release.

Introduction of new accessions from the same country: sanitation

Usually there are no legal restrictions on the introduction of new accessions from the same country where the germplasm bank is located. However, an uncontrolled introduction of pathogen-infected accessions may cause serious problems for the management, characterization, evaluation and utilization of the genotypes included in the repository. To avoid these problems, it is advisable that all materials introduced from the same country will be subject in germplasm bank facilities to a sanitation programme for pathogen elimination (Navarro, 1993). Even if there are no regulations for within-country movement of propagative material, sometimes the risk of introducing new severe pathogens in the areas where the collection is located, or new and more severe strains of existing pathogens, is even higher with local material than with imported material. This is the case in countries or areas with severe endemic diseases. In some cases, obtaining propagative materials from a reliable source of pathogen-tested material outside the country would result in less risk than obtaining material from within the country. For these reasons, and despite the higher cost, some citrus germplasm banks, like those of the USDA-ARS and IVIA, have among their objectives to maintain only healthy genotypes, even for those accessions selected in the country.

Quarantine and sanitation technology: pathogen elimination

The quarantine and sanitation programmes share practically the same technical needs and facilities. The relationships between these various areas of plant health are diagrammed in Fig. 4.2. The two main areas are pathogen testing (see following section) and pathogen elimination.

Pathogens can be eliminated from citrus by nucellar embryony *in vivo* or *in vitro* (Weathers and Calavan, 1959; Navarro and Juárez, 1977, 1981a, b; Roistacher, 1977), by thermotherapy (Roistacher, 1977) and by shoot-tip grafting *in vitro* (Navarro *et al.*, 1975; Navarro, 1992). This last technique is recommended because it is effective in eliminating all citrus pathogens, and recovered plants do not have juvenile characters.

Thermotherapy is the older of these pathogen elimination techniques. Initial attempts to use high temperatures to eliminate viruses from citrus materials were unsuccessful. Finally, Grant (1957) was

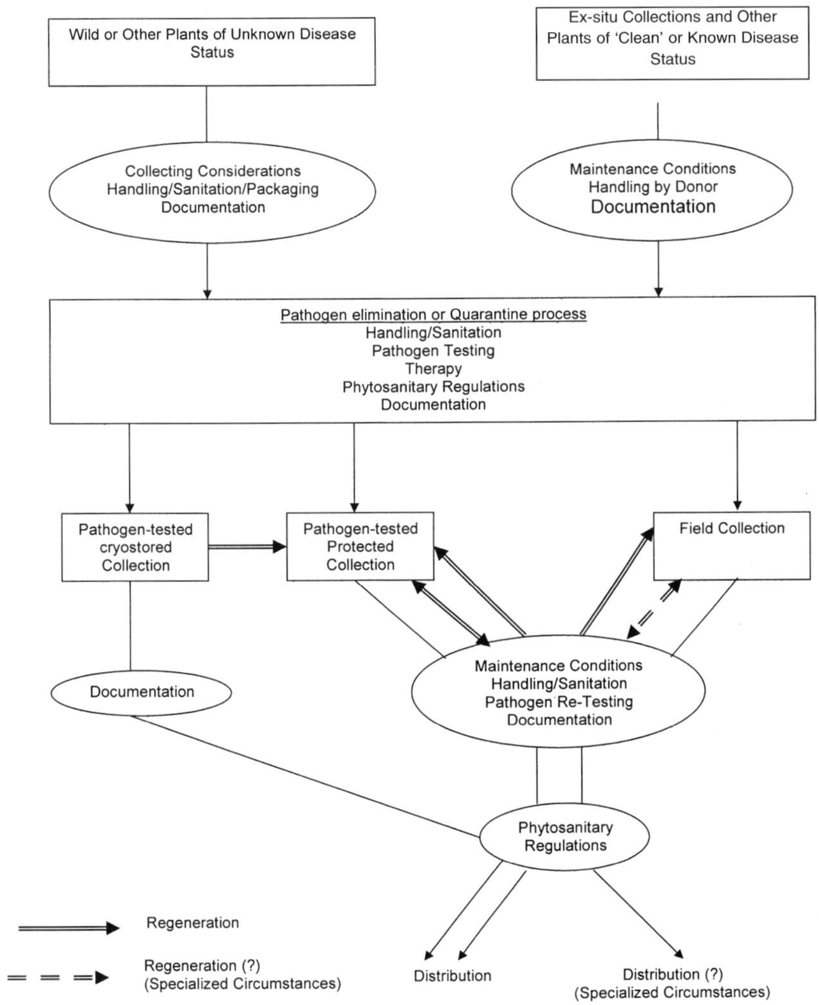

Fig. 4.2. Phytosanitary considerations in the management of citrus genetic resources.

able to eliminate tristeza and psorosis from infected budwood using hot air chambers. Later, various other viruses were eliminated using similar techniques (see a review in Roistacher, 1995). Calavan *et al.* (1972) further developed the use of thermotherapy for elimination of viruses by identifying appropriate rootstocks that were better able to withstand the high temperatures, and Roistacher and Calavan (1972) demonstrated the importance of pre-conditioning prior to heat treatment. There are many variations in thermotherapy techniques, but a commonly used diurnal temperature regime is 40/30°C (16/8 h). While thermotherapy has proven to be very effective in eliminating viruses from citrus, the work of Calavan *et al.* (1972) and that of various other workers (summarized in Roistacher, 1977, 1995) indicated that in many instances non-viral infectious agents (e.g. viroids and mycoplasmas) were not eliminated using this technique. This shortcoming led, in part, to the development of the other major method for pathogen elimination used with citrus, shoot-tip micrografting, which is at this time more widely used (see Chapter 17).

Following therapy, the material is indexed (see below). If the results indicate that one or more graft-transmissible pathogens are still present, the material must be subjected to another cycle of therapy. The alternate cycling of therapy and indexing continues until the material tests negative for all known graft-transmissible pathogens. Infected materials from previous cycles are destroyed by incineration or autoclaving.

Quarantine and sanitation technology: pathogen testing

Pathogen testing of citrus for graft-transmissible pathogens is one of the central points of the entire quarantine process. Detection of graft-transmissible pathogens is based primarily upon biological indexing on specific citrus indicator plants, supplemented with laboratory tests. Specific graft-transmissible pathogens have been reported from certain countries. Indexing for these pathogens should be added to the normal indexing programme when germplasm is introduced from these countries. New citrus pathogens are reported periodically. As they are identified and assay procedures developed, these diseases should be added to the indexing programme. In addition, as additional diseases are newly reported from different countries or geopolitical areas, indexing for these pathogens should be added to the normal indexing programme.

Biological indexing is based upon distinctive pathogen-specific symptoms (usually foliar) which develop in specific plants (indicators) when infected tissue is inoculated therein. This is generally done by the grafting of bark tissue (phloem). Biological indexing is done by comparing symptom expression in indicators inoculated with tissue from the plants to be tested with the symptoms developed in indicators inoculated with tissue from plants known to be infected with specific pathogens and with uninoculated, virus-free asymptomatic controls. Most graft-transmissible pathogens produce symptoms best at cool temperatures, whereas a lesser number produce symptoms best at warm temperatures. Growth chambers are inadequate for indexing, and therefore a well-constructed, properly functioning greenhouse facility is critical (see the following section).

The first report of the use of seedling indicator plants for the diagnosis of citrus diseases was apparently the report of Wallace (1945) on the use of sweet orange for the detection of psorosis. Previous to this, diagnosis was made either on the basis of symptoms on field trees or inoculation and long-term observations. Since then, a number of additional citrus diseases have been reported, demonstrated to be caused by graft-transmissible pathogens, and biological indexing methods developed. For a historical perspective on these matters for most major graft-transmissible diseases of citrus, see Roistacher (1995).

A comprehensive set of protocols for biological indexing of graft-transmissible pathogens is beyond the scope of this chapter. Rather than listing a complete list of published reports, it is better to refer to the published compendiums or handbooks in this area: International Organization for Citrus Virologists (1968, n.d.), Calavan *et al.* (1978), Frison and Taher (1991), Roistacher (1991, 1998) and Lee *et al.* (1999). Additional information on the application of indexing to importation and certification programmes can be found in Navarro (1993), Roistacher (1977) and Roistacher *et al.* (1977).

More recently, various types of laboratory tests have become available for graft-transmissible pathogens. These include serologically based tests, such as enzyme-linked immunosorbent assay (ELISA), dot immunobinding assay (DIBA) and immunospecific electron microscopy (ISEM); nucleic acid-based assays, such as sequential polyacrylamide gel electrophoresis (sPAGE) and polymerase chain reaction (PCR)-based assays; culture of pathogens on media; and microscopy. A complete discussion of the theoretical bases behind these techniques is beyond the scope of this chapter, and the reader should refer to such works as Hampton *et al.* (1990), Dhingra and Sinclair (1995), Singh and Singh (1995) and Schaad *et al.* (2001).

In some cases, laboratory-based assays offer advantages over biological indexing. They are more rapid, require fewer human resources and greenhouse space, and are therefore generally cheaper, and a large number of samples can be processed in a short time. A disadvantage is the need for a well-equipped laboratory with specialized equipment and additional training for the technicians. In some cases, there is the potential for the production of false-positive and false-negative reactions, particularly when using PCR methods on a large scale with different genotypes.

In some cases, laboratory tests can provide additional information, such as titre levels or (to some extent) isolate identity. However, laboratory tests do not so far provide information as to the severity of the isolate, the reaction of the indicator to the isolate, and other biological characteristics. Furthermore, if a disease is of unknown aetiology, it may be possible to detect it with an indicator but not with a biological assay. Only analysis of double-stranded RNA (dsRNA) will detect unknown viruses that produce dsRNA during their replication cycle. In this way, greenhouse and laboratory tests complement but do not replace each other. As mentioned, in some cases, laboratory tests can be performed when it is not possible to perform biological indexing for some reason. For the purpose of a complete index, if a biological test is available for a pathogen, a laboratory test can be used as an adjunct to the biological test but should not entirely replace it. In many cases, laboratory-based tests are not accepted by the regulatory authorities, and the 'index of record' is the biological index. On the other hand, for purposes of re-testing a large number of pathogen-tested germplasm accessions, laboratory tests are often the only feasible method.

Most laboratory tests specific for citrus pathogens have been developed since the 1970s. The earliest true 'laboratory-based' tests for citrus pathogens were serological tests (ELISAs) developed in the middle to late 1970s for CTV (Bar-Joseph *et al.*, 1979, 1980; Garnsey *et al.*, 1979, 1981). Serological tests offer the advantage of permitting the processing of large amounts of samples in a relatively short time. As such, this technique is often preferred for routine re-testing of accessions that have tested 'clean' and for large-scale surveys of productions areas. One agency in California routinely performs over 400,000 ELISA tests a year for CTV. This is the pathogen most commonly assayed in these types of large-scale surveys. The ELISA test for CTV, particularly the immunoprinting ELISA modification, remains one of the standard and most widely accepted tests for a citrus pathogen. It is accepted in many instances where all other pathogens must be assayed by biological indexing. Probably the only other serological test that has been proved to be of real value for diagnosis at this point is that developed recently for psorosis (García *et al.*, 1997; Martín *et al.*, 2002).

More recently, assays based upon nucleic acid analysis have been developed. The first of these were developed in the 1980s for the characterization and later detection of various viroids of citrus. In particular, the sPAGE assay for citrus viroids (Rivera-Bustamante *et al.*, 1986; Duran-Vila *et al.*, 1991, 1993) is widely used and accepted by regulatory agencies.

Since the development of PCR-based techniques in the early 1990s, these have become more common as they are simpler to perform than hybridizations. PCR-based techniques have the advantages of being more sensitive than serological techniques; however, they are more complicated to perform and require a more expensively equipped laboratory and personnel, and sometimes they have problems of producing false-negative and false-positive results. However, PCR-based technologies are in constant development, and performing these types of assays has become easier in recent years due to the development of kits and more automated instruments. At this point, however, nucleic acid-based techniques are less suited to large-scale and rapid testing than are serological techniques.

Other laboratory techniques include hybrid serological–nucleic acid techniques, classical culturing and electron microscopy.

The culture technique for detection of *Spiroplasma citri*, the causal agent of stubborn disease, is probably the most widely used and accepted assay for this pathogen, although biological assays and other laboratory-based techniques are available. Appropriate culturing is also useful for detection of other bacterial diseases.

Facilities

Quarantine and sanitation programmes for citrus have some very specific requirements with regard to facilities. Any facility used for these purposes must be inspected and approved by the regulatory authorities. A thorough discussion of the technical aspects of design for these facilities is beyond the scope of this chapter, but a few general comments will be made.

In classical quarantine procedures, the basic facilities are greenhouses for propagation and indexing, and laboratories for pathogen diagnosis, although facilities for therapy are highly advisable. In the biotechnology quarantine procedure, a tissue culture laboratory is also a requirement. These facilities are also used for elimination of pathogens from local accessions. The design of facilities for quarantine programmes is very important. General considerations in the design of quarantine facilities are reviewed by Mears and Kahn (1999) and Roosjen *et al.* (1999), while some considerations specific to quarantine of citrus are discussed by Roistacher (1991, 1998) and Gumpf (1999).

Ideally, the facilities should be located in an area with a climate suitable for growing citrus, but remote from areas of commercial production. This is almost imperative when the classical procedure is used, but as a practical matter is not always possible. It is therefore important that the facilities are designed and built in a manner which minimizes the risk of pathogens 'escaping' from the quarantine facilities and infecting commercial (or even dooryard) citrus. This may be extremely expensive. The biotechnology procedure has the advantage that pathogens (particularly pests, fungus and bacteria which are the more difficult organisms to contain) are eliminated at the initial step of introduction. This allows the quarantine facility to be located in citrus research stations where trained personnel and facilities are available. Of course, facilities design is insufficient for this purpose unless personnel are well trained in handling plants, soil, etc., so as to minimize this risk. This includes support personnel as well as the scientists actively engaged in the indexing.

The size of facilities will vary depending upon the size and scope of the quarantine or sanitation programme. However, certain general principles apply whether the facilities are large or small. Most of these concern exclusion of pests, prevention of the 'escape' of pathogens, and maintenance of environmental conditions suitable for indexing. In some cases, regulatory agencies will have specific requirements for facilities design. In general, the facilities belong to federal or state agencies involved in plant protection; alternatively, these agencies may have an agreement with a research institution. The facilities, processes and records must be available for inspection by the phytosanitary authorities at any time.

Ideally, facilities should include greenhouses, a screenhouse and a laboratory. The greenhouses are used for the production of indicator plants, rootstocks, etc., and for biological indexing; the screenhouse is used for protected maintenance of plants; and the laboratory is used for tests such as ELISA, PCR, culturing, etc.

The greenhouse facilities are a very important part of the quarantine facilities. However, a detailed discussion of greenhouse facilities for indexing of citrus pathogens is outside the scope of this chapter. A few general points will be made below. If a more thorough treatment is needed, the reader is referred to Roistacher (1991, 1998) and Gumpf (1999). Greenhouses used for indexing must have at least two and preferably three chambers with independently controlled temperatures. A room maintained at cool temperatures is used for the detection of most graft-transmissible pathogens. A room maintained at higher temperatures is used mainly for the detection of viroids and *S. citri*. If at all possible, a third room of intermediate temperature should be used for the production of indicator plants and propagation of trees for other purposes.

Because of the importance of the maintenance of these temperatures, the heating and cooling systems are vital to a properly designed greenhouse. The cooling system is particularly critical due to the great effect that a higher than optimal temperature can have on symptom expression. In some cases, cooling systems will not be able to maintain the appropriate temperatures during the hottest months of the year, and this can dictate the time of the year in which indexing takes place. The construction of the greenhouse is less critical than the environmental control systems. Greenhouses should be constructed so that they are insect proof, and should have double doors with vestibules for this reason.

Screenhouses may be constructed of metal or wooden frames and screened with nylon or stainless steel. It should be emphasized that 'screen' refers to anti-insect screen, not shade screen. The roof of the screenhouse should be at least 3 m from the ground (preferably 4.0–5.0 m) for adequate headroom for vigorously growing trees. Screenhouses should also have a vestibule with double doors.

The design and construction of the laboratory depends on the extent of its use. If only a limited number of laboratory tests are carried out (e.g. ELISA and sPAGE), a fairly small laboratory may be suitable. If larger numbers of materials are handled, or a larger number of laboratory tests carried out, a larger laboratory will be needed. The types of tests carried out will determine the equipment needed, as well. A minimalist indexing laboratory would include an ELISA plate reader and the apparatus for sPAGE. If PCR-based tests are utilized, a thermal cycler and perhaps a gel imaging system will be necessary. Culturing tests for bacteria and fungi necessitate the appropriate supplies and incubators. The biotechnology approach also requires a facility or portion of the facility devoted to tissue culture. This would include a laminar flow hood, incubation facility, etc.

In addition to the construction of the facilities, measures should be taken to minimize the phytosanitary risks such as escape of pathogens that may be present in the quarantine material; the access of insects to the interior of the facilities; and the spread of fungal pathogens within the facilities. Except under controlled situations, plants should not be introduced into the quarantine facilities from other facilities and especially from the field; this is particularly true of other citrus plants of unknown disease status. Normal phytosanitary precautions regarding sterilization of soil and equipment, avoiding contact with soil, maintaining the overall cleanliness of the facilities, etc. are also important.

Certain steps concerning access to the facilities can also reduce phytosanitary risk. These basically include restricting access to the facilities to all except those who are authorized and trained in proper phytosanitary conduct. The facilities should be kept locked as much as possible, and should also be behind a locked fence. Appropriate behaviour regarding ingress and egress is also vital.

Maintenance of citrus germplasm

The germplasm of citrus, as well as of many other perennial crops, is generally maintained in the form of living trees. This is in contrast to germplasm of most field crops and vegetables, which is generally maintained in a seed bank, distributed as seed, and regenerated by grow-outs as needed. The term 'maintenance' should be contrasted with 'preservation', which refers to long-term conservation and is discussed below.

The accessions of living trees are the source of propagative materials for distribution as well as often being able to serve as resources for a limited amount of characterization and evaluation. As such, these collections of living trees are often very similar to working or active collections of germplasm. However, depending on the circumstances and use of the collections, they may also serve as base or even inactive collections. Citrus materials which are preserved cryogenically or *in vitro* are not generally used as active collections since they are costly to regenerate, but embryogenic callus is being exchanged after thawing and regrowth.

All citrus germplasm banks have a field collection that is needed for characterization and preliminary evaluation. A few germplasm banks have a duplicate protected collection that is mainly used to maintain healthy genotypes and to diminish risks of losses of plants due to biotic or abiotic stresses. On rare occasions the germplasm bank also holds a cryostored collection for long-term maintenance.

There are certain principles that apply to both protected and field collections. Collections should include at least two copies of each accession. This creates at least one replicate for observations made directly from trees in the collections as well as providing a back-up in the event that one of the trees dies. However, for genotypes of low use, only one plant can be maintained in the protected collection. This reduces cost, and trees in the field collection can be used as back-up. Accurate maps or lists of locations should be maintained and updated in a timely manner. Trees should be observed periodically for signs of pests or pathogens, abiotically induced problems, injuries, off-types or budsports, etc. Problems should be rectified or the tree repropagated and replaced if appropriate. Proper phytosanitary precautions should be taken. This includes the use of a bleach solution on instruments or tools used on all trees of the collections.

Accuracy of identification is critical. Despite best efforts, accessions are occasionally misidentified. Sometimes this is immediately apparent and sometimes not. This points out the need for constant review of records and checking them for accuracy. Misidentified accessions should be removed from the collection and either discarded (if identity is known and is duplicated) or re-propagated for further study and (it is hoped) identification.

Field collection

Most collections of germplasm of citrus (and, for that matter, most perennial crops) have as their base a field collection or planting. These are necessary for characterization and evaluation activities, and may or may not serve as sources of germplasm for distribution purposes. From the conservation point of view, the field collections have the risk of losing genotypes due to diseases and abiotic stresses such as freezes, floods or strong winds. Because of these threats, but more because of the possibility of becoming infected with a graft-transmissible disease, field collections of citrus germplasm do not normally serve as sources of vegetative propagative material.

Propagation of trees for field collections generally follows practices for production of commercial trees for use in the area. The same is true of cultural practices in most instances. Generally, these practices were developed in order to produce a healthy and thrifty tree, and this type of tree is desirable in a field collection. However, there are some differences regarding field collections of citrus germplasm as compared with commercial plantings.

Commercial spacings are not necessarily needed, but there should be sufficient space between and within rows that a relatively normal or natural form of growth may be observed and documented. Cultural practices which may affect fruit set, quality, etc. should be avoided if one of the goals is to observe or characterize flowering, fruiting and other similar characteristics. Likewise, pruning and other manipulations should be avoided if one goal is to characterize normal growth habit. Commercial harvesting practices are often not possible or desirable. In many instances, fruit is left on the trees as long as possible so that it is available for observation.

Cross-pollination is a common feature of these collections and this does not allow proper characterization of the production of seeds, particularly for self-incompatible but fertile genotypes. Similarly, proper characterization of parthenocarpic genotypes is not possible in these collections. The only way to solve these problems is to exclude pollination insects (mainly bees), for example by covering trees with nets during the flowering period.

Some type of fenced enclosure is often desirable to prevent both theft of fruit and taking of possibly contaminated budwood. Access to the field collection, like access to quarantine facilities and protected collections, should be restricted to those having legitimate business in the collection. Visitors should be accompanied by scientists or workers.

When possible, field collections should be replicated in more than one location. This helps ensure the perpetuation of the accessions since they are backed up at more than one location. In addition, establishment of field collections in locations with different environmental or climatic conditions can help with the evaluation of genotype × environment interactions.

Even if field collections are periodically re-tested for diseases, as a general rule budwood should not be distributed from field trees unless the recipient has an adequate phytosanitary programme in place to prevent introduction of pathogens from this source. Seeds and pollen may be distributed in most instances. Propagative material for replacement trees should if at all possible be taken from a pathogen-tested source of budwood rather than directly from a field tree. If such a source is not available, budwood may be taken from the field tree. Trees should not be placed in a collection when propagated from non-local budwood unless the phytosanitary health of the budwood is acceptable.

Protected collection

A protected collection is maintained under screen or (rarely) in a greenhouse. The protected and field plantings may or may not be located near each other. Ideally, a protected collection, particularly one connected with a certification programme, would be located away from commercial areas with their associated higher risk of contamination or infection. However, this is not always possible in practice.

Protected collections are costly to establish due to the investments in screenhouses, and more expensive to maintain than field collections, since they need to be cared for by labour with higher specialization. However, they are highly recommended for germplasm conservation. They are the only way to maintain healthy genotypes in areas with naturally spreading diseases. This is needed in order to release material for certification programmes and for a safe exchange of germplasm. A protected collection is the only way to guarantee germplasm conservation in areas with severe naturally spreading diseases that may kill field trees. In addition, plants can be much more easily protected against abiotic stresses. An additional advantage of these collections is that parthenocarpy can be easily evaluated.

Protected collections vary somewhat. Some general information on suitable facilities and phytosanitary and security practices was presented above. Within the facilities, the manner in which the trees are grown may vary somewhat. Maintaining trees in pots has some obvious advantages with regard to disease prevention, fertilization, frost protection, the ability to manipulate and move the trees, etc. However, use of pots limits the size of the tree to a certain extent. The largest practical size for pots for this use is about 25 l. Pot sizes larger than this are more difficult to move, and trees maintained therein would be more difficult to trim and train. In other words, at a certain point, it is probably more efficient to maintain large trees planted directly in the soil rather than in extremely large containers. Spacing of trees under screen needs to balance the need for adequate space for growth (mainly, providing enough space such that crowding and shading is not a problem) with the need to maintain accessions in a limited area. Spacing will depend on tree size and vigour.

Trees maintained in pots do not have to be propagated exactly as if they were to be planted in the field. The increased phytosanitary standard that is possible in a protected facility makes a suitable tree for a protected collection somewhat different from a tree destined for field planting. Rootstock selection is not as important as with field trees, but in many, if not most, cases, the rootstocks of choice for field trees are also very well suited and perhaps the best choice for a protected collection. A rootstock that produces a compact, long-lived tree is preferred. Excessive distance between bud union and crown is undesirable in a protected collection both due to height limitation from the structure and from the standpoint of ease of manipulation of the trees. Therefore, the crown of the tree should be formed closer to the bud union than for a tree for the field. The use of a sterilized soil mix makes practical a lower budding height. Regular pruning creates a more compact tree as well as stimulating production of budwood. Pruning has to be done so as to allow fruit production, which is necessary to ascertain trueness to type. Re-potting to larger pots may be necessary as the trees increase in size. Even when pot size is no longer increased, periodic re-potting is useful in order to prune excessive root growth and observe abnormalities in the roots.

With training such as that described above, it is possible to maintain for a considerable period of time a citrus tree at a size suitable for maintenance under screen. Currently, the oldest and largest trees maintained under screen at Riverside are approximately 15 years old and are maintained in 20 l pots. In Spain, the oldest trees are 22 years old; they have been maintained in 25 l containers without re-potting and they grow and produce fruits normally. Of course, at some point, it will be necessary to re-propagate these trees.

Trees maintained under screen are generally not suitable as sources of fruit for accurate characterization and evaluation purposes. However, they help greatly in ascertaining the trueness to type of genotypes, detecting possible misidentified accessions and detecting possible chimeras. These are very important considerations when releasing budwood for certification. Similarly, due to the relatively large amount of manipulation to which they are subjected, trees in a protected collection are not suitable for characterization of most vegetative characteristics. These types of observations should be made on field-grown trees. Trees in a protected collection should only be propagated or repropagated from pathogen-tested budwood from a reliable source.

Long-term preservation

Genetic resources of most plant species are stored in seedbanks. Seeds are dehydrated and stored in sealed containers at low temperatures (usually about –20°C). This is a very safe and inexpensive procedure but requires that seeds be tolerant to desiccation and freezing temperatures. These are the so-called 'orthodox' seeds, as opposed to 'recalcitrant' seeds that lose viability when the moisture content is reduced. For many years, citrus seeds were considered recalcitrant because they could be stored only for short periods of time. As a consequence, citrus genetic resources are maintained as living plants in field or protected collections, with the associated high cost and the risk of losing genotypes due to climatic or biological hazards.

This situation is common to many vegetatively propagated crops, particularly fruit trees. During the last two decades, there have been investigations into the use of tissue culture methods, including slow growing vegetative shoots and cryopreservation of several tissues, for dealing with the problem of long-term conservation of genetic resources of these crops (Withers, 1992; Engelmann, 1997). In citrus, most approaches have been investigated with variable results (Duran-Vila, 1995, 1997) and they are briefly reviewed in the following sections.

SLOW GROWTH OF *in vitro* VEGETATIVE TISSUES. With some species, it is possible to establish procedures for slow growth of shoots, usually using low temperatures and decreased light intensity in conjunction with some modifications in the culture medium. Re-culture of the explants is done at intervals of about one year. This procedure allows a routine conservation of germplasm of several species *in vitro* (Engelmann, 1997). Attempts to follow this approach in citrus were not successful, because the procedure developed for juvenile material involved a complicated cycle of producing shoots from nodal segments, rooting of the shoots and re-culturing of nodal segments (Marín and Duran-Vila, 1991). In addition, there was a low efficiency in both rooting and shoot production from nodal explants, and this procedure could not be used with adult material.

CRYOPRESERVATION OF APICAL SHOOT-TIPS. Shoot-tips of adult material are the best tissue for cryopreservation of citrus genetic resources, because regenerated plants will not have juvenile characteristics and will be readily available for breeding. Recently the successful cryopreservation of shoot-tips from juvenile plants of *P. trifoliata* using the encapsulation dehydration technique has been reported (González-Arnao *et al.*, 1988). However, this work could not be applied at IVIA to other species or adult material (M. T. González-Arnau *et al.*, unpublished results). The main limitation is that following freezing and thawing, only a portion of the shoot-tip may survive. In citrus, small shoot-tips do not regenerate *in vitro* (with the exception of some species such as *P. trifoliata*) and it is necessary to regenerate plants by shoot-tip grafting, which cannot be done with only a portion of the shoot-tip.

CRYOPRESERVATION OF SEEDS. Although citrus seeds were considered recalcitrant, some limited studies have demonstrated that there are *Citrus* species tolerant to desiccation, others partially tolerant and still others recalcitrant (i.e. they lose viability very quickly with desiccation) (Mumford and Grout, 1979; King *et al.*, 1981; Pérez, 1995). Seeds of species tolerant to desiccation, such as lemon, Mexican lime and sour orange, can be cryopreserved with good survival rates by direct immersion in liquid nitrogen after desiccation. Seeds of species partially tolerant to desiccation (e.g. sweet orange and common mandarin) have variable survival rates depending on the degree of dehydration and are difficult to establish on a large scale. No survival was obtained with recalcitrant species (e.g. Cleopatra mandarin, *P. trifoliata*) (Pérez, 1995). Cryopreservation of seeds has not been used in practice for conservation of citrus genetic resources.

CRYOCONSERVATION OF OVULES. Underdeveloped ovules are possibly a good tissue for conservation of genetic resources of citrus polyembryonic genotypes. They are easily excised from immature fruits and efficient plant regeneration can be achieved. There was an early report claiming survival of ovules by direct immersion in liquid nitrogen (Bajaj, 1984). However, survival of ovules has not been achieved in several experiments done at IVIA using different freezing protocols, including fast and slow cooling rates, and vitrification (unpublished results). Following the encapsulation–dehydration technique, occasionally a few ovules have survived (1–16%) but not at a rate high enough for practical application (González-Arnao *et al.*, 2003).

CRYOPRESERVATION OF SOMATIC EMBRYOS. Cryopreservation of somatic embryos produced by ovules of sweet orange cultured *in vitro* was achieved some time ago (Marín and Duran-Vila, 1988; Marín *et al.*, 1993). However, survival rates were low and erratic, and a careful selection of embryos in early developmental stages was needed to achieve survival. The procedure was discarded as a practical method for germplasm preservation. Recently, cryopreservation of somatic embryos of several *Citrus* species produced by ovule culture *in vitro* has been accomplished by the encapsulation–dehydration technique with high survival rates (75–100%) (Gonzalez-Arnao *et al.*, 2003). This method does not require a careful selection of embryos. Recovery of plants was rapid, since the whole embryo survived the treatment and germinated readily. In addition, somatic embryos of polyembryonic varieties are easily produced by ovule culture *in vitro*. Although additional extensive work with a wider range of genotypes needs to be done, this procedure is very promising for the conservation of citrus germplasm. However, there is still the limitation that regenerated plants have juvenile characters.

CRYOPRESERVATION OF EMBRYOGENIC CALLUS. Embryogenic callus produced by ovules cultured *in vitro* of several citrus species and some related genera have been successfully cryopreserved, either directly or in cell suspension cultures, in a number of laboratories using several procedures, including slow cooling and vitrification (Kobayashi *et al.*, 1990; Sakai *et al.*, 1990, 1991a, b; Aguilar *et al.*, 1993; Engelman *et al.*, 1994; Pérez *et al.*, 1997, 1999; Xiancai, 1997). Survival rates are very high, and normally 100% of the calluses included in the cryotubes produce embryos and plants after thawing and reculture. The disadvantage of this procedure is that the production of embryogenic callus by ovules cultured *in vitro* is genotype dependent and time consuming (Pérez *et al.*, 1998) and that recovered plants have juvenile characters. The advantage is that germplasm is maintained as embryogenic callus, a very valuable and difficult to obtain tissue that is the basic material for somatic hybridization (Grosser *et al.*, 2000). Maintaining embryogenic callus under normal growing conditions requires monthly subcultures. This is a time-consuming procedure that may result in problems associated with somaclonal variation, decrease or even loss of their embryogenic potential, and risk of loss by contamination due to *in vitro* manipulations. After thawing, cryopreserved callus has been proved to be excellent material for protoplast isolation (Olivares-Fuster *et al.*, 2000) and has been used for exchange with other laboratories. Embryogenic callus also can be used for genetic transformation (Fleming *et al.*, 2000). Taking into account these advantages and the fact that somatic hybridization and genetic transformation programmes are being carried out at IVIA, this method has been adopted to establish a collection of cryopreserved embryonic callus of more than 40 accessions.

The standard cryopreservation protocol used at the IVIA germplasm bank was described by Pérez *et al.* (1997). Briefly, 150–200 mg of loose cells are cryoprotected in 1.8 ml of liquid basal medium supplemented with 10% (v/v) dimethylsulphoxide (DMSO) and maintained at 4°C for 30 min. Cryoprotected cultures are frozen by slow cooling in 2 ml cryotubes using a programmable freezing unit. Samples are cooled from 4 to –40°C at –0.5°C/min in order to obtain a smooth cooling curve. The samples are fast cooled from –40 to –150°C at –20°C/min and then placed in liquid nitrogen and stored. Cultures are thawed by immersion of the cryotubes in a 37°C water bath for 5 min. The cryoprotectant solution is removed from the cryotubes and cells are washed three times with 1.8 ml of liquid basal medium and transferred to solid basal medium for production of embryos and plants.

Characterization and evaluation

The efficient and effective utilization of citrus germplasm requires sound and accurate knowledge and documentation of its traits, i.e. it entails a description of what is in a collection. Descriptions of a germplasm resource are conveyed by descriptors based upon passport data, characterization, and evaluation of the germplasm. Passport data include basic information on the origin and type of the germplasm. Management data trace the history of an accession, the handling of its propagative units, its distribution, regeneration, etc. This ensures that users of germplasm are handling the materials that they believe they are.

A distinction between characterization and evaluation is sometimes made, although this is somewhat arbitrary and the boundaries somewhat blurred. Characterization refers to documentation of characters that are highly heritable, are easily identified (usually qualitative) and are expressed in all environments, while evaluation consists of documentation of additional characters (often quantitative) which are thought desirable by a consensus of users of the crop. Traditional phenotypic and modern molecular characterization of citrus is discussed by Gmitter *et al.* (1999), and some of the extensive documentation of characterization and evaluation data of citrus and related genera that has been generated over the last century is shown in Table 4.3.

Responsibility for characterization and evaluation varies. The curator is usually involved with characterization (usually the more basic attributes), while advanced or complex evaluations may be beyond the curator's capabilities and/or resources. Curators have the primary responsibility for documentation, which increasingly is via computerized databases, such as the Germplasm Resources Information Network (GRIN) system (Germplasm Resources Information Network, 1995; Mowder and Stoner, 1989), or the specific citrus databases GERMO developed at IVIA (<http://www.ivia.es/deps/biot/germop.htm>) and EGID developed at CIRAD, France (<http://www.corse.inra.fr/sra/>).

The descriptors most widely used for citrus are those developed by the International Board for Plant Genetic Resources (IBPGR) (1988; recently revised as International Plant Genetic Resources Institute (IPGRI), 1999), which are a slightly modified and expanded version of the 'Fruit Description Outline for Citrus' developed many years ago by H. J. Webber of the University of California Citrus Research Center (Webber, 1943; Hodgson, 1967). These descriptors are chiefly concerned with documentation of passport data and basic morphological traits. Other systems of descriptors developed independently are often fairly similar to these standardized descriptors, since the basic attributes to be described are fairly intuitive and obvious. These basic descriptors fall under 'characterization'.

The descriptors are adequate for describing the basic morphology of citrus. However, they do not address some very basic characteristics (e.g. growth rate), and their treatment of important physiological, pathological, horticultural and genetic characteristics is limited to a few items tacked on to the bottom of the morphological descriptors. One obvious example is that, except for one item dealing with one specific scion/rootstock compatibility, rootstock characteristics or other traits that might influence an accession's suitability as a rootstock are not dealt with. Yet, for many accessions, these would be the traits of most interest and importance. Another issue with the IBPGR/IPGRI descriptors is their utility for the citrus relatives.

Some of the shortcomings of the descriptors are due more to their application than to the descriptors themselves. There are two major considerations in assessing the value of descriptor data: geographical area and time. Geographical area can affect citrus morphology and growth via environmental (climatic) effects, soil conditions, water quality, air quality, pest and/or disease endemism, etc. Climatic factors (including temperature, photoperiod, rainfall, humidity and soil temperature) affect vegetative growth, flowering, fruit set, fruit composition, fruit growth and fruit morphology (Reuther *et al.*, 1969; Reuther, 1973; Germaná and Sardo, 1988). Soil conditions interact with water quality in affecting citrus morphology via fertility level (Embleton *et al.*, 1973a; Reitz and Embleton, 1986); presence or absence of vesicular-arbuscular mycorrhizae (Menge *et al.*, 1977); salinity (Bernstein, 1968; Maas, 1993); and water relations (Kriedemann and Barrs, 1981; Syvertsen and Lloyd, 1994). Air pollution can also affect the morphology of citrus (Thompson and Taylor, 1969; Olszyk *et al.*, 1988; Yelenosky, 1991). Time can also affect morphological observations. Due to patterns of seasonal growth, the time of the season in which an observation is made will affect such characteristics as fruit maturity and composition, flowering, percentage fruit set, size of various organs, etc. There are probably also less obvious effects on such parameters as pest and disease resistance, physical properties, etc. that are mediated more by environmental conditions than the time of season *per se*. The age of the trees upon which the observations are made are another facet of the effects of time which should be taken into account.

Another weakness of much descriptive work is that it is performed in one (or a limited number of) site(s) and for one (or a limited number of) year(s). Repetition of observations over time is easily done given adequate resources. However, dealing with the effects of location (i.e. climate) is less easily accomplished due to the fact that germplasm collections usually exist in only one location, and due to cost and other constraints it is often not feasible to replicate collections in a number of diverse climates. In any case, evaluation of perennial crop germplasm is a long-term endeavour. It would also be difficult to account for differences in cultural practices, pests and diseases, etc.

Another constraint caused by resource limitation is the small number of trees that can be maintained and hence evaluated. Many germplasm collections maintain only a few trees of each accession. This is probably sufficient to make general morphological evaluations, but not for more complex types of evaluations that require replicated trials. These include such important traits as pest and disease resistance, rootstock characteristics, responses to environmental conditions or cultural practices, etc. These types of trials, while important in evaluating germplasm and determining its value, are sometimes outside the scope of what can be evaluated within a germplasm collection. Most such collections consist of only a few specimens of each accession, and destructive (or potentially destructive) trials compromise the integrity of a collection. Although these investigations are outside the scope of what can be investigated within a germplasm collection, they may not be outside the responsibility of the repository scientists, depending upon the areas of expertise. However, they would have to be performed on trees planted specifically for trials and not on trees in the collection. More complex evaluations (disease resistance, physical properties, etc.) are very important but often have to be investigated as stand-alone research projects outside of an established repository or germplasm conservation system. These types of investigations require more resources than the initial characterizations, since they are complex, intensive, multiyear projects in many different areas. These types of investigations are by nature open ended and often yield new questions to investigate, all requiring adequate resources. Obviously a complete evaluation of genotypes in germplasm banks is a huge task that usually requires replication trials over a period of time and that is far beyond the responsibilities and resources of germplasm banks. Evaluation is the responsibility of all the citrus scientific community. The main problem is how to coordinate and place all the scattered information together in order to make it easily available. Evaluation is the black hole of genetic conservation.

Despite the limitations of phenotypic descriptors, they are quite useful for management of germplasm. They are currently the only means to differentiate accessions of certain groups of species that have been produced by natural budsport mutations (e.g. sweet oranges, satsumas or clementines), and likewise to identify duplications in the collection. Elimination of duplicates is very important because this reduces the cost of maintenance. Molecular markers have not at this time been developed to the point that they can be utilized for these purposes.

The last several decades have seen the evolution of biochemical and molecular markers as tools with great potential application to germplasm characterization. In contrast to morphologically based phenotypic characterization, molecular markers are generally unaffected by the many factors able to influence plant or organ characteristics. This allows comparisons between accessions within a collection or among collections at different locations at any time of year, while phenotypic characteristics can be masked by environmental or cultural affects.

Molecular characterization has a number of applications in the management of germplasm collections. These include elucidating systematic relationships between accessions; assessing gaps and redundancies in the collection; development of core subsets; characterizing newly acquired germplasm; maintaining trueness to type; monitoring shifts in population genetic structure in heterogeneous germplasm; monitoring genetic shifts caused by differential viability in storage or *in vitro* culture; exploiting associations among traits of interest and genetic markers; and genetic enhancement (Bretting and Widrlechner, 1995; Ayad *et al.*, 1997). One of the most important potential uses of molecular markers is their use in breeding programmes. Identification of genes and markers associated with quantitative traits will greatly increase the efficiency of a breeding programme.

The use of molecular markers for characterization and management of citrus germplasm is currently in its early stages. There are many reports on the development or use of molecular markers in citrus breeding, phylogenetic studies, etc., but many of these do not deal with citrus germplasm collections *per se*. Some of the work that has been done in this area includes the use of ISSR markers to analyse trifoliate accessions (Fang *et al.*, 1997), certain mandarin accessions (Fang *et al.*, 1998b) and miscellaneous accessions (Fang and Roose, 1997), and the use of isozymes, ISSR and simple sequence repeat (SSR) markers to analyse lemon germplasm (Gulsen and Roose, 2001) maintained by the University of California and USDA in Riverside, California. Isozymes have also been used to analyse a large number of accessions of the IVIA germplasm bank (Herrero *et al.*, 1996a, b) and *copia*-like retrotransposon sequences to study clementine accessions of this bank (Bretó *et al.*, 2001). Some of these studies suggested that there is less diversity among certain groups of accessions than previously believed, as noted in the discussion of *Poncirus* above. However, at this point, the technology of molecular analysis of germplasm accessions is not sufficiently developed to 'fingerprint' accessions. Many methodologies yield inconsistent results.

These reports deal with an analysis of genetic diversity among groups of accessions. Potentially, larger subsets and even entire collections can be analysed. In Riverside, Barkley *et al.* (2003) surveyed approximately 400 apparently sexually derived accessions using 25 SSR markers. This is an extremely robust data set that, when fully analysed, revealed a great deal about various aspects of the genetic diversity and relationships among these accessions. The information generated in some cases confirmed existing ideas derived from other types of observations, and in other cases called this into question. The molecular data have to be looked at in a broad context in combination with other types of observations and data.

Molecular markers potentially have a number of other uses in the management of citrus germplasm collections, but are perhaps less developed for these uses than for other crops. This is due to the small size of the worldwide citrus industry as compared with, for instance, wheat and maize. However, these types of uses are bound to increase for citrus. Krueger *et al.* (2003) and Krueger and Roose (2003) reported the use of various molecular markers to reduce redundancies in potential new accessions received as seed, resolve the identity of mislabelled accessions and detect pathogens. One can conceive of many other potential uses for molecular markers, and their use will undoubtedly increase in the future. They will never replace, but rather will complement, traditional morphological, horticultural and phytopathological evaluations.

The main limitation today of molecular markers for germplasm management is that they are not able to differentiate and fingerprint close genotypes within species, particularly in the case of cultivated sweet oranges, clementines, satsumas or lemons, where cultivars have been produced by spontaneous budsport mutations in the field. Most markers used today in citrus are the so-called neutral or anonymous, which are not related to specifically known genes. As citrus genomic projects taking place in several countries advance, many sequence-based markers will be available and this probably will allow a more efficient molecular genotyping of most accessions, facilitating both germplasm management and utilization.

As molecular methodologies improve, identification of redundancies and gaps in collections will improve. A more complete understanding of the relationships between accessions will make it possible to identify more accurately accessions that are completely or basically the same. There is an interaction here between molecular and 'traditional' data. In some cases, traditional observations may indicate that accessions are very similar or identical, whereas molecular analysis may reveal differences that may or may not be significant. Conversely, molecular analysis may show differences that have no practical significance. This is an area of inquiry in which a great deal of progress remains to be made.

Documentation and databases

There is sometimes a tendency to think of documentation of germplasm resources being associated only with *ex situ* collections, but this is too narrow a viewpoint. Documentation of germplasm resources in the broad sense includes documentation both of the overall status of a particular species and its local status, as well as formal collections. Various publications by entities such as IPGRI, IUCN, various governmental and non-governmental agencies and organizations, etc. address some of these broader issues. It is these broader issues that may make a formal effort at conserving genetic resources necessary or desirable. If a species is well represented in a stable environment, it is less vulnerable to extinction or genetic erosion than if it is represented by only a few specimens in an area undergoing rapid degradation.

The methods as well as the style of documentation are undergoing rapid change due to the influence of technology. Early documentation of citrus germplasm was based on personal observation and handwritten notes, with accessions listed in an accession book. Today, observations may be supplemented by information gathered from a global positioning device, with visual information provided by a digital camera and entered into a portable computer. Evaluation of accessions has also been enhanced by advances in instrumentation and electronic data analysis and maintenance. This statement is made with the realization that, while the cost of computer technology is decreasing and use of computers is increasing, in any individual instance, a 'database' may indeed still consist of handwritten records. Furthermore, just as a handwritten observation may be lost or deteriorate, electronic data may become deleted or their format or medium become obsolete. Thus, both hard copy and electronic records should be backed up and, in addition, care must be taken that the format of electronic data is updated periodically so that it may be accessed by currently available machines. Even if data are backed up in several locations electronically, original handwritten records should be retained. They are often invaluable in correcting mistakes or solving various mysteries associated with the collection. Even multiple copies of the same printed form, with notes taken by different workers, should be retained.

Databases are useful for maintaining, manipulating and reporting data. However, today, most database records are still based upon handwritten observations or notes. There are a few scientists who may take a portable computer into the field or greenhouse and make observations directly into it there, but most still make their observations on a datasheet or in a notebook. So, the transcription of original data to the maintained form still offers a potential step for miscopying, whether to a written or an electronic record. Thus, each step in the record-keeping process must be done carefully and with an eye towards quality control.

Often, data are maintained in both local and public databases. The public database generally contains non-confidential data that are of interest to the general public and potential users. These often consist of accession information, passport data and a limited amount of descriptor data. Many important types of data, particularly management data, are critical to local users but may not be important to potential collaborators. Therefore, complete and well-organized local databases are critical to the functioning of a germplasm collection. However, some types of local data may be of interest to colleagues. For instance, a scientist requesting germplasm will often request data from pathogen testing that may have taken place. Accessions or potential accessions should be assigned accession (or pre-accession) numbers that are maintained in the local database. In addition, individual trees within each accession number should have inventory or tracking numbers assigned so that unique properties, problems, etc. associated with individual trees will not become associated with the accession in general or with other trees of the same accession. These numbers should be assigned and recorded as early in the life of the plant as possible. In addition to general information on accessions, local databases should precisely and thoroughly document such attributes of individual trees as rootstock, propagation source, date of propagation, pathogen status, location and notes.

Utilization

Utilization of genetic resources is a primary goal of germplasm banks and in many instances is the justification for investing in genetic resource conservation. Citrus germplasm banks are or can be used for breeding and research, and directly for commercial propagation through certification programmes.

Utilization of citrus genetic resources for breeding

Citrus is affected by important abiotic stresses, such as acidic, alkaline and saline soils, flooding and drought, freezes and high temperatures; by arthropod, vertebrate and nematode pests; and diseases caused by fungi, bacteria, spiroplasmas, phytoplasmas, virus, viroids and virus-like pathogens. In addition, markets demand fruits of high quality that are difficult to produce because of a lack of adequate varieties. Existing high-quality varieties may be impossible to grow in many areas due to susceptibility to the above-mentioned biotic and abiotic stresses.

The consequence of this situation is that the number of varieties and rootstocks that can be used in different citrus-growing areas is very limited and the production is not well adapted to market demands. Genetic improvement has a very high priority to solve these problems, and several countries have been carrying out breeding programmes since the end of the 19th century (Soost and Cameron, 1975; Soost and Roose, 1996). However, the results so far have been very limited as a consequence of the complex reproductive biology of citrus, including several factors such as apomixis, total or partial pollen or ovule sterility of many genotypes of interest, autoincompatibility and incompatibility among genotypes, high heterozygosity and long juvenile periods (Cameron and Frost, 1968; Frost and Soost, 1968; Soost and Cameron, 1975; Soost and Roose, 1996). For these reasons, the number of active breeding programmes for citrus has been much more limited than for other species, and consequently the citrus genetic resources have not had an intensive utilization for breeding.

In recent years, several biotechnology techniques have been established for citrus breeding that solve some of the problems of traditional breeding (Navarro *et al.*, 2004). This includes symmetric and asymmetric protoplast fusion procedures that overcome the heterozygosity problem and allow the recovery of somatic hybrids among sexually incompatible parents, including citrus relatives (Grosser *et al.*, 2000). The optimization of technologies for embryo rescue and ploidy analysis by flow cytometry is allowing establishment of large triploid breeding programmes to recover seedless cultivars (Navarro *et al.*, 2003; Chapter 8). Citrus transformation technologies are opening up completely new and expanding possibilities for citrus improvement (Peña and Navarro, 1999; Chapter 15). The development of molecular markers for qualitative and quantitative traits is also advancing very quickly (Chapter 12) and this will accelerate breeding programmes through marker-assisted selection. In addition, the development of large citrus genomic projects that is already underway in some countries such as Japan, Brazil, Spain and the USA will undoubtedly produce a lot of information that will contribute to citrus improvement. Utilization of genetic resources maintained in germplasm banks for citrus breeding is increasing significantly with the initial application of these new technologies. In the near future, it is expected that utilization for this purpose will be expanded along with further developments in biotechnology.

Utilization of citrus genetic resources for research

Probably the major use of collections of citrus genetic resources is for breeding. However, there are other research uses for materials from these collections. Germplasm collections are continuously being used by physiologists, biologists, plant pathologists and investigators in other disciplines. For instance, the identification of 'Dweet' tangor as an indicator for psorosis (Wallace, 1945) and 'Parson's Special' mandarin for cachexia (Calavan and Christiansen, 1965) was possible because of citrus genetic resources maintained in Riverside.

However, the distinction between research to identify materials for utilization in breeding programmes and other types of research is not always clear. For instance, some of the most common investigations carried out are screening of accessions for useful traits, such as disease resistance/tolerance or adaptation to soil conditions. This information may be utilized in an improvement programme as well as more directly. An example is the trifoliate orange. Early on, its reaction to CTV was known (Costa *et al.*, 1949; Grant and Costa, 1948; Grant *et al.*, 1949, 1951). This led to its direct use both as a rootstock and as a source of CTV resistance in breeding programmes. More recently, the genetic basis of this reaction has been characterized (Gmitter *et al.*, 1996; Mestre *et al.*, 1997a, b, c; Fang *et al.*, 1998a; Fang and Roose, 1999; Deng *et al.*, 2001; Yang *et al.*, 2001). This information may be useful in improvement as well as being interesting as purely basic information.

Many traits documented in Table 4.3 have overlapping bases and results. Not all of these documentations are the result of investigations utilizing formal collections of citrus genetic resources, but some are. Conversely, investigation into and documentation of these types of traits from sources outside a collection can lead to inclusion of certain accessions in collections of genetic resources.

Utilization of citrus genetic resources for certification programmes

The subject of certification programmes for citrus is linked to the area of germplasm resources and will be briefly addressed. Certification programmes are primarily designed to control graft-transmissible and other important diseases. They only use advanced commercial varieties and in most cases they operate independently of germplasm banks. In citrus, there are only a few cases where there is a direct connection between certification programmes and germplasm banks, although some of the certified varieties may come from germplasm banks. For more information on certification programmes in general, a recent review has been produced by Waterworth (1998). For citrus in general, Lee *et al.* (1999) and Navarro (1993) offer good overviews.

The devastating nature of citrus diseases has been noted above. Roistacher (1993), in arguing for mandatory certification programmes for citrus, quite rightly notes that 'Citrus stands alone among the tree and fruit crops of the world for having the largest number of virus and virus-like diseases which have the capacity to debilitate or destroy its industry. There are very few or no comparable diseases of stone, pome, or vine fruits which can match the rapacity of ... citrus diseases ...'. Therefore, certification programmes are highly recommended to support citrus production and help maintain the phytosanitary status of the citrus industry.

Citrus was one of the first crops for which various types of clean stock and certification programmes were devised. Probably the first proto-certification programme for citrus was the 'Psorosis Free' programme started in California in 1937 (Calavan *et al.*, 1978). As additional diseases were shown to be caused by viruses or other graft-transmissible pathogens, the programme evolved to the form in which it is found today (Gumpf *et al.*, 1997; Bash, 1999; Krueger, 2001). In its current form, clean source trees are established by the Citrus Clonal Protection Program (CCPP), which is run by the University of California. After release to the industry, propagation and distribution of trees are regulated by the California Department of Food and Agriculture (CDFA).

The California programme is the oldest and longest-running certification programme for citrus. Other well-established programmes include those of Spain (Pina and Navarro, 2001), Australia (Forsyth, 1990), South Africa (Lee and von Broembsen, 1990) and Florida (Castle *et al.*, 2001). The protected collection of the Spanish germplasm bank is totally integrated in the certification programme, since it is the Protected Foundation Block of Initial Material for certification (Navarro *et al.*, 2002; see also Chapter 17). Many other countries are establishing certification programmes or taking the first step in starting phytosanitation programmes. Many of these programmes are supported technically by the FAO and often start foundation plantings with materials obtained from established programmes in California or Spain. Frequently, expert consultations from the FAO or other entities provide the impetus for start up of local indexing and sanitation programmes to complement the introduced germplasm. In addition to the FAO, various other NGOs have taken an active role in promoting or establishing regional certification programmes. These notably include IPGRI, which through its relationship with the International Agricultural Research Centers and the Consultative Group of International Agricultural Research, seeks to promote phytosanitary standards in the exchange of germplasm for research purposes (Frison and Diekmann, 1998), and the European/Mediterranean Plant Protection Organization. More recently, the North American Plant Protection Organization has become involved in this area.

Navarro (1993) and Lee *et al.* (1999) differentiate certification programmes from quarantine and clean stock (or sanitation) programmes. In this schema, quarantine programmes (as discussed above) are designed to ensure the safe introduction of new types of germplasm, while clean stock programmes have the aim of producing and maintaining healthy, true-to-type local cultivars. In reality, most of the techniques and technology used in a quarantine programme are also utilized in clean stock programmes, and sub-federal level regulations or quarantines make handling local cultivars similar from a security standpoint to handling foreign introductions even if the risk of introducing exotic pests or pathogens is lower.

Certification programmes, as the term is used by Navarro (1993) and Lee *et al.*

(1999), introduce regulatory or legal requirements and restrictions that enforce to some degree the use of clean material. These regulations are based upon local conditions and may not regulate all pests and pathogens. For instance, in California the CCPP provides a quarantine programme for introduction of new varieties from outside California, and also functions as a clean stock programme for cultivars already established in California. In providing these functions, CCPP tests for all known graft-transmissible pathogens. However, after release from state (and where applicable, federal) quarantine(s), the propagation of these materials is regulated by the CDFA, and testing is done for only a limited range of endemic diseases.

The most important components of certification programmes are the sources of propagative material. The bases for all further propagations are Foundation Blocks. These include both Protected Foundation Blocks of Initial Material and field-planted or protected Foundation Blocks usually located at the nurseries. The tendency today is to protect as much as possible all blocks of the programme. In Spain, IVIA holds the Protected Foundation Block of initial material, which is the protected collection of the germplasm bank, and the nurseries hold nine foundation blocks that are also protected. The multiplication blocks located in 39 nurseries are also all protected. Protected Foundation Blocks offer greater security as far as avoidance of infection by an insect vector, but infrastructure costs are greater. Foundation plantings should be periodically re-tested for the pathogens included in the regulations, which always include at least the endemic pathogens. Although this is obviously more important for a field-planted Foundation Block than for a Protected Foundation Block, the latter should also be re-tested although regulations usually establish longer testing intervals. The most important diseases from a re-testing standpoint are endemic diseases that are naturally spread.

Due to the expenses associated with maintaining and re-testing foundation trees, only a small number of trees from each variety are generally maintained (typically 2–6). This is not enough to supply budwood directly to propagate nursery trees. Therefore, material from the Foundation Block is normally used to establish Budwood Increase Blocks, allowing the rapid and efficient multiplication of buds. Increase Blocks may be protected or established in the field. The final step in a certification programme is the production of certified nursery trees from either foundation materials or Increase Blocks. Production may be protected or in the field. Again, the actual requirements for certification are mandated by regulatory agencies and vary from location to location.

Certification programmes may be either mandatory or voluntary. Mandatory programmes are of course more coercive, but in most cases are more effective than voluntary programmes. In voluntary programmes, the initially higher investment in certified materials may lead some nurserymen to use non-certified materials despite the long-term negative economic impact of using this inferior material. This can lead to greater disease pressure than if a programme was mandatory and, if endemic vector-transmitted diseases exist, these can be transmitted to trees produced under a voluntary certification programme. Thus, a few non-cooperative nurseries seeking a short-term economic advantage can cause long-term phytosanitary and economic damage to an industry.

Certification programmes are among the best established means of increasing phytosanitary health, and some of those for citrus are among the oldest in the world. In conjunction with quarantine and clean stock programmes, they remain important weapons in the ongoing fight against citrus diseases.

Germplasm banks may have a very important role in certification programmes by supplying the initial materials for propagation, particularly if they maintain healthy plants of commercial varieties. In some cases, such as in Spain, the germplasm bank is an integrated part of the certification programme and also is in charge of the post-entry quarantine station. Nurseries have already released 100 million plants to growers that originated in the germplasm bank. Healthy material from some germplasm banks, like those located in the USA, Spain and France (Corsica), have been used by many countries to start certification programmes when resources to establish their own clean stock programmes were not available. This is an example of international collaboration that has produced large benefits to the citrus industry worldwide and that shows the importance of maintaining citrus genetic resources.

References

Adikaram, N.K.B., Abhayawardhane, Y., Ratanayake Bandara, B.M., Gunatilaka, A.A.L. and Wijeratne, E.M.K. (1989) Antifungal activity, acid and sugar content in the wood apple (*Limonia acidissima*) and their relation to fungal development. *Plant Pathology* 38, 258–265.

Aguilar, M.E., Engelmann, F. and Andmichaux-Ferrière, N. (1993) Cryopreservation of cell suspensions of *Citrus deliciosa* Tan. and histological study. *Cryo-Letters* 14, 217–228.

Airy-Shaw, H.K. (1939) Notes on the nomenclature of certain Rutaceae–Aurantioideae. *Kew Bulletin of Miscellaneous Information* 1939, 290–293.

Albrigo, L.G. (ed.) (1997) *Identification and Conservation of Genetic Resources of Citrus and its Relatives.* Proceedings of a Symposium held during the VIII Congress of the International Society of Citriculture, Sun City, South Afica.

Albrigo, L.G. (1999) Networking to accomplish citrus germplasm conservation goals. In: Broadbent, P., Sykes, S.R., Bevington, K.B. and Hailstones, D. (eds) *Proceedings of the Citrus Germplasm Conservation Workshop.* Brisbane, Australia, pp. 44–45.

Albrigo, L.G. (ed.) (2001). *Proceedings of the Global Citrus Germplasm Network.* Network meeting held during the Ninth Congress of the International Society of Citriculture, International Society of Citriculture, Orlando, Florida.

Albrigo, L.G., Gmitter, F.G. and Menini, U.G. (1997) Requirements and priorities for citrus germplasm conservation. In: Albrigo, L.G. (ed.) *Identification and Conservation of Genetic Resources of Citrus and Its Relatives.* Proceedings of a Symposium held during the VIII Congress of the International Society of Citriculture, Sun City, South Africa, pp. 6–10.

Allen, B.M. (1967) *Malayan Fruits.* Donald Moore Press Ltd, Singapore.

Al-Saadawi, I.S. and Al-Rubeaa, A.J. (1985) Allelopathic effects of *Citrus aurantium* L. I. Vegetational patterning. *Journal of Chemical Ecology* 11, 1515–1525.

Al-Saadawi, I.S., Arif, M.B. and Al-Rubeaa, A.J. (1985) Allelopathic effects of *Citrus aurantium* L. II. Isolation, characterization, and biological activities of phytotoxins. *Journal of Chemical Ecology* 11, 1527–1534.

Anderson, C. (2001) Citrus germplasm resources and their use in Argentina, Brazil, Chile, Cuba, and Uruguay. In: Albrigo, L.G. (ed.) *Proceedings of the Global Citrus Germplasm Network.* Network meeting held during the Ninth Congress of the International Society of Citriculture, Orlando, Florida, pp. 49–56.

Anderson, E.F. (1986a) Ethnobotany of hill tribes of northern Thailand. I. Medicinal plants of Akha. *Economic Botany* 40, 38–53.

Anderson, E.F. (1986b) Ethnobotany of hill tribes of northern Thailand. II. Lahu medicinal plants. *Economic Botany* 40, 442–450.

Anupunt, P. (1999) Citrus germplasm collection and conservation in Thailand. In: Broadbent, P., Sykes, S.R., Bevington, K.B. and Hailstones, D. (eds) *Proceedings of the Citrus Germplasm Conservation Workshop.* Brisbane, Australia, pp. 139–145.

Armstrong, J. (1975) The current status of the sub-family Aurantioideae. *Australian Plants* 8, 226–236.

Asíns, M.J., Herrero, R., Pina, J.A., Carbonell, E.A. and Navarro, L. (1996) Genetic relationship in Citrus and related genera. *Proceedings of the International Society of Citriculture* 1, 248–253.

Asthana, A., Larson, R.A., Marley, K.A. and Tuveson, R.W. (1992) Mechanisms of citrus phototoxicity. *Photochemistry and Photobiology* 56, 211–222.

Attaway, J.A. (1994) Citrus flavonoids with anticarcinogenic and antitumor properties. In: Huang, M.T. (ed.) *Food Phytochemicals for Cancer Prevention. I. Fruits and Vegetables.* Symposium Series, American Chemical Society, Washington DC, pp. 241–248.

Aubert, B. (1987) *Le Greening, une Maladie Infectieuse des Agrumes d'Origine Bactérienne, Transmis par des* Homopteres psyllides. IRFA/CIRAD, St Pierre.

Awasthi, A.K. (1991) Ethnobotanical studies of the negrito islanders of Andaman Islands, India – the Great Andamanese. *Economic Botany* 45, 274–280.

Ayad, W.G., Hodgkin, T., Jaradat, A. and Rao, V.R. (1997) *Molecular Genetic Techniques for Plant Genetic Resources*. International Plant Genetic Resources Institute, Rome.

Azab, S.A. and Hegazy, A.K. (1995) Growth performance and plant water relations of seven citrus rootstocks under the arid environment of Qatar. *Arab Gulf Journal of Scientific Research* 13, 605–620.

Back, E.A. and Pemberton, C.E. (1915) Susceptibility of citrus fruits to the attack of the Mediterranean fruit fly. *Journal of Agricultural Research* 3, 311–330.

Bailey Hortatorium, L.H. (1976) *Hortus Third*. Macmillan, New York.

Baines, R.C., Bitters, W.P. and Clarke, O.F. (1960) Susceptibility of some species and varieties of Citrus and some other Rutaceous plants to the citrus nematode. *Plant Disease Reporter* 44, 281–285.

Baines, R.C., Cameron, J.W., Soost, R.K., Klotz, L.J., DeWolfe, T.A. and Small, R.H. (1967) Citrus nematode and phytophthora resistant rootstocks. *California Citrograph* 52, 294–295.

Bajaj, Y.P.S. (1984) Induction and growth in frozen embryos of coconut and ovules of citrus. *Current Science* 53, 1215–1216.

Baker, R. (1994) Potential dietary benefits of citrus pectin and fiber. *Food Technology* 48, 133–138.

Bandara, K.A.N.P., Peries, I.D.R., Kumar, V., Larunaratne, V. and Ranasinghe, M.A.S.K. (1990) Insecticidal activity of *Acorus calamus* L. and *Glycosmis mauritiana* (Lam.) Tanaka against *Aphis craccivora* (Homoptera: Aphididae). *Tropical Agriculture (Trinidad)* 67, 223–228.

Bar-Joseph, M., Garnsey, S.M., Gonsalves, D., Moskovitz, M., Purcifull, D.E., Clark, M.F. and Loebenstein, G. (1979) The use of enzyme-linked immunosorbent assay for the detection of citrus tristeza virus. *Phytopathology* 69, 190–194.

Bar-Joseph, M., Garnsey, S.M., Gonsalves, D. and Purcifull, D.E. (1980) Detection of citrus tristeza virus. I. Enzyme-linked immunosorbent assay (ELISA) and SDS-immunodiffusion methods. In: *Proceedings of the Eighth Conference of the International Organization of Citrus Virologists*. IOCV, Riverside, California, pp. 1–8.

Barkley, N.A., Roose, M.L. and Krueger, R.R. (2003) Assessing genetic diversity in citrus by utilizing molecular markers. In: *Proceedings, International Society of Citriculture IX Congress*. Orlando, Florida, December 2000, pp. 126–127.

Barrett, H.C. and Rhodes, A.M. (1976) A numerical taxonomic study of the affinity relationships in cultivated Citrus and its close relatives. *Systematic Botany* 1, 105–136.

Bash, J. (1999) The California Citrus Clonal Protection Program. In: Aubert, B (ed.) *Proceedings of the Fifth International Congress of the International Society of Citrus Nurserymen*. Montpellier, France, pp. 323–327.

Beavers, J.B. and Hutchison, D.J. (1985) Evaluation of selected *Citrus* spp. and relatives for susceptibility to root injury by *Diaprepes abbreviatus* larvae (Coleoptera: Cuculionidae). *Florida Entomologist* 68, 222–223.

Bender, G.S. and Menge, J.A. (1986) Evaluation of tolerance of common citrus rootstocks to dry root rot caused by *Fusarium solani*. In: *Proceedings of the Second World Congress of the International Society of Citrus Nurserymen* pp. 164–172.

Benevente-García, O., Castillo, J., Marín, F.R., Ortuño, A. and Del Río, J.A. (1997) Uses and properties of *Citrus* flavonoids. *Journal of Agriculture and Food Chemistry* 34, 4505–4515.

Beretta, M.J.G., Pompeu, J., Jr, Derrick, K.S., Lee, R.F., Barthe, G.A. and Hewitt, B.G. (1992) Evaluation of rootstocks in Brazil for resistance to declinio. In: *Proceedings of the International Society of Citriculture* pp. 841–843.

Berhow, M.A., Hasegawa, S., Kwan, K. and Bennett, R.D. (2000) Limonoids and the chemotaxonomy of Citrus and the Rutaceae family. In: Berhow, M.A., Hasegawa, S. and Manners, G.D. (eds) *Citrus Limonoids: Functional Chemicals in Agriculture and Food*. American Chemical Society, Washington DC, pp. 212–228.

Bernstein, L. (1968) Salinity factors and their limits for citrus culture. *Proceedings of the 1st International Citrus Symposium* 3, 1779–1782.

Bettencourt, E., Hazekamp, T. and Perry, M.C. (1992) *Directory of Germplasm Collections. 6.1. Tropical and Subtropical Fruits and Tree Nuts*. International Board for Plant Genetic Resources, Rome.

Bevington, K.B. and Castle, W.S. (1982) Development of the root system of young 'Valencia' orange trees on

rough lemon and Carrizo citrange rootstocks. *Proceedings of the Florida State Horticultural Society* 95, 33–37.
Bhattacharya, S.C. and Dutta, S. (1956) *Classification of Citrus Fruits of Assam.* Monograph No. 20. Indian Council for Agricultural Research, Delhi, 110 pp.
Bhumannavar, B.S., Singh, S.P. and Sulladmath, V.V. (1988) Evaluation of citrus germplasm resistance to the oriental red mite, *Eutetranychus orientalis* (Klein) under tropical humid South Indian conditions. *Topical Pest Management* 34, 193–198.
Bhumannavar, B.S., Singh, S.P. and Sulladmath, V.V. (1989) Field evaluation of citrus germplasm for resistance to the black aphid, *Toxoptera aurantii* (Boy.) under tropical humid South Indian conditions. *Insect Science and its Applications* 10, 81–88.
Bitters, W.P. (1950) Rootstocks with dwarfing effect. *California Agriculture* 4 (2), 5–14.
Bitters, W.P. (1959) Rootstocks in relation to control of tristeza. In: Wallace, J.M. (ed.) *Citrus Virus Diseases.* University of California, Division of Agricultural Sciences, Berkeley, California, pp. 203–207.
Bitters, W.P. (1972) Reaction of some new citrus hybrids and introductions as rootstocks to inoculations with tristeza virus in California. In: *Proceedings of the Fifth Conference of the International Organization of Citrus Virologists* pp. 112–120.
Bitters, W.P. and Batchelor, L.D. (1951) Effect of rootstocks on the size of orange fruits. *Proceedings of the American Society for Horticultural Science* 57, 133–141.
Bitters, W.P., McCarty, C.D. and Cole, D.A. (1973a) Evaluation of trifoliate orange selections as rootstocks for Washington navel and Valencia orange. In: *I Congreso Mundial de Citricultura.* Murcia-Valencia, Spain, pp. 127–131.
Bitters, W.P., Cole, D.A. and McCarty, C.D. (1973b) Citrus rootstocks from the papeda group. *Citrograph* 58, 419–420, 438.
Bitters, W.P., Cole, D.A. and McCarty, C.D. (1977) Citrus relatives are not irrelevant as dwarfing stocks or interstocks for citrus. In: *Proceedings of the Second World Congress of the International Society of Citriculture* 2, 561–567.
Bonavia, E. (1890) *The Cultivated Oranges and Lemons etc. of India and Ceylon.* 2 volumes. A.W. Allen and Co., London.
Bové, J.M., Navarro, L., Bonnet, P., Zreik, L. and Garnier, M. (1996) Reaction of citrus cultivars to graft-inoculation of *Phytoplasma aurantifolia*-infected lime shoots. In: *Proceedings of the 13th Conference of the International Organization of Citrus Virologists.* Riverside, California, pp. 249–251.
Bowman, K.D. (1994) 'Cipo' sweet orange and its unique growth habit. *Fruit Varieties Journal* 48, 230–234.
Bowman, K.D., Shapiro, J.P. and Lapointe, S.L. (2001) Sources of resistance to *Diaprepes* weevil in subfamily Aurantioideae, Rutaceae. *HortScience* 36, 332–336.
Bracke, M.E., Bruyneel, E.R., Vermuelen, S.J., Vennekens, K., Van Mark, V. and Mareel, M.M. (1994) Citrus flavonoid effect on tumor invasion and metastasis. *Food Technology* 48, 121–124.
Bretó, M.P., Ruiz, C., Pina, J.A. and Asíns, M.J. (2001) The diversification of *Citrus clementina* Hort. ex Tan., a vegetatively propagated crop species. *Molecular Phylogenetics and Evolution* 21, 285–293.
Bretting, P.K. and Widrlechner, M.P. (1995) Genetic markers and plant genetic resource management. *Plant Breeding Reviews* 13, 11–86.
Brizicky, G.K. (1962a) The genera of the Rutaceae in the southeastern United States. *Journal of the Arnold Arboretum* 43, 1–22.
Brizicky, G.K. (1962b) Taxonomic and nomenclatural notes on Xanthoxylum and Glycosmis (Rutaceae). *Journal of the Arnold Arboretum* 43, 80–93.
Broadbent, P. (1999) Phytosanitary practices to safely collect and exchange citrus germplasm. In: Broadbent, P., Sykes, S.R., Bevington, K.B. and Hailstones, D. (eds) *Proceedings of the Citrus Germplasm Conservation Workshop.* Brisbane, Australia, October 1997, pp. 33–39.
Broadbent, P., Sykes, S.R., Bevington, K.B. and Hailstones, D. (eds) (1999) *Proceedings of the Citrus Germplasm Conservation Workshop.* Brisbane, Australia, October 1997.
Broadbent, P., Sykes, S.R., Bevington K.B., Zhusheng, C., Hailstones, D. and Lewis, W.J., (2003) Variance among *Poncirus* accessions from the People's Republic of China. In: *Proceedings of the International Society of Citriculture IX Congress* pp. 209–210.
Brooks, H.J. and Barton, D.W. (1977) A plan for national fruit and nut germplasm repositories. *HortScience* 12, 298–300.

Brophy, J.J., Goldsack, R.J. and Forster, P.I. (2001) The leaf oils of the Australian species of *Citrus* (Rutaceae). *Journal of Essential Oil Research* 13, 264–268.

Brown, A.H.D., Frankel, O.H., Marshall, D.R. and Williams, J.T. (eds) (1989) *The Use of Plant Genetic Resources*. Cambridge University Press, Cambridge.

Brown, W.H. (1954) *Useful Plants of the Philippines*. Republic of the Philippines, Department of Agriculture and Natural Resources, Technical Bulletin 10, Manila.

Burger, D.W. (1985) Pollination effects of fruit production of 'Star Ruby' grapefruit (*Citrus paradisi* Macf.). *Scientia Horticulturae* 25, 71–76.

Burkill, I.H. (1935) *A dictionary of the Economic Products of the Malay Peninsula*. Governments of the Straits Settlements and Federated Malay States, London.

Busch, L., Lacy, W.B., Burkhardt, J., Hemken, D., Moraga-Rojel, J., Koponen, T. and de Souza Silva, J. (1995) *Making Nature, Shaping Culture: Plant Biodiversity in Global Context*. University of Nebraska Press, Lincoln, Nebraska.

Button, J. and Bornman, C.H. (1971) Development of nucellar plants from unpollinated and unfertilized ovules of the Washington navel orange *in vitro*. *South African Journal of Botany* 37, 127–134.

Ca, D.D. (1999) Present situation of germplasm resources in Vietnam. In: Broadbent, P., Sykes, S.R., Bevington, K.B. and Hailstones, D. (eds) *Proceedings of the Citrus Germplasm Conservation Workshop*. Brisbane, Australia, pp. 146–158.

Calavan, E.C. and Christiansen, D.W. (1965) Variability of cachexia reactions among varieties of rootstocks and within clonal propagations of citrus. In: *Proceedings of the Third Conference of the International Organization of Citrus Virologists*. University of Florida Press, Gainesville, Florida, pp. 76–85.

Calavan, E.C., Mather, S.M. and McEachern, E.H. (1978) Registration, certification, and indexing of citrus trees. In: Reuther, W., Calavan, E.C. and Carman, G.E. (eds) *The Citrus Industry. Vol. IV. Crop Protection*, revised edn. University of California Division of Agricultural Sciences, Berkeley, California, pp. 185–222.

Calavan, E.C., Roistacher, C.N. and Nauer, E.M. (1972) Thermotherapy of citrus for inactivation of certain viruses. *Plant Disease Reporter* 56, 976-980.

Calomme, M.L., Peters, A., Vlietninck, D. and Vanden Berghe, D. (1996) Inhibition of bacterial mutagenesis by *Citrus* flavonoids. *Planta Medica* 62, 222–226.

Cambra, M., Gorris, M.T., Marroquín, C., Román, M.P., Olmos, A., Martínez, M.C., Hermoso de Mendoza, A., López, A. and Navarro, L. (2000) Incidence and epidemiology of Citrus tristeza virus in the Valencian Community of Spain. *Virus Research* 71, 85–95.

Cameron, J.W. (1974) Present status and future outlook for germplasm reserves in citrus. *Fruit Varieties Journal* 28, 86–89.

Cameron, J.W. and Frost, H.B. (1968) Genetics, breeding and nucellar embryony. In: Reuther, W., Batchelor, L.D., Webber, H.J. (eds) *The Citrus Industry. Vol. II. Anatomy, Physiology, Genetics, and Reproduction*. Division of Agricultural Sciences. University of California, Berkeley, California, pp. 325–381.

Cameron, J.W., Baines, R.C. and Soost, R.K. (1969a) Development of rootstocks resistant to the citrus nematode, by breeding and selection. *Proceedings of the 1st International Citrus Symposium* 2, 949–954.

Cameron, J.W., Carman, G.E. and Soost, R.K. (1969b) Differential resistance of *Citrus* species hybrids to infestation by the California red scale, *Aonidiella aurantii* (Mask.). *Journal of the American Society for Horticultural Science* 94, 694–696.

Cameron, J.W., Klotz, L.J., DeWolfe, T.A. and Soost, R.K. (1972) Estimates of the resistance of *Citrus* × *Poncirus* hybrids to feeder root infection by *Phytophthora* spp., by a greenhouse seedling test. *Plant Disease Reporter* 56, 927–931.

Campbell, C.W. (1974) The Wampee, a fruit well adapted to southern Florida. *Proceedings of the Florida State Horticultural Society* 87, 390–392.

Campbell, C.W. and Goldweber, S. (1979) Rootstocks for *Citrus* in the limestone soils of southern Florida. *Proceedings of the Florida State Horticultural Society* 92, 290–291.

Caro, M., Cerdá, A., Fernández, F.G. and Guillén, M.G. (1973) Tolerancia a la salinidad de portainjertos cítricos durante la germinación. *I Congreso Mundial de la Citricultura* 1, 313–316.

Carpenter, J.B. and Furr, J.R. (1962) Evaluation of tolerance to root rot caused by *Phytophthora parasitica* in seedlings of citrus and related genera. *Phytopathology* 52, 1277–1285.

Carpenter, J.B. and Furr, J.R. (1967) Susceptibility of *Citrus macrophylla* 'Alemow' to cachexia virus. *Plant Disease Reporter* 51, 925–927.

Carpenter, J.B. and Reece, P.C. (1969) *Catalog of Genera, Species, and Subordinate Taxa in the Orange*

Subfamily Aurantioideae (Rutaceae). ARS 34-106. United States Department of Agriculture, Agricultural Research Service, Washington DC.
Carpenter, J.B., Burns, R.M. and Furr, J.R. (1975) Phytophthora-tolerant rootstocks for lemons. *Plant Disease Reporter* 59, 54–56.
Carpenter, J.B., Burns, R.M. and Sedlacek, R.F. (1981) Phytophthora resistant rootstocks for Lisbon lemons in California. *Citrograph* 66, 287–288, 291–292.
Carpenter, J.B., Burns, R.M. and Sedlacek, R.F. (1982) Performance of rootstocks inoculated with virus. *Citrograph* 67, 101–106.
Cassady, J.M., Ojima, N., Chang, C.J. and McLaughlin, J.L. (1979) An investigation of the anti-tumor activity of *Micromelum integerrimum* (Rutaceae). *Journal of Natural Products* 42, 274–278.
Castle, W.S. (1980) Citrus rootstocks for tree-size-control in Florida. *Proceedings of the Florida State Horticultural Society* 93, 24–27.
Castle, W.S. (1987) Citrus rootstocks. In: Rom, R.C. and Carlson, R.F. (eds) *Rootstocks for Fruit Crops*. John Wiley and Sons, New York, pp. 361–399.
Castle, W.S. and Krezdorn, A.H. (1975) Effect of citrus rootstocks on root distribution and leaf mineral content of 'Orlando' tangelo trees. *Journal of the American Society for Horticultural Science* 100, 1–4.
Castle, W.S. and Krezdorn, A.H. (1977) Soil water use and apparent root efficiencies of citrus trees on four rootstocks. *Journal of the American Society for Horticultural Science* 102, 403–406.
Castle, W.S. and Phillips, R.H. (1977) Potentially dwarfing rootstocks for Florida citrus. *Proceedings of the Second World Congress of the International Society of Citriculture* 2, 558–561.
Castle, W.S. and Youtsey, C.O. (1977) Root system characteristics of citrus nursery trees. *Proceedings of the Florida State Horticultural Society* 90, 39–44.
Castle, W.S., Kesinger, M. and Rouse, R.E. (2001) The current situation in the Florida citrus nursery industry. In: *Proceedings of the Sixth World Congress of the International Society of Citrus Nurserymen* pp. 17–25.
Chadha, K.L. (1995) Status report on tropical fruit species in South Asia. In: Arora, R.K. and Ramanatha Rao, V. (eds). *Expert Consultation on Tropical Fruit Species of Asia*. International Plant Genetic Resources Institute, Office for South Asia, New Delhi, pp. 45–60.
Chadha, K.L. and Singh, H.B. (1996) Status report of genetic resources of Citrus in India. In: *Status Reports on Genetic Resources of Citrus: East Asia, South Asia*. I.P.G.R.I. Project No. B06. International Plant Genetic Resources Institute Regional Office for Asia, the Pacific, and Oceania, Singapore, pp. 1–88.
Chapman, W.B. and Brown, J.R. (1993) An illusion of cold-hardiness of *Eremocitrus*. *Fruit Gardener* 25, 15.
Chapot, H. (1963) Le 'Mal secco'. *Al Awamia* 9, 89–125.
Chase, M.W., Morton, C.M. and Kallkuni, J.A. (1999) Phylogenetic relationships of Rutaceae: a cladistic analysis of the subfamilies using evidence from RBCL and ATPB sequence variation. *American Journal of Botany* 86, 1191–1199.
Chaudhary, U.L. (1999) Citrus germplasm conservation in Nepal. In: Broadbent, P., Sykes, S.R., Bevington, K.B. and Hailstones, D. (eds) *Proceedings of the Citrus Germplasm Conservation Workshop*. Brisbane, Australia, pp. 131–135.
Cheng, F.S. and Roose, M.L. (1995) Origin and inheritance of dwarfing by the *Citrus* rootstock *Poncirus trifoliata* 'Flying Dragon'. *Journal of the American Society for Horticultural Science* 120, 286–291.
Childs, J.F.L. (1951) Cachexia, a bud-transmitted disease and the manifestation of phloem symptoms in certain varieties of citrus, citrus relatives and hybrids. *Proceedings of the Florida State Horticultural Society* 64, 47–51.
Childs, J.F.L. and Johnson, R.E. (1966) Preliminary report of seed transmission of psorosis virus. *Plant Disease Reporter* 50, 81–83.
Chowdhury, D.N. and Das, D.P. (1979) Investigation on the toxicity of some carbazole derivatives and plant extracts. *Current Science* 48, 344–346.
Cohen, M. (1974) Effect of exocortis inoculation on performance of Marsh grapefruit trees on various rootstocks. In: *Proceedings of the Sixth Conference of the International Organization of Citrus Virologists* pp. 117–121.
Cooper, W.C. (1952) Influence of rootstock on injury and recovery of young citrus trees exposed to freezes of 1950–51 in the Rio Grande Valley. In: *Proceedings of the Rio Grande Valley Horticultural Society* 6, 16–24.

Cooper, W.C. (1961) Toxicity and accumulation of salts in citrus trees on various rootstocks in Texas. *Proceedings of the Florida State Horticultural Society* 74, 95–104.

Cooper, W.C. and Gorton, B.S. (1951) Salt tolerance of various citrus rootstocks. In: *Proceedings of the Fifth Annual Rio Grande Horticultural Institute*. Weslaco, Texas, pp. 46–52.

Cooper, W.C. and Gorton, B.S. (1952) Toxicity and accumulation of chloride salts in citrus on various rootstocks. *Proceedings of the American Society for Horticultural Science* 59, 143–146.

Cooper, W.C. and Peynado, A. (1954) Screening citrus rootstock seedlings for tolerance to calcareous soils. *Proceedings of the Eighth Annual Rio Grande Valley Horticultural Institute*. Weslaco, Texas, pp. 100–105.

Cooper, W.C. and Peynado, A. (1955) Boron accumulation in citrus as influenced by rootstock. In: *Proceedings of the Ninth Annual Rio Grande Valley Horticultural Institute*. Weslaco, Texas, pp. 86–94.

Cooper, W.C. and Olson, E.O. (1951) Influence of rootstock on chlorosis of young Red Blush grapefruit trees. *Proceedings of the American Society for Horticultural Science* 57, 125–132.

Cooper, W.C. and Shull, A.V. (1953) Salt tolerance of and accumulation of sodium and chloride ions in grapefruit on various rootstocks grown in a naturally saline soil. In: *Proceedings of the Seventh Annual Rio Grande Valley Horticultural Institute*. Weslaco, Texas, pp. 107–117.

Cooper, W.C., Gorton, B.S. and Olson, E.O. (1952) Ionic composition of citrus as influenced by rootstock and scion and concentrations of salts and boron in the substrate. *Plant Physiology* 31, 256–257.

Cooper, W.C., Peynado, A. and Shull, A.V. (1955) Boron accumulation as influenced by rootstock. *Proceedings of the Rio Grande Horticultural Society* 7, 107–117.

Cooper, W.C., Peynado, A. and Olson, E.O. (1958) Response of grapefruit on two rootstocks to calcium additions to high-sodium, boron-contaminated, and saline irrigation water. *Soil Science* 86, 180–189.

Corner, E.J.H. (1988) *Wayside Trees of Malaya*, Vol. 2, 3rd edn. Malayan Nature Society, Kuala Lumpur.

Coronel, R.E. (1995) Status report on fruit species germplasm conservation and utilization in South-east Asia. In: Arora, R.K. and Ramanatha Rao, V. (eds) *Expert Consultation on Tropical Fruit Species of Asia*. International Plant Genetic Resources Institute, Office for South Asia, New Delhi, pp. 85–102.

Cossman, K.F. (1939) Citrus roots: their anatomy, osmotic pressure and periodicity of growth. *Palestine Journal of Botany, Rehovot Series* 3, 3–104.

Costa, A.S., Grant, T.J. and Moreira, S. (1949) Investigações söbre a tristeza dos *Citrus*. II. Conceitos e dados sôbre a reação das plantas cítricas à tristeza. *Bragantia* 9, 59–80.

Crescimanno, F.G., Somma, V. and Calabrese, F. (1973) Preliminary research on resistance of some rootstocks to Mal secco. In: *Proceedings, I Congreso Mundial de Citricultura*. Murcia-Valencia, Spain, pp. 119–120.

Dagar, H.S. and Dagar, J.C. (1991) Plant folk medicines among the Nicobarese of Katchal Island, India. *Economic Botany* 45, 114–119.

Dashen, L. and Fangcong, W. (1989) Distribution and varieties of kumquats in China. In: *International Symposium on Horticultural Germplasm, Cultivated and Wild. Part 1. Fruit Trees*. Beijing, pp. 279–282.

Dass, H.. (1990) Citrus germplasm of India. In: Bangyan, H. and Qian, Y. (eds) *Proceedings of the International Citrus Symposium*, Guangzhou, PR China, pp. 65–69.

Davenport, S.B. and Rouse, R.E. (1992) Survey of young citrus in the Immokalee Budwood Foundation Grove for foot rot susceptibility. *Proceedings of the Florida State Horticultural Society* 105, 38–42.

Davis, R.M. (1984) Comparison of numbers of *Tylenchulus semipenetrans* Cobb on six citrus rootstocks in South Texas. *Journal of the Rio Grande Valley Horticultural Society* 37, 95–98.

De Cicco, V., Paradies, M. and Salerno, M. (1984) Behaviour of some lemon rootstocks towards Mal secco root infections: preliminary results. In: *Proceedings of the International Society for Citriculture* pp. 428–430.

Deng, X.X., Guo, W.W. and Zhang, W.C. (1997a) Citrus germplasm and its utilization. In: *Proceedings of the International Society of Citriculture* pp. 1217–1221.

Deng, X.X., Guo, W.W. and Zhang, W.C. (1997b) Citrus germplasm and its utilization. In: Albrigo, L.G. (ed.) *Identification and Conservation of Genetic Resources of Citrus and Its Relatives*. Proceedings of a Symposium held during the VIII Congress of the International Society of Citriculture, Sun City, South Africa, pp. 18–25.

Deng, Z., Huang, S., Ling, P., Yu, C., Tao, Q., Chen, C., Wendell, M.K., Zhang, H.B. and Gmitter, F.G., Jr

(2001) Fine genetic mapping and BAC contig development for the citrus tristeza virus resistance gene locus in *Poncirus trifoliate* (Raf.). *Molecular Genetics and Genomics* 265, 739–747.

Dhingra, O.D. and Sinclair, J.B. (1995) *Basic Plant Pathology Methods*, 2nd edn. CRC Press, Boca Raton, Florida.

Ding, S.Q., Zhang, X.N., Bao, Z.R. and Ling, M.Q. (1984) A new species of Poncirus from China. *Acta Botanica Yunnanica* 6, 292–293.

Ding, S.Q., Zhang, X.N., Bao, Z.R. and Liang, M.Q. (1990) Citrus ichangensis in Yunnan. In: Bangyan, H. and Qian, Y. (eds) *Proceedings of the International Citrus Symposium*. Guangzhou, PR China, pp. 170–173.

Donadio, L.C. and Banzato, D.A. (1988) Relationship of citrus cultivars and declinio. In: *Proceedings of the Tenth Conference of the International Organization of Citrus Virologists* pp. 384–387.

Dornelles, C.M.M. (1976) Stem pitting in citrus varieties and nucellar clones in Rio Grande do Sul, Brazil. In: *Proceedings of the Seventh Conference of the International Organization of Citrus Virologists* pp. 71–74.

Douglas, T.J. and Walker, R.R. (1983) 4-Desmethylsterol composition of citrus rootstocks of different salt exclusion capacity. *Physiologum Plantarum* 58, 69–74.

Dowell, R.V. (1989) Toxicity of water extracts of *Murraya paniculata* Jack leaves to immature citrus blackfly, *Aleurocanthus woglumi* Ashby (Homoptera: Aleyrodidae). *Pan-Pacific Entomologist* 65, 163–165.

Duan, Z.Y. (1990) Primary study of fuming evergreen trifoliate orange (*P. trifoliata* Raf.). In: *Proceedings of the International Citrus Symposium*. Guangzhou, PR China, pp. 166–169.

Duran-Vila, N. (1995) Cryoconservation of germplasm of citrus. *Biotechnology in Agriculture and Forestry* 32, 70–86.

Duran-Vila, N. (1997) Cryoconservation of citrus genetic resources. In: Razdan, M.K. and Cocking, E.C. (eds) *Conservation of Plant Genetic Resources* In Vitro. Science Publishers, Inc., Enfield, New Hampshire, pp. 175–199.

Duran-Vila, N., Pina, J.A., Mollins, M.I. and Navarro, L. (1991) A new indexing method for cachexia. In: *Proceedings of the 11th Conference of the International Organization of Citrus Virologists*. IOCV, Riverside, California, pp. 219–223.

Duran-Vila, N., Pina, J.A. and Navarro, L. (1993) Improved indexing of citrus viroids. In: *Proceedings of the 12th Conference of the International Organization of Citrus Virologists*. IOCV, Riverside, California, pp. 202–211.

Dutta, S. (1958) Origin and history of citrus fruits of Assam. *Indian Journal of Horticulture* 15, 146–153.

Eldridge, J. (1975) Bush medicine in the Exumas and Long Island, Bahamas: a field study. *Economic Botany* 29, 307–332.

El-Otmani, M. (1996) Tolerance of seedlings of six citrus rootstocks to high soluble calcium carbonate content. In: *Proceedings of the International Society of Citriculture* pp. 290–295.

El-Otmani, M., Coggins, C.W., Jr and Duymovic, A. (1990) Citrus cultivars and production in Morocco. *HortScience* 25, 1343–1346.

El-Shazly, S.M., Alcaraz, C.F. and Carpena, O. (1992) Effect of some citrus rootstocks on leaf mineral contents of young and adults lemons trees. *Proceedings of the International Society of Citriculture* pp. 275-279.

Embleton, T.W., Labanauskas, C.K. and Bitters, W.P. (1962) The influence of certain rootstocks on the concentration of boron, iron, manganese, and other elements in lemon leaves, and on boron toxicity symptoms. *Proceedings of the American Society for Horticultural Science* 80, 285–290.

Embleton, T.W., Reitz, H.J. and Jones, W.W. (1973a) Citrus fertilization. In: Reuther, W. (ed.) *The Citrus Industry. Vol. III. Production Technology*. University of California Division of Agricultural Science, Berkeley, California, pp. 122–182.

Embleton, T.W., Jones, W.W., Labanauskas, C.K. and Reuther, W. (1973b) Leaf analysis as a diagnostic tool and guide to fertilization. In: Reuther, W. (ed.) *The Citrus Industry. Vol. III. Production Technology*. University of California, Division of Agricultural Sciences, Berkeley, California, pp. 183–210, 447–495.

Engelman, F. (1997) *In vitro* conservation methods. In: Ford-Lloyd, B.V., Newburry, J.H. and Callow, J.A. (eds) *Biotechnology and Plant Genetic Resources: Conservation and Use*. CABI, Wallingford, UK, pp. 119–162.

Engelmann, F., Dambier, D. and Ollitrault, P. (1994) Cryopreservation of embryogenic cell suspensions and calluses of citrus using a simplified freezing process. *Cryo-Letters* 15, 53–58.

Engler, A. (1931) Rutaceae. In: Engler, A. and Prantl, E. (eds) *Die natürlichen Pflanzenfamilien*, 2nd edn., Vol. 19a. Engelman, Leipzig, pp. 187–359.

Fallahi, E. (1992) Tree canopy volume and leaf mineral concentrations of 'Redblush' grapefruit on twelve rootstocks. *Fruit Varieties Journal* 46, 44–48.

Fallahi, E. and Rodney, D.R. (1992) Tree size, yield, fruit quality, and leaf mineral nutrient concentration of 'Fairchild' mandarin on six rootstocks. *Journal of the American Society for Horticultural Science* 117, 28–31.

Fang, D.Q. (1993) Intra and intergeneric relationships of *Poncirus polyandra*: investigation by leaf isozymes. *Journal of Wuhan Botanical Research* 11, 34–39.

Fang, D.Q. and Roose, M.L. (1997) Identification of closely related citrus cultivars with inter-simple sequence repeat markers. *Theoretical and Applied Genetics* 95, 408–417.

Fang, D.Q. and Roose, M.L. (1999) A novel gene conferring citrus tristeza virus resistance in *Citrus maxima* (Burm.) Merrill. *HortScience* 34, 334–335.

Fang, D.Q., Roose, M.L., Krueger, R.R. and Federici, C.T. (1997) Fingerprinting trifoliate orange germplasm accessions with isozymes, RFLPs, and inter-simple sequence repeat makers. *Theoretical and Applied Genetics* 95, 211–219.

Fang, D.Q., Federici, C.T. and Roose, M.L. (1998a) A high-resolution linkage map of the citrus tristeza virus resistance gene region in *Poncirus trifoliate* (L.) Raf. *Genetics* 150, 883–890.

Fang, D.Q., Krueger, R.R. and Roose, M.L. (1998b) Phylogenetic relationships among selected *Citrus* germplasm accessions revealed by inter-simple sequence repeat (ISSR) markers. *Journal of the American Society for Horticultural Science* 123, 612–617.

Fantz, P.R. (1988) Nomenclature of the Meiwa and Changshou kumquats, intrageneric hybrids of *Fortunella*. *HortScience* 23, 249–250.

Feder, W.A. (1968) Differential susceptibility of selections of *Poncirus trifoliata* to attack by the citrus nematode, *Tylenchulus semipenetrans*. *Israel Journal of Agricultural Research* 18, 175–179.

Federici, C.T., Fang, D.Q., Scora, R.W. and Roose, M.L. (1998) Phylogenetic relationships within the genus *Citrus* (Rutaceae) and related genera as revealed by RFLP and RAPD analysis. *Theoretical and Applied Genetics* 96, 812–822.

Feldmesser, J. and Hannon, C.I. (1969) Susceptibility of two citrus rootstocks to *Pratylenchus* spp. *Plant Disease Reporter* 53, 603–607.

Ferguson, L., Van Gundy, S.D. and Roose, M.L. (1996) Assessment of citrus rootstocks for citrus nematode resistance. In: *Proceedings of the International Society of Citriculture* pp. 95–99.

Fleming, G.H., Olivares-Fuster, O., Fatta del-Bosco, S. and Grosser, J.W. (2000) An alternative method for the genetic transformation of sweet orange. *In Vitro Cellular and Developmental Biology-Plant* 36, 450–455

Food and Agriculture Organization of the United Nations. (1996) *The State of the World's Plant Genetic Resources for Food and Agriculture*. Food and Agriculture Organization of the United Nations, Rome. Also available online: <http://www.fao.org/WAICENT/OIS/PRESS_NE/PRESSENG/H8F.HTM>.

Ford, H.W. (1952) The distribution of feeder roots of orange and grapefruit trees on rough lemon rootstock. *Citrus Magazine* 14, 22–23.

Ford, H.W. (1954) The influence of rootstock and tree age on root distribution in citrus. *Proceedings of the American Society for Horticultural Science* 63, 137–142.

Ford, H.W. (1964) The effect of rootstock, soil type, and soil pH on citrus root growth in soils subject to flooding. *Proceedings of the Florida State Horticultural Society* 77, 41–45.

Ford, H.W. and Feder, W.A. (1958) Procedures used for rapid evaluation of citrus for resistance to certain endoparasitic nematodes. *Proceedings of the American Society for Horticultural Science* 71, 278–284.

Ford, H.W. and Feder, W.A. (1960) *Citropsis gilletiana*, a citrus relative resistant to the burrowing nematode in laboratory tests. *Proceedings of the Florida State Horticultural Society* 73, 60–64.

Ford, H.W. and Feder, W.A. (1962) Laboratory evaluation of certain citrus rootstock selections for resistance to burrowing nematode. *Journal of the Rio Grande Valley Horticultural Society* 16, 35–39.

Ford, H.W. and Feder, W.A. (1964) *Three Citrus Rootstocks Recommended for Trial in Spreading Decline Areas*. Circular S-151, University of Florida Agricultural Experiment Station, Gainesville, Florida.

Ford, H.W. and Feder, W.A. (1969) Development and use of citrus rootstocks resistant to the burrowing nematode, *Radopholus similis*. *Proceedings of the First International Citrus Symposium* 2, 941–948.

Ford, H.W. and Hutchins, P.C. (1967) Burrowing nematodes may survive on variants of 'Milam' lemon. *HortScience* 2, 56.

Forster, P.I. (2000) *Clausena smyrelliana* (Rutaceae: Aurantioideae), a new and critically endangered species from south-east Queensland. *Austrobaileya* 5, 715–720.

Forsyth, J.B. (1988) Citrus cultivars in Australia. *HortScience* 23, 684–686.

Forsyth, J.B. (1990) The NSW Citrus Crop Improvement Program. In: *Proceedings of the Third World Congress of the International Society of Citrus Nurserymen* pp. 71–75.

Foster, J.A. and Hadidi, A. (1998) Exclusion of plant viruses. In: Hadidi, A., Khetarpal, R.K. and Koganezawa, H. (eds) *Plant Virus Disease Control.* APS Press, St Paul, Minnesota, pp. 301–319.

Fowler, C. and Mooney, P. (1990) *Shattering: Food, Politics, and the Loss of Genetic Diversity.* University of Arizona Press, Tucson, Arizona.

Fraser, L.R. and Singh, D. (1969) Reactions of Indian citrus varieties to greening. *Proceedings of the First International Citrus Symposium* 1, 365–366.

Frison, E.A. and Diekmann, M. (1998) IPGRI's role in controlling virus diseases in plant germplasm. In: Hadidi, A., Khetarpal, R.K. and Koganezawa, H. (eds) *Plant Virus Disease Control.* APS Press, St Paul, Minnesota, pp. 230–236.

Frison, E.A. and Taher, M.M (eds) (1991) *FAO/IBPGR Technical Guidelines for the Safe Movement of Citrus Germplasm.* Food and Agricultural Organization/International Board for Plant Genetic Resources, Rome.

Frost, H.B. and Soost, R.K. (1968) Seed reproduction: development of gametes and embryos. In: Reuther, W., Batchelor, L.D. and Webber, H.J. (ed.) *The Citrus Industry. II. Anatomy, Physiology, Genetics, and Reproduction.* University of California, Division of Agricultural Sciences, Berkeley, California, pp. 290–324.

Fulton, H.R. (1925) Relative susceptibility of citrus varieties to attack by *Gloeosporium limetticolum* (Clausen). *Journal of Agricultural Research* 30, 629–635.

Furr, J.R. and Armstrong, W.W., Jr (1959) Breeding citrus for cold hardiness. *Proceedings of the Florida State Horticultural Society* 72, 66–71.

Furr, J.R. and Carpenter, J.B. (1961) Program for breeding citrus rootstocks resistant to Phytophthora root rot. *Proceedings of the Florida State Horticultural Society* 74, 18–23.

Furr, J.R., Brown, R.T. and Olson, E.O. (1966) Relative cold tolerance of progenies of some citrus crosses. *Journal of the Rio Grande Valley Horticultural Society* 20, 109–112.

Galipienso, L., Navarro, L., Ballester-Olmos, J.F., Pina, J.A., Moreno, P. and Guerri, J. (2000) Host range and symptomatology of a graft-transmissible pathogen inducing bud union crease on trifoliate rootstocks. *Plant Pathology* 49, 308–314.

Galipienso, L., Vives M.C., Moreno, P., Milne, R.G., Navarro, L. and Guerri, J. (2001) Partial characterization of citrus leaf blotch virus, a new virus from Nagami kumquat. *Archives of Virology* 146, 357–368.

Ganeshan, S. and Sulladmath, V.V. (1983) Pollen storage studies on Citrus limon Burm.: varietal differences and influence of flower type. *Gartenbauwissenschaft* 48, 51–54.

Garces, A. M. (1999) Citrus germplasm conservation in the Philippines. In: Broadbent, P., Sykes, S.R., Bevington, K.B. and Hailstones, D. (eds) *Proceedings of the Citrus Germplasm Conservation Workshop.* Brisbane, Australia, October 1997, pp. 136–138.

García, M.L., Sánchez de la Torre, M.E., Dal Bó, E., Dejlouah, K., Rouag, N., Luisoni, E., Milne, R.G. and Grau, O. (1997) Detection of citrus psorosis-ringspot virus using RT-PCR and DAS-ELISA. *Plant Pathology* 48, 830–836.

Gardner, F.E. and Horanic, G.E. (1963) Cold tolerance and vigor of young citrus trees on various rootstocks. *Proceedings of the Florida State Horticultural Society* 76, 105–110.

Gardner, J.M., Konos, Y. and Chandler, J.L. (1986) Bioassay and host-selectivity of *Alternaria citri* toxins affecting rough lemon and mandarins. *Physiological and Molecular Plant Pathology* 29, 293–304.

Garnier, M., Zreik, L. and Bové, J.M. (1991a) Witches' broom, a lethal mycoplasmal disease of lime trees in the Sultanate of Oman and the United Arab Emirates. *Plant Disease* 75, 546–551.

Garnsey, S.M. (1992) Tolerance of Gou tou rootstock to severe exotic isolates of citrus tristeza virus (CTV). *Proceedings of the Florida State Horticultural Society* 105, 35–38.

Garnsey, S.M. (1999) Systemic diseases. In: Timmer, L.W. and Duncan, L.W. (eds) *Citrus Health Management.* American Phytopathological Society Press, St Paul, Minnesota, pp. 95–106.

Garnsey, S.M., Gonsalves, D. and Purcifull, D.E. (1979) Rapid diagnosis of citrus tristeza virus infections by sodium dodecyl sulfate SDS-immunodiffusion procedures. *Phytopathology* 69, 88–95.

Garnsey, S.M., Bar-Joseph, M. and Lee, R.F. (1981) Applications of serological indexing to develop control strategies for citrus tristeza virus. *Proceedings of the International Society for Citriculture* 1, 448–452.

Georgiou, A. (2001) Citrus biodiversity in Cyprus. In: Albrigo, L.G. (ed.) *Proceedings of the Global Citrus*

Germplasm Network. Network meeting held during the Ninth Congress of the International Society of Citriculture. Orlando, Florida, December 2000, pp. 43–48.

Germaná, C. and Sardo, V. (1988) Correlations among some physiological and climatic parameters in orange trees. *Proceedings of the International Society of Citriculture* 1, 525–534.

Germplasm Resources Information Network (1995) *GRIN Manual.* USDA-ARS Germplasm Resources Laboratory, GRIN Database Management Unit, Beltsville, Maryland.

Ghazanfar, S.A. and Al-Sabahi, A.M.A. (1993) Medicinal plants of northern and central Oman (Arabia). *Economic Botany* 47, 89–98.

Ghosh, S.P. (1999) Indigenous citrus germplasm in India and conservation strategy. In: Broadbent, P., Sykes, S.R., Bevington, K.B. and Hailstones, D. (eds) *Proceedings of the Citrus Germplasm Conservation Workshop.* Brisbane, Australia, October 1997, pp. 96–105.

Given, D.R. (1994) *Principles and Practices of Plant Conservation.* Timber Press, Portland, Oregon.

Global Citrus Germplasm Network (1998) Conclusions of the International Technical Meeting on Globalization of the Work on Genetic Resource Conservation and Utilization of *Citrus* and *Citrus* Relatives. Acireale, Italy, December 1997. Also available online at: www.lal,ufl.edu/CONGRESS/Gcnrept.html.

Gmitter, F.G. (1995) Origin, evolution, and breeding of the grapefruit. *Plant Breeding Reviews* 13, 345–363.

Gmitter, F. and Hu, X. (1989) The role of Yunnan Province, China, in the origin of contemporary Citrus species (Rutaceae). In: Chinese Society of Horticultural Science (ed.) *International Symposium on Horticultural Germplasm.* Beijing, September 1988, pp. 146–153.

Gmitter, F.G. and Hu, X. (1990) The possible role of Yunnan, China, in the origin of contemporary Citrus species (Rutaceae). *Economic Botany* 44, 237–277.

Gmitter, F.G., Jr, Xiao, S.Y., Huang, S., Hu, X.L., Garnsey, S.M. and Deng, Z. (1996) A localized linkage map of the citrus tristeza virus resistance gene region. *Theoretical and Applied Genetics* 92, 688–695.

Gmitter, F.G., Krueger, R.R. and Roose, M.L. (1999) Citrus germplasm characterization by phenotype and molecular markers. In: Broadbent, P., Sykes, S.R., Bevington, K.B. and Hailstones, D. (eds) *Proceedings of the Citrus Germplasm Conservation Workshop.* Brisbane, Australia, October 1997, pp. 46–62.

Gondell, M.A. (1946) *La Susceptibilidad de Diferentes Especies y Variedades Citricas a la* Phytophthora citrophthora *(Sm. y Sm.) Leon.,* P. parasitica *Dastur y* P. megasperma *Leon. en la Zona de Concordia (Entre Ríos).* Republica Argentina, Ministerio de Agricultura de la Nación, Dirección General de Investigaciones, Instituto de Sanidad Vegetal, Publicaciones Ser. A, Buenos Aires.

González-Arnao, M.T., Engelmann, F., Urra, C., Morenza, M. and Rios, A. (1988) Cryopreservation of citrus apices using the encapsulation-dehydratation technique, *CryoLetters* 19, 177–182.

González-Arnao, M.T., Juárez, J., Ortega, C., Navarro L. and Duran-Vila, N. (2003) Cryopreservation of ovules and somatic embryos of citrus using the encapsulation-dehydration technique. *CryoLetters* 24, 85–94.

Goor, A. and Nurock, M. (1968) *The Fruits of the Holy Land.* Hebrew University Press, Jerusalem, pp. 152–181.

Gottwald, T.R., Graham, J.H., Civerolo, E.L., Barrett, H.C. and Hearn, C.J. (1993) Differential host range reaction of citrus and citrus relatives to citrus canker and citrus bacterial spot determined by leaf mesophyll susceptibility. *Plant Disease* 77, 1004–1009.

Grant, T.J. (1957) Effect of heat treatments on tristeza and psorosis viruses in citrus. *Plant Disease Reporter* 41, 232–234.

Grant, T.J. and Costa, A.S. (1948) A progress report on studies of tristeza disease of citrus in Brazil. I. Behavior of a number of citrus varieties as stocks for sweet orange and grapefruit, and as scions over sour orange rootstocks, when inoculated with the tristeza virus. *Proceedings of the Florida State Horticultural Society* 61, 20–33.

Grant, T.J., Costa, A.S. and Moreira, S. (1949) Studies of tristeza disease of citrus in Brazil. III. Further results on the behavior of citrus varieties as rootstocks, scions, and seedlings when inoculated with the tristeza virus. *Proceedings of the Florida State Horticultural Society* 62, 72–79.

Grant, T.J., Costa, A.S. and Moreira, S. (1951) Variations in stem pitting on tristeza-inoculated plants of different citrus groups. *Proceedings of the Florida State Horticultural Society* 64, 42–46.

Grimm, G.R. and Hutchison, D.J. (1977) Evaluation of *Citrus* spp., relatives and hybrids for resistance to *Phytophthora parasitica* Dastur. *Proceedings of the Second World Congress of the International Society of Citriculture* 2, 863–865.

Grosser, J.W. and Gmitter, F.G., Jr (1990) Protoplast fusion and citrus improvement. *Plant Breeding Reviews* 8, 339–374.

Grosser J.W., Ollitrault P. and Olivares-Fuster O. (2000) Somatic hybridization in citrus: an effective tool to facilitate variety improvement. *In Vitro Cellular and Developmental Biology-Plant* 36, 434–449.

Guarino, L., Ramanatha Rao, V. and Reid, R. (1995) *Collecting Plant Genetic Resources: Technical Guidelines*. CAB International, Wallingford, UK.

Guerra, M., Galvão Bezerra dos Santos, K., Barros e Silva, A.E. and Ehrendorfer, F. (2000) Heterochromatin banding patterns in Rutaceae–Aurantoideae – a case of parallel chromosomal evolution. *American Journal of Botany* 87, 735–747.

Guerri, J., Pina, J.A., Vives, M.C., Navarro, L. and Moreno, P. (2004) Seed transmission of Citrus leaf blotch virus: implications in quarantine and certification programs. *Plant Diseases* 88, 906.

Gulsen, O. and Roose, M.L. (2001) Lemons: diversity and relationships with selected *Citrus* genotypes as measured with nuclear genome markers. *Journal of the American Society for Horticultural Science* 126, 309–317.

Gumpf, D.J. (1999) Citrus quarantine, California. In: Kahn, R.P. and Mathur, S.B. (eds) *Containment Facilities and Safeguards for Exotic Plant Pathogens and Pests*. American Phytopathological Society Press, St Paul, Minnesota, pp. 151–156.

Gumpf, D.J., Semancik, J.S., Bash, J.A., Greer, G., Diaz, J., Serna, R. and Gonzáles, R. (1997) The California Citrus Clonal Protection Program. *Proceedings of the Congress of the International Society for Citriculture* 1, 445–447.

Gupta, S.K. and Banerjee, A.B. (1972) Screening of selected West Bengal plants for antifungal activity. *Economic Botany* 26, 255–259.

Haas, A.R.C. (1945a) Boron content of citrus trees grown on various rootstocks. *Soil Science* 59, 465–479.

Haas, A.R.C. (1945b) Influence of rootstock on the composition of citrus leaves and rootlets. *Soil Science* 60, 445–461.

Halma, F.F. (1929) Quantitative differences in palisade tissue in citrus leaves. *Botanical Gazette* 87, 319–324.

Hampton, R., Ball, E. and De Boer, S. (eds) (1990) *Serological Methods for Detection and Identification of Viral and Bacterial Plant Pathogens*. American Phytopathological Society Press, St Paul, Minnesota.

Hearn, C.J., Cooper, W.C., Register, R.O. and Young, R. (1963) Influence of variety and rootstock upon freeze injury to citrus trees in the 1962 Florida freeze. *Proceedings of the Florida State Horticultural Society* 76, 75–81.

Hearn, C.J., Reece, P.C. and Fenton, R. (1969) Effects of pollen source on fruit characteristics and set of four *Citrus* hybrids. *Proceedings of the First International Citrus Symposium* 1, 183–187.

Herrero, R., Asins, M.J., Carbonell, E.A. and Navarro, L. (1996a) Genetic diversity in the orange subfamily Aurantioideae. I. Intraspecies and intragenus genetic variability. *Theoretical and Applied Genetics* 92, 599–609.

Herrero, R., Asins, M.J., Pina, J.A., Carbonell, E.A. and Navarro, L. (1996b) Genetic diversity in the orange subfamily Aurantioideae. II. Genetic relationship among genera and species. *Theoretical and Applied Genetics* 93, 1327–1334.

Hilgeman, R.H., Rodney, D.R., Dunlap, J.A. and Hales, T.A. (1966) Rootstock evaluation for lemons on two soil types in Arizona. *Proceedings of the American Society for Horticultural Science* 88, 280–290.

Hirai, M., Mitsue, S., Kita, K. and Kajiura, I. (1990) A survey and isozyme analysis of wild mandarin, tachibana (*Citrus tachibana* (Mak.) Tanaka) growing in Japan. *Journal of the Japanese Society for Horticultural Science* 59, 1–7.

Hirano, E. (1931) Relative abundance of stomata in Citrus and some related genera. *Botanical Gazette* 92, 296–310.

Hodgson, R.W. (1961) Taxonomy and nomenclature in citrus. *Proceedings of the Second Conference of the International Organization of Citrus Virologists*. University of Florida Press, Gainesville, Florida, pp. 1–7.

Hodgson, R.W. (1967) Horticultural varieties of citrus. In: Reuther, W., Webber, H.J. and Batchelor, L.D. (eds) *The Citrus Industry. Vol. I. History, World Distribution, Botany, and Varieties*. University of California, Division of Agricultural Science, Berkeley, California, pp. 431–591.

Hodgson, R.W. and Eggers, E.R. (1938) Rootstock influence on the composition of citrus fruits. *California Citrograph* 23, 499–531.

Holden, J.H.W. and Williams, J.T. (1984) *Crop Genetic Resources: Conservation and Evaluation*. George Allen and Unwin, London.

Holden, J., Peacock, J. and Williams, T. (1993) *Genes, Crops, and the Environment*. Cambridge University Press, Cambridge.

Honda, K. (1990) Identification of host-plant chemicals stimulating oviposition by swallowtail butterfly, *Papilio protenor*. *Journal of Chemical Ecology* 16, 325–337.

Hooker, J.D. (1875) *Flora of British India.* Reeve and Co., London. (Rutaceae, 1, 484–517).
Hor, Y.L., Chan, Y.K. and Ramanatha Rao. V. (1999) Sustainable conservation and use of crop genetic resources for food security and agribusiness in Malaysia. *Plant Genetic Resources Newsletter* 117, 21–26.
Howard, F.W. (1979a) Comparación de seis especies de *Citrus* como plantas hospederas de *Aleurocanthus woglumi* Ashby. *Folia Entomológica Mexicana* 41, 33–40.
Howard, F.W. (1979b) Studies of the host plant suitability of *Ardisi solanacea* and *Citrus jambhiri* for citrus blackfly and citrus whitefly. *Florida Entomologist* 62, 380–382.
Hu, X. (1989) Fruit germplasm resources in Yunnan China. In: Chinese Society of Horticultural Science (ed.) *International Symposium on Horticultural Germplasm.* Beijing, PR China, pp. 208–214.
Huang C.C. (1959) Preliminary study on Chinese Rutaceae (3). *Acta Phytotaxonomica Sinica* 8, 69–124 plus plates.
Huang C.C. (1978) Materials of Chinese Rutaceae. *Acta Phytotaxonomica Sinica* 16, 81–85.
Huang, C.C. (1987) New taxa and notes on *Glycosmis* (Corr.) (Materials for Chinese Rutaceae IV). *Guihaia* 7, 115–124.
Hume, H.H. (1902) *The Kumquats.* Florida Agricultural Experiment Station, Bulletin No. 65, DeLand, Florida, pp. 551–566.
Hung, T.H., Wu, M.L. and Su, H.J. (2001) Identification of the Chinese box orange (*Severinia buxifolia*) as an alternative host of the bacterium causing citrus Huanglongbing. *European Journal of Plant Pathology* 107, 183–189.
Hutchison, D.J. and Grimm, G.R. (1972) Variation in *Phytophthora* resistance of Florida rough lemon and sour orange clones. *Proceedings of the Florida State Horticultural Society* 85, 38–39.
Hutchison, D.J. and O'Bannon, J.H. (1972) Evaluating the reaction of citrus selections to *Tylenchulus semipenetrans. Plant Disease Reporter* 56, 747–751.
Hutton, R.J., Broadbent, P. and Bevington, K.B. (2000) Viroid dwarfing for high density citrus plantings. *Horticultural Reviews* 24, 277–317.
Ieki, H. (1981) Resistance of citrus to scab. *Proceedings of the International Society of Citriculture* pp. 340–344.
Ingram, J. and Moore, H.E., Jr (1975) Rutaceae. *Baileya* 19, 169–171.
Inoue, K., Iguchi, I., Hara, S. and Tajeda, Y. (1985) Resistance of nucellar seedlings and seedlings of citrus crosses to scab (*Elsinoe fawcettii* Bitanc. & Jenkins). *Bulletin of the Shizuoka Prefectural Citrus Experiment Station* 21, 27–33.
International Board for Plant Genetic Resources (1982) *Genetic Resources of Citrus.* International Board for Plant Genetic Resources, Rome
International Board for Plant Genetic Resources (1988) *Descriptors for Citrus.* International Board for Plant Genetic Resources, Rome.
International Organization of Citrus Virologists (1968) *Indexing Procedures for 15 Virus Diseases of Citrus Trees.* Agriculture Handbook No. 333. USDA-ARS, Washington DC.
International Organization of Citrus Virologists (n.d.) *A Supplement to the Handbook 'Indexing Procedures for 15 Virus Diseases of Citrus Trees'.* Photocopied.
International Plant Genetic Resources Institute (1999) *Descriptors for Citrus.* International Plant Genetic Resources Institute, Rome.
Intrigliolo, F. and Starrantino, A. (1988) Nutritional features of 16 clones of lemons. In: *Proceedings of the International Society of Citriculture* pp. 673–679.
Ippoloti, A., Nigro, F. and Lima, G. (1997) Influence of the scion on the susceptibility of sour orange rootstock to *Phytophthora gummosis* and root rot. *Phytopathologia Mediterranea* 36, 81–86.
Iwamasa, M. (1988) Citrus cultivars in Japan. *HortScience* 23, 687–690.
Iwanami, T., Omura, M. and Ieki, H. (1993) Susceptibility of several citrus relatives to satsuma dwarf virus. In: *Proceedings of the 12th Conference of the International Organization of Citrus Virologists.* IOCV, Riverside, California, pp. 352–356.
Iwasa, S. and Shiraishi, S. (1957) Studies on the trifoliate orange. *Studies from the Institute of Horticulture, Kyoto (Japan)* 8, 59–63.
Iwasaki, N. and Oogaki, C. (1985) Photosynthetic characteristics of some citrus species under various temperatures and light conditions. *Journal of the Japanese Horticultural Society* 54, 315–322.
Iwasaki, T. (1943) On the big leaf and small leaf strain of trifoliate orange (*Poncirus trifoliata*). *Journal of the Horticultural Association of Japan* 14, 302–305.
Iwasaki, T. and Nishiura, M. (1963) On the tetraploid, large leaf, and small leaf stains of trifoliate orange. II.

Growth of seedlings and their practical value as citrus rootstocks. *Bulletin of the Horticultural Research Station, Okitsu (Japan), Series B* 2, 15–24.
Iyer, U.M. and Mani, U.V. (1990) Studies on the effect of curry leaves supplementation *(Murraya koenigii)* on lipid profile, glycated proteins and amino acids in non-insulin-dependent diabetic patients. *Plant Foods for Human Nutrition* 40, 275–282.
Jacobson, M. (1989) Botanical insecticides. *Symposium Series of the American Chemical Society* 387, 1–10.
Jain, S.K. (1965) Medicinal plant lore of the tribals of Bastar. *Economic Botany* 19, 236–250.
Jain, S.K. and Borthakur, S.K. (1980) Ethnobotany of the Mikars of India. *Economic Botany* 34, 264–272.
Jones, D.T. (1982) Lelang: limau of the mangrove. *Nature Malaysiana* 7 (3), 26–29.
Jones, D.T. (1985) The Aurantioideae of South-east Asia with special reference to germplasm collection in Malaysia and Singapore. *International Board for Plant Genetic Resources Regional Committee for South-east Asia Newsletter* 9 (2), 6–10.
Jones, D.T. (1987) Rare plant profile no. 1: *Merrillia caloxylon. Botanic Gardens Conservation News* 1 (1), 38–42.
Jones, D.T. (1989) Collection, utilization, and conservation of citrus genetic resources in Malaysia. In: Zakri, A.H. (ed.) *Genetic Resources of Under-utilised Plants in Malaysia.* Malaysian National Committee on Plant Genetic Resources, Kuala Lumpur, pp. 141–160.
Jones, D.T. (1990) A background for the utilization of citrus genetic resources in South-east Asia, I: classification of the Aurantioideae. *Proceedings of the Fourth International Asia Pacific Conference on Citrus Rehabilitation.* Chiang Mai, Thailand, pp. 31–37.
Jones, D.T. (1991) A background for the utilization of citrus genetic resources in South-east Asia. II. Germplasm collection, documentation and research. In: *Proceedings of the International Citrus Symposium.* Guangzhou, PR China, pp. 61–64.
Jones, D.T. (1995) Rutaceae. In: Soepadmo, E. and Wong, K.M. (eds) *Tree Flora of Sabah and Sarawak,* Vol. 1. Sabah Forestry Department/Forest Research Institute Malaysia/Sarawak Forestry Department, Sabah, Malaysia, pp. 351–419.
Jones, D.T. and Ghani, F.D. (1987) Citrus research and germplasm collection in Malaysia. *Malaysian Applied Biology* 16, 139–144.
Joseph, S. and Peter, K.V. (1985) Curry leaf (*Murraya koenigii*), perennial, nutritious, leafy vegetable. *Economic Botany* 39, 68-73.
Kahn, R.P. (1999a) Regulatory concepts. In: Kahn, R.P. and Mathur, S.B. (eds) *Containment Facilities and Safeguards for Exotic Plant Pathogens and Pests.* American Phytopathological Society Press, St Paul, Minnesota, pp. 3–7.
Kahn, R.P. (1999b) Biological concepts. In: Kahn, R.P. and Mathur, S.B. (eds) *Containment Facilities and Safeguards for Exotic Plant Pathogens and Pests.* American Phytopathological Society Press, St Paul, Minnesota, pp. 8–16.
Kahn, R.P., Waterworth, H.E., Gillaspie, A.G., Jr, Foster, J.A., Goheen, A.C., Monroe, R.L., Povich, W.L., Mock, R.G., Luhn, C.F., Calavan, E.C. and Roistacher, C.N. (1979) Detection of viruses or viruslike agents in vegetatively propagated plant importations under quarantine in the United States, 1968–1978. *Plant Disease Reporter* 63, 775–779.
Kalt, W. (2001) Health functional phytochemicals of fruit. *Horticultural Reviews* 27, 269–315.
Kandaswami, C., Perkins, E., Solonick, D.S., Drzewiccki, G. and Middleton, E., Jr (1991) Anti-proliferative effects of citrus flavonoids on a human squamous cell carcinoma. *Cancer Letters* 56, 390–393.
Kaplan, D.T. (1981) Characterization of citrus rootstock responses *to Tylenchulus semipenetrans* (Cobb). *Journal of Nematology* 13, 492–498.
Kaplan, D.T. and O'Bannon, J.H. (1981) Evaluation and nature of nematode resistance in Swingle citrumelo. *Proceedings of the Florida State Horticultural Society* 94, 33–36.
Kefford, J.F. and Chandler, B.V. (1961) The influence of rootstocks on the composition of oranges, with special reference to bitter principles. *Australian Journal of Agricultural Research* 12, 56–68.
Khairi, M.M. and Hall, A.E. (1976) Comparative studies of net photosynthesis and transpiration of some citrus species and relatives. *Physiologum Plantarum* 36, 35–39.
Khan, I.A. and Roose, M.L. (1988) Frequency and characteristics of nucellar and zygotic seedlings in three cultivars of trifoliate orange. *Journal of the American Society for Horticultural Science* 113, 105–110.
Kiang, C.K. and Zu-Zhao, J. (1984) Types of *Citrus ichangensis* in south-west China. *HortScience* 19, 539–540.
King, M.W., Soetisna, U. and Roberts, E.H. (1981) The dry storage of *Citrus* seeds. *Annals of Botany* 48,

865–872.

Kirkpatrick, J.D. and Bitters, W.P. (1969) Physiological and morphological response of various rootstocks to salinity. *Proceedings of the First International Citrus Symposium* 1, 391–399.

Klotz, L.J. and Fawcett, H.S. (1930) The relative resistance of varieties and species of citrus to pythiacystis and gummosis and other bark diseases. *Journal of Agricultural Research* 41, 415–425.

Klotz, L.J., Bitters, W.P., deWolfe, T.A. and Garber, M.J. (1967) Orchard tests of citrus rootstocks for resistance to Phytophthora. *Citrograph* 53, 38–55.

Klotz, L.J., DeWolfe, T.A. and Baines, R.C. (1968) Resistance of trifoliate orange stocks to gummosis. *California Citrograph* 54, 259–260.

Knorr, L.C. (1956) Suscepts, indicators, and filters of tristeza virus, and some differences between tristeza in Argentina and in Florida. *Phytopathology* 46, 557–560.

Knorr, L.C. (1977) Citrus. In: Hewitt, W.B. and Chiarappa, L. (eds) *Plant Health and Quarantine in International Transfer of Genetic Resources.* CRC Press Inc., Cleveland, pp. 111–118.

Ko, W.W. (1999) Conservation of citrus and citroid germplasm in Malaysia. In: Broadbent, P., Sykes, S.R., Bevington, K.B. and Hailstones, D. (eds) *Proceedings of the Citrus Germplasm Conservation Workshop.* Brisbane, Australia, October 1997, pp. 112–130.

Kobayashi, S., Sakai, A. and Oiyama, L. (1990) Cryopreservation in liquid nitrogen of cultured navel orange (*Citrus sinensis* Osb.) nucellar cells and subsequent plant regeneration. *Plant Cell Tissue and Organ Culture* 23, 15–20.

Kobayashi, S., Uchimiya, H. and Ikeda, I. (1983) Plant regeneration from Trovita orange protoplasts. *Japanese Journal of Breeding* 3, 119–122.

Kohomoto, K., Akimitsu, K. and Otani, H. (1991) Correlation of resistance and susceptibility of citrus to *Alternaria alternata* with sensitivity to host-specific toxins. *Phytopathology* 81, 719–722.

Koizumi, M. (1981) Resistance of citrus plants to bacterial canker disease: a review. In: *Proceedings of the International Society of Citriculture* pp. 402–405.

Kong, Y.C., Cheng, K.F., Ng, K.H., But, P.P.H., Li, Q., Yu, S.X., Chang, H.T., Cambie, R.C., Kinoshita, T., Kan, W.S. and Waterman, P.G. (1986) A chemotaxonomic division of *Murraya* based on the distribution of the alkaloids yuehchukene and girinimbine. *Biochemical Systematics and Ecology* 14, 491–497.

Kong, Y.-C., But, P.P.-H., Ng, K.-H., Li, Q., Cheng, K.-F. and Waterman, P.G. (1988) *Micromelum*: a key genus in the chemosystematics of the Clauseneae. *Biochemical Systematics Ecology* 16, 485–489.

Kriedemann, P.E. and Barrs, H.D. (1981) Citrus orchards. In: Kozlowski, T.T. (ed.) *Water Deficits and Plant Growth.* Vol. VI. Academic Press, New York, pp. 325–417.

Krueger, R.R. (1997a) An estimation of the annual cost to maintain a citrus germplasm repository. In: *Proceedings of the International Society of Citriculture* pp. 1235–1240.

Krueger, R.R. (1997b) An estimation of the annual cost to maintain a citrus germplasm repository. In: Albrigo, L.G. (ed.) *Identification and Conservation of Genetic Resources of Citrus and Its Relatives.* Proceedings of a Symposium held during the VIII Congress of the International Society of Citriculture, Sun City, South Africa, pp. 56–67.

Krueger, R.R. (1999) National clonal repository at Riverside, California, protects and provides unique citrus and date palm collections. *Diversity* 15 (3), 19–21.

Krueger, R.R. (2001) The California citrus certification program supports the industry. In: *Proceedings of the Sixth World Congress of the International Society of Citrus Nurserymen*, pp. 198–201.

Krueger, R.R. and Roose, M.L. (2003) Use of molecular markers in management of citrus germplasm resources *Journal of the American Society for Horticultural Science* 128, 827–837.

Krueger, R.R., Gulsen, O. and Roose, M.L. (2003) Use of molecular markers in management of citrus germplasm resources. In: *Proceedings of the International Society of Citriculture IX Congress.* Orlando, Florida, December 2000, p. 167.

Labanauskas, C.K. and Bitters, W.P. (1974) The influence of rootstocks and interstocks on the nutrient concentrations in 'Valencia' orange leaves. *Journal of the American Society for Horticultural Science* 99, 32–33.

Lal, S.D. and Lata, K. (1980) Plants used by the Bhat community for regulating fertility. *Economic Botany* 34, 273–275.

Lal, S.D. and Yadav, B.K. (1983) Folk medicines of Kirikshretra District (Haryana), India. *Economic Botany* 37, 299–305.

Lam, L.K.T., Zhang, J. and Hasegawa, S. (1994) Citrus limonoid reduction of chemically induced tumorigenesis. *Food Technology* 48, 104-109.

Laranjeira, F.F., Pompeu, J., Jr, García, A., Jr, Vieira, M., Harakava, R. and Beretta, M.J.G. (1998) Screening for tolerance to *Xylella fastidiosa*, the causal agent of citrus variegated chlorosis. *Fruits* 53, 345–349.

Laville, E. and Blondel, L. (1979) Comportement de differents clones de *Poncirus trifoliata* aux attaques racinaires a *Phytophthora* sp. *Fruits* 34, 175–178.

Le, H.T., Hancock, J., Trinh, T.T. and Ho, P.H. (1999) Germplasm resources in Vietnam: major horticultural and industrial crops. *HortScience* 34, 175–180.

Lee, A.T.C. and von Broembsen, L.A. (1990) The practical implementation of the South African Citrus Improvement Program. In: *Proceedings of the Third World Congress of the International Society of Citrus Nurserymen*, pp. 77–82.

Lee, H.A. (1918) Further data on the susceptibility of rutaceous plants to citrus-canker. *Journal of Agricultural Research* 15, 661–665.

Lee, H.A. (1925) The comparative resistance to foot rot of various citrus species as root stocks. *Philippine Journal of Science* 27, 243–256.

Lee, R.F., Lehman, P.S. and Navarro, L. (1999) Nursery practices and certification programs for budwood and rootstocks. In: Timmer, L.W. and Duncan, L.W. (eds) *Citrus Health Management.* American Phytopathological Society Press, St Paul, Minnesota, pp. 35–46.

Leite Jr, R.P. and Mohan, S.K. (1984) Evaluation of citrus cultivars for resistance to canker caused by *Xanthomonas campestris* pv. *citri* (Hasse) Dye in the state of Paraná, Brazil. In: *Proceedings of the International Society of Citriculture* pp. 385–389.

Levy, Y., Lifshitz, J., de Malach, Y. and David, Y. (1999) The response of several *Citrus* genotypes to high-salinity irrigation water. *HortScience* 34, 878–881.

Li, Q, Zhu, L.-F., But, P.P.-H., Kong, Y.-C., Chang, H.-T. and Waterman, P.G. (1988) Monoterpene and sesquiterpene rich oils from the leaves of *Murraya* species: chemotaxonomic significance. *Biochenical Systematics and Ecology* 16, 491–494.

Li, W., Donadio, L.C., Beretta, M.J.G., Sempionato, O.R. and Miranda, V.S. (1996a) Resistance of tolerance of citrus species and cultivars to citrus variegated chlorosis. In: *Proceedings of the International Society of Citriculture* pp. 216–218.

Li, W., Donadio, L.C., Sempionato, O.R., Rossetti, V. and Beretta, M.J.G (1996b) Clone selection of sweet orange 'Pera' (*C. sinensis* (L.) Osbeck) to citrus variegated chlorosis. In: *Proceedings of the International Society of Citriculture* pp. 280–282.

Li, W., Donadio, L.C., Beretta, M.J.G., Rossetti, V., Stuchi, E.S., Sempionato, O.R., Coutinho, A. and Miranda, V.S. (1996c) Seedling resistance of 20 citrus rootstocks to citrus variegated chlorosis. In: *Proceedings of the International Society of Citriculture* pp. 283–285.

Li, W., Donadio, L.C., Sempionato, O.R., Rossetti, V., Stuchi, E.S. and Beretta, M.J.G., (1996d) Effect of 20 rootstocks on the severity of citrus variegated chlorosis (cvc) of sweet orange 'Pera' (*C. sinensis* (L.) Osbeck). In: *Proceedings of the International Society of Citriculture* pp. 286–289.

Li, W.B., Ayres, A.J., He, C.X. and Donadio, L.C. (2000) Susceptibility of tangerines to citrus variegated chlorosis. *Acta Horticulturae* 353, 253–257.

Li, W., Pria, W., Jr, Lacava, P.M. and Hartung, J.S. (2003) Transmission of *Xylella fastidiosa* through seed to seedlings of *Citrus sinensis.* In: *Eighth International Congress of Plant Pathology.* New Zealand, March 2003 (abstract).

Lin, Z. and Myhre, D.L. (1991a) Differential response of citrus rootstocks to aluminum levels in nutrient solutions: I. Plant growth. *Journal of Plant Nutrition* 14, 1223–1238.

Lin, Z. and Myhre, D.L. (1991b) Differential response of citrus rootstocks to aluminum levels in nutrient solutions: II. Plant mineral concentrations. *Journal of Plant Nutrition* 14, 1239–1254.

Linnaeus, C. (1753) *Species Plantarum. Vol. II.* Holmiae.

Luro, F., Laigret, F., Bové, J.M. and Ollitrault, P. (1992) Application of random amplified polymorphic DNA (RAPID) to citrus genetics and taxonomy. *Proceedings of the International Society of Citriculture* 1, 225–228.

Luthria, D.L., Ramakrishnan, V., Verma, G.S., Prabhu, B.R. and Banerji, A. (1989) Insect antifeedants from *Atalantia racemosa. Journal of Agriculture Food and Chemistry* 37, 1435–1437.

Luxmore, R.J., Stolzy, L.H., Joseph, H. and DeWolfe, T.A. (1971) Gas space porosity of citrus roots. *HortScience* 6, 447–448.

Maas, E.V. (1993) Salinity and citriculture. *Tree Physiology* 12, 195–216.

Mabberley, D.J. (1997) A classification for edible *Citrus* (Rutaceae). *Telopea* 7, 167–172.

Mabberley, D.J. (1998) Australian Citrae with notes on other Aurantioideae (Rutaceae). *Telopea* 7, 333–344.

Mabesa, L.B. (1990) Calamansi or calamondin. In: Nagy, S., Shaw, P.F. and Wardowski, W.F. (eds) *Fruits of Tropical and Subtropical Origin: Composition, Properties, and Uses.* Florida Science Source Inc., Lake Alfred, Florida, pp. 348 372.

Machado, M.A. (1997) Citrus germplasm collections in Brazil: evaluation and characterization. In: Albrigo, L.G. (ed.) *Identification and Conservation of Genetic Resources of Citrus and its Relatives.* Proceedings of a Symposium held during the VIII Congress of the International Society of Citriculture. Sun City, South Africa, pp. 68–69.

Maggs, D.H. and Alexander, D.McE. (1969) The quantitative growth of young seedlings of the citrus rootstocks *Citrus sinensis* (L.) Osbeck, *Poncirus trifoliata* (L.) Raf. and their hybrid, *Carrizo citrange. Annals of Botany* 133, 981–990.

Manandhar, N.P. (1995) An inventory of some herbal drugs of Myagdi District, Nepal. *Economic Botany* 49, 371–379.

Manners, G.D. and Hasegawa, S. (1999) Squeezing more from citrus fruits. *Chemistry and Industry* 1999, 542–545.

Manthey, J.A., Grohmann, K., Montanari, A., Ash, K. and Manthey, C.L. (1999) Polymethoxylated flavones derived from *Citrus* suppress tumor necrosis factor-α expression by human monocytes. *Journal of Natural Products* 62, 441–444.

Marais, L.J. and Lee, R.F. (1991) Screening of rootstocks for citrus blight tolerance in South Africa – preliminary results. In: *Proceedings of the 11th Conference of the International Organization of Citrus Virologists* pp. 303–305.

Marais, L.J., Lee, R.F. and Breytenbach, J.H.J. (1993) Screening of rootstocks for blight tolerance in South Africa. In: *Proceedings of the IV World Congress of the International Society of Citrus Nurserymen* pp. 118–124.

Marín, M.L. and Duran-Vila, N. (1988) Survival of somatic embryos and recovery of plants of sweet orange (*Citrus sinensis* (L.) Osb.) after immersion in liquid nitrogen. *Plant Cell Tissue and Organ Culture* 14, 51–57.

Marín, M.L. and Duran-Vila, N. (1991) Conservation of citrus germplasm *in vitro. Journal of the American Society for Horticultural Sciences* 116, 740–746.

Marín, M.L., Gogorcena, Y., Ortiz, J. and Duran-Vila, N. (1993) Recovery of whole plants of sweet orange from somatic embryos subjected to freezing thawing treatments. *Plant Cell Tissue and Organ Culture* 34, 27–33.

Martín, S., Alioto, D., Milne, R.G., Guerri, J. and Moreno, P. (2002) Detection of citrus psorosis virus in field trees by direct tissue blot immunoassay in comparison with ELISA, symptomology, biological indexing and cross-protection tests. *Plant Pathology* 51, 134–141.

McCarty, C.D., Bitters, W.P. and Van Gundy, S.D. (1979) Susceptibility of 25 citrus rootstocks to the citrus nematode. *HortScience* 14, 54–55.

McClean, A.P.D. (1961) Transmission of tristeza virus to *Aeglopsis chevalieri* and *Afraegle paniculata. South African Journal of Agricultural Science* 4, 83–94.

McClean, A.P.D. and Schwarz, R.E. (1970) Greening or blotchy-mottle disease of citrus. *Phytophylactica* 2, 177–194.

McNeely, J.A., Gadgil, M., Levèque, C., Padoch, C. and Redford, K. (1995) Human influences on biodiversity. In: Heywood, V.H. (ed.) *Global Biodiversity Assessment.* Cambridge University Press, Cambridge, pp. 711–821.

Mears, D.R. and Kahn, R.P. (1999) New concepts and designs for quarantine greenhouses. In: Kahn, R.P. and Mathur, S.B. (eds) *Containment Facilities and Safeguards for Exotic Plant Pathogens and Pests.* American Phytopathological Society Press, St Paul, Minnesota, pp. 79–92.

Mehra, K.L. and Sastrapadja, S. (eds) (1985) *Proceedings of the International Symposium on South-east Asian Plant Genetic Resources.* International Board for Plant Genetic Resources, Jakarta.

Mendel, M.J., Alford, A.R., Rajab, M.S. and Bentley, M.D. (1991) Antifeedant effects of citrus limonoids differing in A-ring structure on Colorado potato beetle (Coleoptera: Chrysomelidae) larvae. *Journal of Economic Entomology* 884, 1158–1162.

Menge, J.A., Lembright, H. and Johnson, E.L.V. (1977) Utilization of mycorrhizal fungi in citrus nurseries. *Proceedings of the International Society of Citriculture* 1, 129-132.

Merrill, E.D. (1927) Swinglea, a new generic name in the Rutaceae. *Journal of the Arnold Arboretum* 8, 131–132.

Mestre, P.F., Asíns, M.J., Pina, J.A., Carbonell, E.A. and Navarro, L. (1997a) Molecular markers flanking citrus tristeza virus resistance gene from *Poncirus trifoliate* (L.) Raf. *Theoretical and Applied Genetics* 94, 458–464.

Mestre, P.F., Asíns, Carbonell, E.A. and Navarro, L. (1997b) New gene(s) involved in the resistance of

Poncirus trifoliate (L.) Raf. to citrus tristeza virus. *Theoretical and Applied Genetics* 95, 691–695.

Mestre, P.F., Asíns, M.J., Pina, J.A. and Navarro, L. (1997c) Efficient search for new resistant genotypes to the citrus tristeza closterovirus in the orange subfamily Aurantioideae. *Theoretical and Applied Genetics* 95, 1282–1288.

Middleton, E., Jr, and Kandaswami, C. (1994) Potential health-promoting properties of *Citrus* flavonoids. *Food Technology* 48, 115–119.

Miller, E.G., Gonzáles-Sanders, A.P., Couvillon, A.M., Wright, J.M., Hasegawa, S. and Lam, L.K.T. (1992) Inhibition of hamster buccal pouch carcinogenesis by limonin 17-β-D-glucopyranoside. *Nutrition and Cancer* 17, 1–7.

Miller, E.G., Gonzáles-Sanders, A.P., Couvillon, A.M., Binnie, W.H., Hasegawa, S. and Lam, L.K.T. (1994) *Citrus* limonoids as inhibitors of oral carcinogenesis. *Food Technology* 48, 110–114.

Minessy, F.A. and Higazy, M.K. (1957) Polyembryony in different citrus varieties in Egypt. *Alexandria Journal of Agricultural Research* 5, 89–96.

Mitra, R.L. and Subramanyam, K. (1969) *Glycosmis pentaphylla* (Rutaceae) and related Indian taxa. *Journal of the Arnold Arboretum* 50, 153–157.

Molino, J.F. (1993) History and botany of *Clausena anisum-olens* (Blanco) Merr. cv. 'Clausanis' (Rutaceae), a promising essential oil plant. *Acta Horticulturae* 331, 183–190.

Molino, J.F. (1994) Révision du genre *Clausena* Burm. f. (Rutaceae). *Bulletin du Museum National d'Histoire Naturellé, section B, Adansonia* 16, 105–153.

Moll, J.N. and van Vuuren, S.P. (1982) Observations on the occurrence and annual variations of greening in the citrus. *Subtropica* 3 (10), 18–20.

Montanari, A., Widmer, W. and Nagy, S. (1997) Health promoting phytochemicals in citrus fruit and juice products. *Recent Advances in Phytochemistry* 31, 31–52.

Moore, G. (2001) Oranges and lemons: clues to the taxonomy of *Citrus* from molecular markers. *Trends in Genetics* 17, 536–540.

Moreira, S., Gurgel, J.T.A. and de Arruda, L.F. (1947) Poliembrionia em *Citrus*. *Bragantia* 7, 69–106.

Morinaga, K. and Ikeda, F. (1990) The effects of rootstocks on photosynthesis, distribution of photosynthetic product, and growth of young Satsuma mandarin trees. *Journal of the Japanese Society for Horticultural Science* 59, 29–34.

Morton, J.F. (1984) The curryleaf tree (*Murraya koenigii* Spreng.) is attracting attention in Florida. *Proceedings of the Florida State Horticultural Society* 97, 314–317.

Mowder, J. and Stoner, A. (1989) Information systems. *Plant Breeding Reviews* 7, 57–66.

Mumford, P.M. and Grout, B.W.W. (1979) Desiccation and low temperature (–196°C) tolerance of *Citrus limon* seed. *Seed Science and Technology* 7, 407–410.

Müller, G.W. and Garnsey, S.M. (1984) Susceptibility of citrus varieties, species, citrus relatives, and non-rutaceous plants to slash-cut mechanical inoculation with citrus tristeza virus (CTV). In: *Proceedings of the Ninth Congress of the International Organization of Citrus Virologists* pp. 33–40.

Narayanaswamy, V. (1941) A revision of the Indo-Malaysian species of Glycosmis. *Record of the Botanical Survey of India* 14(2), 1–72.

National Research Council (1991) *Managing Global Genetic Resources: the U.S. National Plant Germplasm System.* National Academy Press, Washington DC.

National Research Council. (1993) *Managing Global Genetic Resources: Agricultural Crop Issues and Policies.* National Academy Press, Washington DC.

Navarro, L. (1976) The Citrus Variety Improvement Program in Spain. In: Calavan, E.C. (ed.), *Proceedings of the Seventh Conference of the International Organization of Citrus Virologists.* IOCV, Riverside, California, pp. 198–203.

Navarro, L. (1992) Citrus shoot-tip grafting *in vitro*. In: Bajaj, J.P.S. (ed.) *Biotechnology in Agriculture and Forestry, Vol. 18. High-Tech and Micropropagation.* Springer-Verlag, Berlin, pp. 327–338.

Navarro, L. (1993) Citrus sanitation, quarantine and certification programs. In: Moreno, P., da Graça, J.V. and Timmer, L.W. (eds) *Proceedings of the 12th Conference of the International Organization of Citrus Virologists.* IOCV, Riverside, California, pp. 383–391.

Navarro, L. and Juárez, J. (1977) Elimination of Citrus pathogens in propagative budwood. II. *In vitro* propagation. *Proceedings of the International Society of Citriculture* 3, 973–987.

Navarro, L. and Juárez, J. (1981a) Nucellus culture *in vitro* to obtain virus-free nucellar plants of monoembryonic citrus cultivars. In: Bové, J.M. and Vogel, J.M. (eds) *Virus and Virus-like Diseases of Citrus*, 2nd edn. SETCO-IRFA, Paris.

Navarro, L. and Juárez, J. (1981b) Ovule culture *in vitro* to obtain virus-free nucellar plants of seedless polyembryonic citrus cultivars. In: Bové, J.M. and Vogel, J.M. (eds) *Virus and Virus-like Diseases of Citrus*, 2nd edn. SETCO-IRFA, Paris.

Navarro, L., Roistacher, C.N. and Murshige, T. (1975) Improvement of shoot-tip grafting *in vitro* for virus-free citrus. *Journal of the American Society for Horticultural Science* 100, 471–479.

Navarro, L., Ballester, J.F., Juárez, J., Pina, J.A., Arregui, J.M., Bono, R., Fernandez de Cordova, L. and Ortega, C. (1980) The Citrus Variety Improvement Program in Spain (CVIPS) after four years. In: Calavan, E.C., Garnsey, S.M. and Timmer, L.W. (eds) *Proceedings of the Eighth Conference of the International Organization of Citrus Virologists.* IOCV, Riverside, California, pp. 289–294.

Navarro, L., Ballester, J.F., Juárez, J., Pina, J.A., Arregui, J.M[a]. and Bono, R. (1981) Development of a program for disease-free citrus budwood in Spain. *Proceedings of the International Society of Citriculture* 1, 70–73.

Navarro, L., Juárez, J., Pina, J.A. and Ballester, J.F. (1984a) The citrus quarantine station in Spain. In: Garnsey, S.M., Timmer, L.W. and Dodds, J.A. (eds) *Proceedings of the Ninth Conference of the International Organization of Citrus Virologists.* IOCV, Riverside, California, pp. 365–370.

Navarro, L., Pina, J.A., Ballester, J.F., Moreno, P. and Cambra, M. (1984b) A new graft transmissible disease found in Nagami kumquat. In: Garnsey, S.M., Timmer, L.W. and Dodds, J.A. (eds) *Proceedings of the Ninth Conference of the International Organization of Citrus Virologists.* IOCV, Riverside, California, pp. 234–240.

Navarro, L., Juárez, J., Pina, J.A., Ballester, J.F. and Arregui, J.M[a]. (1988) The citrus variety improvement program in Spain (CVIPS) after eleven years. In: Timmer, L.W., Garnsey, S.M. and Navarro, L. (eds) *Proceedings of the Tenth Conference of the International Organization of Citrus Virologists.* IOCV, Riverside, California, pp. 400–406.

Navarro, L., Civerolo, E.L., Juárez, J. and Garnsey, S.M. (1991) Improving therapy methods for citrus germplasm exchange. In: Brlansky, R.H., Lee, R.F. and Timmer, L.W. (eds) *Proceedings of the 11th Conference of the International Organization of Citrus Virologists.* IOCV, Riverside, California, pp. 400–408.

Navarro, L., Pina, J.A., Juárez, J., Ballester-Olmos, J.F., Arregui, J.M., Ortega, C., Navarro, A., Duran-Vila, N., Guerri, J., Moreno, P., Cambra, M. and Zaragoza, S. (2002). The Citrus Variety Improvement Program in Spain in the Period 1975–2001. In: *Proceedings of the 15th Conference of the International Organization of Citrus Virologists*, IOCV, Riverside, California, pp. 306–316.

Navarro, L., Juárez, J., Aleza, P. and Pina, J.A. (2003) Recovery of triploid seedless mandarin hybrids from 2n × 2n and 2n × 4n crosses by embryo rescue and flow cytometry. In: Vasil, I.K. (ed.) *Plant Biotechnology 2002 and Beyond.* Kluwer Academic Publishers, Dordrecht, The Netherlands, pp. 541–544.

Navarro, L., Olivares-Fuster, O., Juárez, J., Aleza, P., Pina, J.A., Ballester-Olmos, J.F., Cervera, M., Fagoaga, C., Duran-Vila, N. and Peña, L. (2004) Applications of biotechnology to citrus improvement in Spain. *Acta Horticulturae* 632, 221–234.

Nguyen, R. and Fraser, S. (1989) Lack of suitability of commercial limes and lemons as hosts of *Anastrepha suspensa* (Diptera: Tephritidae). *Florida Entomologist* 72, 718–720.

Nicolosi, E., Deng, Z.N., Gentile, A., La Malfa, S., Continella, G. and Tribulato, E. (2000) Citrus phylogeny and genetic origin of important species as investigated by molecular markers. *Theoretical and Applied Genetics* 100, 1155–1166.

Niles, R.K., Freckman, D.W. and Roose, M.L. (1995) Use of trifoliate orange as a comparative standard for assessing the resistance of citrus rootstocks to citrus nematode. *Plant Disease* 79, 813–818.

Nishida, M., Nishiura, M., Iwamasa, M., Tsuchiya, S. and Kozaki, I. (1981) The problems of citrus genetic resources in Japan. Photocopy.

Nito, N., Katayama, Y., Yamaguchi, S. and Fukushima, H. (1999) Germplasm conservation of citrus relatives at Saga University. In: Broadbent, P., Sykes, S.R., Bevington, K.B. and Hailstones, D. (eds) *Proceedings of the Citrus Germplasm Conservation Workshop.* Brisbane, Australia, October 1997, pp. 106–111.

O'Bannon, J.H. and Ford, H.W. (1977) Resistance in citrus rootstocks to *Radopholus similis* and *Tylenchulus semipenetrans* (Nematoda). *Proceedings of the Second Congress of the International Society of Citriculture* 2, 544–549.

Okamoto, Y. (1935) Comparative experiment on the strains of trifoliate orange as rootstocks for Satsuma orange. *Annual Report of the Oita Agricultural Experiment Station (Citrus Branch) for 1933*, p. 123.

Olivares-Fuster, O., Asíns, M.J., Duran-Vila, N. and Navarro, L. (2000) Cryopreserved callus, a source of pro-

toplasts for citrus improvement. *Journal of Horticultural Sciences and Biotechnology* 75, 635-640.
Olson, E.O. (1960) Evaluation of rootstocks for Valencia orange trees following inoculation under screen-house conditions with a severe strain of tristeza virus. *Journal of the Rio Grande Horticultural Society* 14, 29–32.
Olson, E.O. and Godfrey, G.H. (1953) Cotton root rot of various citrus rootstock seedlings. *Proceedings of the Seventh Annual Rio Grande Valley Horticultural Institute.* Weslaco, Texas, pp. 25–30.
Olson, E.O., Cooper, W.C., Maxwell, N. and Shull, A.V. (1962) Survival, size and yield of xyloporosis- and exocortis-infected old-line red grapefruit trees on 100 rootstocks. *Journal of the Rio Grande Valley Horticultural Society* 16, 44–51.
Olszyk, D.M., Cabrera, H. and Thompson, C.R. (1988) California statewide assessment of the effects of ozone on crop productivity. *Journal of the Air Pollution Control Association* 38, 928–931.
Omura, M. (1996) Status report on genetic resources of citrus in Japan. In: *Status Reports on Genetic Resources of Citrus: East Asia, South Asia.* International Plant Genetic Resources Institute Project No B06. International Plant Genetic Resources Institute Regional Office for Asia, the Pacific, and Oceania, Singapore.
Omura, M. (1997) Genetic resources of citrus in Japan: a summary. In: Albrigo, L.G. (ed.) *Identification and Conservation of Genetic Resources of Citrus and Its Relatives.* Proceedings of a Symposium held during the VIII Congress of the International Society of Citriculture. Sun City, South Africa, pp. 70–72.
Ortiz, J.M., Zaragoza, S. and Bono, R. (1988) The major citrus cultivars of Spain. *HortScience* 23, 691-692.
Osman, M.B., Mohamed, Z.A., Idris, S. and Aman, R. (1995) Tropical fruit production and genetic resources in South-east Asia: identifying the priority fruit species. In: Arora, R.K. and Ramanatha Rao, V. (eds). *Expert Consultation on Tropical Fruit Species of Asia.* International Plant Genetic Resources Institute, Office for South Asia, New Delhi, pp. 61–84.
Panigrahi, G. (1975) Nomenclatural notes on *Hesperethusa crenulata* (Rutaceae). *Bulletin of the Botanical Survey of India* 17, 195–197.
Panigrahi, G. (1977) Proposal to reject the name *Limonia acidissima* L. Nom. Ambig. Rejic. in favour of *L. elephantum* (Corr.), Comb. Nov. *Taxon* 26, 576–577.
Panigrahi, G. (1985) Proposal to conserve 4087 Glycosmis against Panel (Rutaceae). *Taxon* 34, 715–716.
Parmar, C. and Kaushal, M.K. (1982) *Wild fruits of the Sub-Himalayan Region.* Kalyani Publishers, Delhi.
Paul, A. and Cox, P.A. (1995) An ethnobotanical survey of the uses for *Citrus aurantium* (Rutaceae) in Haiti. *Economic Botany* 49, 549–256.
Peltier, G.L. (1918) Susceptibility and resistance to citrus-canker of the wild relatives, citrus fruits, and hybrids of the genus citrus. *Journal of Agricultural Research* 14, 337–357.
Peltier, G.L. and Frederich, W.J. (1920) Relative susceptibility to citrus-canker of different species and hybrids of the genus Citrus, including the wild relatives. *Journal of Agricultural Research* 19, 339–362.
Peltier, G.L. and Frederich, W.J. (1923) Relative susceptibility of citrus fruits and hybrids to *Cladosporium citri* Massee. *Journal of Agricultural Research* 24, 955–959.
Peltier, G.L. and Frederich, W.J. (1924) Further studies on the relative susceptibility to citrus canker of different species and hybrids of the genus citrus, including the wild relatives. *Journal of Agricultural Research* 28, 227–239.
Peña, L. and Navarro, L. (1999) Transgenic citrus. In: Bajaj, Y.P.S. (ed.) *Biotechnology in Agriculture and Forestry, Vol. 44, Transgenic Trees.* Springer Verlag, Berlin, pp. 39–54.
Pérez, R.M. (1995) Crioconservación de recursos genéticos de cítricos. Tesis doctoral. Universidad Politécnica de Valencia.
Pérez, R.M., Navarro, L. and Duran-Vila, N. (1997) Cryoconservation and storage of embryogenic callus cultures of several citrus species and cultivars. *Plant Cell Reports* 17, 44–49.
Pérez, R.M., Galiana, A.M., Navarro, L. and Duran-Vila, N. (1998) Embryogenesis *in vitro* of several *Citrus* species and cultivars. *Journal of Horticultural Science and Biotechnology* 73, 796–802.
Pérez, R.M., Mas, O., Navarro, L. and Duran-Vila, N. (1999) Production and cryoconservation of embryogenic cultures of mandarin and mandarin hybrids. *Plant Cell Tissue and Organ Culture* 55, 71–74.
Phillips, R.L. (1969) Dwarfing rootstocks for citrus. *Proceedings of the First International Citrus Symposium* 1, 401–406.
Phillips, R.L. and Castle, W.S. (1977) Evaluation of twelve rootstocks for dwarfing citrus. *Journal of the American Society for Horticultural Science* 102, 526–528.
Pina, J.A. and Navarro, L. (2001) Certification program for citrus nursery plants in Spain. In: *Proceedings of*

the Sixth World Congress of the International Society of Citrus Nurserymen pp. 26-31.

Pio, R.M., Pompeu, J., Jr, Figueiredo, J.O., Barbin, D. and Demetri, C.G.B. (1984) Polyembryony in 26 trifoliata-type rootstocks. In: *Proceedings of the International Society of Citriculture* pp. 24–25.

Pollard, G.V., Riley, R. and Wattie, E. (1983) Preliminary investigations on the selection of citrus species by the leaf-cutting ant, *Acromyrmex octospinosus* (Reich) (Formicidae, Attini). *Tropical Agriculture (Trinidad)* 60, 282–285.

Raeburn, P. (1995) *The Last Harvest: the Genetic Gamble that Threatens to Destroy American Agriculture.* University of Nebraska Press, Lincoln, Nebraska.

Raghavendra Rao, N.N. and Prasad, M.B.N.V. (1983) Evaluation of strains of *Poncirus trifoliata* and trifoliate orange hybrids for resistance to *Phytophthora* root rot. *Scientia Horticulturae* 20, 85–90.

Rai, M., Chandel, K.P.S. and Gupta, P.N. (1997a) Occurrence, distribution, and diversity in the genus Citrus in the Indian gene centre. In: *Proceedings of the International Society of Citriculture,* pp. 1228–1234.

Rai, M., Chandel, K.P.S. and Gupta, P.N. (1997b) Occurrence, distribution, and diversity in the genus Citrus in the Indian gene centre. In: Albrigo, L.G. (ed.) *Identification and Conservation of Genetic Resources of Citrus and its Relatives.* Proceedings of a Symposium held during the VIII Congress of the International Society of Citriculture. Sun City, South Africa, pp. 26–39.

Ramanatha Rao V. and Arora, R.K. (1999) Conservation strategy. In: Broadbent, P., Sykes, S.R., Bevington, K.B. and Hailstones, D. (eds) *Proceedings of the Citrus Germplasm Conservation Workshop.* Brisbane, Australia, October 1997, pp. 7–21.

Ranade, S.A., Rana, T.S., Srivastava, A.P. and Nair, K.N. (2006) Molecular differentation in *Murraya* Koenig ex L. species in India inferred through ITS, RAPD and DAMD analysis. *Current Science* 90, 1253–1258.

Rasmussen, G.K. and Smith, P.F. (1958) Effects of fertilizer rate, rootstock and leaf age on the level of sulfur in citrus leaves. Proceedings of the American Society for Horticultural Science 71, 241–247.

Ream, C.L. and Furr, J.R. (1976) Salt tolerance of some *Citrus* species, relatives, and hybrids tested as rootstocks. *Journal of the American Society for Horticultural Science* 101, 265-267.

Reitz, H.J. and Embleton, T.W. (1986) Production practices that influence fresh fruit quality. In: Wardowski, W.F., Nagy, S. and Grierson, W. (eds) *Fresh Citrus Fruits.* AVI, New York, pp. 49–77.

Reuther, W. (1973) Climate and citrus behavior. In: Reuther, W. (ed.) *The Citrus Industry. Vol. III. Production Technology.* University of California Division of Agricultural Science, Berkeley, California, pp. 280–337.

Reuther, W. (1977) Genetic resources conservation of citrus species and near relatives from the international viewpoint. *Proceedings of the International Society of Citriculture* 2, 604–606.

Reuther, W. (1988) Major commercial citrus varieties in the United States. *HortScience* 23, 693–697.

Reuther, W., Rasmussen, G.K., Hilgeman, R.H., Cahoon, G.A. and Cooper, W.C. (1969) A comparison of maturation and composition of 'Valencia' oranges in some major subtropical zones of the United States. *Journal of the American Society for Horticultural Science* 94, 144–157.

Ríos-Castaño, D. and Tafut, R.R. (1972) Tabog, *Swinglea glutinosa* (Blanco) Merr. como patron para kumquats, *Fortunella* sp. *Proceedings of the Tropical Region of the American Society for Horticultural Science* 16, 115–124.

Rivera-Bustamante, R., Gin, R. and Semancik, J.S. (1986) Enhanced resolution of circular and linear molecular forms of viroid and viroid-like RNA by electrophoresis in a discontinuous-pH system. *Analytical Biochemistry* 156, 91–95.

Roia, F.C., Jr (1966) The use of plants in hair and scalp preparations. *Economic Botany* 20, 17–30.

Roistacher, C.N. (1977) Elimination of citrus pathogens in propagative budwood. I. Budwood selection, indexing and thermotherapy. *Proceedings of the International Society for Citriculture* 3, 965–972.

Roistacher, C.N., Calavan, E.C. and Navarro, L. (1977) Concepts and procedures for importation of citrus budwood. *Proceeding of the International Society of Citriculture* 1, 133–136.

Roistacher, C.N. (1991) *Graft-transmissible Diseases of Citrus. Handbook for Detection and Diagnosis.* International Organization of Citrus Virologists/Food and Agriculture Organization of the United Nations, Rome.

Roistacher, C.N. (1993) Arguments for establishing a mandatory certification program for citrus. In: *Proceedings of the Fourth World Congress of the International Society of Citrus Nurserymen.* Johannesburg, South Africa, pp. 18–34.

Roistacher, C.N. (1995) *A Historical Review of the Major Graft-transmissible Diseases of Citrus.* FAO Regional Office for the Near East, Cairo, Egypt.

Roistacher, C.N. (1998) Indexing for viruses in citrus. In: Hadidi, A., Khetarpal, R.K. and Koganezawa, H. (eds) *Plant Virus Disease Control.* American Phytopathological Society Press, St Paul, Minnesota, pp. 301–319.

Roistacher, C.N. and Calavan, E.C. (1972) Heat tolerance of preconditioned citrus budwood for virus inactivation. In: *Proceedings of the Fifth Conference of the International Organization of Citrus Virologists.* University of Florida Press, Gainesville, Florida, pp. 256–261.

Roistacher, C.N. and Nauer, E.M. (1968) Frequency of virus infection in citrus budwood introduced into the United States. In: *Proceedings of the Fourth Conference of the International Organization of Citrus Virologists.* University of Florida Press, Gainesville, Florida, pp. 386–391.

Roistacher, C.N., Calavan, E.C. and Navarro, L. (1977) Concepts and procedures for importation of citrus budwood. *Proceedings of the International Society of Citriculture* 1, 133–136.

Roose, M. and Kupper, R. (1992) Effects of citrus rootstocks on freeze tolerance in California. In: *Proceedings of the International Society of Citriculture* pp. 256–258.

Roose, M.L., Bitters, W.P., Cole, D.A. and Traugh, S.A. (1985) Progress report on rootstock studies. *Citrograph* 70, 127-131.

Roosjen, M.G., Stitger, M. and Oomen, P.A. (1999) Greenhouse pest exclusion and eradication safeguards and principles. In: Kahn, R.P. and Mathur, S.B. (eds) *Containment Facilities and Safeguards for Exotic Plant Pathogens and Pests.* American Phytopathological Society Press, St Paul, Minnesota, pp. 182–185.

Rossetti, V., Nakadaira, J.T., Calza, R. and Miranda, C.A.B. (1965) Estudos sobre a clorose zonada dos citros. *Arquivos do Instituto Biológico (São Paulo)* 332, 111–125.

Rouse, R.E. (1988) Major citrus cultivars of the world as reported from selected countries. *HortScience* 23, 680–684.

Rouse, R.E. and Wutscher, H.K. (1985) Heavy soil and bud union crease with some grapefruit clones limit use of Swingle citrumelo rootstock. *HortScience* 20, 259–261.

Rouseff, R.L. and Nagy, S. (1994) Health and nutritional benefits of *Citrus* fruit components. *Food Technology* 48, 125–132.

Roy, S.K. (1998) Bael fruit. In: Shaw, P.E., Chan, H.T., Jr and Nagy, S. (eds) *Tropical and Subtropical Fruits.* AgScience Inc., Auburndale, Florida, pp. 217–260.

Ruíz, C., Bretó, M.P. and Asíns, M.J. (2000) A quick methodology to identify sexual seedlings in citrus breeding programs using SSR markers. *Euphytica* 112, 89-94.

Russo, F. (1976/1977) Il miglioramento genetico per la resistenza al 'Mal secco' de limone in Italia. *Annali dell'Instituto Sperimentale per l'Agrumicoltura* 9/10, 231–243.

Saamin, S. and Ko, W.W. (1997a) Biodiversity and conservation of citrus and its relatives in Malaysia and vicinity. In: *Proceedings of the International Society for Citriculture* pp. 1221–1227.

Saamin, S. and Ko, W.W. (1997b) Biodiversity and conservation of citrus and its relatives in Malaysia and vicinity. In: Albrigo, L.G. (ed.) *Identification and Conservation of Genetic Resources of Citrus and Its Relatives.* Proceedings of a Symposium held during the VIII Congress of the International Society of Citriculture. Sun City, South Africa, pp. 40–49.

Sadana, G.L. and Joshi, R. (1979) Comparative susceptibility of different varieties of citrus to the attack of the phytophagous mite, *Brevipalpus californicus. Science & Culture* 45, 413–414.

Sagee, O., Hasdai, D., Hamou, M. and Shaked, A. (1992) Screenhouse evaluation of new citrus rootstocks for tolerance to adverse soil conditions. In: *Proceedings of the International Society of Citriculture* pp. 299–303.

Sakai, A., Kobayashi, S. and Oiyama, I. (1990) Cryopreservation of nucellar cells of navel orange (*Citrus sinensis* Osb. var. *brasiliensis* Tanaka) by vitrification. *Plant Cell Reports* 9, 30–33.

Sakai, A., Kobayashi, S. and Oiyama, I. (1991a) Cryopreservation of nucellar cells of navel orange (*Citrus sinensis* Osb.) by a simple freezing method. *Plant Science* 74, 243–248.

Sakai, A., Kobayashi, S. and Oiyama, I. (1991b) Survival of nucellar cells of navel orange (*Citrus sinensis* Osb. Var. *brasiliensis* Tanaka) cooled to –196°C by vitrification. *Journal of Plant Physiology* 137, 465–470.

Salibe, A.A. (1977) The stem-pitting effects of tristeza on different citrus hosts and their economic significance. *Proceedings of the Second Congress of the International Society of Citriculture* 2, 953–955.

Salibe, A.A. and Moreira, S. (1965) Reaction of types of citrus as scion and as rootstock to xyloporosis virus. In: *Proceedings of the Third Conference of the International Organization of Citrus Virologists* pp. 70–75.

Samuel, R., Ehrendorfer, F., Chase, M.W. and Greger, H. (2001) Phylogenetic analyses of Aurantioideae (Rutaceae) based upon con-coding plastid DNA sequences and phytochemical features. *Plant Biology* 3, 77–87.

San Martín A., J. (1983) Medicinal plants in central Chile. *Economic Botany* 37, 216–227.
Santiago, A. and Sarkawi, S.B. (1962) *An Illustrated Guide to Malayan Citrus Species and Varieties.* Bulletin No. 111, Division of Agriculture, Federation of Malaysia, Kuala Lumpur.
Saunt, J. (2000) *Citrus Varieties of the World,* 2nd edn. Sinclair International Limited, Norwich, UK.
Savage, E.M., Cooper, W.C. and Piper, R.B. (1945) Root systems of various citrus rootstocks. *Proceedings of the Florida State Horticultural Society* 58, 44–48.
Schaad, N.W., Jones, J.B. and Chun, W. (eds) (2001) *Laboratory Guide for Identification of Plant Pathogenic Bacteria,* 3rd edn. American Phytopathological Society Press, St Paul, Minnesota.
Scora, R.W. (1975) On the history and origin of citrus. *Bulletin of the Torrey Botanical Club* 102, 369–375.
Scora, R.W. (1989) Biochemistry, taxonomy, and evolution of modern cultivated citrus. *Proceedings of the International Society of Citriculture* 1, 277–289.
Scora, R.W. and Kumamoto, J. (1983) Chemotaxonomy of the genus *Citrus.* In: Waterman, P.G. and Grundon, M.F. (eds) *Chemistry and Chemical Taxonomy of the Rutales.* Academic Press, London, pp. 343–351.
Scora, R.W., Kumamoto, J., Soost, R.K. and Nauer, E.M. (1982) Contribution to the origin of the grapefruit, *Citrus paradisi* (Rutaceae). *Systematic Botany* 7, 170–177.
Serit, M., Ishida, M., Hagiwara, N., Kim, M., Yamamoto, T. and Takahashi, S. (1992) Meliaceae and Rutaceae limonoids as termite antifeedants evaluated using *Reticulitermes speratus* Kolbe (Isoptera: Rhinotermitidae). *Journal of Chemical Ecology* 18, 593–603.
Shah, N.C. and Joshi, M.C. (1971) An ethnobotanical study of the Kumaon region of India. *Economic Botany* 25, 414–422.
Shaked, A., Goell, A. and Hamou, M. (1988) Screening citrus rootstocks and rootstock–scion combinations for tolerance to calcareous soils. In: *Proceedings of the International Society of Citriculture* pp. 83–86.
Shands, H.L., Fitzgerald, P.J. and Eberhart, S. (1988) Program for plant germplasm preservation in the United States. In: Knutson, L. and Stoner, A.K. (eds) *Biotic Diversity and Germplasm Preservation, Global Imperatives.* Kluwer Academic Publishing, Waaginengen, The Netherlands, pp. 97–116.
Shands, H.L. (1995) The U.S. national plant germplasm system. *Canadian Journal of Plant Science* 75, 9–15.
Shannon, L.M. and Zaphrir, J. (1958) The relative influence of two citrus rootstock species upon plant growth and upon inorganic composition of the scion. *Proceedings of the American Society for Horticultural Science* 71, 257–264.
Shannon, L.M., Frolich, E.F. and Cameron, S.H. (1960a) Characteristics of *Poncirus trifoliata* selections. *Proceedings of the American Society for Horticultural Society* 76, 163–169.
Shannon, L.M., Frolich, E.F. and Cameron, S.H. (1960b) Characteristics of *Poncirus trifoliata* selections. *Journal of the Rio Grande Horticultural Society* 14, 108–118.
Shapiro, J.P. and Gottwald, T.R. (1995) Resistance of eight cultivars of citrus rootstock to a larval root weevil, *Diaprepes abbreviatus* L. (Coleoptera: Curculionidae). *Journal of Economic Entomology* 88, 148–154.
Shapiro, J.P., Bowman, K.D. and Smith, H. (1997) Resistance of citrus rootstocks and *Glycosmis pentaphylla* against larval citrus root weevil, *Diaprepes abbreviatus,* in live root or diet-incorporation assays. *Florida Entomologist* 80, 471–477.
Shapiro, J.P., Bowman, K.D. and Lapointe, S.L. (2000) Dehydrothalebanin, a source of resistance from *Glycosmis pentaphylla* against the citrus root weevil *Diaprepes abbreviatus. Journal of Agriculture and Food Chemistry* 48, 4404–4409.
Sharples, G.C. and Hilgeman, R.H. (1972) Leaf mineral composition of five rootstock cultivars grown on sour orange and rough lemon rootstock. *Journal of the American Society for Horticultural Science* 97, 427–430.
Sinclair, W.B. and Bartholomew, E.T. (1944) Effects of rootstock and environment on the composition of oranges and grapefruit. *Hilgardia* 3, 125–176.
Singh, B. (1981) *Establishment of First Gene Sanctuary in India for Citrus in Garo Hills.* Concept Publishing Co., New Delhi.
Singh, H.P. and Chadha, K.L. (1993) Genetic resources of citrus. In: Chadha, K.L. and Pareek, O.P. (eds) *Advances in Horticulture. Vol. 1. Fruit Crops Part 1.* Malhotra Publishing, New Delhi, pp. 95–121.
Singh, H.P. and Uma, S. (1995) *Genetic Resources of Tropical Fruits in India.* Technical Document No. 59, Indian Institute of Horticultural Research, Bangalore.
Singh, R. (1985) Indigenous citrus germplasm in India. In: Nijjar, G.S. (ed.) *Fruit Breeding in India.* 2nd edn. Oxford and I.B.H., New Delhi, pp. 53–63.

Singh, R. and Nath, N. (1969) Practical approach to the classification of citrus. In: *Proceedings of the First International Citrus Symposium* pp. 435–440.
Singh, R.N. and Roy, S.K. (1984) *The Bael: Cultivation and Processing.* Indian Council of Agricultural Research, New Delhi.
Singh, R.P. and Singh, U.S. (eds) (1995) *Molecular Methods in Plant Pathology.* CRC Press, Boca Raton, Florida.
Sinha-Roy, S.P. and Chakraborty, D.P. (1976) Psoralen, a powerful germination inhibitor. *Phytochemistry* 15, 2005–2006.
Small, J.K. (1933) *Manual of the Southeastern Flora.* Science Press, New York, pp. 756–761.
Smith, G.S., Hutchison, D.J. and Henderson, C.T. (1987) Screening sweet orange citrus cultivars for relative susceptibility to Phytophthora root rot. *Proceedings of the Florida State Horticultural Society* 100, 64–66.
Smith, P.F. (1975) Effect of scion and rootstock on mineral composition of mandarin-type citrus leaves. *Journal of the American Society for Horticultural Science* 100, 368–369.
Smith, P.F., Reuther, W. and Specht, A.W. (1949) The influence of rootstock on the mineral composition of Valencia orange leaves. *Plant Physiology* 24, 455–461.
Smith, R.L. and Wallace, A. (1954) Preliminary studies of some physiological root characteristics in citrus and avocado. *Proceedings of the American Society for Horticultural Science* 63, 143–145.
Solel, Z. and Kimchi, M. (1997) Susceptibility and resistance of citrus genotypes to *Alternaria alternata* pv. *citri. Journal of Phytopathology* 145, 389–391.
Solel, Z. and Spiegel-Roy, P. (1978) Methodology of selection of lemon clones for tolerance to Mal secco (*Phoma tracheiphila*). *Phytoparasitica* 6, 129–134.
Soost, R.K. and Cameron, J.W. (1975) Citrus. In: Janick, J. and Moore, J.N. (eds) *Advances in Fruit Breeding.* Purdue University Press, West Lafayette, Indiana, pp. 507–540.
Soost, R.K. and Roose, M.L. (1996) Citrus. In: Janick, J. and Moore, J.N. (eds) *Fruit Breeding. Vol. I. Tree and Tropical Fruits.* John Wiley & Sons, New York, pp. 257–323.
Spitler, G.H., Armstrong, J.W. and Couey, H.M. (1984) Mediterranean fruit fly (Diptera: Tephritidae) host status of commercial lemon. *Journal of Economic Entomology* 77, 1441–1444.
Stange, R.R., Jr, Midland, S.L., Eckert, J.W. and Sims, J.J. (1993) An antifungal compound produced by grapefruit and valencia orange after wounding of the peel. *Journal of Natural Products* 56, 1627–1629.
Stone, B.C. (1978a) Studies in Malesian Rutaceae. I. Notes toward a revision of the genus *Glycosmis* Correa. *Federation Museums Journal* 23, 75–109.
Stone, B.C. (1978b) Studies in Malesian Rutaceae. II. New records of taxa and new taxa or Aurantioideae from Borneo. *Federation Museums Journal* 23, 111–116.
Stone, B.C. (1985a) New and noteworthy paleotropical species of Rutaceae. *Proceedings of the Academy of Natural Sciences of Philadelphia* 137 (2), 213–228.
Stone, B.C. (1985b) A conspectus of the genus *Glycosmis* Correa: studies in Malesian Rutaceae, III. *Proceedings of the Academy of Natural Sciences of Philadelphia* 137 (2), 1–27.
Stone, B.C. (1985c) Rutaceae. In: Dassanayake, M.D. (ed.) *A Revised Handbook to the Flora of Ceylon.* Vol. V. Smithsonian Institution/National Science Foundation, Washington DC, pp. 406–478.
Stone, B.C. (1994a) Citrus fruits of Assam: a new key to species, and remarks on *C. assamensis* Bhattacharya and Dutta, 1956. *Gardens' Bulletin (Singapore)* 46, 105–112.
Stone, B.C. (1994b) Additional notes on the genus *Glycosmis* Correa (Rutaceae). *Gardens' Bulletin (Singapore)* 46, 113–119.
Stone, B.C. and Jones, D.T. (1988) New and noteworthy Rutaceae-Aurantioideae from northern Borneo. Studies in Malesian Rutaceae, V. *Proceedings of the Academy of Natural Sciences of Philadelphia* 140, 267–274.
Stone, B.C. and Nicolson, D.H. (1978) Arguments for *Limonia acidissima* L. (Rutaceae) and against its rejection as a *Nomen Abiguum. Taxon* 27, 551–552.
Stone, B.C., Lowry, J.B., Scora, R.W. and Jong, K. (1973) *Citrus halimii*: a new species from Malaya and peninsular Thailand. *Biotropica* 5, 102–110.
Storey, R. and Walker, R.R. (1999) Citrus and salinity. *Scientia Horticulturae* 78, 39–81.
Stubbs, L.L. (1963) Tristeza-tolerant strains of sour orange. *F.A.O. Plant Protection Bulletin* 11, 8–10.
Sudahono, Byrne, D.H. and Rouse, R.E. (1994) Greenhouse screening of citrus rootstocks for tolerance to bicarbonate-induced iron chlorosis. *HortScience* 29, 113–116.

Swingle, W.T. (1909) *The Limitation of the Satsuma Orange to Trifoliate-Orange Stock.* Circular No. 46, United States Department of Agriculture, Bureau of Plant Industry, Washington DC.

Swingle, W.T. (1912) Le genre Balsamocitrus et un nouveau genre voisin, Aeglopsis. *Journal of the Washington Academy of Science* 2, 353–354.

Swingle, W.T. (1913a) Citrus ichangensis, a promising, hardy, new species from south-western China and Assam. *Journal of Agricultural Research* 1, 1–15.

Swingle, W.T. (1913b) Chaetospermum, a new genus of hard-shelled citrous fruits. *Journal of the Washington Academy of Science* 3, 99–102.

Swingle, W.T. (1914–1917) Citrus and related genera. Extracted from: Bailey, L.H. (ed.) *The Standard Cyclopedia of Horticulture,* vols. I – IV. The MacMillan Co., New York.

Swingle, W.T. (1914a) Citropsis, a new tropical African genus allied to Citrus. *Journal of Agricultural Research* 1, 419–436.

Swingle, W.T. (1914b) Eremocitrus, a new genus of hardy, drought-resistant citrous fruits from Australia. *Journal of Agricultural Research* 2, 85–100.

Swingle, W.T. (1914c) The name of the wood-apple, Feronia limonia. *Journal of the Washington Academy of Science* 4, 325–328.

Swingle, W.T. (1915a) Merope angulata, a salt-tolerant plant related to Citrus, from the Malay Archipelago. *Journal of the Washington Academy of Sciences* 5, 420–425.

Swingle, W.T. (1915b) Microcitrus, a new genus of Australian citrous fruits. *Journal of the Washington Academy of Sciences* 5, 569–578.

Swingle, W.T. (1915c) A new genus, Fortunella, comprising four species of kumquat oranges. *Journal of the Washington Academy of Sciences* 5, 165–176.

Swingle, W.T. (1916a) Pamburus, a new genus related to Citrus, from India. *Journal of the Washington Academy of Science* 6, 335–337.

Swingle, W.T. (1916b) Pleiospermium, a new genus related to Citrus, from India, Ceylon and Java. *Journal of the Washington Academy of Science* 6, 426–431.

Swingle, W.T. (1916c) *Severinia buxifolia,* a Citrus relative native to southern China. *Journal of the Washington Academy of Science* 6, 651–657.

Swingle, W.T. (1918) Merrillia, a new Rutaceous genus of the tribe Citrae from the Malay Peninsula. *Philippine Journal of Science* 13, 335–343.

Swingle, W.T. (1938) A new taxonomic arrangement of the orange subfamily, Aurantiodeae. *Journal of the Washington Academy of Sciences* 28, 530–532.

Swingle, W.T. (1939) Clymenia and Burkillanthus, new genera also three new species of Pleiospermium (Rutaceae-Aurantiodeae). *Journal of the Arnold Arboretum* 20, 250–253.

Swingle, W.T. (1940a) Limnocitrus, a new genus, also new species of Wenzelia, Paramignya and Atalantia (Rutaceae-Aurantioideae). *Journal of the Arnold Arboretum* 21, 1–24.

Swingle, W.T. (1940b) Three new species of Citropsis, also new varieties of Atalantia and Fortunella (Rutaceae-Aurantioideae). *Journal of the Arnold Arboretum* 21, 115–132.

Swingle, W.T. (1940c) New varieties and new combinations in the genera Clausena, Oxanthera, and Triphasia of the orange subfamily, Aurantioideae. *Journal of the Washington Academy of Science* 30, 79–83.

Swingle, W.T. (1942) Three new varieties and two new combinations in *Citrus* and related genera of the orange subfamily. *Journal of the Washington Academy of Sciences* 32, 24–26.

Swingle, W.T. (1943) The botany of citrus and its wild relatives of the orange subfamily. In: Webber, H.J. and Batchelor, L.D. (eds) *The Citrus Industry. Vol. I. History, Botany, and Breeding.* University of California Press, Berkeley, California, pp. 129–474.

Swingle, W.T, and Reece, P.C. (1967) The botany of citrus and its wild relatives. In: Reuther, W., Webber, H.J. and Batchelor, L.D. (eds) *The Citrus Industry. Vol. I. History, World Distribution, Botany, and Varieties.* California Division of Agriculture Science, Berkeley, California, pp. 190–423.

Sykes, S.R. (1993) Evaluation of germplasm for *Citrus* breeding in Australia. In: Hayashi, T., Omura, M. and Scott, N.S. (eds) *Techniques on Gene Diagnosis and Breeding in Fruit Trees.* Fruit Tree Research Station, Japanese Ministry of Agriculture, Forestry, and Fisheries, Tsukuba, Japan, pp. 219–230.

Sykes, S.R. (1997) Citrus germplasm resources in Australia. In: Albrigo, L.G. (ed.) *Identification and Conservation of Genetic Resources of Citrus and Its Relatives.* Proceedings of a Symposium held during the VIII Congress of the International Society of Citriculture. Sun City, South Africa, pp. 50–55.

Sykes, S.R. (1999) Citrus germplasm In Australia with special reference to indigenous members of the subfamily Aurantioideae. In: Broadbent, P., Sykes, S.R., Bevington, K.B. and Hailstones, D. (eds) *Proceedings of the Citrus Germplasm Conservation Workshop.* Brisbane, Australia, October 1997, pp. 76–84.

Sykes, S.R. (2001) Native citrus hits the limelight. <http://www.pi.csiro.au/Media/MediaReleases/2001/MR05-01-01.htm>

Syvertsen, J.P. (1981) Hydraulic conductivity of four commercial citrus rootstocks. *Journal of the American Society for Horticultural Science* 106, 378–381.

Syvertsen, J.P. and Graham, J.H. (1985) Hydraulic conductivity of roots, mineral nutrition, and leaf gas exchange of citrus rootstocks. *Journal of the American Society for Horticultural Science* 110, 865–869.

Syvertsen, J.P. and Lloyd, J.J. (1994) Citrus. In: Schaffer, B. and Anderson, P.C. (eds) *Handbook of Environmental Physiology of Fruit Crops. Vol. 2. Sub-tropical and Tropical Crops.* CRC Press, Boca Raton, Florida, pp. 65–99.

Tanaka, T. (1932) Philippine Rutaceae-Aurantiodeae (Revisio Aurantiacearum. VII). *Transactions of the Natural History Society of Formosa* 22, 418–433.

Tanaka, T. (1933) General remarks on the genus *Fortunella* (2). *Studia Citrologica* 6, 19–40.

Tanaka, T. (1937) Further revision of Rutaceae-Aurantioideae of India and Ceylon. Revisio Aurantiacearum VIII. *Journal of the Indian Botanical Society* 16, 227–240.

Tanaka, T. (1954) *Species Problem in Citrus. Revisio Aurantiacearum IX.* Japanese Society for the Promotion of Science, Tokyo.

Tanaka, T. (1966) Hodgson's citrus classification discussed. *Bulletin of University of Osaka Prefecture, Series B* 18, 25–29.

Tanaka, T. (1969a) Taxonomic problems of citrus fruits in the Orient. *Bulletin of University of Osaka Prefecture, Series B* 21, 133–138.

Tanaka, T. (1969b) Misunderstanding with regards to citrus classification and nomenclature. *Bulletin of University of Osaka Prefecture, Series B* 21, 139–145.

Tanaka, T. (1977) Fundamental discussion of citrus classification. *Studia Citrologica* 14, 1–6.

Tanaka, Y. (1969) Citrus rootstock problems in Japan. *Proceedings of the International Citrus Symposium,* Riverside, California, pp. 407–410.

Thanassoulopoulos, C.C. (1991) 'Ermioni' a new lemon cultivar resistant to Mal secco disease (*Phoma tracheiphila*). *Journal of Phytopathology* 131, 234–242.

Thompson, C.R. and Taylor, O.C. (1969) Effects of air pollutants on growth, leaf drop, fruit drop, and yield of citrus trees. *Environmental Science and Technology/American Chemical Society* 3, 934–940.

Timmer, L.P., Garnsey, S.M. and Graham, J.H. (2000) *Compendium of Citrus Diseases,* 2nd edn. American Phytopathological Society Press, St Paul, Minnesota.

Traub, H.P. and Robinson, T.R. (1937) *Improvement of Subtropical Fruit Crops: Citrus.* United States Department of Agriculture, Yearbook Separate No. 1589, Washington DC, pp. 749–826.

Turrell, F.M. (1947) Citrus leaf stomata: structure, composition, and pore size in relation to penetration of liquids. *Botanical Gazette* 108, 476–483.

Turrell, F.M. (1961) Chlorophyll content and reflection spectra of citrus leaves. *Botanical Gazette* 123, 10–15.

Tuzcu, O. (2001) Citrus diversity situation in Turkey. In: Albrigo, L.G. (ed.) *Proceedings of the Global Citrus Germplasm Network.* Network meeting held during the Ninth Congress of the International Society of Citriculture. Orlando, Florida, pp. 34–42.

Tuzcu, O., Cinar, A. Gksedef, M., Ozsan, M. and Bcici, M. (1984) Resistance of citrus rootstocks to *Phytophthora citrophora* during winter dormancy. *Plant Disease* 68, 502–505.

Ueno, I., Iwamasa, M. and Nishiura, M. (1967) Embryo number of various varieties of citrus and its relatives. *Bulletin of the Horticultural Research Station, Japan, Series B* 7, 11–21.

Uhe, G. (1974) Medicinal plants of Samoa: a preliminary survey of the use of plants for medicinal purposes in the Samoan Islands. *Economic Botany* 28, 1–30.

Van Gundy, S.D. and Kirkpatrick, J.D. (1964) Nature of resistance in certain citrus rootstocks to citrus nematode. *Phytopathology* 54, 419–427.

Vanderweyen, A. (1980) Influence de la variete d'agrume sur la sensibilite du porte-greffe a la gommose a phytophthora. *Al Awamia* 60, 67–80.

Verzera, A., Russo, C., La Rosa, G., Bonaccorsi, I. and Cotroneo, A. (2001) Influence of cultivar on lemon oil composition. *Journal of Essential Oil Research* 13, 343–347.

Verheij, E.W.M. and Coronel, R.E. (eds) (1991) *Plant Resources of South-East Asia. No 2. Edible Fruits and Nuts.* Pudoc, Wageningen, The Netherlands.

Vives, M.C., Galipienso, L., Navarro, L., Moreno, P. and Guerri, J. (2001) The nucleotide sequence and genomic organization of citrus leaf blotch virus: candidate type species for a new virus genus. *Virology* 287, 225–233.

Vogel, R. and Bové, J.M. (1971) Réactions de quelques porte-greffe a l'exocortis. *Fruits* 26, 295–300.

Wallace, A. and Mueller, R.T. (1972) Resistance of rough lemon and trifoliate orange seedlings to high levels of iron chelates. *HortScience* 7, 121–122.

Wallace, A., Naude, C.J., Mueller, R.T. and Zidan, Z.I. (1952) The rootstock–scion influence on the inorganic composition of citrus. *Proceedings of the American Society for Horticultural Science* 59, 133–142.

Wallace, A., North, C.P. and Frolich, E.F. (1953) Interaction of rootstock, soil pH, and nitrogen on the inorganic composition of citrus. *Proceedings of the American Society for Horticultural Science* 62, 75–78.

Wallace, H.M. and Lee, L.S. (1999) Pollen source, fruit set and xenia in mandarins. *Journal of Horticultural Science and Biotechnology* 74, 82–86.

Wallace, J.M. (1945) Technique for hastening foliage symptoms of psorosis of citrus. *Phytopathology* 35, 535–541.

Waterworth, H.E. (1993) Processing foreign plant germplasm at the National Plant Germplasm Quarantine Center. *Plant Disease* 77, 854–860.

Waterworth, H.E. (1998) Certification for plant viruses: an overview. In: Hadidi, A., Khetarpal, R.K. and Koganezawa, H. (eds) *Plant Virus Disease Control.* American Phytopathological Society Press, St Paul, Minnesota, pp. 325–345.

Weathers, L.G. and Calavan, E.C. (1959) Nucellar embryony as a means of freeing citrus clones of virus diseases. In: Wallace, J.M. (ed.) *Citrus Virus Diseases.* University of California Division of Agricultural Sciences, Berkeley, California, pp. 197–202.

Webber, H.J. (1943) Cultivated varieties of citrus. In: Webber, H.J. and Batchelor, L.D. (eds). *The Citrus Industry. Vol. I. History, World Distribution, Botany, and Varieties.* University of California Press, Berkeley, California, pp. 475–668.

Weidong, G., Fang, J., Zheng, D., Li, Y., Lu, X., Rao, R.V., Hodgkin, T. and Zongwen, Z. (2000) Utilization of germplasm conserved in Chinese national genebanks – a survey. *Plant Genetic Resources Newsletter* 123, 1–8.

Westwood, M.N. (1989) Maintenance and storage: clonal germplasm. *Plant Breeding Reviews* 7, 111–128.

White, C.T. (1926) A new species of Paramignya from Papua with notes on two other Papuan Rutaceae. *Journal of the Arnold Arboretum* 7, 231–233.

White, G.A., Shands, H.L. and Lovell, G.R. (1989) History and operation of the National Plant Germplasm System. *Plant Breeding Reviews* 7, 5–54.

Wiersma, J.H. and León, B. (1999) *World Economic Plants. A Standard Reference.* CRC Press, Boca Ratón, Florida.

Wijnands, D.O. (1984) Nomenclatural note on the calamondin (Rutaceae). *Baileya* 22, 134–136.

Wilkes, G. (1988) Germplasm preservation: objectives and needs. In: Knutson, L. and Stoner, A.K. (eds) *Biotic Diversity and Germplasm Preservation. XIII Beltsville Symposium.* USDA-ARS, Beltsville, Maryland. pp. 13–42.

Williams, T.E. (1991) Preservation and maintenance of virus-free clonal citrus germplasm at the National Clonal Germplasm Repository for Citrus, Riverside, California. In: *Proceedings of the 26th Annual Meeting of the Caribbean Food Crops Society.* Mayagüez, Puerto Rico, pp. 24–30.

Williams, T.E. (1992a) *Ex-situ* preservation, evaluation, and maintenance of pathogen-free clonal germplasm of Citrus and related genera: the USDA-ARS National Germplasm Repository for Citrus. *Proceedings of the International Society of Citriculture* 1, 44–47.

Williams, T.E. (1992b) The USDA-ARS National Germplasm Repository for clean-stock citrus germplasm. In: *Proceedings of the Workshop on CTV and* Toxoptera citricidus *in Central America and the Caribbean Basin,* Maracay, Venezuela, pp. 148–154.

Winston, J.R. (1923) *Citrus Scab: its Cause and Control.* United States Department of Agriculture, Bulletin No. 1118, Washington DC.

Winston, J.R., Bowman, J.J. and Bach, W.J. (1927) *Citrus melanose and its Control.* United States Department of Agriculture, Bulletin No. 1474, Washington DC.

Winters, H.F. (1976) *Microcitrus papuana,* a new species from Papua New Guinea [Rutaceae]. *Baileya* 20, 19–24.

Withers, L.A. (1992) *In vitro* conservation. In: Hammerschlag, F.A. and Litz, R.E. (eds) *Biotechnology of Perennial Fruit Crops.* CAB International, Wallingford, UK, pp. 169–200.

Woodruff, R.E. and Olson, E.O. (1960) Effects of rootstocks on physical and chemical composition of fruit of six citrus varieties in Texas. *Journal of the Rio Grande Valley Horticultural Society* 14, 77–84.

World Wide Fund for Nature and International Union for Conservation of Nature and Natural Resources (1994–1997) *Centres of Plant Diversity. A Guide and Strategy for their Conservation.* Three volumes. International Union for Conservation of Nature and Natural Resources Publications Unit, Cambridge, UK.

Wutscher, H.K. (1979) Citrus rootstocks. *Horticultural Reviews* 1, 237–269.

Wutscher, H.K. (1986) Effect of three rootstocks on concentration and distribution of 14 elements in young 'Hamlin' orange trees. *Proceedings of the Congress of the Interamerican Society for Tropical Horticulture* 30, 95–103.

Wutscher, H.K. (1989) Alternation of fruit tree nutrition through rootstocks. *HortScience* 24, 578–584.

Wutscher, H.K. and Bistline, F.W. (1988) Rootstock influences juice colour of 'Hamlin' orange. *HortScience* 23, 724–725.

Wutscher, H.K. and Dube, D. (1977) Performance of young nucellar grapefruit on 20 rootstocks. *Journal of the American Society for Horticultural Science* 102, 267–270.

Wutscher, H.K. and Shull, A.V. (1972) Performance of 13 citrus cultivars as rootstocks for grapefruit. *Journal of the American Society for Horticultural Science* 97, 778–781.

Wutscher, H.K. and Shull, A.V. (1975) Yield, fruit quality, growth and leaf nutrient levels of 14-year-old grapefruit, *Citrus paradisi* Macf., trees on 21 rootstocks. *Journal of the American Society for Horticultural Science* 100, 290–294.

Wutscher, H.K. and Shull, A.V. (1976a) Performance of 'Orlando' tangelo on 16 rootstocks. *Journal of the American Society for Horticultural Science* 101, 88–91.

Wutscher, H.K. and Shull, A.V. (1976b) Performance of 'Marr's' early orange on eleven rootstocks in South Texas. *Journal of the American Society for Horticultural Science* 101, 158–161.

Wutscher, H.K., Olson, E.O., Shull, A.V. and Peynado, A. (1970) Leaf nutrient levels, chlorosis, and growth of young grapefruit trees on 16 rootstocks grown on calcareous soil. *Journal of the American Society for Horticultural Science* 95, 259–261.

Wutscher, H.K., Peynado, A., Cooper, W.C. and Hill, H. (1973) Method of irrigation and salt tolerance of citrus rootstocks. *Proceedings of the First World Congress of Citriculture* 1, 299–309.

Xiancai, L. (1997) Cultivo *in vitro* de especies afines a los cítricos. Tesis doctoral, Universidad Politecnica de Valencia.

Xiao, L., Lovatt, C.J., Bertling, I. and Liao, Z. (2001) Differences in endogenous hormones between normal and dwarfing trifoliate orange rootstocks. *HortScience* 36, 602–603.

Xuannan, Q. and Dehui, H. (1989) A study of the dwarfing characters of *Severinia buxifolia* as a rootstock for citrus. In: Chinese Society for Horticultural Science (ed.) *International Symposium on Horticultural Germplasm, Cultivated and Wild. Part 1. Fruit Trees.* Chinese Society for Horticultural Science, Beijing, pp. 350–352.

Xueqin, Z. (1995) Genetic resources of tropical fruit species in China. In: Arora, R.K. and Ramanatha Rao, V. (eds) *Expert Consultation on Tropical Fruit Species of Asia.* International Plant Genetic Resources Institute, Office for South Asia, New Delhi, pp. 103–111.

Xueyuan, Z., Changyhong, Z., Yuanhui, J., Xinghua, H., Jiezhong, C. and Zhusheng, C. (1993) Evaluation of tristeza tolerance of 45 citrus types. *Proceedings of the 12th Conference of the International Organization of Citrus Virologists.* IOCV, Riverside, California, pp. 73–77.

Yamada, S. and Sawamura, K. (1952) *Studies on the Dwarf Disease of Satsuma Orange,* Citrus unshiu *Marcovich (Preliminary Report).* Horticulture Division, National Tokai-kinki, Agricultural Experiment Station, Bulletin No. 1, Okitsu, Japan, pp. 61–71.

Yamamoto, M., Kobayashi, S., Nakamura, Y. and Yamada, Y. (1993) Phylogenetic relationships of *Citrus* revealed by diversity of cytoplasmic genes. In: Hayashi, T., Omura, M. and Scott, N.S. (eds) *Techniques on Gene Diagnosis and Breeding in Fruit Trees.* Fruit Tree Research Center, Tsukuba, Japan, pp. 39–46.

Yang, R.Z. and Tang, C.S. (1988) Plants used for pest control in China: a literature review. *Economic Botany* 42, 376–402.

Yang, Z.N., Ye, X.R., Choi, S., Molina, J., Moonan, F., Wing, R.A., Roose, M.L. and Mirkov, T.E. (2001) Construction of a 1.2-Mb contig including the citrus tristeza virus resistance gene locus using a bacterial artificial chromosome library of *Poncirus trifoliata* (L.) Raf. *Theoretical and Applied Genetics* 44, 382–393.

Yelenosky, G. (1985) Cold hardiness in citrus. *Horticultural Reviews* 7, 201–238.

Yelenosky, G. (1991) Responses and adaptations of citrus trees to environmental stresses. *Israeli Journal of Botany* 40, 259–270.

Yelenosky, G. and Hearn, C.J. (1976) Cold damage to young mandarin hybrid trees and different rootstocks on flatwood soil. *Proceedings of the Florida State Horticultural Society* 80, 53–56.

Yelenosky, G. and Young, R. (1977) Cold hardiness of orange and grapefruit trees on different rootstocks during the 1977 Florida freeze. *Proceedings of the Florida State Horticultural Society* 90, 49–53.

Yelenosky, G., Hearn, C.J. and Cooper, W.C. (1968) Relative growth of trifoliate orange selections. *Proceedings of the American Society for Horticultural Science* 93, 205–209.

Yelenosky, G., Brown, R.T. and Hearn, C.J. (1973) Tolerance of trifoliate orange selections and hybrids to freezes and flooding. *Proceedings of the Florida State Horticultural Society* 86, 99–104.

Yelenosky, G., Barrett, H. and Young, R. (1978) Cold hardiness of young hybrid trees of *Eremocitrus glauca* (Lindl.) Swing. *HortScience* 13, 257–258.

Yin-min, Y. (1985) The status of Fortunella genetic resources in China. *Fruit Varieties Journal* 39, 17–20.

Yinmin, Y., Xiaodong, L., Suqin, D. and Mingqing, L. (1976) Honghe papeda, a new species of citrus from Yunnan. *Acta Phytotaxonomica Sinensis* 14, 57–60.

Yoshida, T. (1996) Graft compatibility of *Citrus* with plants in the *Aurantiodeae* and their susceptibility to citrus tristeza virus. *Plant Disease* 80, 414–417.

Youtsey, C. (1997) Rootstocks for calamondins and kumquats. *Newsletter of the International Society of Citrus Nurserymen* No. 16, p. 2.

Young, R. (1963a) Climate – cold hardiness – citrus. *Journal of the Rio Grande Valley Horticultural Society* 17, 3–14.

Young, R. (1963b) Freeze injury to four- and nine-year-old citrus cultivars and related species in the Lower Rio Grande Valley of Texas. *Journal of the Rio Grande Valley Horticultural Society* 17, 30–36.

Young, R. (1963c) Freeze injury to young seedlings of citrus and related species in the Rio Grande Valley. *Journal of the Rio Grande Valley Horticultural Society* 17, 37–42.

Young, R. (1966) Freezing points and lethal temperatures of citrus leaves. *Proceedings of the American Society for Horticultural Science* 88, 272–279.

Young, R. and Hearn, C.J. (1972) Screening citrus hybrids for cold hardiness. *HortScience* 7, 14–18.

Young, R.H. (1977) The effect of rootstocks on citrus cold hardiness. *Proceedings of the Second Congress of the International Society of Citriculture* 2, 518–522.

Young, R.H. and Olson, E.O. (1963a) Freeze injury to citrus varieties in the Lower Rio Grande Valley of Texas. *Proceedings of the American Society for Horticultural Science* 83, 333–336.

Young, R.H. and Olson, E.O. (1963b) Freeze injury to citrus trees on various rootstocks in the Lower Rio Grande Valley of Texas. *Proceedings of the American Society for Horticultural Science* 83, 337–343.

Young, R.H., Peynado, A. and Cooper, W.C. (1960) Effect of rootstock–scion combination and dormancy on cold hardiness of citrus. *Journal of the Rio Grande Valley Horticultural Society* 14, 58–65.

Zhang, W.C., Deng, X.X. and Deng, Z.A. (1992) Citrus germplasm preservation and varietal improvement works in China. *Proceedings of the International Society of Citriculture* 1, 67–71.

Zhaomin, S. (1989) The present situation and future of citrus production in China. *HortScience* 24, 904–905.

Zhusheng, C. (1997) Conservation and identification of genetic resources of citrus and its relatives. In: Albrigo, L.G. (ed.). *Identification and Conservation of Genetic Resources of Citrus and its Relatives.* Proceedings of a Symposium held during the VIII Congress of the International Society of Citriculture, Sun City, South Africa, pp. 11–17.

Zhusheng, C. (1999) Citrus germplasm in China. In: Broadbent, P., Sykes, S.R., Bevington, K.B. and Hailstones, D. (eds) *Proceedings of the Citrus Germplasm Conservation Workshop*. Brisbane, Australia, October 1997, pp. 85–95.

Zhusheng, C., Yinmin, Y., Liangzhen, W., Tianchi, G. and Jiangdong. (1996) Status report on genetic resources of citrus in China. In: *Status Reports on Genetic Resources of Citrus: East Asia, South Asia.* IPGRI Project No. B06. International Plant Genetic Resources Institute Regional Office for Asia, the Pacific, and Oceania, Singapore.

Zubrzycki, H.M. and Diamante de Zubrzycki, A. (1980) Resistencia a *Xanthomonas citri* (Hasse) Dowson, en diferentes especies y cultivares citricos. *Proceedings, 2nda Congreso Nacional de Citricultura* 2, 39–50.

5 Nucellar Embryony

J. L. Kepiro and M. L. Roose

Department of Botany and Plant Sciences, University of California, Riverside, CA 92521, USA

Nucellar embryony is one of the more unusual features of seed reproduction in citrus. The term refers to development of embryos from the maternal tissue called the nucellus that surrounds the embryo sac. This chapter describes the trait, how it influences citrus rootstock and scion breeding, how the trait is inherited, and methods to manipulate it in breeding programmes.

Background of Nucellar Embryony (Apomixis) in Citrus

The trait of nucellar embryony in citrus has long been a subject of investigation. Webber (1900) described this trait in an article entitled, 'Complications in citrus hybridization caused by polyembryony'. Polyembryony (multiple embryos in one seed) is associated with nucellar embryony because it frequently results in polyembryonic seeds from which multiple seedlings germinate (Fig. 5.1). Refined definitions of the trait based on developmental studies led to the term nucellar embryony, since the majority of the multiple embryos arise from cells of the somatic nucellus tissue surrounding the embryo sac (Kobayashi *et al.*, 1979). A broader term is apomixis, generally defined as asexual reproduction through seed. Apomixis includes a variety of different types of asexual reproduction and is reviewed thoroughly in Asker and Jerling (1992).

Nucellar Embryony in Citrus Rootstock Breeding

In citrus, nucellar embryony is generally essential in citrus rootstocks because it

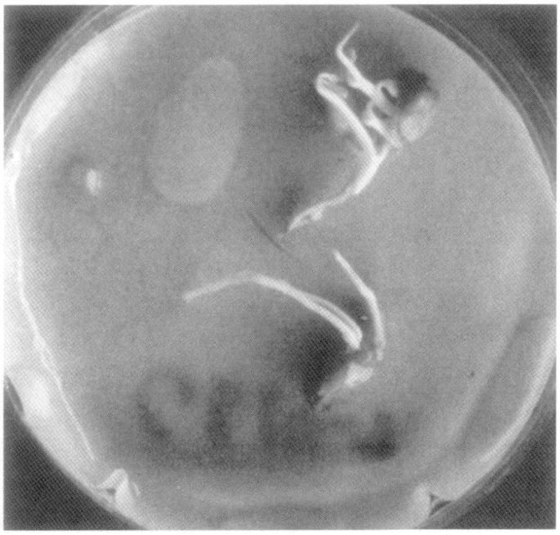

Fig. 5.1. Multiple seedlings emerging from a citrus seed.

allows nurseries to propagate trees on highly heterozygous, but genetically uniform seedling populations. This reduces variation in rootstock and scion performance in comparison with that which would occur if rootstock seedlings resulted from sexual reproduction. Nucellar embryony allows citrus rootstock breeders to produce F_1 hybrids that are highly heterozygous, but produce seedlings that are genetically uniform and identical to the mother tree. If nucellar embryony were not present, citrus breeders and nurseries could take several different approaches to produce genetically uniform rootstocks. Rootstocks could be vegetatively propagated as rooted cuttings or propagated using *in vitro* culture methods. Another alternative would be to use seedling populations of genotypes that are homozygous for all important traits. Although these strategies are used in other tree crops and have been attempted in citrus, none has been commercially successful, perhaps because they are more difficult or costly to implement than growing nucellar seedling populations of citrus rootstocks. Therefore, nucellar embryony is considered an essential trait in citrus rootstocks. The rootstock breeder must therefore choose parents that produce progeny having the trait, and must select progeny that produce a high proportion of nucellar seedlings and few sexual seedlings.

Nucellar Embryony in Citrus Scion Breeding

In contrast to rootstock breeding, nucellar embryony impedes progress in scion breeding. When using a female parent with the trait, a breeder must invest monetary and labour resources to generate a population, but the breeder has few new hybrids to use for selection. The process is inefficient and costly in a long generation tree crop. The other alternative is to choose female parents that lack nucellar embryony. This is possible and frequently done, but many important scion cultivars have nucellar embryony so this strategy limits the choice of crosses. If the gene(s) responsible for nucellar embryony can be identified, it may be possible to silence it by transformation with an antisense or RNA interference (RNAi) construct, or perhaps turn it off and on with an inducible promoter system.

Nucellar Embryony

Nucellar embryony is a heritable trait found in some citrus varieties. Nucellar embryony is an adventitious form of apomictic reproduction, wherein the somatic cells of the nucellus tissue are initialized to enter into an embryonic pathway of development. Development of nucellar embryos was studied in detail by Koltunow *et al.* (1995), Kobayashi *et al.* (1979) and Wakana and Uemoto (1988). The nucellar embryos develop from nucellar initial cells that originate in the nucellus that surrounds the embryo sac (Fig. 5.2). These cells occur in all portions of the nucellus, but many migrate to the micropylar end so in many varieties nearly all nucellar embyos are found at the micropylar end (Wakana and Uemoto, 1988). Nucellar embryos generally initiate development before fertilization, about ten days before anthesis, but this may vary with the genotype. Many embryo sacs contain several to many nucellar initial cells. As these begin to divide and grow, the number of growing embryos diminishes. At later stages, the larger embryos tend to be located at the micropylar end of the embryo sac (Fig. 5.2).

In citrus, nucellar embryony does not prevent normal sexual reproduction (Esan and Soost, 1977; Wilms *et al.*, 1983). This results in trees bearing fruit with seeds containing two classes of embryos: (i) zygotic embryos from a fertilization event; and (ii) nucellar embryos that are genetically identical to the maternal parent. Many genotypes with nucellar embryony also have a high frequency of multiple embryos, since many cells in the nucellus initiate embryogenesis and several embryos may mature. A seed may contain two or more seedlings that germinate, but not every seed produced

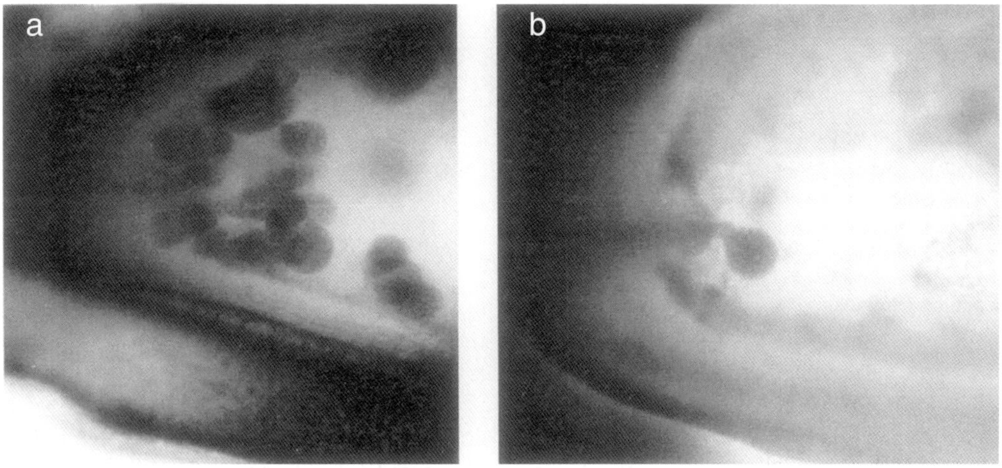

Fig. 5.2. Citrus ovules, about 60 days after pollination, of (a) a variety with a high level of nucellar embryony and (b) a monoembryonic (sexual) variety.

by a tree with nucellar embryony has multiple mature embryos.

In a genotype that produces seeds by nucellar embryony, normal sexual reproduction can also occur. Such a genotype can produce several different types of seeds (Wakana and Uemoto, 1988): (i) seeds with one mature, sexual embryo; (ii) seeds with one mature nucellar embryo; (iii) seeds with multiple mature nucellar embryos; and (iv) seeds with one mature sexual and one or more mature nucellar embryos. Seeds may also contain immature embryos that cannot germinate normally. This is because not only does adventitious apomixis occur without preventing normal sexual reproduction, but also its success depends on fertilization for endosperm formation (Esan and Soost, 1977).

Nucellar embryo growth is arrested at the late globular stage without endosperm development (Koltunow *et al.*, 1995). Most fertilization events involve a double fertilization: (i) of the polar nuclei to form the endosperm; and (ii) of the egg to form a zygote. Developmental competition between the zygotic and nucellar embryos, as well as the genotype of the zygote, will affect maturation and, in seeds with mature nucellar embryos, immature zygotic and nucellar embryos are often present but do not germinate (Ueno *et al.*, 1967; Esan and Soost, 1977; Koltunow *et al.*, 1995).

The number and type of embryos produced may vary from tree to tree and also at different positions on a single tree (Nasharty, 1945; Parlevliet and Cameron, 1959). This variation has been suggested to be controlled by minor genes, pollen source and environmental conditions (Frost, 1926; Parlevliet and Cameron; 1959, Khan and Roose, 1988). A study specifically designed to determine the origin of seedlings produced by open-pollinated *Poncirus trifoliata* cultivars with the genotype of nucellar embryony found that the majority of the mature embryos that germinate are genetically identical to the maternal parent (Khan and Roose, 1988). The proportion of nucellar seedlings has been estimated in many citrus rootstock cultivars using isozymes and other genetic markers, with most cultivars having more than 90% nucellar seedlings (Table 5.1; Roose and Kupper, 1992), but this sample is biased toward genotypes producing a high proportion of nucellar embryos because citrus rootstocks are selected to have a high percentage of nucellar seedlings.

Table 5.1. Percentage of nucellar seedlings in citrus rootstocks as detected by isozyme analysis.

Rootstock	Percentage zygotic	References
Pomeroy trifoliate	14.1	Khan and Roose, 1988
Rubidoux trifoliate	4.6	Khan and Roose, 1988
Trifoliate orange	2.6, 21.6	Anderson *et al.*, 1991; Moore and Castle, 1988
Flying Dragon trifoliate orange	18.0, 29.7	Moore and Castle, 1988; Khan and Roose, 1988
Swingle citrumelo	9.3, 9.6, 15.2, 17.7	Ashari *et al.*, 1988; Moore and Castle, 1988; Xiang and Roose, 1988
Sacaton citrumelo	38.7, 39.7	Xiang and Roose, 1988; Xiang and Roose, 1988
C-32 citrange	5.7, 19.7	Xiang and Roose, 1988; M. L. Roose and R. S. Kupper, personal communication
Carrizo citrange	0.0	Anderson *et al.*, 1991;
Troyer citrange	0.0, 0.8	Moore and Castle, 1988; Anderson *et al.*, 1991
Uvale citrange	1.3	Moore and Castle, 1988
Yuma citrange	36.1, 50.6	Moore and Castle, 1988, Xiang and Roose, 1988
Amblycarpa	4.9	Moore and Castle, 1988
Cleopatra mandarin	0.8	Anderson *et al.*, 1991
Cuban shaddock	12.0	Moore and Castle, 1988;
Rough lemon	2.1, 5.0, 5.5	Anderson *et al.*, 1991; Moore and Castle, 1988; Xiang and Roose, 1988
Milam	0.0	Moore and Castle, 1988
Sour orange	0.0	Moore and Castle, 1988
Sweet orange	0.0, 0.8	Moore and Castle, 1988; Anderson *et al.*, 1991
Taiwanica	34.8	Xiang and Roose, 1988
Volkamer lemon	16, 23.7, 27.3	Moore and Castle, 1988; Xiang and Roose, 1988
Yuma Ponderosa	20.0	Xiang and Roose, 1988

At least 94% of zygotic seedlings from self-pollination should have been detected based on the number of enzyme loci studied, except in sweet orange (87%) and Cleo (50%).

Nucellar Embryony and Polyembryony

The term polyembryonic has in the past been used to refer to a genotype that produces multiple embryos in a single seed by the process of nucellar embryogenesis. Use of this term to indicate a genotype capable of nucellar embryogenesis causes confusion when attempting to distinguish the trait of nucellar embryony from sexual twinning. While sexual twins may be identical to each other, they originate by sexual reproduction and therefore are genetically distinct from the maternal genotype. A study of twins in hybrids of *Citrus* × *Poncirus* from strictly sexual seed parents demonstrated that supernumerary embryos can bud off from a zygotic embryo and produce seed with multiple embryos (Cameron and Garber, 1968). It is worth making a special note of this and reiterating that polyembryonic seeds do not by their mere existence indicate the tree producing such seed has the trait of nucellar embryony. However, because twinning is rare (Frost, 1926; Parlevliet and Cameron, 1959), trees that produce seed with a moderate or high frequency of multiple embryos can be classified as probably having nucellar embryony, and those that produce very few (less than ~6%) seeds with multiple embryos can be classified mostly as lacking nucellar embryony.

The term monoembryonic has been used to refer to a single seed that contains one embryo or to describe a strictly sexual seed parent. If used as a descriptor of a tree that is a strictly sexual seed parent, the term can lead to confusion when multiple embryos arise by twinning. Trees called 'monoembryonic' do produce seeds with two or more embryos by twinning at a low frequency, but these embryos will be genetically different from the maternal parent (Cameron and Garber, 1968). For example, Cameron and Soost (1979) classified trees that produced seed with multiple embryos in no more than 7% of the seed as monoembryonic and trees with 8 or 9% as near-monoembryonic. Iwamasa *et al.* (1967) defined varieties producing 1–6% polyembryonic seed as genetically monoembryonic. It is preferable to use the term 'strictly sexual' to describe trees lacking the trait of nucellar embryony.

Inheritance of Nucellar Embryony

The work of Parlevliet and Cameron (1959) suggests that nucellar embryony is controlled by a single major dominant gene that is heterozygous in trifoliate and absent in 'Chandler' pummelo. They also suggest minor genes may control the level of expression. Other work suggests that several genes control nucellar embryony and that polyembryony is an independent trait (Garcia *et al.*, 1999; Asins *et al.*, 2002). Garcia *et al.* (1999) evaluated the progeny of a cross between two parents known to have nucellar embryony, *Citrus volkameriana* ('volkamer' lemon) and *Poncirus trifoliata* var 'Rubidoux'. The cross produced 50 fruit-yielding hybrid progeny from which seed samples were germinated. Twenty-five random seedlings from each individual were genotyped with isozymes to determine the seedlings' origin as either nucellar or zygotic. Eight to ten seeds were scored for multiple embryos and the percentage of polyembryonic seed calculated for 38 of the 50 individuals in the test population. A variety of marker types were used in their mapping analysis, with 73 polymorphic markers in *P. trifoliata* and 97 in *C. volkameriana*. They propose a model with two quantitative trait loci (QTLs) in *P. trifoliata* and four QTLs in *C. volkameriana* controlling apomixis, with individual QTLs contributing up to 24% of the total variation. Markers linked to polyembryony were found at different positions in each parent. TAA15 (a *C. volkameriana* marker for polyembryony) was linked to *Apo2*, the QTL with the strongest effect on apomixis.

Kepiro (2003) studied inheritance of nucellar embryony in 88 progeny of a cross of Chandler pummelo × trifoliate orange. They scored the trait by counting the number of seedlings that germinated from an average of 283 seeds per progeny tree. Only 17 progeny produced more than 8.5% polyembryonic seeds. Mapping and QTL analysis identified a major, dominant QTL that was heterozygous in trifoliate orange and which appears essential for production of more than about 2% polyembryonic seeds. DNA marker studies on a subset of seedlings supported a correlation between production of nucellar seedlings and production of a high proportion of polyembryonic seeds. Some trees that produced 1–2% polyembryonic seed did produce nucellar embryos, but in others no apparently nucellar seedlings were found among seedlings from polyembyronic seeds. Therefore, the proportion of polyembryonic seeds could not be used to determine whether or not such trees have nucellar embryony. This major locus and markers associated with it showed significant segregation distortion, with only about 32% of progeny producing at least 1.4% polyembryonic seed instead of the expected 50%. In those progeny having polyembryony, a second unlinked QTL accounted for much of the variation in the percentage of polyembyronic seeds. These QTLs were confirmed in a population of open-pollinated (mostly selfed) progeny of trifoliate orange. No QTLs influencing polyembryony were detected in the Chandler pummelo parent. Thus far, it has not been possible to compare the map locations of QTLs detected by Garcia *et al.* (1999) with those studied by Kepiro because the maps developed do not share any common markers. While the results of Kepiro suggest that inheritance of polyembryony from *Poncirus* involves only a few genes, it is quite possible that additional polymorphic genes involved in nucellar embryony occur in *Citrus* or other *Poncirus* genotypes.

The relationship between the proportion of polyembryonic seeds and the proportion of nucellar seedlings has not been adequately tested in any segregating population. Garcia *et al.* (1999) tested this relationship, but they examined only 25 seedling per progeny tree, too few to have a high probability of detecting polyembryony in genotypes that produce only 1–5% polyembryonic seeds.

Screening Hybrid Populations for Nucellar Embryony

When the rootstock breeder selects parents for hybridization, if the rootstock cultivar will be propagated as seedlings, then the parents must be selected to produce nucellar progeny. The frequency of progeny that produce mostly nucellar seedlings will be higher if both parents have nucellar embryony, but this cross configuration will produce mostly progeny identical to the mother. This dilemma is unresolved, although screening for DNA markers that predict nucellar embryony would greatly reduce the cost of propagating hybrids that lack the trait. The appropriate decision is determined by the importance, genetic control and ease of selection for the other traits that can be contributed by each potential parent.

Screening hybrids for adequate levels of nucellar embryony can be done in several ways. The first question is 'what is an unacceptable number of zygotic (sexual) seedlings?' Among existing commercial rootstocks, the proportion of zygotic seedlings ranges from less than 1% up to 50% (Table 5.2). The acceptable proportion will depend on a number of factors. One issue is the degree of difficulty nursery personnel would encounter in rogueing the seedling population (eliminating off-types). This will depend on the level of heterozygosity of the population for easily recognized morphological traits such as the trifoliate leaf morphology present in most trifoliate × *Citrus* hybrids. If the parents are fairly closely related (e.g. two mandarins), then it will be more difficult to identify the zygotic off-types than if the parents are distantly related. On the other hand, if the parents are closely related, the performance of a tree budded on a zygotic seedling may be

Table 5.2. Percentage of polyembryonic and nucellar seedlings in major citrus groups.

Seed parent	Seedlings/seed	% nucellar
Lemon: Eureka, Lisbon, etc.	1.05–1.06	32–33
Rough lemon	1.24–1.96	54–98
Mexican Lime	1.29	78
Mandarin: Dancy, Kara	1.37–1.71	100
Mandarin: Satsuma	1.44	90
Mandarin: Kishiu	1.00	0
Mandarins: King, Ponkan	1.01-1.42	21–98
Grapefruit: Marsh	1.08	96
Pummelo: 11 cultivars	1.00	0
Sweet orange: four cultivars	1.09–2.00	39–97
Sour orange	1.21	85
Tangelo: Orlando, Minneola	1.31–1.49	83–97
Trifoliate orange	1.03–1.26	13–73

Source: after Frost and Soost (1968).

fairly similar to that of a tree budded on a nucellar seedling. Evidence (Xiang and Roose, 1988) suggests that, at least in some locations, most zygotic seedlings result from selfing rather than outcrossing. This makes zygotic seedlings more difficult to identify, but also makes their characteristics more predictable. Nurseries dislike using rootstocks that are excessively variable, so we suggest that 10–15% zygotic seedlings is the maximum permissible level in most cases.

In some rootstocks, particularly *Citrus* × *Poncirus* hybrids, the proportion of nucellar seedlings can be estimated fairly well from a visual screen of 100 or more seedlings. Each seedling should be carefully examined for leaf morphology, internode spacing, thorn length, stem morphology and any other visible traits (Fig. 5.3). A good practice is to compare the new hybrid with an established rootstock, such as Troyer or Carrizo, which produces a high proportion of nucellar seedlings. Another protocol would be estimate the proportion of polyembryonic seeds from a germination test or dissection of mature seeds, but this indirect method is less desirable because the correlation between the proportion of polyembryonic seeds and the proportion of nucellar seedlings is not well established in most populations. Before release of a rootstock or scion variety, the proportion of nucellar seedlings can be estimated more precisely by 'fingerprinting' a large (100–200) sample of seedlings with many heterozygous DNA markers (unlinked simple sequence repeats (SSRs), amplified fragment length polymorphisms (AFLPs) or inter-simple sequence repeats (ISSRs) are appropriate). Many loci that are heterozygous in the female parent should be screened because, for each heterozygous locus, the probability of a selfed seedling having the same genotype as the mother is one in two. Testing five unlinked, co-dominant, heterozygous loci gives an overall probability of misclassifying a hybrid as nucellar of $(0.50)^5 = 1/32 = 0.031$, so that about 3% of actual hybrids would be dis-

Fig. 5.3. Zygotic (identified by isozyme analysis) and nucellar seedlings of trifoliate orange. Note that some zygotic seedlings are dwarf, but many are not.

carded. Twelve dominant loci would be needed to achieve the same accuracy. This procedure is too laborious and costly to apply to seedlings from each of a large population of hybrids as would occur in the early stages of a breeding programme. Again, easily screened DNA markers for the genes that govern the presence and proportion of nucellar embyros would be particularly valuable because of the cost of propagating many hybrids to fruiting.

Distinguishing Hybrid from Nucellar Seedlings

If a scion breeder uses a female parent that has nucellar embryony, then the breeding plan should include a protocol for separating hybrid from nucellar seedlings. As indicated above, it is often possible to identify hybrids morphologically based on a character contributed by the male, but not the female parent. When this is not possible, a genetic marker screen can be used. However, the problem is easier than identification of the proportion of nucellar seedlings. To identify a hybrid, the breeder need only find a single co-dominant locus for which the female and male parents do not share alleles, i.e. crosses of the configuration aa × bb, ab × cc or ab × cd. In each of these situations, the hybrid can be unambiguously distinguished from a nucellar seedling. Dominant markers can also be used if the female is 00 (band absent) and the male is 11 (homozygous band present). Note that in either case only a single locus is required, although confirmation with a second locus is always desirable.

Future Strategies

Nucellar embryony is currently considered to be an essential trait in citrus rootstocks because it provides a low-cost, low-technology method for propagation of genetically uniform citrus rootstocks. However, in a larger context, the requirement for nucellar embryony presents rootstock breeders with a significant additional trait that must be selected. Many potential rootstocks are discarded because they produce inadequate numbers of nucellar seedlings. Breeders could hybridize strictly sexual parents to produce large populations of novel combinations (pummelo × mandarin), select seedlings based on desired traits or markers for them, and use rooted cuttings to propagate selected hybrids vegetatively for rootstock trials. If such a programme produced rootstocks with clearly superior traits, then an innovative nursery might develop a method to propagate trees on such a rootstock economically, and growers may be willing to pay a premium for such trees. While this scenario may never become reality, if breeders never attempt certain crosses because of lack of nucellar embryony, their potential will never be known.

References

Anderson, C.M., Castle, W.S. and Moore, G.A. (1991) Isozymic identification of zygotic seedlings in Swingle citrumelo *Citrus paradisi* × *Poncirus trifoliata* nursery and field populations. *Journal of the American Society for Horticultural Science* 116, 322–326.

Ashari, S., Aspinall, D. and Sedgley, M. (1988) Discrimination of zygotic and nucellar seedlings of five polyembryonic rootstocks by isozyme analysis and seedling morphology. *Journal of Horticultural Science* 63, 695–703.

Asins, M.J., Garcia, M.R., Ruiz, C. and Carbonell, E.A. (2002) Molecular markers for the genetic analysis of apomixis. In: Jain, S.M., Brar, D.S. and Ahloowalia, B.S. (eds) *Molecular Techniques in Crop Improvement*. Kluwer Academic Publishers, Dordrecht, The Netherlands, pp. 265–281.

Asker, S.E. and Jerling L. (1992) *Apomixis in Plants*. CRC Press, Boca Raton, Florida.

Cameron, J.W. and Garber, M.J. (1968) Identical-twin hybrids of *Citrus* × *Poncirus* from strictly sexual seed parents. *American Journal of Botany* 55, 199–205.

Cameron, J.W. and Soost R.K. (1979) Sexual and nucellar embryony and F1 hybrids in advance crosses of *Citrus* with *Poncirus*. *Journal of the American Society for Horticultural Science* 104, 408–410.
Esan, A. and Soost, R.K. (1977) Adventive embryogenesis in citrus and its relation to pollination and fertilization. *American Journal of Botany* 64, 607–614.
Frost, H.B. (1926) Polyembryony, heterozygosis and chimeras in *Citrus*. *Hilgardia* 1, 365–402.
Frost, H.B. and Soost, R.K. (1968) Seed reproduction: development of gametes and embryos. In: Reuther, W., Batchelor, L.D. and Webber, H.J. (eds) *The Citrus Industry*, Vol. 1. University of California Press, Berkeley, California, pp. 290–324.
Garcia, R., Asins, M.J., Forner, J. and Carbonell, E.A. (1999) Genetic analysis of apomixis in *Citrus* and *Poncirus* by molecular markers. *Theoretical and Applied Genetics* 99, 511–518.
Iwamasa, M., Ueno, I. and Nishiura, M. (1967) Inheritance of nucellar embryony in citrus. *Bulletin of the Horticultural Research Station of Japan, Series B* 7, 1–8.
Kepiro, J.L. (2003) Molecular genetic analysis of nucellar embryony (apomixis) in *Citrus maxima* × *Poncirus trifoliata*. PhD dissertation, The University of California, Riverside, California.
Khan, I.A. and Roose, M.L. (1988) Frequency and characteristics of nucellar and zygotic seedlings in three cultivars of trifoliate orange. *Journal of the American Society for Horticultural Science* 113, 105–110.
Kobayashi, S., Ieda, I. and Nakantani, M. (1979) Studies on the nucellar embryogenesis in *Citrus* II. Formation of the primordium cell of the nucellar embryo in the ovule of the flower bud, and its meristematic activity. *Journal of the Japanese Society of Horticultural Science* 48, 179–185.
Koltunow, A.M., Soltys, K., Nito, N. and McClure, S. (1995) Anther, ovule, seed, and nucellar embryo development in *Citrus sinensis* cv. Valencia. *Canadian Journal of Botany* 73, 1567–1582.
Moore, G.A. and Castle, W.S. (1988) Morphological and isozymic analysis of open-pollinated *Citrus* rootstock populations. *Journal of Heredity* 79, 59–63.
Nasharty, A.H. (1945) Some morphological and physiological aspects of polyembryony in *Citrus*. PhD dissertation. University of California, Los Angeles.
Parlevliet, J.E. and Cameron, J.W. (1959) Evidence on the inheritance of nucellar embryony in citrus. *Proceedings of the American Society for Horticultural Science* 74, 252–260.
Roose, M.L. and Kupper, R.S. (1992) Causes and consequences of variability in rootstocks. In: Khan, I.A. (ed.) *Proceeding of the First International Seminar on Citriculture in Pakistan*. University of Agriculture, Faisalabad, Pakistan, pp. 122–130.
Ueno, I., Iwamasa, M. and Nishiura, M. (1967) Embryo number of various varieties of *Citrus* and its relatives. *Bulletin of the Horticultural Research Station of Japan, Series B* 7, 11–21.
Wakana, A. and Uemoto, S. (1988) Adventive embryogenesis in *Citrus* (Rutaceae). II. Post-fertilization development. *American Journal of Botany* 75, 1031–1047.
Webber, H.J. (1900) Complications in citrus hybridization caused by polyembryony. *Science* 11, 308.
Wilms, H.J., Van Went, J.L., Cresti, M. and Ciampolini, F. (1983) Adventive embryogenesis in *Citrus*. *Caryologia* 36, 65–78.
Xiang, C. and Roose, M.L. (1988) Frequency and characteristics of nucellar and zygotic seedlings in 12 citrus rootstocks. *Scientia Horticulturae* 37, 47–59.

6 Cytogenetics

MUHAMMAD JAFAR JASKANI[1], MITSUO OMURA[2] AND IQRAR A. KHAN[3]

[1]*Institute of Horticultural Sciences, University of Agriculture, Faisalabad 38040, Pakistan;* [2]*National Agricultural Research Organization (NARO), Okitsu 458-6, Shimizu, Shizuoka, 424-0292 Japan;* [3]*Nuclear Institute of Agriculture & Biology, Faisalabad, Pakistan/Sultan Qaboos University, Muscat, Oman*

Introduction

Citrus and related genera have 18 chromosomes in diploid somatic cells (Frost, 1925a; Nakamura, 1929). The basic chromosome number, x=9, is invariant in *Citroideae/Aurantioideae* (Stace *et al.*, 1993) and largely conserved in *Rutaceae* (Smith-White, 1954). Variant poloidies of different types have been reported (Guerra, 1984).

There have been spontaneously occurred tetraploids from nucellar seedling (Frost, 1925b). Tetraploid clones were generated by artificial induction using colchicine (Barrett, 1974; Oiyama and Okudai, 1986). Naturally occurring triploid 'Tahiti' lime, triploids between tetraploid and diploid crosses, and spontaneously occurring triploids from diploid crosses have been reported (Krug and Bacchi, 1943; Oiyama *et al.*, 1991; 1980). The percentage of triploid obtained was higher in the smaller seed with less than 0.1 g weight. The triploid seedless grapefruit cultivars, 'Oroblanco' and 'Melogold' have been released (Soost and Cameron, 1980, 1985).

Pentaploid, hexaploid as well as tetraploids were obtained from crosses between triploid and diploid (Esen and Soost, 1972a). These unexpected ploidy levels may have arisen from the functioning of doubly unreduced female gametes (Esen and Soost, 1973). Oiyama and Kobayashi (1993) obtained haploids from diploid 'Clementine' and 'Lee' pollinated with pollen from a triploid plant.

Haploid plantlet regeneration through gynogenesis in *Citrus clementina* Hort. ex Tan., cv. Nules, has been induced by *in vitro* pollination with pollen from 'Oroblanco', a triploid grapefruit hybrid (Gemanà and Chiancone, 2001).

Interspecific hybridization, ploidy level and the mono/polyembryonic nature of the variety in question may also contribute to the frequency of polyploid progenies (Cameron and Soost, 1969; Wakana *et al.*, 1981).

Aneuploids such as trisomics and monosomics are usually convenient tools for the genome mapping of the specific chromosomes. From the crossing between diploid and tetraploid, Esen and Soost (1972b) detected various aneuploids including 2n=22, 24, 25, 28, 29, 30, 31, 33,

37, 38, 39, and 41. In the crosses between diploid with triploid pollen, a number of trisomics (2n=19) were obtained (Sharma and Bal, 1957; Oiyama and Kobayashi, 1993). Several types of meiotic irregularities capable of producing aneuploid gametes have been reported by researchers (Raghuvanshi, 1962a,b; Naithani and Raghuvanshi, 1963).

Meiotic behavior of somatic hybrids provide valuable information for their practical utilization in citrus breeding programs. Generally abnormal tetrad formation and irregular chromosome behavior with univalent or multivalent pairing occur in somatic hybrid plants. Meiotic abnormalities such as chromosome bridges and chromosomes orientated away from the equatorial plate are frequently observed in somatic hybrids resulting in different sizes of pollen grain (Chen *et al.*, 2004).

Karyotype analysis in most plants has been identified by the length, and the positions of centromere and secondary constriction of each chromosome. Kandaelaki (1938) attempted to classify the metaphase chromosomes into 3 groups i.e. distinctively unequal-armed chromosomes, chromosomes with only slightly unequal-armed chromosomes, and with satellites. However, the staining methods used traditionally with aceto-carmine, aceto-orcein or Feulgen's solution were less informative to reveal detailed structure under the usual optical microscope because the mitotic chromosomes are very small (1.0-4.0 μm) and most of them are similar in morphology (Krug, 1943).

Recent developments in genomics require high resolution karyotype analysis so that chromosome map could provide the integrated information of physical mapping in *Citrus*. The techniques of karyotype analysis have been highly improved by the development of enzymatic maceration of specimens, fluorochrome staining for the structure analysis of chromosomes, and FISH on the mapping of the specific DNA sequences on chromosomes.

Citrus breeding programs need improved chromosomal analysis. Cultivated *Citrus* species have been hybridized with some wild relatives such as *Murraya, Severinia, Atalantia,* and *Swinglea,* in order to introduce desirable traits, mainly resistance to pests and pathogens (Barrett, 1977; Motomura *et al.*, 1995). The identification of chromosomes of different genomes could be a simple method of identifying citrus hybrids and is thus important for future work (Cameron and Frost, 1968).

Tissue Preparation Techniques

A general protocol for cytogenetic investigations is summarized in Table 1. To study the tissue by light microscopy the specimen is generally sliced into thin sections or

Table 1. A slide preparation protocol for observation of *Citrus* mitotic figures.

1. Collect root tips with 3-5 mm in length with vigorous growth and soak in water in refrigerator at 6 °C for 24 hours.
2. Fix with ice-cold fluid of methanol:acetic acid (3:1) for 1 day.
3. Transfer roots to ice-cold citrate buffer (0.01 M citric acid/sodium citrate, pH4.6).
4. Excise 10 root tips with @ 1mm long and transfer into enzyme solution (7-8% Pectinase Sigma + 2% Cellulase Onozuka in the citrate buffer) in an Eppendorf tube.
5. Spin down specimens by centrifugation and incubate at 37°C for 1.5 to 2.5 hours.
6. Centrifuge at 2500 rpm to discard supernatant, add new citrate buffer and suspend with gentle pipetting.
7. After washing with citrate buffer 3 times, add 100-200 μl cold methanol: acetic acid (3:1) and resuspend.
8. Drop cell maceration mixture on slide glass and air-dry.
9. Confirm the preparation by phase contrast microscopy.

Modified after Ito *et al.*, (1992)

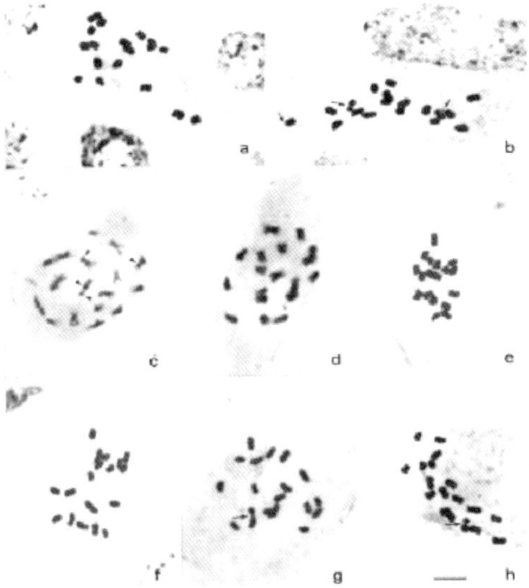

Fig. 6.1. Chromosomal variability in *Citrus* species. (a and b) Metaphases of *Citrus depressa* showing (a) no secondary constrictions (SECs) and (b) one proximal and two terminal SECs. Prophase (c), prometaphase (d) and metaphase (e) chromosome complement of *C. deliciosa*. Metaphase of *C. pennivesiculata* (f), *C. volkameriana* (g) and *Poncirus trifoliata* (h). Larger arrows, proximal SECs; small arrows, terminal SECs; arrowheads, heteropycnotic blocks. The bar in (h) represents 5 mm for all figure parts (Guerra *et al.*, 1997).

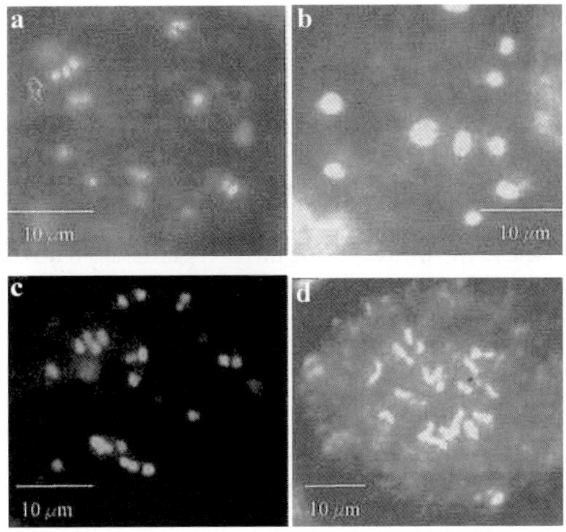

Fig. 6.2. (a) Diploid cell (2n = 18) of *Swinglea glutinosa* with chromosomes scattered; (b) multinucleated protoplast of 'Ruby Red' grapefruit; (c) cell from 'Ruby Red' grapefruit + 'Succari' sweet orange with 22 chromosomes; (d) a cell from *Swinglea glutinosa* + sour orange with 22 chromosomes (Louzada *et al.*, 2002).

squashed and contrast within tissues is induced using dyes or fluorochromes. Citrus chromosome preparations for routine examination can be made from pollen mother cells, root or shoot meristem and embryos (Guerra *et al.*, 1997). The metaphase stage is arrested by pre-treatment solutions before the fixation (Fig. 6.1). Mitotic analysis of the samples can be improved by pretreatments with 0.002 M 8-hydroxyquinoline (8HQ) for 1 h at room temperature and then 20-23 h at 10°C (Oiyama and Okudai, 1986). Also, 1,4-dichlorobenzene has been employed successfully as a pretreatment reagent in citrus karyotype studies (Jaskani, 1998). It straightens of the chromosome arms and prevents excessive clumping. Louzada *et al.* (2002) improved chromosome scattering and micronuclei visualization by treating suspension cells for 2 h with 1,4-dicholorobenzene (Fig. 6.2). Storage of root tips at 6°C enhanced mitotic index in different *Citrus* species (Ito *et al.*, 1993). The combination of 8HQ and 4°C treatment resulted in increased resolution of mitotic figures in the root tips (Roose *et al.*, 1998).

Fixation

The aim of fixation is to preserve the tissue in a state that most closely reflects a living cell. The choice of fixative is generally dependent on the tissue of interest as different fixatives can preserve particular tissue elements. However, the most commonly used fixatives for general plant histology are buffered aldehyde and formalin/acid/alcohol mixtures (FAA). FAA fixes nucleic acids very well but gives poor morphological preservation and makes the tissue hard (Spence, 2001). Chen *et al.* (2004) fixed flower buds for 48 h in Carnoy's solution of ethanol-acetic acid (3:1) at room temperature and kept in 70%

ethanol at 4°C until use, while, Louzada *et al.* (2002) fixed suspension cells in the same fixative but for 24 h.

Maceration

Softening of meristematic tissues is traditionally obtained by soaking in 1N HCl at 60°C for 10 min. Enzymatic digestion has been reported to improve karyotype analysis (Jamieson *et al.*, 1986). Fixed anthers macerated in a 100 µl enzyme mixture (2% cellulase and 0.5% pectolyase) in 75 mM KCl, pH 4.8, for 20 min at 37°C digested cell wall and improved mitotic index. Compared to the traditionally used maceration by soaking samples in 1N HCl at 60°C and squashing them on a glass slide, the enzymatic digestion of cell wall produces stable specimen for the karyotype analysis including *in situ* hybridization (Jamieson *et al.*, 1986). Citrus root tips fixed with in Carnoy's solution, are rinsed with water or 0.1 M citrate buffers and then macerated with the combination of cellulase and pectinase (Ito *et al.*, 1992). The optimum combinations of enzymes were different among experiments. For example, the root tip cells of *C. junos* seedling macerated in 2% cellulase Onozuka RS and 6% pectinase (Sigma) in citrate buffer, generated better preparation index with digestion at 37 °C for 2-2.5 hr (Ito *et al.*, 1992). A combination of 3% cellulase RS and 1% pectinase Y-23 were used for the observation of young leaf mitosis (Befu *et al.*, 2000).

Staining

Specific staining is obtained by using dye which has an affinity for a particular cell type or tissue element. The root meristems are squashed in a drop of 45% acetic acid and then stained in a 1% aceto-carmine or aceto-orcine (Rao *et al.*, 1992) or 2% Giemsa solution (Guerra *et al.*, 1997). Staining root tips in Schiff"s solution (leuco-basic fucsin) for 45 min in the dark enhances chromosome visualization. Another dye reported to stain anther squashes is Carbol Fuchsin solution (Chen *et al.*, 2004). To visualize karyotypes under fluorescence microscope, Louzada *et al.* (2002) stained digested cell suspensions with 4,6-diamidino2-phenylindole dihydrochloride (DAPI) @ 0.4 µg/ml.

In Situ Hybridization

Gall and Pardue (1969) and John *et al.* (1969) were the first to localize nucleic acids directly to biological material. They used radioactively labeled RNA to detect rDNA in cytological preparations. Important applications of *in situ* hybridization are to display gene expression in a particular organ, tissue, or cell layer of interest (RNA:RNA *in situ* hybridization), and to localize particular DNA sequences to chromosomes, interphase nuclei, or DNA fibre spreads (DNA:DNA *in situ* hybridization). The probes used for *in situ* hybridization can be classified into four categories.

1. Chromosome paints (DNA/DNA *in situ* hybridization): The probes are organized from flow-sorted chromosomes and are PCR amplified using degenerate oligonucleotide primers and labeled nucleotides. Chromosome paints are used to label and identify individual chromosomes in a cytological preparation. These probes have been used for phylogenetic studies and medical diagonosis of gross chromosomal abnormalities.
2. Total genomic DNA (DNA/DNA *in situ* hybridization): As total genomic DNA is labeled for hybridization, the technique is called as GISH. It can localize the parental origin of chromatin in hybrid organisms and can used to identify alien chromosomes and chromosomal segments. The technique can also help to determine the ancestry of natural allopolyploids, and detects intergenomic translocations in hybrids.
3. Cloned DNA fragments (DNA/DNA *in situ* hybridization): These are the most usually used types of probe, with sequences

cloned in bacteria as plasmids, cosmids, or bacterial artificial chromosomes (BACs), or in yeast as yeast artificial chromosomes (YACs). These probes are normally used to map genes and repetitive sequences on the chromosomes.

4. RNA probes (RNA/RNA *in situ* hybridization): The sequence of interest is frequently cloned into a vector containing the bacteriophage RNA polymerase promotor sequences. By using appropriate RNA polymerases, it is possible to generate single-stranded RNA probes complementary to the coding (sense) or non-coding (antisense) DNA strands.

Fluorescence *In Situ* Hybridization (FISH)

Fluorescence *in situ* hybridization (FISH) has considerably contributed to a better understanding of plant genome structure and evolution. Using probes for total genomic DNA, the technique facilitated identification of parental genomes in hybrids and individual chromosomes in chromosome complements (Jiang and Gill, 1994; Taketa *et al.*, 2000), analysis of genomic distribution of mobile genetic elements (Balint-Kurti *et al.*, 2000), integration of genetic and physical maps with marker tagged BAC clones (Yuan *et al.*, 2000), physical mapping of genes on chromosomes (Dolezelova *et al.*, 1998) and other repetitive DNA sequences (Schmidt and Heslop-Harrison, 1996). The theory is same as for southern hybridization, except that the DNA to which the probe will hybridize is the actual chromosome. The probe is labeled using fluorescently tagged nucleotides, added to a chromosomal preparation from the species of interest and viewed using a fluorescent microscope. The probe hybridizes to the complementary sequences. Since the technique uses a fluorescent probe, it is called fluorescence *in situ* hybridization or FISH. The protocol for FISH analysis is given in Table 2.

Fluorescent *in situ* hybridization has been used for gene mapping, for integrating genetic and physical maps, and for cytogenetic studies of citrus (Roose *et al.*, 1998). The rDNA probes (18S-5.8S-25S) labeled with biotin or rhodamine and 5S rDNA probes labelled with digoxigenin were applied to locate rDNA sites on root-tip metaphase chromosomes of *Citrus sinensis* L., *Poncirus trifoliata* L. Raf., and *Citrus* x *Poncirus* hybrids. Counterstaining with the fluorochromes chromomycin A3 and DAPI uniquely identified many but not all chromosomes. *C. sinensis* had five 18S-25S rDNA sites, *P. trifoliata* had seven, and three different *Citrus* x *Poncirus* hybrids had five or six sites. Four 5S rDNA sites were detected as linked to 18S-25S rDNA sites. Karyotype and molecular analysis of Trovita sweet orange chromosomes showed three CMA+/DAPI- heterochromatic regions (Fig. 6.3) positive to the 26S rDNA probe (Matsuyama *et al.*, 1996). Telomere arrays consisting of Tm(A)Gn were detected on the extreme ends of each chromosome and most of the CMA+/DAPI- heterochromatic regions lied close inside the telomere-specific repeated sequences. DNA fingerprinting of *Citrus* genomes using a satellite sequence composed of TGG repeats

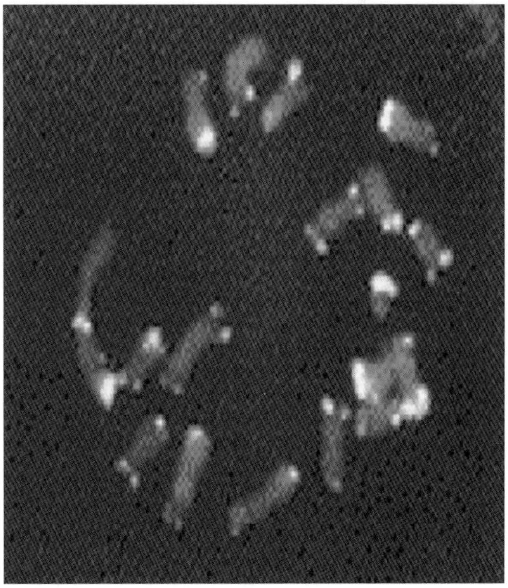

Fig. 6.3. FISH doublet signals on a sweet orange (Trovita) chromosome probed with a telomeric repeated sequence (Matsuyama *et al.*, 1996).

Table 2. A general protocol for FISH analysis.

1. Treatment to remove substances inhibiting the hybridization
   - i Wash slide with the mixture of 2% cellulase RS, 1.5% pectolyase, and 0.3% macerozyme R-200 with 1mM EDTA (pH 4.2) for 15 min at 37°C, following washing with twice of 2x SSC for 5 min .
   - ii Digest proteins by proteinase K (1µg/ml) solution in TE buffer (pH 8.0) for 15 min at 37°C, followed by washing with 2x SSC for 5 min twice.
   - iii Acidify with 45% acetic acid for 5 min at 37°C.
   - iv RNase digestion 0.1 mg/ml in 2x SSC for 1 hour at 37°C, then 2x SSC washing, and dehydration with 70%, 95% and 99% EtOH series before final air-drying.
2. Probe preparation (PCR labeling method)
   - i PCR random labeling with biotynylated dUTP (0.14 mM)+0.06 mM dTTP, 0.2 mM each dATP, dCTP, and dGTP, along with 0.5 µM primer pairs by 2.5 U Ampli TaqGold on 5 µl (20pmol) template DNA amplified 30 cycles of 94°C for 1 min, 55°C for 2 min and 72°C for 2 min after 3 min denaturation at 94°C, and followed with 7 min extension at 72°C.
   - ii Denaturation with formamide at 95°C for 10 min and finally in 2x SSC.
3. Hybridization
   - i Denaturation of chromosome specimen in the probe solution at 70°C for 6 min and incubate overnight (10 hr) at 38°C
   - ii Wash specimen by a series of solution of 2x SSC at 40°C for 10 min, 50% formamide in 2x SSC at 40°C for 10 min, 2x SSC at 40°C for 10 min, then 4x SSC at 40°C for 10 min.
   - iii Blocking by 5% BSA in BT buffer at 37°C for 5 min
   - iv Hybridization with avidin-FITC in 1% BSA containing 4x SSC for 60 min in the dark at 37°C, and then wash 3 times with BT buffer at 40°C for each 10 min.
   - v Blocking with 5% goat serum in BT at 37°C for 5 min.
   - vi Reaction with 2% Bio-Anti avidin solution in BT at 37°C for 30 min, and then wash 3 times with BT at 40°C of each 5 min.
   - vii Blocking again with 5% BSA in BT at 37°C for 5 min.
   - viii Reaction with 2% extra avidin –FITC solution in BT at 37°C for 30 min, and wash twice with BT at 40°C for 10 min.
   - ix Stain with PI added with 1% DABCO in phosphate buffer, and mount for observation.

and sequence tagged microsatellite site (STMS) markers has been reported (Matsuyama *et al.*, 1992; Kijas *et al.*, 1995). However, TGG repeat-related sequences were not found in *Citrus*-specific CMA+/DAPI- heterochromatic regions located at the extreme ends of each chromosome (Matsuyama *et al.*, 1999). This suggests that the TGG-repeated sequences are evolutionarily conserved and that the CMA+/DAPI- heterochromatic regions are added to the chromosome ends at a recent stage in *Citrus* evolution.

Multicolor FISH (MCFISH) using 5S and 45S rDNA specific probes simultaneously have provided valuable information on the evolution of rDNA sites and the relationships between wild and cultivated polyploid species (Taketa *et al.*, 1999; Schrader *et al.*, 2000; Mishima *et al.*, 2002). FISH using a 45S rDNA probe was found useful to elucidate the chromosomal location and the variation in the number of sites of 45S rDNA in 10 *Diospyros* species (Choi *et al.*, 2003).

Mitotic metaphase chromosomes are usually selected for FISH but pachytene (meiotic) chromosomes could be better substrates. Two homologous chromosomes are present in each pachytene combination that is joined along their entire length by a proteinaceous scaffold called the synaptonemal complex (SC) (Moses, 1968). Because each homologue contains two chromatids, there are four closely associated copies of each locus available for hybridization on a bivalent. In comparison, there are only two nearby copies of each locus available for FISH on a metaphase chromosome. In spreads of pachytene, chromosomes that have been prepared to reveal SCs (SC spreads), chromatin extends as a diffuse cloud around each SC. The loops of DNA extending from the SC appear to be more accessible to FISH probes than the DNA of condensed metaphase chromosomes (Moens and Pearlman, 1989; Heng *et al.*, 1994), and SC spreads can be prepared relatively free of overlying debris. Additionally, pachytene chromosomes are 5–15 times longer than corresponding metaphase chromosomes (Stack, 1984). The closely associated loci that are not resolvable by FISH on metaphase chromosomes should be resolvable on pachytene chromosomes.

Although the FISH techniques has high potential to identify the specific gene or regions of chromosome, the information of genetic linkage map has not been integrated onto the cytological chromosome map. Large insert genomic sequences in BAC have been constructed. The future development to increase the resolution of FISH signals will contribute to assign each chromosome to the consensus linkage group and also to the development of Citrus genomics to bridge between the linkage map and physical map.

Genomic *In Situ* Hybridization (GISH)

The genomic *in situ* hybridization (GISH) technique provides a direct and visual method for effectively determining the number and position of parental chromosomes. It has been extensively and successfully applied to the genetic identification of numerous interspecific and intergeneric plant hybrids (Gavrilenko *et al.*, 2001; Zhou *et al.*, 2001; Xia *et al.*, 2003).

Somatic hybrids can be analyzed by genomic *in situ* hybridization Fu *et al.* (2004) analyzed somatic hybrids combining Goutou sour orange (*Citrus aurantium* L.) with trifoliate orange [*Poncirus trifoliata* (L.) Raf]. GISH analysis confirmed that 18 chromosomes came from trifoliate orange and the remaining 18 from Goutou sour orange, as with most symmetric somatic hybrid plants; moreover, chromosome translocations were also observed in one plant.

Even though GISH is a powerful tool for parental genome analysis of citrus somatic hybrids (Matsuyama *et al.*, 1996; Pedrosa *et al.*, 2000),yet, few GISH studies in citrus somatic hybrids have been reported. Partly, because the citrus chromosomes being small and morphologically indistinguishable. Accomplishment of high-quality well-dispersed chromosome preparation need skills and practice. Fu *et al.* (2004) described an enzyme-macerating-flame method to prepare distinct and countable mitotic chromosomes to avoid intricacy in chromosome preparation.

Heterochromatin banding

Banding techniques for chromosomes can reveal the detailed structure of karyotype based on molecular features such as constitutive heterochromatic region with highly repetitive sequences or nucleolar organizing regions. Amongst such banding techniques, C-banding has been applied to *Citrus* chromosomes by Giemsa staining (Guerra, 1985; Wei et al., 1988). Liang (1988) identified Giemsa C-bands after the treatment with 5% $Ba(OH)_2$ at 50°C for 15 minutes. The heterochromatin blocks in *Poncirus* were predominantly telomeric and centromeric bands which were composed of highly repeated DNA sequences. They showed the heteromorphism among possible homologous chromosome pairs, including 3 pairs in *Fortunella margarita*, 4 pairs in *C. sinensis*, 3 pairs in *C. paradisi*, 0-4 pairs in mandarin species.

Compared to such staining as Giemsa and Feulgen methods, fluorochromes produce clear chromosomes because many of them directly stain DNAs revealed under epi-fluoresent microscope system. The staining of chromosome specimens by a fluorochrome 4',6-diamidino-2-phenylindole staining (DAPI) revealed clear chromosome figure in late prophase to early metaphase as well as typical metaphase cells (Ito *et al.*, 1992). DAPI strongly combines with AT rich sequence region of DNA and allows to visualize chromosomes clearly under UVepifluorescence microscopy. The method is convenient to count chromosome numbers instead of the conventional aceto-carmine or aceto-orcein staining methods. However, the simple staining with DAPI can not show the specific banding patterns in *Citrus* chromosome. The chromomycin A3 (CMA) stains strongly GC rich region of chromosome. To emphasize the fluoresence by CMA staining, the counter staining is recommended as described in Table 3. By the application of double staining method (Schweizer, 1976; Hizume *et al.*, 1989) with DAPI and CMA, Guerra (1993) identified 6 main CMA positive banding types of *Citrus* chromosome (Fig. 6.4): A type chromosome has two telomeric and one proximal band, B has one telomeric and one major proximal band, C has two telomeric bands, D has one large telomeric band, E has one small telomeric band, and F has bands absent or only very fine and not always visible. Abkenar *et al.,* (2007) described intergeneric and trigeneric hybrids using CMA banding patterns.

In detailed karyotype analyses of young leaves from various *Citrus* species, many variations have been identified as summarized in Table 4. Proximal heterchromatin with CMA+ were conspicuous in *Citrus* and they were highly variable among species, even distinguishing homoeologous pair in chromosome I. Six pairs of chromosomes in *Citrus* and *Poncirus* showed no or very small heterochromatin regions while all chromosomes have proximal bands in *Fortunella* (Miranda *et al.*, 1997). The amount of GC rich CMA-positive regions is high in *Fortunella* compared to other species/genera. The proportion of CMA-positive region to total chromosome length is 24.5% in *Fortunella hindsii* and 34.2% in *F. crassifolia*, and 21.4% in *Citrus grandis*, 20.5% in *C. sinensis*, and 22.0% in *C. succosa*. The similarities between *Citrus* and *Poncirus* suggests little heterochromatin diversification among karyotypes of these genera.

Table 3. Protocols for DAPI and CMA double staining for flurochrome banding of *Citrus* chromosome

A. Chromomycin A_3 (CMA) staining	
1. Rinse with McIlvaine buffer (pH 7.0) (0.63 g citric acid- 6.19 g Na_2HPO_4/500 ml)	30 min
2. Pre-treat with 2 or 3 drops of 0.1 mg/ml Distamycin A in buffer and cover with parafilm and keep in moistened box	10 min
3. Rinse briefly with buffer supplemented with 5mM $MgSO_4$	10 min
4. Stain with 2 or 3 drops of CMA staining solution 0.1 mg/ml in 5mM $MgSO_4$ buffer and covered with parafilm and keep in moistened box	10 min
5. Rinse with 5mM $MgSO_4$-buffer	10 min
6. Stain with 0.1 µg/ml DAPI in buffer	5 min
7. Wash with buffer	10 min
8. Cover with non-fluorescent glycerin mounted with glass/cover slip and store overnight in refrigerator	
9. Observe with epi-fluorescent B or BV excitation (420 nm)	
B. 4',6-diamidino-2-phenylindole (DAPI) staining	
1. Soaking with MacIlvaine buffer	30 min
2. Drop 0.25 mg/ml Actinomycin D and covered with parafilm in moistened box	10 min
3. Rinse with buffer	10 min
4. Staining with DAPI solution in buffer	5 min
5. Rinse with buffer 10 min	
6. Mount with non-fluorescent glycerin by glass slip and observe with UV excitation (355 nm)	

*2% Dabco (1,4-diazabicyclo[2.2.2] octane) for anti-fading in glycerol: MacIlvaine buffer 1:1 Modified after Hizume *et al.*, (1989).

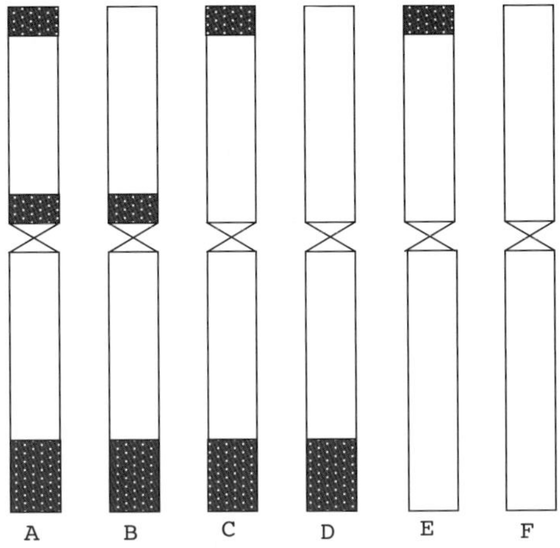

Fig. 6.4. Classification of karyotypes in CMA banding pattern in *Citrus* [after Guerra (1993), modified by Befu *et al.* (2000)].

Guerra *et al.* (2000) stained citrus chromosomes with the fluorochromes chromomycin A3 (CMA) and 4'-6-diamidino-2-phenylindol (DAPI) and examined variable number of regions that appeared bright or positive with CMA, and faint or negative with DAPI (Fig. 6.5). Following three types of banding patterns were recognized:

1. CMA^+ heterochromatin associated with nucleolus organizing regions (NORs). Only one pair of CMA^+ blocks per monoploid complement is examined in this type of heterochromatin. Sometimes it appears partially decondensed and as a secondary constriction. Generally, it represents a particular type of heterochromatin and the chromatin linked to the nucleolus is CMA^+ and GC rich (Schweizer, 1976).
2. CMA^+ heterochromatin not associated with the nucleolus organizing regions (NORs). The species having large number of CMA^+ blocks show that the most of them are not nucleolus-associated bands. In *Citrus sinensis* only three of the 16-18 CMA^+ blocks corresponded to rDNA sites by *in situ* hybridization (Matsuyama *et al.*, 1996) indicating that the most of the CMA^+ blocks were constituted by another DNA sequence.
3. CMA^- heterochromatin observed in proximal regions of some species. After conventional staining the proximal region of prophase chromosomes is heteropycnotic and appears to correspond to small, DAPI-brilliant chromocenters (Guerra, 1993) and was identified in a few chromosomes with the C-banding method (Guerra, 1985). However, Ito *et al.* (1993) examined occurrence of proximal chromatin in every chromosome of Trovita orange.

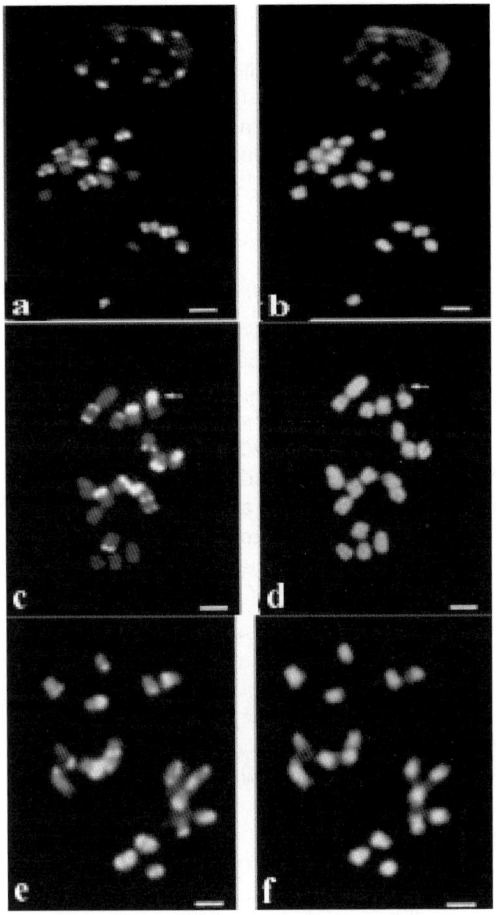

Fig. 6.5. Banding patterns of some Citrinae and Balsamocitrinae species. (a and b) Metaphase and interphase nucleus of *Poncirus trifoliata*; (c and d) *Citrus reticulata*. Arrows show a telomeric $CMA^+/DAPI^-$ band; (e and f) *Fortunella crassifolia*. (a), (c) and (e) were stained with CMA, and (b), (d) and (f) were stained with DAPI. Bars = 2.5 µm (Guerra *et al.*, 2000).

Table 4. Karyotypes of Citrus on the CMA banding pattern.

Species	Cultivar	Karyotype	Tissue	Reference
C. grandis	–	3A+3C+4D+8E	Root	Guerra (1993)
	Shadenyu	3A+2C+7D+6E	Root	Miranda *et al.* (1997)
	Tosa buntan	1A+1B+5C+2D+9E	Young leaf	Befu *et al.* (2000)
	Suisho-buntan	3A+3C+3D+9E	Young leaf	Befu *et al.* (2001)
C. medica	–	1A+1B+1C+9D+6E(F)	Root	Guerra *et al.* (1993)
	Fingered citron	2B+8D+8E	Young leaf	Befu *et al.* (2001)
C. sinensis	–	2B+2C+7D+7E(F)	Root	Guerra (1993)
	Trovita	2B+2C+7D+7E	Root	Matsuyama *et al.* (1996)
			Root	Miranda *et al.* (1997)
			Young leaf	Befu *et al.* (2000)
C. aurantifolia	–	2A+1C+7D+8E(F)	Root	Guerra (1993)
C. succosa	Honchiso	1A+1B+10D+6E	Root	Miranda *et al.* (1997)
C. unshiu	Okitsu wase	1A+1C+8D+8E	Young leaf	Befu *et al.* (2001)
C. leiocarpta	Koji	1C+8D+9E	Young leaf	Befu *et al.* (2001)
C. paradisi	Duncan GF	1A+2B+2C+4D+8E	Young leaf	Befu *et al.* (2001)
Poncirus trifoliata	–	2B+10D+6E	Root	Miranda *et al.* (1997)
	–	4B+8D+6E	Young leaf	Befu *et al.* (2000)
Fortunella crassifolia	–	2A+2C+14D	Root	Miranda *et al.* (1997)

Type E in this classification includes small telomeric bands and F for absent bands.

CMA banded chromosomes proved to be a very useful tool for carrying out a more comprehensive analysis of *Citrus* phylogeny. Chromosomes of type A and B were found only in a few mandarin hybrids and in the lime-lemon-citron-pummelo group (Guerra, 1993; Befu *et al.*, 2001). Therefore, it can be hypothesized that all mandarin karyotypes with these chromosome types are hybrids. Chromosomes of type C, present in Mediterranean mandarins and in most other accessions, may be part of the original mandarin karyotype. On the contrary, type E chromosome, found only in *C. depressa*, seems to be restricted to a small group of mandarin species. Cornelio *et al.* (2003) also revealed highly differentiated banding patterns in mandarin accessions and hybrids, and classified in four groups according to the presence/absence of CMA banded chromosomes. CMA banding patterns of chromosomes compared in 17 accessions of mandarins classified chromosomes into six types (A-F) based on the number and position of CMA positive bands (Yamamoto and Tominaga, 2003). Type F chromosomes are present only in some mandarins originating in Japan, which could thus be distinguished from mandarins originating in other areas. This is the first report of type F chromosome in citrus. Amounts of relative heterochromatin per karyotype vary largely (Guerra *et al.*, 2000). CMA-banded karyotype estimates a heterochromatin proportion of 20.58-22.74% in *Citrus* species (Miranda *et al.*, 1997).

Wide hybridization in *Citrus* affects karyotype stability. Chromosome organization at prophase show heteropycnotic blocks with Giemsa or Feulgen staining in many chromosomes at the proximal region but at terminal region of few (Guerra, 1985, 1987). Heteropycnotic regions show variation (Guerra *et al.*, 2000) and commonly maintained in species with a low DNA amount and small chromosome size (Guerra, 1987). Diploid genomes of *Citrus* appear relatively small (between 0.73 and 0.82 pg/2C for 2n=2x=18) and variations were observed within species as *C. reticulata* has the smallest nuclear genome while *C. medica* has the largest (Ollitrault *et al.*, 1994).

Cyto-taxonomy

Polyploid levels, chromosome number, their size and shape etc., are given due weightage for the classification in addition to morphological characters in cyto-taxonomic investigations.

Lapin (1937) confirmed polyploidy and noted peculiar configurations (quadrivalents) at metaphase attached with slender threads and rods, particularly the trivalents. Moreover, Kandelaki (1938) noted characteristic differences in citrus chromosome morphology. Cytological studies with respect to mitosis and meiosis of different species and biotypes of *Citrus* plants revealed variable chromosome size and shape (Rao *et al.*, 1992). Therefore, grouping all the species into *Citrus* according to Reece (1969) and Tanaka (1969) needs to be debated in view of their salient cyto-taxonomic characteristics.

Flow Karyotyping

A quick and accurate approach to look at changes in genome size during evolution and differentiation is flow cytometry (FCM). Polyploidy often accompanies differentiation, an important part of plant development, with different cell types having characteristic ploidies (Galbraith *et al.*, 1991; Bino *et al.*, 1993) where flow cytometry can be used as a rapid screening tool. The genome size of citrus is small with 382 Mb (Arumuganathan and Earle, 1991). FCM has shown that mean nuclear DNA content of the species of the genera *Fortunella* and *Citrus*, was 0.81pg/2C (Kayim *et al.*, 1998).

Flow cytogenetics can be used for detection of aberrant cell cycles, changes in nuclear DNA amounts and sorting of chromosomes for gene mapping and library construction (Heslop-Harrison, 1995). The sensitivity of the flow karyotyping can detect numerical and structural chromosome changes in plants, including chromosome polymorphism (Dolezel *et al.*, 1994). The numerical changes in chromosome have been demonstrated in barley, where trisomy of chromosome 6 was identified (Lee *et al.*, 2000). An alien chromosome presence was detected in an oat–maize addition line (Li *et al.*, 2001) and six wheat-rye lines, where flow karyotyping monitored the frequency of alien chromosomes in the population (Kubalakova *et al.*, 2003). Recognition of chromosome translocations and deletions has also been known in field bean, garden pea, barley, rye and wheat (Dolezel and Lucretti, 1995; Neumann *et al.*, 1998; Gill *et al.*, 1999; Lysak *et al.*, 1999; Vrana *et al.*, 2000; Kubalakova *et al.*, 2002, 2003). Detection of chromosomes is generally dependent on change in chromosome size and difference from the remaining chromosomes. Kubalakova *et al.* (2002) reported that a weakly differentiated chromosome can be detected by a characteristic change in the flow karyotype. Similarly, flow karyotyping is sensitive enough to detect polymorphism in relative DNA content of chromosomes in some agronomic crops, and the fingerprint patterns of flow karyotypes characteristic for certain cultivars are heritable (Lee *et al.*, 2000, 2002; Kubalakova *et al.*, 2002, 2003).

The results obtained todate mainly focus on several agronomic and vegetable crop species and confirmed the usefulness of flow cytogenetics. About 17 species have been flow karyotyped (Dolezel *et al.*, 2004). Flow cytometry is routinely used in identification of triploids and other polyploids of citrus at the Citrus Research and Education Center-University of Florida, Lake Alfred, USA. It is expected that the technology will play a greater role in the study of nuclear genome, gene isolation and mapping in citrus.

References

Abkenar, A.A., Matsumoto, R. and Tanaka, S. 2007. Parental choromosome differentiation in citrangequat [*Fortunella* sp. X (*Citrus sinensis* x *Poncirus trifoliate*)] using CMA banding patterns. *Scientia Horticulturae (in press).*

Arumuganathan, K. and Earle, E.D. (1991) Nuclear DNA content of some important plant species. *Plant Molecular Biology Reporter* 9, 208-218.

Balint-Kurti, P.J., Clendennen, S.K, Dolezelova, M., Valarik, M., Dolezel, J., Beetham, P.R. and May, G.D. (2000) Identification and chromosomal localization of the monkey retrotransposon in *Musa* sp. *Molecular and General Genetics* 263, 908-915.

Barrett, H.C. (1974) Colchicine induced polyploidy in *Citrus. Botanical Gazette* 135, 29-41.

Barrett, H.C. (1977) Intergeneric hybridization of *Citrus* and other genera in citrus cultivar improvement. *Proceedings of the International Society of Citriculture* 2, 586-589.

Befu, M., Kitajima, A. and Hasegawa, K. (2001) Chromosome composition of some *Citrus* species and cultivars based on the chromomycin A (3) (CMA) banding patterns. *Journal of the Japanese Society for Horticultural Science* 70, 83-88.

Befu, M., Kitajima, A., Ling, Y.X. and Hasegawa, K. (2000) Classification of 'Tosa-Buntan' pummelo (*Citrus grandis* [L.] Osb.), 'Washington' navel orange (*C. sinensis* [L.] Osb.) and trifoliate orange (*Poncirus trifoliata* [L.] Raf.) chromosomes using young leaves. *Journal of the Japanese Society of Horticultural Sciences 69*, 22-28.

Bino, R.J., Lanter, S., Verhoeven, H.A. and Kraak, H.L. (I993) Flow cytometric determination of nuclear replication stages in seed tissues. *Annals of Botany* 72, 181-187.

Cameron, J.W. and Frost, H.B. (1968) Genetics, breeding and nucellar embryony. In: Reuther, W., Batchelor, L.D. and Webber, H.J. (eds) *The Citrus Industry*, Vol. II. University of California Press, California, USA. pp. 325-370.

Cameron, J.W. and Soost, R.K. (1969) Characters of new populations of *Citrus* polyploids, and the relation between tetraploidy in the pollen parent and hybrid tetraploid progeny. *Proceedings of the first International Citrus Symposium* 1, 199-205.

Chen, C.L., Guo, W.W., Yi, H.L. and Deng, X.X. (2004) Cytogenetic analysis of two interspecific Citrus *allotetraploid somatic hybrids and their diploid fusion parents.* Plant Breeding *123, 332-337.*

Choi, Y.A., Tao, R., Yonemori, K. and Sugiura, A. (2003) Simultaneous visualization of 5S and 45S rDNAs in persimmon (*Diospyros kaki*) and several wild relatives (*Diospyros* spp.) by fluorescent in situ hybridization (FISH) and multiColor FISH (MCFISH). *Journal of the American Society of Horticultural Sciences* 128, 736-740.

Cornelio, M.T.M., Figueirôa, A.R.S., Santos, K.G.B., Carvalho, R., Soares Filho W.S. and Guerra, M. (2003) Chromosomal relationships among cultivars of *Citrus reticulata* Blanco, its hybrids and related species. *Plant Systematics and Evolution* 240, 149-161.

Dolezel, J. and Lucretti, S. (1995) High-resolution flow karyotyping and chromosome sorting in *Vicia faba* lines with standard and reconstructed karyotypes. *Theoretical Applied Genetics* 90, 797-802.

Dolezel, J., Kubalakova, M., Bartos, J. and Macas, J. (2004) Flow cytogenetics and plant genome mapping. *Chromosome Research* 12, 77-91.

Dolezel, J., Lucretti, S. and Schubert, I. (1994) Plant chromosome analysis and sorting by flow cytometry. *CRC Critical Reviews in Plant Sciences* 13, 275-309.

Dolezelova, M., Valarik, M., Swennen, R. and Dolezel, J. (1998) Physical mapping of the 18S-25S and 5S ribosomal RNA genes in diploid bananas. *Biologia Plantarum* 41, 497-505.

Esen, A. and Soost, R.K. (1972a) Tetraploid progenies from 2x X 4x crosses of *Citrus* and their origin. *Journal of the American Society of Horticultural Sciences* 97, 410-414.

Esen, A. and Soost, R.K. (1972b) Aneuploid in *Citrus. American Journal of Botany* 59, 473-477.

Esen, A. and Soost, R.K. (1973) Precocious development and germination of spontaneous triploid seed in citrus. *Journal of Heredity* 64, 147-154.

Frost, H. B. (1925a) The chromosomes of *Citrus. Journal of the Washington Academy of Sciences* 15, 1-3.

Frost, H. B. (1925b) Tetraploidy in Citrus. *Proceedings of the National Academy of Sciences USA* 11, 535-7.

Fu, C.H., Chen, C.L., Guo, W.W. and Deng, X.X. (2004) GISH, AFLP and PCR-RFLP analysis of an intergeneric somatic hybrid combining Goutou sour orange and *Poncirus trifoliata. Plant Cell Reports* 23, 391-396.

Galbraith, D.W., Harkins, K.R. and Knapp, S. (1991) Systemic endopolyploidy in *Arabidopsis thaliana. Plant Physiology* 96, 985-989.

Gall, J.G. and Pardue, M.L. (1969) Formation and detection of RNA-DNA hybrid molecules in cytological preparations. *Proceeding National Academy of Sciences USA* 63, 378-383.

Gavrilenko, T., Thieme, R. and Rokka V.M. (2001) Cytogenetic analysis of *Lycopersicon esculentum* + *Solanum etuberosum* somatic hybrids and their androgenetic regenerants. *Theoretical Applied Genetics* 103, 231-239.

Germanà M.A. and Chiancone B. (2001) Gynogenetic haploids of Citrus after *in vitro* pollination with triploid pollen grains. Plant Cell, Tissue and Organ Culture. 66: 59-66.

Gill, K.S., Arumuganathan, K. and Le, J.H. (1999) Isolating individual wheat (Triticum aestivum) chromosome arm by flow cytometric analysis of ditelosomic lines. *Theoretical Applied Genetics* 98, 1248-1252.

Guerra, M. (1984) Cytogenetics of *Rutaceae.* II. Nuclear DNA content. *Caryologia* 37, 219-226.

Guerra, M. (1985) Cytogenetics of *Rutaceae.* III. Heterochromatin patterns. *Caryologia* 38, 335-346.

Guerra, M. (1987) Cytogenetics of *Rutaceae.* IV. Structure and systematic significance of the interphase nuclei. *Cytologia* 52, 213-222.

Guerra, M. (1993) Cytogenetics of *Rutaceae.* V. High chromosomal variability in *Citrus* species revealed by CMA/DAPI staining. *Heredity* 71, 234-241.

Guerra, M., dos Santos, K.C.B., Silva, A.E.B. and Ehrendorfer, F. (2000) Heterochromatin banding patterns in *Rutaceae-Aurantioideae*-a case of parallel chromosomal evolution. *American Journal of Botany* 87, 735-747.

Guerra, M., Pedrosa, A. and Silva, A.E.B. (1997) Chromosome number and secondary constriction variation in 51 accessions of a citrus germplasm bank. Brazilian Journal of Genetics *120, 489-496.*

Heng, H.H.Q., Tsui, L.C. and Moens, P.B. (1994) Organization of heterologous DNA inserts on the mouse meiotic chromosome core. *Chromosoma 103,* 401-407

Heslop-Harrison, J.S. (1995) Flow cytometry and genome analysis. *Probe* 5, 14-17.

Hizume, M., Ohgiku, A. and Tanaka, A. (1989) Chromosome banding in the genus *Pinus.* II. Interspecific variation of fluorescent banding patterns in *P. densiflora* and *P. thunbergii. Botanical Magazine, Tokyo* 102, 25-36.

Ito, Y., Omura, M. and Nesume, H. (1993) Improvement of chromosome observation methods for *Citrus.* In: Hayashi, T., Omura, M. and Scott, N.S. (eds) *Techniques on gene diagnosis and breeding.* FTRS, Tsukuba, Ibaraki, Japan, pp. 31-38.

Ito, Y., Omura, M., Nesumi, H. and Yoshida, T. (1992) Improvement of preparation and observation methods for *Citrus* chromosomes. *Bulletin of Fruit Tree Research Station* 23, 57-66.

Jamieson, G., Evans, I.J. and Barnes, S.R. (1986) An enzymatic method of preparing plant chromosomes for *in situ* hybridization. *Stain Technology* 61, 21-25.

Jaskani, M.J. (1998) Interploid hybridisation and regeneration of Kinnow mandarin. PhD thesis, Department of Horticulture, University of Agriculture, Faisalabad, Pakistan.

Jiang, J. and Gill, B.S. (1994) Non-isotopic in situ hybridization and plant genome mapping: the first 10 years. *Genome* 37, 717-725.

John, H.A., Birnstiel, M.L. and Johns, K.W. (1969) RNA-DNA hybrids at the cytological level. Nature 223, 582-587.

Kandaelaki, G.V. (1938) Karyological investigations of some cultivated species and varieties of the genus Citrus. Tbiliss Botanichesk Institute (Tbilis) Trudy 4, 109-120 (*Biological Abstracts* 14, 4086; 1940).

Kayim, M., Koc, N.K. and Rokka, V.M. (1998) Variation of the nuclear DNA content of species of subtribe Citrinae (Rutaceae). *HortScience* 33, 1247-1250.

Kijas, J.M.H., Fowler, J.C.S. and Thomas, M.R. (1995) An evolution of sequence tagged microsatellite site markers for genetic analysis within *Citrus* and related species. *Genome* 38, 346-355.

Krug, C.A. (1943) Chromosome number in the subfamily *Aurantioideae* with special reference to the genus *Citrus. Botanical Gazette* 104, 602-611.

Krug, C.A. and Bacchi, O. (1943) Triploid varieties of *Citrus. Journal of Heredity* 34, 277-283.

Kubalakova, M, Vrana, J., Cihalikova, J., Simkova, H. and Dolezel, J. (2002) Flow karyotyping and chromosome sorting in bread wheat (*Triticum aestivum* L.). *Theoretical Applied Genetics* 104, 1362-1372.

Kubalakova, M., Valarik, M. and Bartos, J. (2003) Analysis and sorting of rye (Secale cereale L.) chromosomes using flow cytometry. *Genome* 46, 893-905.

Lapin, W.K. (1937) Investigations on polyploidy in citrus. *USSR. All-Union Science Research Institute Humid Subtropics Workshop* 1, 1-68. [In Russian, with English summary]

Lee, J.H., Arumuganathan, K. and Chung, Y.S. (2000) Flow cytometric analysis and chromosome sorting of barley (*Hordeum vulgare* L.). *Molecules and Cells* 10, 619-625.

Lee, J.H., Arumuganathan, K. and Kaeppler, S.M. (2002) Variability of chromosomal DNA contents in maize (*Zea mays* L.) inbred and hybrid lines. *Planta* 215, 666-671.

Li, L.J., Arumuganathan, K. and Rines, H.W. (2001) Flow cytometric sorting of maize chromosome 9 from an oat–maize chromosome addition line. *Theoretical Applied Genetics* 102, 658-663.

Liang, G. (1988) Studies on the giemsa C-banding patterns of some *Citrus* and its related genera. *Acta Genetica Sinica* 15, 409-415.

Louzada, E.S., del Rio, H.S., Xia, D. and Moran-Mirabal, J.M. (2002) Preparation and fusion of *Citrus* sp. Microprotoplasts. *Journal of the American Society of Horticultural Sciences* 127, 484-488.

Lysak, M.A., Cihalikova, J., Kubalakova, M., Simkova, H., Kunzel, G. and Dolezel, J. (1999) Flow karyotyping and sorting of mitotic chromosomes of barley (*Hordeum vulgare* L.). *Chromosome Research* 7, 431-444.

Matsuyama, T., Akihama, T., Ito, Y., Omura, M. and Fukui, K. (1996) Characterization of heterochromatic regions in 'Trovita'orange (*Citrus sinensis* Osbeck) chromosomes by the fluorescent staining and FISH method. *Genome* 39, 941-945.

Matsuyama, T., Akihama, T., Ito, Y., Omura, M. and Fukui, K. (1999) D*istribution of TGG repeat-related sequences in 'Trovita' orange (*Citrus sinensis *Osbeck) chromosomes. Genome* 42, 1251-1254.

Matsuyama, T., Motohashi, R., Akihama, T. and Omura, M. (1992) DNA fingerprinting in Citrus cultivars. *Japanese Journal of Breeding* 42, 155-159.

Miranda, M., Ikeda, F., Endo, T., Moriguchi, T. and Omura, M. (1997) Comparative analysis on the distribution of heterochromatin in *Citrus, Poncirus* and *Fortunella* chromosomes. *Chromosome Research* 5, 86-92.

Mishima, M., Ohmido, N., Fukui, K. and Yahara, T. (2002) Trends in site number change of rDNA loci during polyploid evolution in *Sanguisorba* (Rosaceae). *Chromosoma* 110, 550-558.

Moens, P.B. and Pearlman, R.E. (1989) Satellite DNA I in chromatin loops of rat pachytene chromosomes and in spermatids. *Chromosoma 98,* 287-294.

Moses, M.J. (1968) Synaptinemal complex. *Annual Review of Genetics 2,* 363-412.

Motomura, T., Hidaka, T., Moriguchi, T., Akihama, T. and Omura, M. (1995) Intergeneric somatic hybrids between *Citrus* and *Atalantia* or *Severinia* by electrofusion, and recombination of mitochondrial genomes. *Breeding Science* 45, 309-314.

Naithani, S.P. and Raghuvanshi, S.S. (1963) Cytogenetical studies in *Citrus*. Part I. *Genetica* 33, 301-312.

Nakamura, M. (1929) Cytological studies in the genus *Citrus*. I. On the Wase satsuma originated through bud variation. *Studia Citrologia* 3, 1-14 [In Japanese, with English summary].

Neumann, P., Lysak, M., Dolezel, J. and Macas, J. (1998) Isolation of chromosomes from *Pisum sativum* L. hairy root cultures and their analysis by flow cytometry. *Plant Science* 137, 205-215.

Oiyama, I. and Kobayashi, S. (1993) Haploids obtained from diploid x triploid crosses of citrus. *Journal of the Japanese Society of Horticultural Sciences* 62, 89-93.

Oiyama, I., Kobayashi, S., Yoshinaga, S., Ohgawara, T. and Ishii, S. (1991) Use of pollen from a somatic hybrid between *Citrus* and *Poncirus* in the production of triploids. *HortScience* 26, 1082.

Oiyama, I. and Okudai, N. (1986) Production of Colchicine-induced autotetraploid plants through micrografting in monoembryonic citrus cultivars. *Japanese Journal of Breeding* 36, 371-376.

Oiyama, I., Okudai, N. and Takahara, T. (1980) Studies on polyploid breeding in citrus. (3) Triploid progenies from small seeds of diploid sweet orange cultivars. *Journal of the Japanese Society of Horticultural Sciences* 52-53.

Ollitrault, P., Dambier, D., Luro, F. and Duperray, C. (1994) Nuclear genome size variation in Citrus. *Fruits* 49, 390-393.

Pedrosa, A., Schweizer, D. and Guerra, M. (2000) Cytological heterozygosity and the hybrid origin of sweet orange [*Citrus sinensis* (L.) Osbeck]. *Theoretical Applied Genetics* 100, 361-367.

Raghuvanshi, S.S. (1962a) Cytogenetical studies in genus *Citrus*. *Caryologia* 15, 143-149.

Raghuvanshi, S.S. (1962b) Cytogenetical studies in genus *Citrus*. IV. Evolution in genus *Citrus*. *Cytologia* 27, 172-188.

Rao, A.R., Khan, M.D., Hameed, M. and Mahmood, T. (1992) Polyploidy in *Citrus* species and biotypes in Punjab. In: Khan, I.A. (ed.) *Proceedings of the First International Seminar on Citriculture.* University of Agriculture Faisalabad, Pakistan, pp. 98-102.

Reece, P.C. (1969) Classification of Citrus. In: Chapman, H.D. (ed.) *Proceedings of the First International Citrus Symposium.* University of California, Riverside, 1, 429-434.

Roose, M.L., Schwarzacher, T. and Heslop-Harrison, J.S. (1998) The chromosomes of *Citrus* and *Poncirus* species and hybrids: Identification of characteristic chromosomes and physical mapping of rDNA loci using *in situ* hybridization and fluorochrome banding. *Journal of Heredity* 89, 83-86.

Schmidt, T. and Heslop-Harrison, J.S. (1996) The physical and genomic organization of microsatellite in sugar beet. *Proceedings of the National Academy of Sciences, USA* 93, 8761-8765.

Schrader, O., Budahn, H. and Ahne, R. (2000) Detection of 5S and 25S rRNA genes in *Sinapis alba, Raphanus sativus* and *Brassica napus* by double fluorescence *in situ* hybridization. *Theoretical Applied Genetics* 100, 665-669.

Schweizer, D. (1976) Reverse fluorescent chromosome banding with chromomycin and DAPI. *Chromosoma* 58, 307-324.

Sharma, A.K. and Bal, A.B. (1957) Chromosome studies in *Citrus.* I. *Agronomia Lusitana* 19, 101-126.

Smith-White, S. (1954) Chromosome numbers in the Boronieae (Rutaceae) and their bearing on the evolutionary development of tribe in the Australian flora. *Australian Journal of Botany* 2, 287-303.

Soost, R.K. and Cameron, J.W. (1969) Tree and fruit characters of *Citrus* triploids from tetraploid by diploid crosses. *Hilgardia* 39, 569-579.

Soost, R.K. and Cameron, J.W. (1980) `Oroblanco', a triploid pummelo-grapefruit hybrid. *HortScience* 15, 667-669.

Soost, R.K. and Cameron, J.W. (1985) `Melogold', a triploid pummelo-grapefruit hybrid. *HortScience* 20, 1134-1135.

Spence, J. (2001) Plant histology. In: Hawes, C. and Satiat-Jeunemaitre, B. (eds) *Plant Cell Biology: A Practical Approach,* Second Edition. Oxford University Press Inc., New York, pp. 189-206.

Stace, H.M., Armstrong, J.A. and James, S.H. (1993) Cytoevolutionary patterns in Rutaceae. *Plant Systematics and Evolution* 187, 1-28.

Stack, S. M. (1984) Heterochromatin, the synaptonemal complex, and crossing over. *Journal of Cell Science 71,* 159-176.

Taketa, S., Ando, H., Takeda, K., Harrison, G.E. and Heslop-Harrison, J.S. (2000) The distribution, organization and evolution of two abundant and widespread repetitive DNA sequences in the genus *Hordeum. Theoretical Applied Genetics* 100, 169-176.

Taketa, S., Harrison, G.E. and Heslop-Harrison, J.S. (1999) Comparative physical mapping of the 5S and 18-25S rDNA in nine wild *Hordeum* species and cytotypes. *Theoretical Applied Genetics* 98, 1-9.

Tanaka, T. (1969) Taxonomic problems of citrus fruits. *Orient. Bulletin of the University of Osaka Prefecture,* Series B 21, 133-138.

Vrana, J., Kubalakova, M., Simkova, H., Cihalikova, J., Lysak, M.A. and Dolezel, J. (2000) Flow-sorting of mitotic chromosomes in common wheat (*Triticum aestivum* L). *Genetics* 156, 2033-2041.

Wakana, A., Iwamasa, M. and Uemoto, S. (1981) Seed development in relation to ploidy of zygotic embryo and endosperm in polyembryonic citrus. *Proceedings International Society of Citriculture* 1, 35-39.

Wei, W., Cheng, Y. and Duan, Y. (1988) Studies on the evolution of *Citrus* based on karyotype and C-banding patterns. *Acta Horticultural Sinica* 15, 223-228.

Xia, G., Xiang, F., Zhou, A., Wang, H. and Chen, H. (2003) Asymmetric somatic hybridization between wheat (*Triticum aestivum* L.) and *Agropyron elongatum* (Host) Nevishi. *Theoretical Applied Genetics* 107, 299-305.

Yamamoto, M. and Tominaga, S. (2003) High chromosomal variability of mandarins (*Citrus* spp.) revealed by CMA banding. *Euphytica 129,* 267-274.

Yuan, Q., Liang, F., Hsiao, J., Zismann, V., Benito, M.I., Quackenbush, J., Wing, R. and Buell, R. (2000) Anchoring of rice BAC clones to the rice genetic map in silico. *Nucleic Acids Research* 28, 3636-3641.

Zhou, A., Xia, G., Zhang, X., Chen, H. and Hu, H. (2001) Analysis of chromosomal and organellar DNA of somatic hybrids between *Triticum aestivum* and *Haynaldia villosa* Schur. *Molecular Genetics and Genomics* 265, 387-393.

7 Haploidy

Maria Antonietta Germanà

Dipartimento SENFIMIZO, Sezione di Frutticoltura Mediterranea, Tropicale e Sutropicale, Facoltà di Agraria, Università degli Studi di Palermo, Viale delle Scienze, 11, 90128 Palermo, Italy

Introduction

Citrus, with 105.4 Mt in 2005 (FAOSTAT, Database), are the most widely cultivated fruit trees in the world.

The most commercially important *Citrus* species are oranges (*C. sinensis* L. Osbeck) and tangerines (*C. unshiu* Marc., *C. nobilis* Lour., *C. deliciosa* Ten., *C. reticulata* Blanco and their hybrids) (>80%), followed by lemons (*C. limon* L. Burm. f.), limes (*C. aurantifolia* Christm. Swing.) and grapefruits (*C. paradisi* Macf.) in almost equal proportions.

All cultivated forms of *Citrus* and related genera (*Poncirus*, *Fortunella*, etc.) are diploid with a monoploid number of chromosomes ($n = x = 9$) (Frost, 1925). Triploid and tetraploid forms of *Citrus* also exist. Most species of the genus *Citrus* are characterized by polyembryony, which consists of the production of 1–40 adventive embryos by the nucellus (Fusurato, 1957) so that two or more embryos develop in a single seed.

The main goals of *Citrus* breeding are to obtain new varieties with a shorter juvenile non-fruiting period, an increased yield, a longer ripening season, regular fruit bearing, seedlessness and improved external and internal quality of the fruits. To make available new scions and rootstocks selected for resistance or tolerance to biotic and abiotic stresses is another important aim in *Citrus* improvement.

Citrus breeding is based either on conventional methods (hybridization, selection, mutation) or biotechnological methods employing *in vitro* tissue culture, regeneration from protoplasts, somatic hybridization, *in vitro* mutant selection, genetic transformation and haploid production. An integrated approach between innovative and conventional tools is fundamental to obtaining large improvements in a short time.

The Importance of Haploids and Doubled Haploids

The life of higher plants comprises a sporophytic ($2n$) and a gametophytic (n) generation, where the gametes are products of meiosis, and the sporophytic generation, usually diploid, develops from the zygote produced by fertilization. In the course of evolution, the gametophytic generation has been progressively reduced in terms of both size and lifespan, in comparison with the sporophytic generation. For this reason, the influence of the gametophytic phase on the sporophyte is usually underestimated and gametes are only considered as vectors for transmission of the genome to the next sporophytic generation; Heslop-Harrison (1979) termed this step of angiosperm life the 'forgotten generation'.

More recently, both basic and applied studies have improved the knowledge of pollen biology and pollen biotechnology, making the manipulation of pollen development and function a reliable tool for crop improvement (Mulcahy, 1986). The most important application of pollen biotechnology in breeding and genetic studies is the ability to obtain haploids and doubled haploids.

Haploids are sporophytic plants with the gametophytic chromosome number because they originate from a single gamete.

The importance of haploids in plant breeding and genetic research was recognized with the discovery of the first natural haploid in *Datura stramonium* and *Nicotiana* (Blakeslee *et al.*, 1922, 1924; Kostoff, 1929), but long before techniques for producing haploids by *in vitro* gametic embryogenesis became available.

The discovery made by Guha and Maheshwari (1964, 1966) that, by *in vitro* culture of immature anthers of *Datura innoxia*, a change in the normal gametophytic development into sporophytic development can be induced and embryoids with a haploid chromosome number can be obtained, led to further and extensive research on anther culture.

The interest of breeders in haploids or, by doubling the chromosome numbers, doubled haploids, lies in the possibility of shortening the time needed to produce completely homozygous lines compared with conventional breeding. In fact, haploidiploidization through gametic embryogenesis allows the single-step development of complete homozygous lines from heterozygous parents. In a conventional breeding programme, pure lines are developed after several generations of selfing and still may not be 100% homozygous. In the case of woody plants, such as *Citrus*, generally characterized by a long reproductive cycle, a high degree of heterozygosity, large size and, sometimes, self-incompatibility, there is no other chance to obtain haploidization through conventional methods. Actually, the absence of pure lines in woody plants makes genetic studies rather difficult.

Haploid plants, with a gametophytic set of chromosomes in the sporophyte, and homozygous doubled haploid plants aroused interest in the fields of genetic and developmental studies, as well as for plant breeding. In fact, they have a potential use in mutation research, selection, genetic analysis, the production of inbred lines required to utilize hybrid vigour (heterosis), and genetic transformation.

New superior cultivars produced via gametic embryogenesis (above all through anther culture) have been reported for rice, wheat, tobacco, maize and pepper (Evans, 1989), and doubled haploids are being routinely used in breeding programmes for new cultivar development in many crops such as aubergine, pepper, barley and rape (Veilleux, 1994).

Often, *in vitro* regenerated plants show differences in their morphological and-biochemical characteristics, as well in chromosome number and structures. 'Gametoclonal variation', the variation observed among plants regenerated from cultured gametic cells (Evans *et al.*, 1984; Morrison and Evans, 1987), is another opportunity to use haploids in crop improvement. Unlike 'somaclonal variation' which is related to the variation among plants regenerated from cultured cells or tissue (Larkin and Scowcroft, 1981), gametoclonal variation results from both meiotic and mitotic division. Moreover, because of their homozygosity, in the gametoclones it is possible to observe the direct expression of both dominant and recessive mutations. Several different sources of variation have to be considered in order to explain gametoclonal variation, such as new genetic variation induced by the cell culture procedures, new variation resulting

Table 7.1. Haploids and doubled haploids obtained in Citrus by gynogenesis.

Genotypes	Response	References
Citrus clementina Hort. ex Tan.	(×) plantlets	Oiyama and Kobayashi, 1993
Lee mandarin (clementine × Orlando tangelo)	(×) plantlets	Oiyama and Kobayashi, 1993
Citrus clementina Hort. ex Tan.	(×) plantlets	Ollitrault *et al.*, 1996
Citrus clementina Hort. ex Tan.	(×) plantlets	Germanà and Chiancone, 2001

from segregation and independent assortment, new variation induced by the chromosome doubling procedure and new variation induced at diploid level, resulting in heterozygousity (Morrison and Evans, 1987; Huang, 1996).

Doubled haploids can also increase the efficiency of crop breeding programmes because they are important in genome mapping, providing excellent material to obtain reliable information on the location of major genes and qualitative trait loci (QTLs) for economically important traits (Khush and Virmani, 1996).

Haploids and Haploid Production in Citrus

In *Citrus natsudaidai*, haploid seedlings were first obtained by the application of gamma rays (Karasawa, 1971).

One haploid embryo was obtained in an immature seed from a diploid (Clementine mandarin) × diploid (Pearl tangelo) cross (Esen and Soost, 1972).

Haploids can be induced mainly through two strategies: from the female gamete, and from the male gamete. Although a lot of research has been carried out on gametic embryogenesis in *Citrus* spp. and their relatives (Germanà, 1997), not much of it has been successful (Tables 7.1 and 7.2).

Anther culture technique has been regularly employed to recover haploids in *Citrus*. Nevertheless, since many *Citrus* species proved to be recalcitrant to this method, other methods have been and could be attempted to produce haploids.

Moreover, a comparison of doubled haploid plants coming from male or female gametes could be interesting because the method of producing doubled haploids seems to produce differences in doubled haploid field performance. The maternally derived homozygous regenerants are generally more vigorous after *in vivo* transfer and have less gametoclonal variation than doubled haploids obtained from male gametes (Snape *et al.*, 1988; Wernsman *et al.*, 1989).

Table 7.2. Haploids and doubled haploids (DH) obtained in *Citrus* and its relatives by anther culture.

Genotypes	Response	References
Poncirus trifoliata L. Raf.	(×) plantlets	Hidaka *et al.*, 1979
C. madurensis Lour.	(×) (2×) plantlets	Chen *et al.*, 1980
Hybrid No. 14 of *C. ichangensis* × *C. reticulata*	(×) plantlet	Deng *et al.*,1992
Citrus clementina Hort. ex Tan.	(×) DH (3×) plantlets	Germanà *et al.*, 1994, 2000a, 2005, Germanà and Chiancone, 2003
C. limon L. Burm. f.	(×) calli	Germanà *et al.*, 1991
Mapo tangelo (*C. deliciosa* × *C.* paradisi)	(×) embryoids	Germanà and Reforgiato, 1997
Clausena excavata	(×) embryoids	Froelicher and Ollitrault, 2000

Gynogenesis

Selection of seedlings

Parthenogenesis (the production of an embryo from an egg cell without the participation of the male gamete) and apogamy (the production of an embryo from a gametophytic cell other than the ovum) are the causes of spontaneously occurring haploids.

Spontaneously developed haploids were reported in over 100 angiosperm species (Kasha, 1974). Most of the haploids in fruit trees (especially pome and stone fruits) are of spontaneous origin (Zhang *et al.*, 1990). Spontaneous haploid plants were obtained in apple, pear, peach, plum, apricot, etc., but in very low number and they were not very viable (Zhang *et al.*, 1990).

In situ *parthenogenesis induced by irradiated pollen followed by* in vitro *culture of embryos*

Parthenogenesis induced *in vivo* by irradiated pollen, followed by *in vitro* culture of embryos, can be an alternative method of obtaining haploids. Gynogenesis by *in situ* pollination with irradiated pollen has been successfully used for *Petunia* (Raquin, 1985), *Cucumis melo* (Sauton and Dumas de Vaulx, 1987), *Daucus carota* (Rode and Dumas de Vaulx, 1987), *Malus domestica* (Zhang and Lespinasse, 1991; Hofer and Lespinasse, 1996), *Pyrus communis* (Bouvier *et al.*, 1993), *Actinidia deliciosa* (Pandey *et al.*, 1990; Chalak and Legave, 1997) and *Citrullus lanatus* (Sari *et al.*, 1994).

The production of nine haploid plantlets, which did not survive, and two embryogenic callus lines was achieved in clementine (*Citrus clementina* Hort. ex Tan.), cv. SRA 63 after *in situ* parthenogenesis induced by pollen of Meyer lemon (*Citrus meyeri* Y. Tan.) irradiated at 300, 600 and 900 Gy from a ^{60}Co source (Ollitrault *et al.*, 1996). Flowers of clementine SRA 63 were pollinated in the field with the irradiated pollen; fruits were picked at maturity and embryos were cultivated *in vitro.*

The method is based on the *in vitro* culture of immature seeds or embryos obtained as a result of pollination with pollen irradiated by γ-rays from ^{60}Co, and it should be tested in those species in which *in vitro* anther culture and gynogenesis have not been successfully applied. Irradiation does not hinder pollen germination, but prevents pollen fertilization, stimulating the development of haploid embryoids from ovules. The success of this technique is dependent on the choice of radiation dose, the developmental stage of the embryos at the time of culture, the culture conditions and the media requirements.

In situ *or* in vitro *parthenogenesis induced by triploid pollen followed by* in vitro *culture of embryos*

Three haploid plants were obtained from two monoembryonic diploid (clementine and 'Lee') × triploid hybrid of 'Kawano natsudaidai' (*C. natsudaidai*) *in vivo* crosses (Oiyama and Kobayashi, 1993). Triploidy of pollen, like irradiation, does not hinder pollen germination, but prevents pollen fertilization and stimulates the development of haploid embryoids from ovules. Haploid and diploid embryoids did not show any difference in their size; however, haploid seedlings grew very slowly in the soil. Restriction endonuclease analyses of nuclear rDNA and of chloroplast DNA determined the maternal origin of these haploids.

Haploid plantlet regeneration through gynogenesis in *C. clementina* Hort. ex Tan., cv. Nules, has been induced by *in vitro* pollination with triploid pollen (Germanà and Chiancone, 2001). The pollen source chosen was 'Oroblanco', a triploid grapefruit-type citrus obtained in 1958 through a cross between an acidless pummelo (*Citrus grandis* Osbeck) and a seedy, tetraploid grapefruit (*C. paradisi* Macf.) (Soost and Cameron, 1980).

The *in vitro* stigmatic pollination technique consists of applying pollen to the apical part of the stigma of an excised gynoecium implanted in solid culture

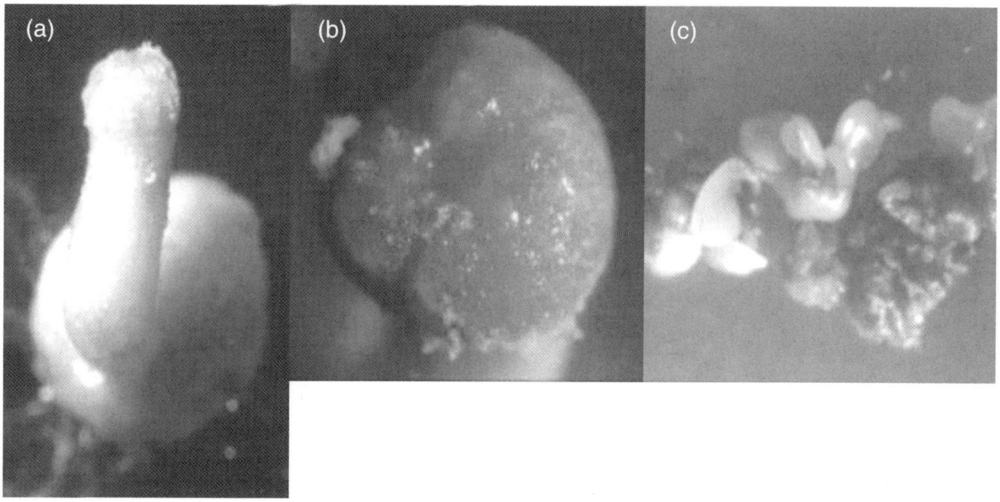

Fig. 7.1. (a) An excised gynoecium of *Citrus clementina*, cv. Nules implanted in solid medium and pollinated with triploid pollen grains. (b) Stigma exudate and 'Oroblanco' pollen grains on the stigma of a pistil taken from a mature bud flower. (c) Gynogenic embryoids breaking through the 'Nules' ovary four months after *in vitro* pollination with triploid 'Oroblanco' pollen grains.

medium (Fig. 7.1a and b). Some ovaries were transformed into brownish and friable callus, sometimes breaking to reveal ovules. From this kind of ovary, the gynogenic embryoids emerged (Fig. 1c) 4–5 months after *in vitro* pollination, which is practically the same time required for androgenesis. The pollination and mature stage of pistils were necessary for gynogenic embryoid regeneration. Although unlike clementine anther culture (Germanà *et al.*, 1994, 2000a), embryogenic calluses were not obtained, parthenogenesis induced *in vitro* by triploid pollen can be usefully employed in attempting to obtain haploids in monoembryonic genotypes of *Citrus* for which androgenesis has not yet been successful in haploid production.

Anther and isolated microspore technique

The *in vitro* anther or isolated microspore culture technique is usually the most effective and widely used method of producing haploids and doubled haploids.

Regeneration from male gametes has been reported in about 200 species belonging to some families, such as *Solanaceae*, *Cruciferae* and *Gramineae*; many other families (*Leguminosae* and woody plants) appear, instead, rather recalcitrant (Dunwell, 1986; Hu and Yang, 1986; Sangwan-Norrel *et al.*, 1986; Bajaj, 1990; Raghavan, 1990; Wenzel *et al.*, 1995). Since 1970, extensive research has been carried out to obtain haploids by anther culture or gynogenesis for perennial species breeding, with not always satisfactory results.

The cellular, biochemical and molecular bases for the transformation of microspores into pollen embryoids have not yet been completely understood. However, it is already possible to indicate some findings. For example, it is known that the capacity to regenerate from a male gamete is genetic and inheritable, and that the stage of microspore development is critical for induction. Usually in the period around the first haploid mitosis (late uninucleate or early bicellular pollen stage), male gametes become competent to differentiate in a different way from the gametophytic pathway with continued growth and division. Moreover, external stresses are necessary to enable competent microspores to undergo embryogenic development. The stress can be physical (also wounding connected to the anther excision and culture), thermal (heat, cold) or chemical (water stress, starvation). The induced microspores are characterized by an altered synthesis and an accumulation of RNA and proteins, and it seems that the genes involved in this reprogramming are stress-related and/or associated with the zygotic embryogenesis.

Anther culture

Anther culture is an easier and more commonly used method than isolated microspore culture, because the isolation of microspores requires a higher degree of skill and better equipment than anther culture (Heberle-Bors, 1989).

Research on haploidization by anther culture has been carried out on several fruit trees, especially in pome and stone fruits, and sometimes produced embryoids that rarely germinated (Zhang *et al*., 1990). The induction of embryogenesis from cultured apple anthers is still low and highly genotype dependent (Hofer, 1995, 1997).

As regards *Citrus* and their relatives, by anther culture, haploid plantlets have been recovered from *Poncirus trifoliata* L. Raf. (Hidaka *et al*., 1979) and *C. madurensis* Lour. (Chen *et al*., 1980), and one doubled haploid plantlet has been obtained from the hybrid No. 14 of *C. ichangensis* × *C. reticulata* (X.X. Deng *et al*., 1992); haploid plantlets and highly embryogenic haploid calli of *C. clementina* Hort. ex Tan. (Germanà *et al*., 1994, 2000a; Germanà, 2003a; Germanà and Chiancone, 2003); haploid, but albino embryoids of Mapo tangelo (*C. deliciosa* × *C. paradisi*) (Germanà and Reforgiato, 1997) were produced; haploid and diploid calli, embryoids and leafy structures but no green plants of *C. limon* L. Burm. f. (Germanà *et al*., 1991); and haploid embryoids of *Clausena excavata* (Froelicher and Ollitrault, 2000) have also been achieved (Table 7.2).

In *Citrus* and its relatives, Hidaka *et al*. (1979) first reported anther culture in *P. trifoliata* and later in *C. aurantium* (1981) and *C. sinensis* (1984a, b); Chen *et al*. (1980) reported anther culture in *C. madurensis*; and Chaturvedi and Sharma (1985) worked on *C. aurantifolia* anther culture. Ling *et al*. (1988) studied *C. madurensis* anther culture. Further studies on anther culture of several *Citrus* genotypes were carried out by Starrantino (1986) and Geraci and Starrantino (1990). X.X. Deng *et al*. (1992) reported on *P. trifoliata* and a hybrid of *C. ichangensis* × *C. reticulata* anther culture. Germanà reported research on *C. limon* (Germanà *et al*., 1991) and *C. clementina* (Germanà *et al*., 1994, 2000a; Germanà, 2003a), *C. reticulata* (Germanà *et al*., 1994) and tangelo Mapo (Germanà and Reforgiato, 1997) anther culture.

Only heterozygous plantlets have been obtained by anther culture in *C. aurantium* (Hidaka *et al*., 1981; Germanà *et al*., 1992; Germanà, 2003b), *C. sinensis* (Hidaka, 1984b), *C. aurantifolia* (Chaturvedi and Sharma, 1985), *C. madurensis* (Ling *et al*., 1988), *C. reticulata* (Germanà *et al*., 1994; Germanà, 2003b) (Fig. 7.2) and *P. trifoliata* (Deng *et al*., 1992). Anther culture produced embryonal structures in *C. sinensis* and *C. paradisi* (Starrantino, 1986), and mixoploid calli were obtained from *C. reticulata*, *C. deliciosa*, *C. sinensis*, *C. limon*, *C. paradisi* and Mapo tangelo (Geraci and Starrantino, 1990).

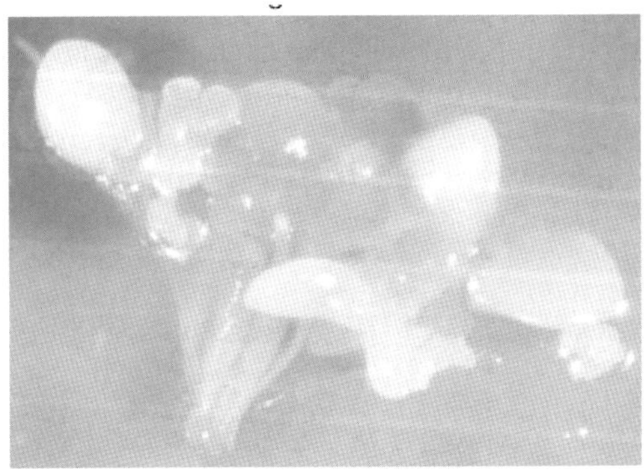

Fig. 7.2. A cluster of somatic embryoids from *Citrus reticulata* (cv. Avana) anther culture.

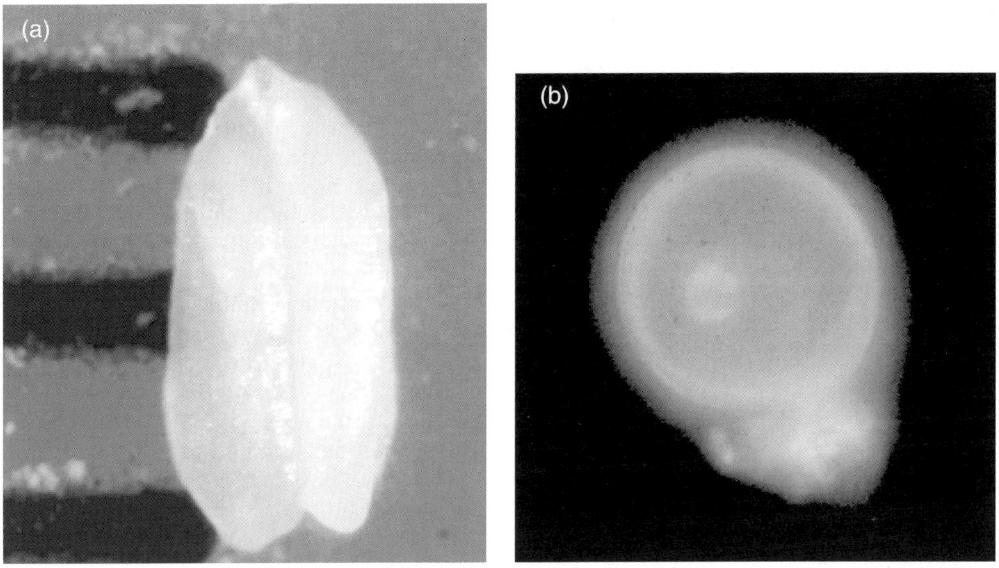

Fig. 7.3. (a) Anther of *C. clementina* Hort. ex Tan. with microspores at the uninucleate stage before culture. (b) A microspore at the uninucleate stage.

In these cases, anther culture could be regarded as a method of obtaining somatic embryogenesis, which is a very efficient method of regeneration. Embryogenic callus is valuable for propagation or genetic improvement and can be used for somatic hybridization by protoplast fusion, genetic transformation, synthetic seed production and germplasm storage.

Anther culture technique in Citrus

Floral buds, with the pollen grains at a specific stage of development, are collected from the donor plant. After pre-treatment, the buds are surface sterilized by immersion for 3 min in 70% (v/v) ethyl alcohol, followed by immersion in sodium hypochlorite solution (~1.5% active chlorine in water) containing a few drops of Tween-20 for 15–20 min, and finally rinsed three times for 5 min with sterile distilled water. Petals are aseptically removed with small forceps, and anthers (Fig. 7.3a) are carefully dissected and placed into the medium.

The stage of pollen development is commonly determined by staining one or more anthers per bud with acetocarmine, Schiff's reagent or 4′,6-diamidino-2-phenylindole dihydrochloride (DAPI) staining (Fig. 7.3b).

Usually, tubes are employed for *Citrus* anther culture (Hidaka *et al.*, 1979, 1981; Hidaka, 1984a; Chatuvedi and Sharma, 1985; Starrantino, 1986; Hidaka and Omura 1989b; Geraci and Starrantino, 1990; X.X. Deng *et al.*, 1992), but Petri dishes (Starrantino, 1986; Germanà *et al.*, 1991, 1992, 1994, 1997, 2000a; Germanà and Chiancone, 2003) and bottles or flasks have also been used (Chen, 1985; Ling *et al.*, 1988).

Isolated microspore culture

Pollen culture is performed by removing somatic anther tissue. This technique, although more difficult and laborious, is ideal for studying the mechanism of pollen embryogenesis, because it eliminates the unknown effects of the sporophytic anther tissue, thereby allowing a greater control over the culture process.

Isolated microspore culture technique in Citrus

Investigation of isolated microspore culture of several *Citrus* species (lemon, orange, clementine, sour orange, grapefruit) and a related genus (*Poncirus*) has been carried out (Germanà *et al.*, 1996). Anthers at the uninucleated stage from cold-pre-treated (4°C for 10 days) and surface-sterilized flower buds are excised and pre-cultured on either liquid or solid medium in Petri dishes. After 5–15 days at 27 ± 1°C in the dark, anthers are gently squeezed with a glass pestle in 2 ml of liquid medium. The anther mixture is filtered through a sterile nylon sieve (40 μm) and microspores are centrifuged (1000 r.p.m. for 10 min) and washed twice with fresh medium. Finally, the clean isolated microspores are resuspended in fresh liquid medium at a density of 10^3–10^4 grains/ml, placed in a thin layer (3 ml) in Petri dishes (6 cm in diameter), sealed with parafilm and incubated at 27 ± 1°C under cool white fluorescent lamps with a photosynthetic photon flux density of 35 μmol/m^2/s for a photoperiod of 16 h of light per day. After various periods of time (1–4 months), the isolated microspores of almost all investigated *Citrus* species produced multinucleated structures and developed into small proembryos, which failed to develop any further, although several media and different methods (double layer, liquid medium, soft agar, solid agar, etc.) were employed. Formation of 'pseudobulbils', white or green spherical bodies, described in *Citrus* by Button and Kochba (1977), has been obtained only in those genotypes (clementine and lemon) that had also produced haploids by anther culture (Fig. 7.4).

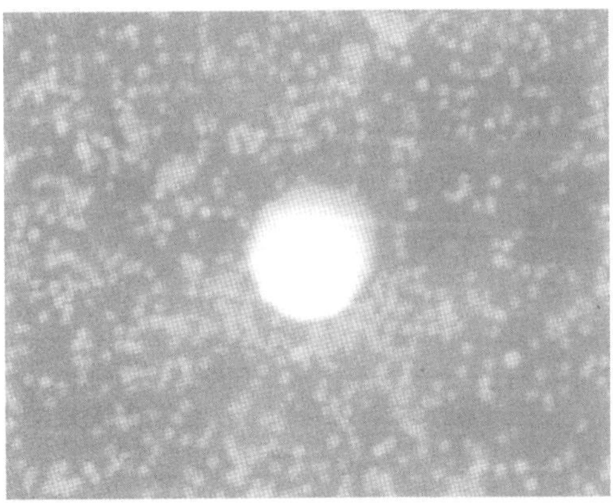

Fig. 7.4. Pseudobulbil produced after about eight months in a *Citrus limon* (L CNR 26) isolated microspore culture.

Medium for isolated microspore culture is more complex than that for anther culture. The NTH (Nitsch and Nitsch, 1969) or N6 (Chu, 1978) liquid filter sterilized medium was supplemented with casein, glutamine, malt extract, biotin, myo-inositol, glycine, pyridoxine, thiamine, serine, coconut water and ascorbic acid, with galactose and sorbitol as carbon source and a complex combination of growth regulators (2,4-dichlorophenoxyacetic acid (2,4-D) + kinetin + zeatin + giberellic acid).

Factors affecting *in vitro* pollen embryogenesis

Although progress in *pollen embryogenesis* has taken many steps forward in recent years, several aspects of this phenomenon remain unclear, particularly the induction process and the factors that control it. The identification of the inhibitory and stimulatory factors is of fundamental importance, especially in recalcitrant species such as *Citrus*.

In vitro pollen embryogenesis is affected by numerous factors: genotype; the pre-treatment applied to anthers or to floral buds; pollen developmental stage; donor plant growth conditions; culture media (macro- and microelements, carbon source and plant growth regulators); and incubation conditions.

Experience in several genotypes has shown that often a particular modification in the procedures (such as the use of maltose in barley, the stress in wheat, and so on) can result in breakthroughs in making the process work. Research in recent years regarding factors affecting gametic embryogenesis in *Citrus* (Germanà *et al.*, 2000a; Germanà and Chiancone, 2003) has resulted in an increase in the number of the cultivars responding to the phenomenon, an improvement in the induction rate of haploid production and an understanding of the ways in which the difficulties of *in vivo* acclimatization can be overcome. As a result, to date, more than 100 homozygous embryogenic callus lines, each coming from single anthers of the different clementine cultivars Nules, Sra 63 and Monreal, have become available.

Genotype

The induction rate of pollen embryos is mainly influenced by the genotype and different culture factors (Heberle-Bors, 1980). 'In most species only a few genotypes can be induced to undergo androgenetic development' (Vasil, 1980).

Because the number of microspores competent to undergo embryogenesis ('E-grains', Sunderland, 1978; or 'P-grains', Heberle-Bors, 1982) depends on the genotype (Heberle-Bors, 1985), it is necessary to know the optimal conditions to turn the development of pollen towards a sporophytic pathway and to avoid embryo abortion.

Almost all research carried out on pollen embryogenesis in *Citrus* recognizes the preponderant influence of the genotype (Khuroshvili *et al.*, 1982; Hidaka, 1984a; Chen, 1985; Germanà *et al.*, 1991, 1994), even though the frequency of P-grains in *Citrus* has not been the object of any research.

The proof that the genotype has a fundamental influence on the success of the gametic embryogenic process is evident from Table 7.2; to date, in the genus *Citrus*, almost all haploids and doubled haploids have been regenerated in different cultivars of the one species *C. clementina*, although research on anther culture of numerous genotypes has been reported (Germanà, 1992, 1997).

Research (Germanà *et al.*, 1994; and unpublished results) carried out simultaneously on several *Citrus* cultivars (four cultivars of clementine, two of mandarin, four of sweet orange, four of sour orange, five of lemons and four of grapefruits) resulted in plantlets being obtained from haploid embryogenic callus only in *C. clementina* cv. Nules, haploid callus in one cultivar of *C. limon,* and diploid and highly embryogenic callus from two cultivars of *C. reticulata* (Avana and Tardivo di Ciaculli) and two cultivars of *C. aurantium* (A.A. CNR 10 and A.A. CNR 23). All experiments were carried out under the same culture conditions and pre-treatments (4°C for 4 days), testing 11 different media. Further research improved the plantlet induction rate of clementine and confirmed its androgenetic response (Germanà *et al.*, 2000a; Germanà and Chiancone, 2003).

Stage of pollen development

The pollen developmental stage is a complex factor that affects the success of anther culture. The suitable stage differs depending on the crop species tested. Generally, pollen grains between the uninucleate and early bicellular stage are cultured. After the pollen grains begin to accumulate storage reserves, they usually lose their embryogenic capacity and follow the gametophytic developmental pathway (Heberle-Bors, 1989; Raghavan, 1990). Usually, the stage of pollen development is tested in one anther per floral bud size by the acetic–carmine method (Sharma and Sharma, 1972). The anthers are collected from flower buds at different stages of development and squashed in 1% acetocarmine in 45% acetic acid for observation under an optical microscope to determine the stage of pollen development. DAPI fluorescent staining has also been used. However, different developmental stages have been observed within a single anther, and between different anthers of the same flower bud in *Citrus* and in *Poncirus* as well as in many other genera (Vasil, 1967; Shull and Menzel, 1977; Hidaka *et al.*, 1979, 1981; Chen, 1985).

Hidaka *et al.* (1979), studying the effect

of different developmental stages of *P. trifoliata* pollen grains on the formation of embryoids, pseudobulbils and calli, indicated the early uninucleate stage as most suitable for embryoid production. Anthers at other developmental stages (from pollen mother cell (PMC) to bicellular stage) produced only calli. Pseudobulbils could not be obtained at PMC, tetrad or bicellular stages. In *C. aurantium*, Hidaka *et al.* (1981) obtained embryoids from anthers only at the late uninucleate stage and callus production from anthers at all other developmental stages except for the PMC in the meiosis and the tetrad stage. Various authors have used different developmental stages in *Citrus* anther culture including: the uninucleate stage (Hidaka, 1984a, b; Hidaka and Omura, 1989b; Germanà *et al.*, 1990, 1991, 1994; X.X. Deng *et al.*, 1992; Germanà and Chiancone, 2003); tetrads at the uninucleate stage (Chaturvedi and Sharma, 1985); the middle anaphase period (Chen, 1985); and the stage of just released spores to the first haploid mitosis (Starrantino, 1986; Geraci and Starrantino, 1990).

Physiological condition of the donor plant

The number of P-grains also depends on the growth conditions of the donor plants (Heberle-Bors, 1985), because their formation *in vivo* and/or *in vitro* seems to be connected with a nitrogen starvation phenomenon (Heberle-Bors, 1983).

Although the physiological condition of the donor plant can dramatically affect the androgenic process, this parameter has been investigated only in herbaceous plants because of the difficulties of determining it in open-air cultivated, perennial woody plants. Significant seasonal variations in anther response have been observed in many genotypes, and it has been noticed that anthers removed from field-grown plants give a better response than those picked from greenhouse-grown plants (Vasil, 1980). In fact, the physiological and growth conditions of the donor plant, which affect the endogenous levels of hormones and the nutritional status of the tissues of the anther (Sunderland and Dunwell, 1977), are important for the success of the embryogenetic process. This parameter has not been considered in the studies regarding *Citrus* anther culture.

The influence of the physiological condition of the donor plant, affected by climatic (temperature, photoperiod and light intensity) and pedological conditions, should be investigated, because this could help to explain the reasons why the response to anther culture is so season-dependent, although the same conditions (pollen development stage, floral bud pretreatments, medium, light and temperature conditions of culture) are employed.

Pre-treatment

It has been observed that stress treatment (such as chilling, high temperature, high humidity, water stress, anaerobic treatment, centrifugation, sucrose and nitrogen starvation) applied to excised floral buds or to anthers before culture acts as a trigger for inducing the sporophytic pathway, preventing the development of fertile pollen (gametophytic pathway) (Sangwan-Norrel *et al.*, 1986; Touraev *et al.*, 1997), although sometimes different results have been obtained (Sunderland, 1974).

The stress, triggering the microspore development from gametophytic to sporophytic, seems to act by altering the polarity of the division at the first haploid mitosis involving reorganization of the cytoskeleton (Nitsch and Norreel, 1973; Reynolds, 1997), delaying and modifying pollen mitosis (two equal-size vegetative-type nuclei instead of one vegetative and one generative), blocking starch production or dissolving microtubules (Nitsch, 1977), or maintaining the viability of the cultured P-grains (Heberle-Bors, 1985).

Cold pre-treatment is routinely employed in anther culture of many crops, and its effect is genotype-dependent (Powell, 1988; Osolnik *et al.*, 1993).

Cold temperature is the most common physiological stress applied in anther culture of *Citrus*. The different cold pre-treatments employed in *Citrus* anther culture are: 5°C for 2 h (Starrantino, 1986), overnight (Geraci and Starrantino, 1990) and 3–5 days (Ling *et al.*, 1988); and 4°C for 4 days (Germanà *et al.*, 1991), for 6 days (Germanà *et al.*, 1994), for 2–6 days (X.X. Deng *et al.*, 1992) and for 10 days (Germanà *et al.*, 1997, 2000a).

In a study carried out by Chen (1985) on the effects of 0–25 days of 3°C cold pre-treatment on *C. madurensis* Lour., a duration of 5–10 days was favourable for inducing callus and embryoids.

A study carried out on factors affecting anther culture in *C. clementina* showed a negative effect on callus production of centrifugation (4000 r.p.m. for 5 min) and high temperature (40°C for 24 h) when compared with chilling (4°C for 10 days) (Germanà *et al.*, 2000a). These and other results (Germanà and Chiancone, 2003) indicated that high temperature applied to the floral buds before culture is not recommended for androgenesis in *Citrus* cv. Nules, contrasting with what has been reported in *Solanum melongena* L. (Rotino, 1996), in *Brassica oleracea* (Keller *et al.*, 1983) and in *Solanum chacoense* (Cappadocia *et al.*, 1984).

Culture medium

The diverse genotypes show very different basal medium requirements to induce pollen-derived plant formation.

The basal media most used in *Citrus* anther culture are: B5 medium (Gamborg *et al.*, 1968; Khuroshvili *et al.*, 1982); DB medium (Drira and Benbadis, 1975); MS medium (Murashige and Skoog, 1962; Hidaka *et al.*, 1979, 1981; Khuroshvili *et al.*, 1982; Hidaka, 1984a, 1984b; Starrantino, 1986; Ling *et al.*, 1988; Hidaka and Omura, 1989b; Germanà, 1994); modified MS medium (Chaturvedi and Sharma, 1985; Chen, 1985; Germanà *et al.*, 1994); Murashige and Tucker (1969) medium (Geraci and Starrantino, 1990; X.X. Deng *et al.*, 1992; Froelicher and Ollitrault, 2000); N_6 medium (Chu, 1978; Chen, 1985; Starrantino, 1986; Germanà *et al.*, 1997, 2000a; Germanà and Chiancone, 2003); SH medium (Schenk and Hildebrandt, 1972; Chaturvedi and Sharma, 1985); Sj-1 medium (Starrantino, 1986); Chaturvedi and Mitra (1974) medium (Germanà *et al.*, 1994); and Nitsch and Nitsch (1969) medium (NTH) (Khuroshvili *et al.*, 1982; Germanà *et al.*, 1997, 2000a).

A study comparing three basal media in the induction media: MS (Murashige and Skoog, 1962); NTH (Nitsch and Nitsch, 1969); and N_6 (Chu, 1978) showed a higher efficiency of the last two media in producing callus (Germanà *et al.*, 2000a).

Carbon source

The carbon source and its concentration is an essential component in the medium for embryo induction. The effect of its concentration is probably related to osmotic pressure regulation during the induction phase (Sunderland and Dunwell, 1977; Sangwan and Sangwan-Norrel 1990). Furthermore, high concentrations of carbohydrate seem to be deleterious (Keller *et al.*, 1975).

Sucrose is the most common carbon source used in the anther culture of *Citrus* and their relatives, at 5% concentration (Hidaka *et al.*, 1979, 1981; Hidaka, 1984b; Hidaka and Omura, 1989b; Geraci and Starrantino, 1990; Froelicher and Ollitrault, 2000), although other concentrations have also been reported: 8% (Starrantino, 1986), 2% (Ling *et al.*, 1988) and 2.5–5% (Drira and Benbadis, 1975; X.X. Deng *et al.*, 1992). Hidaka (1984) studied the effects of sucrose concentration (1, 3, 5, 7 and 9%) on embryoid and callus formation and found that 3% sucrose was the ideal concentration to form embryoids in *P. trifoliata*, 7% in *C. aurantium* and 1% in *C. sinensis* (Trovita orange); 10% was found to be ideal in *C. madurensis* (Chen, 1985); and 3 and 6% in *C. limon* androgenesis (Germanà *et al.*, 1991).

The influence on anther culture of two carbon sources (sucrose and glucose) was tested in two *C. clementina* and two *C. reticulata* cultivars (Germanà *et al.*, 1994). Sucrose (5%) was the best, although with a variable response depending on the species or the variety tested.

Glycerol in combination with sucrose stimulated callus production in *C. clementina* (Germanà *et al.*, 2000a), while galactose was found to be very effective for the production of embryoids from *Citrus* calli (Hidaka and Omura, 1989a). More recent research has shown a positive influence of the combination of lactose and galactose in inducing haploid production in clementine (M. A. Germanà *et al.*, unpublished).

Plant growth regulators

The effects of plant growth regulators have been widely investigated in anther culture of *Citrus*. In *C. limon* and *C. medica* anther culture, Drira and Benbadis (1975) obtained callus on a medium containing 1.0 mg/l 2,4-D, 1.0 mg/l α-naphthaleneacetic acid (NAA) and 0.1 mg/l 6-benzylaminopurine (BA). Hidaka *et al.* (1979) found that a medium containing 0.2 mg/l of both indole-3-acetic acid (IAA) and kinetin (Kin) was the most efficient in embryoid formation, while the addition of 2,4-D increased callus formation in *P. trifoliata* anther culture. Furthermore, Hidaka *et al.* (1981) obtained embryoid production from *C. aurantium* by anther culture in medium supplemented with 0.02 mg/l Kin and 0.02 or 2.0 mg/l IAA.

In *C. sinensis*, Hidaka (1984b) found that the best composition of pollen embryogenesis induction medium was similar to that used for sour orange, having lower concentrations (0.002 or 0.02 mg/l) of IAA and Kin. Chaturvedi and Sharma (1985) supplemented medium with 0.5 mg/l BA and 1.0 mg/l IAA while studying androgenesis in *C. auratifolia*.

Chen (1985) found that 1.0 mg/l BA and 0.1 mg/l 2,4-D was the best hormone combination for embryoid production and that the 2,4-D concentration in the medium is crucial for embryoid production in *C. madurensis*: an increase of its concentration promotes callus formation and inhibits embryoid development.

Starrantino (1986), testing numerous media containing an auxin (0.1, 0.5, 1.0, 5.0 and 10.0 mg/l of IAA or 2,4-D) in combination with a cytokinin (0.1, 0.5, 1.0, 5.0 and 10.0 mg/l of BA or Kin), obtained two embryonal structures from two *C. sinensis* anthers in medium containing 1.0 mg/l 2,4-D and 10.0 mg/l Kin, and one embryonal structure from an anther of *C. paradisi* in medium supplemented with 10 mg/l 2,4-D and 0.5 mg/l Kin.

Ling *et al.* (1988) obtained the highest frequency (0.92%) of embryoid formation in *C. madurensis* by supplementing medium with 2.0 mg/l of both IAA and Kin.

Geraci and Starrantino (1990) succeeded in obtaining the highest percentage of callus proliferation (25%) in the presence of 1 mg/l BA and 0.5 mg/l 2,4-D in *C. reticulata*, *C. deliciosa*, *C. paradisi* and Mapo tangelo (*C. deliciosa* × *C. paradisi*), while in *C. sinensis* and *C. limon*, the highest callus proliferation rate (44.2%) was obtained with 1.0 mg/l of both NAA and BA.

X.X. Deng *et al.* (1992) reported the best response in *P. trifoliata* and an ichang papeda hybrid on medium having 0.1 mg/l NAA.

The anther culture of *C. limon* was successful on medium containing 2.0 mg/l Kin + 1.0 mg/l zeatin (ZI) + 0.1 mg/l NAA (Germanà *et al.*, 1991).

Germanà *et al.* (1994) found the best hormonal combination for callus production (6–28%, depending on the genotype) in anther culture of *C. clementina* and *C. reticulata* in: 0.02 mg/l NAA + 0.5 mg/l ZI + 0.5 mg/l Kin.

In *Clausena excavata*, Froelicher and Ollitrault (2000) obtained the best results with the following hormonal conbination: BA at low concentration (0.1 or 0.3 mg/l), alone or with 2,4-D (0.1 mg/l).

Further research (Germanà and Chiancone, 2003) showed that thidiazuron (TDZ), one of the most active cytokinins in woody plant tissue culture (Huetteman and Preece, 1993), is effective in inducing haploid embryoids and plantlet regeneration in *C. clementina* cv. Nules.

Activated charcoal

The addition of activated charcoal (0.5-2 g/l) to the medium increases the efficiency of androgenesis in several genera. It seems to act by removing inhibitory substances from the medium, and presumably from the anther wall, and by regulating the level of endogenous and exogenous growth regulators (Reinert and Bajaj, 1977; Vasil, 1980; Heberle-Bors, 1985).

X.X. Deng *et al.* (1992) found that medium containing activated charcoal was effective in the medium having *P. trifoliata* embryoid induction. However, no positive effect of activated charcoal addition (0.5 g/l) has been observed in anther culture of several *Citrus* species (Germanà *et al.*, 1994; M.A. Germanà *et al.*, unpublished).

Other substances

The addition of glutamine was necessary to produce callus in *C. limon* anther culture (Drira and Benbadis, 1975). In our laboratory, the best results were obtained by the addition to media of casein, biotin and sometimes coconut water, together with glutamine.

Although the way in which they act is not completely understood, various natural undefined extracts, e.g. coconut water, are used since they improve pollen response. They probably provide one or more substances which stimulate pollen division.

A recent study (Chiancone *et al.*, 2006) showed that the addition to medium of spermidine increased the number of clementine anthers producing haploid embryoids. Polyamines are compounds found in all living organisms, classified as growth regulators and involved in many biological processes, such as growth, development and stress response (Kumar *et al.*, 1997). Moreover, their involvement in *in vitro* organogenesis and embryogenesis and their capacity to inhibit ethylene biosynthesis has been highlighted.

pH

pH is another factor which can influence the gametic embryogenic process (Stuart *et al.*, 1987). In *Citrus* anther culture, the pH of the medium is usually adjusted to 5.7–5.8 before autoclaving. The effect of the pH level (4, 5, 6, 7 and 8) of the medium on anther culture of *P. trifoliata*, *C. aurantium* and *C. sinensis* has been studied by Hidaka (1984a), who found that both pH 5 and 6 were effective in embryoid formation in all genotypes tested.

Solidifying agents

Usually, *Citrus* anther culture media are solidified by adding agar. Chaturvedi and Sharma (1985) obtained diploid plantlet regeneration by floating *C. aurantifolia* anthers on a liquid medium, then embedding them in a semi-solid medium. Generally, we obtained better results in a solid medium rather than a liquid one, also when pre-culturing anthers for pollen isolation. In the liquid medium, anthers initially swell, later turning brown and sometimes shrivelling. Probably the two steps (first liquid and later semi-solid) in the culture are essential for success.

Research on *C. clementina*, Mapo tangelo, *Fortunella margarita* and *C. paradisi* anther culture showed the beneficial effect of potato starch as a gelling agent on callus production (Germanà *et al.*, 1997, 2000a; M.A. Germanà *et al.*, unpublished).

Froelicher and Ollitrault (2000) in anther culture of *Clausena excavata* added gelrite to solidify the medium.

Incubation conditions

The incubation conditions have not received much attention in *Citrus* anther culture (especially light quality and photoperiod).

Hidaka (1984a) reported inducing embryoids at 24 and 28°C in trifoliate orange, sour orange and 'Trovita' orange.

Chen (1985) observed that temperature seems to be more important than light in *Citrus* androgenesis, and obtained embryoids at 20–25°C, especially under dark conditions (2.21% induction rate).

Different temperatures were used in

Citrus: 25 ± 1°C (Ling *et al.*, 1988) and 27–28 ± 1°C (Hidaka *et al.*, 1979, 1981; Hidaka, 1984b; Chaturvedi and Sharma, 1985; Starrantino, 1986; Hidaka and Omura, 1989a; Geraci and Starrantino, 1990; Germanà *et al.*, 1990, 1991, 1994; Deng *et al.*, 1992; Germanà and Reforgiato, 1997; Froelicher and Ollitrault, 2000).

The light intensities used are: 500 lux with a 16 h photoperiod (Hidaka *et al.*, 1979, 1981; Hidaka 1984a, b); 500–800 lux with a 12 h photoperiod (Chen, 1985); 3000 lux with a 15–16 h photoperiod (Chaturvedi and Sharma, 1985; Ling *et al.*, 1988); 1000 lux with a 16 h photoperiod (Starrantino, 1986; Geraci and Starrantino, 1990; X.X. Deng *et al.*, 1992); and 3500 lux with a 16 h photoperiod (Germanà *et al.*, 1991). An inductive period in darkness (10–15 days; 3–4 months Froelicher and Ollitrault, 2000) is usually applied in androgenesis research.

In our experiments, Petri dishes are usually incubated at 27 ± 1°C for 15 days in the dark, and then placed under cool white fluorescent lamps (Philips TLM 30W/84) with a photosynthetic photon flux density of 35 $\mu mol/m^2/s$ and a 16 h photoperiod (Germanà *et al.*, 1994, 2000; Germanà and Reforgiato, 1997; M.A. Germanà *et al.*, unpublished).

Preliminary research on the effect of light quality on anther culture of *C. clementina* Hort. ex Tan., cultivar Nules, testing, after one month of darkness, four light qualities: white, red, far-red and blue, and using as control conditions continuous darkness and white light under a photoperiod of 16 h, showed that gametic embryoids and embryogenic callus were obtained only under photoperiodic conditions of white light, suggesting that the alternation of light and dark is necessary for the gametic embryogenesis process in clementine (Germanà *et al.*, 2005a).

Fig. 7.5. Direct embryogenesis in *Citrus* anther culture.

Embryo development from microspores and the origin of haploids

Gametic embryogenesis (from male or female gametes) can be considered to be one example of cellular totipotency, usually defined as the capacity of the somatic cell to regenerate an entire new plant and, evolutionarily, an important survival adaptation mechanism (Reynolds, 1997).

The developmental process of a plant from a single microspore is referred to as microspore embryogenesis, although the

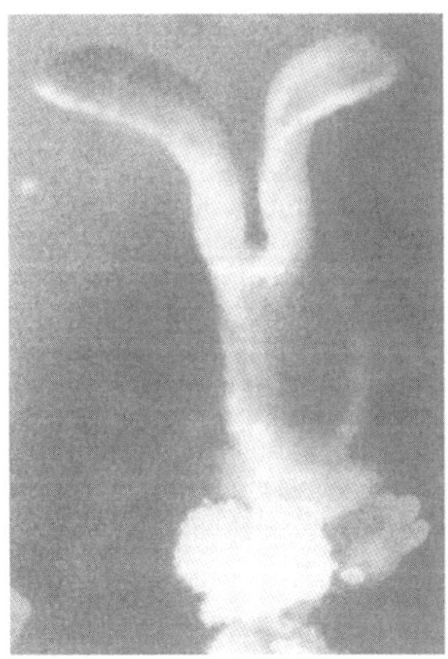

Fig. 7.6. Secondary embryogenesis in *Citrus* anther culture.

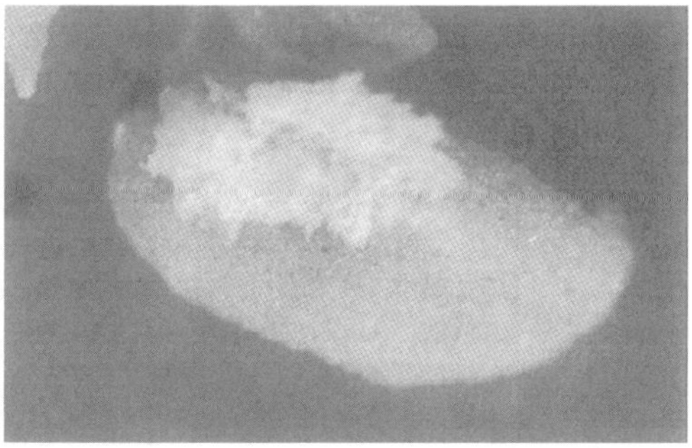

Fig. 7.7. A non-morphogenic callus from anther culture.

route of regeneration may be via direct embryogenesis (Fig. 7.5), secondary embryogenesis (Fig. 7.6) or, less frequently, organogenesis. In other cases, microspores in culture produce undifferentiated calli, instead of embryoids.

After one week of culture, most of the anthers are swollen and after 2–3 months they start to produce calli or embryoids. Most of the calli are non-morphogenic (Fig. 7.7), but many of them appear highly embryogenic and they maintain their potential for a long time. The morphogenic calli appear friable (Fig. 7.8) and white. Sometimes calli develop from two different lobes of an anther (Fig. 7.9). The embryogenic calli differentiate into a clump of embryoids (Fig. 7.10a and b).

The well-structured embryoids develop normally like zygotic embryos, through the globular, the heart, the torpedo and the cotyledonary stages, and often produce secondary embryoids. Often teratomatal structures, cotyledonary fused (Fig. 7.11a and b), pluricotyledonary (Fig. 7.12) and thickened embryoids are observed. Sometimes pseudobulbils, with or without callus, are produced in *Citrus* anther culture (Fig. 7.13a and b).

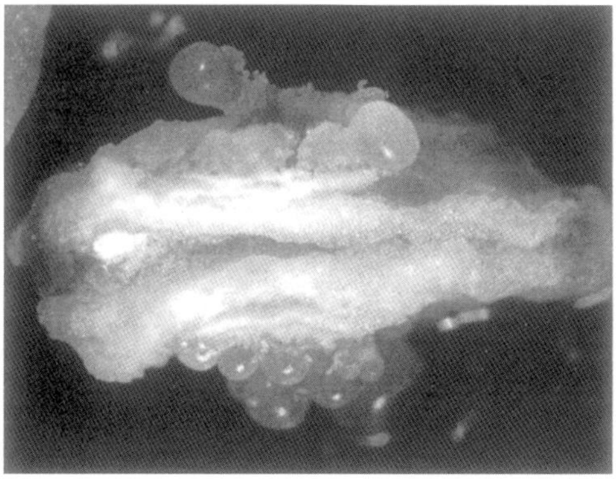

Fig. 7.9. Calli emerging from two different lobes of an anther of cv. Nules clementine.

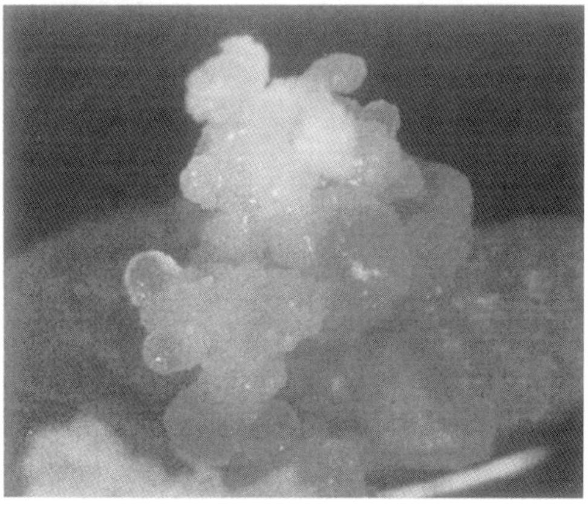

Fig. 7.8. An embryogenic, haploid, friable, white callus emerging after three months of culture from anther culture of the cultivar Nules of *C. clementina* Hort. ex Tan.

Green, compact and non-morphogenic calli emerging from anthers were also observed in *Poncirus*, *C. clementina* and *C. limon* (Fig. 7.14).

Divisions in pollen grains of various species start at different intervals after the first pollen mitosis, and after the trauma of wounding and culture, depending on the degree of repression of the sporophytic gene programme (Heberle-Bors, 1985).

Several studies have been carried out on early nuclear division events of the microspores of herbaceous species. Hidaka and Omura (1989b) described cytologically the development of embryoids from microspores in *C. aurantium* and *P. trifoliata*. They observed three main routes of development. In route A, the microspores

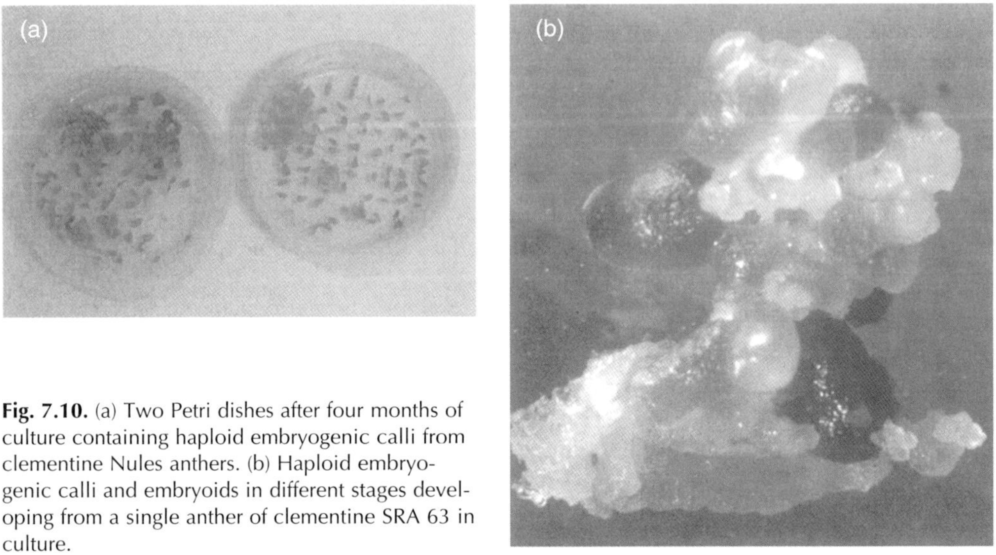

Fig. 7.10. (a) Two Petri dishes after four months of culture containing haploid embryogenic calli from clementine Nules anthers. (b) Haploid embryogenic calli and embryoids in different stages developing from a single anther of clementine SRA 63 in culture.

lose their contents. In route B, microspores develop as *in vivo*, normal nuclear division with one vegetative and one or two generative nuclei; this route is rarely observed. Route C is divided into two routes. In route C1, two types of morphologically similar nuclei are observed: vegetative-type nuclei (C1a) or generative-type nuclei (C1b). Route C2, the repeated division of only the vegetative-type nucleus or the repeated division of both vegetative and generative-type nuclei independent of each other, seems to

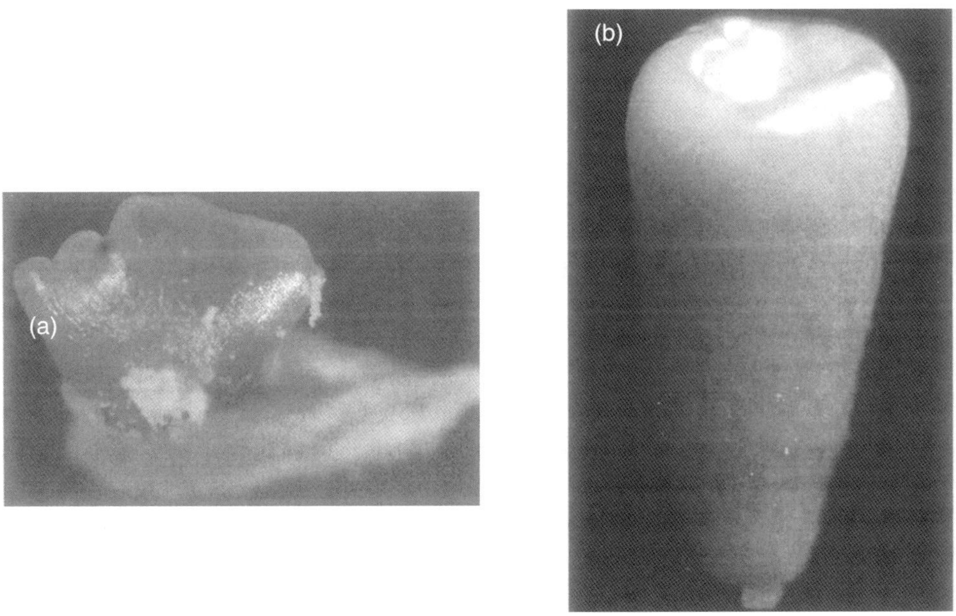

Fig. 7.11. (a and b) A cotyledonary fused embryoid.

Fig. 7.12. A pluricotyledonary embryoid.

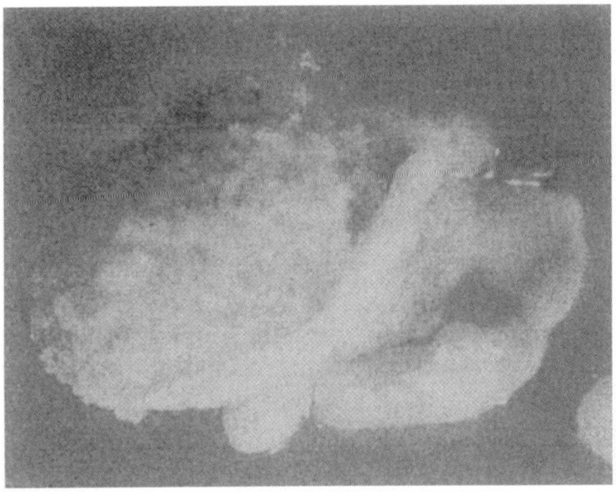

Fig. 7.14. A green and compact callus emerging from inside a *Poncirus trifoliata* anther.

form the most embryoids in *Citrus* anther culture.

When the nucleus divides without cell division, a multinucleate pollen grain is initially formed which later gives rise to a multicellular structure, then developing into a proembryoid (Fig. 7.15) and finally into an embryoid, until the exine rupture. Moreover, nuclear fusion among vegetative and generative nuclei has been observed, and this can explain an increase in ploidy level.

A morphological and ultrastructural study, at the cellular and subcellular level, of early microspore embryogenesis in several embryogenic varieties of *C. clementina* is in progress in a joint project between the University of Palermo (Italy) and the Centro de Investigaciones Biologicas (Spain). Microscopic analysis has revealed very important aspects of this embryogenic

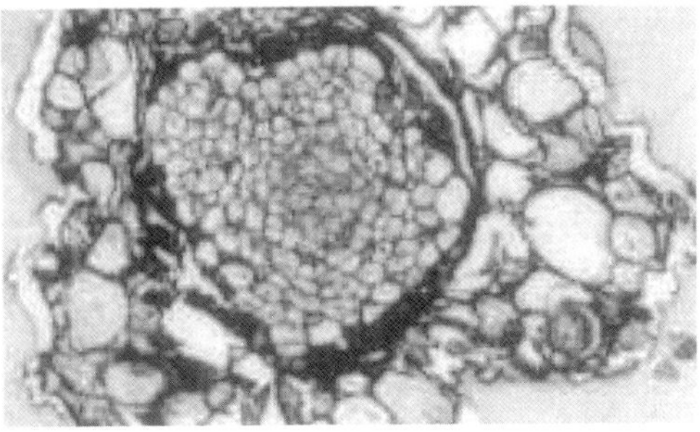

Fig. 7.15. A proembryoid located inside the anther. Fragments of the broken exine are still at its periphery. (2 μm semi-thin section stained with toluidine blue.) This photo was taken at the laboratory of Dr Maria Carmen Risueño, Centro de Investigaciones Biologicas, CSIS, Spain.

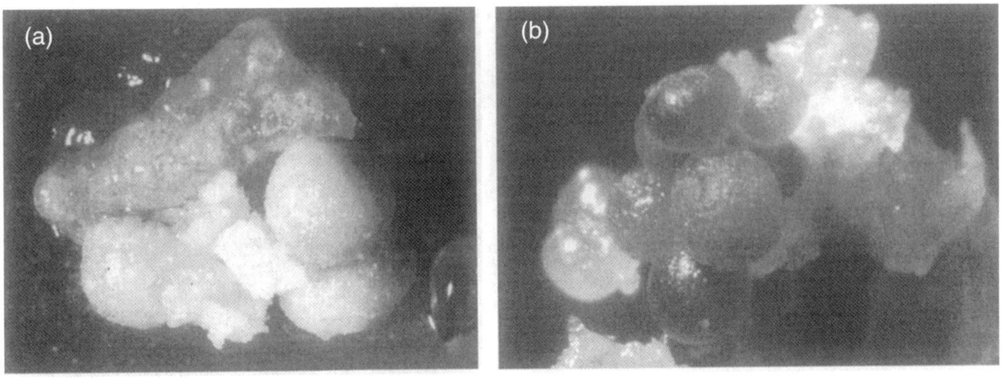

Fig. 7.13. (a and b) Pseudobulbils from *Citrus* anther culture.

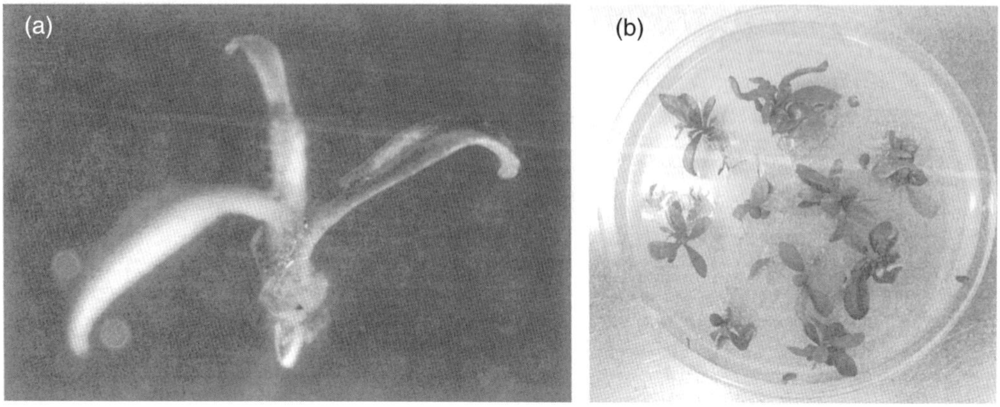

Fig. 7.16. (a and b) Haploid embryoid germination.

process, indicating differences between *Citrus* microspore-derived embryos and those derived from other embryogenic species, such as starch accumulation during the first embryonic stages (Ramirez *et al.*, 2003). Moreover, different cellular types have been observed in these embryos after the exine breakdown.

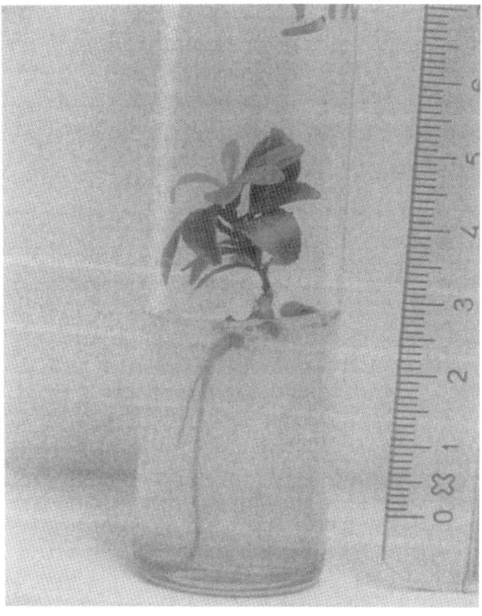

Fig. 7.17. Haploid plantlet of Nules clementine obtained from embryoid germination.

Plant recovery, hardening and characterization

Plantlet formation from cultured anthers may occur either directly through embryogenesis of microspores or indirectly through organogenesis or embryogenesis of microspore-derived callus.

The highly embryogenic haploid callus is multiplied in MS medium supplemented with 5% sucrose, 0.02 mg/l NAA and 0.8% agar, maintaining its embryogenic potential for several years. As the embryos appear, they are germinated in Petri dishes (Fig. 7.16a and b) with MS medium containing 3% (w/v) sucrose, 1 mg/l GA_3, 0.01 mg/l NAA and 0.75% (w/v) agar (germination medium), and they are later transferred to Magenta boxes (Sigma V8505) or to test tubes (Fig. 7.17).

The embryoids develop normally through the globular, heart, torpedo and cotyledonary stages, and often produce secondary embryoids. Haploid embryoids of clementine vigorously germinate *in vitro*; in contrast, haploid plantlets grow slowly in soil, presumably due to recessive harmful genes expressed in homozygosity. These plantlets, when transplanted *in vivo*, usually die as a result of fungal contamination. Better results have been obtained by grafting *in vitro* homozygous small shoots (2–3 mm) on to etiolated 20-day-old Troyer cit-

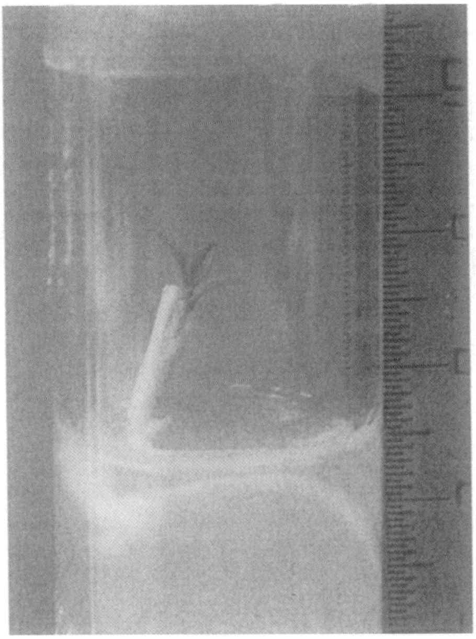

Fig. 7.18. *In vitro* grafting of homozygous *Citrus clementina*, cv. Nules on to etiolated 'Troyer' citrange seedlings to improve the material survival.

range seedlings (Fig. 7.18). After 3–4 months, the grafted plantlets obtained were washed with sterile water to remove the medium from their roots and then transferred to sterilized pots containing peat moss, sand and soil in the ratio 1:1:1 for the acclimation phase (Fig. 7.19). The new scions obtained were later grafted on to 2-year-old sour orange seedlings. They showed a more compact habitus (Fig. 7.20a and b) and a decrease in vigour, with significantly smaller leaves, shorter internodes and more thorns when compared with the heterozygous parent of the same age of grafting (Germanà *et al.*, 2000b).

More vigorous growth has been observed by grafting the second time on to *C. macrophylla* instead of on to *C. aurantium* (unpublished).

Ploidy of androgenetic plants

Haploid, and especially triploid, diploid, aneuploid and mixoploid calli and plantlets have been produced from *Citrus* and its relatives by anther culture. Non-haploids may arise from: (i) somatic tissue of anther walls; (ii) the fusion of nuclei; (iii) endomitosis within the pollen grain; or (iv) irregular microspores formed by meiotic irregularities (D'Amato, 1977; Sunderland and Dunwell, 1977; Narayanaswamy and George, 1982; Sangwan-Norrel, 1983). The developmental stage of the pollen at the time of the culture can cause ploidy variation in regenerated plants: in particular, the older the stage, the higher the ploidy level of the embryoids obtained (Maheshwari *et al.*, 1980).

Besides regeneration from somatic tissue of anther, heterozygosity can also be observed when the plants are regenerated from unreduced microspores or in the case of new variation induced at the diploid level (gametoclonal variation) (Wenzel *et al.*, 1977; Morrison and Evans, 1987).

CYTOLOGICAL CHARACTERIZATION OF REGENERANTS. Chromosome number has been counted in root tip cells from regenerated embryos and plantlets, using the standard Feulgen technique (Lillie, 1951). The explants were pretreated with 0.05% (w/v) aqueous solution of colchicine for 2 h at room temperature, fixed overnight in 3:1 (v/v) ethanol:glacial acetic acid, and stored in 70% ethanol until viewing.

Chromosome counts carried out on root apices of embryos and of plantlets obtained

Fig. 7.19. A haploid plantlet of *Citrus clementina*, cv. Nules transferred to soil.

Fig. 7.20. Doubled haploid Nules grafted on to sour orange seedlings 1 year (a) and 5 years (b) after grafting.

from *in vitro* androgenesis of clementine showed the haploid set of chromosomes ($n = x = 9$) (Fig. 7.21) (Germanà *et al.*, 1991, 1994, 2000a, b; Germanà, 1997). During culture, haploid calli spontaneously diploidize, producing doubled haploid embryoids and plantlets (Germanà, 1997), and sometimes the presence of a triploid number of chromosomes (Fig. 7.22) in homozygous calli cells has also been observed.

FLOW CYTOMETRY. Identification of regenerants from anther culture of clementine and *Clausena excavata* has also been performed by flow cytometry (Ollitrault *et al.*, 1996; Froelicher and Ollitrault, 2000; Germanà *et al.*, 2005b).

Isozyme analyses

Because of the spontaneous diploidization of the haploid calluses, cytological analysis cannot always identify androgenic plants, and isozyme analyses have been employed to decide the gametic origin of calluses and plantlets (Germanà *et al.*, 1991, 1994, 2000a, b; Germanà and Reforgiato, 1997). Isozyme techniques allow the distinction to be made between androgenetic and somatic tissue when the enzyme is heterozygotic in the diploid condition of the donor plant and the regenerants show lack of an allele.

To identify the origin of calli, embryoids and plantlets obtained, their crude extracts are analysed using two enzyme sys-

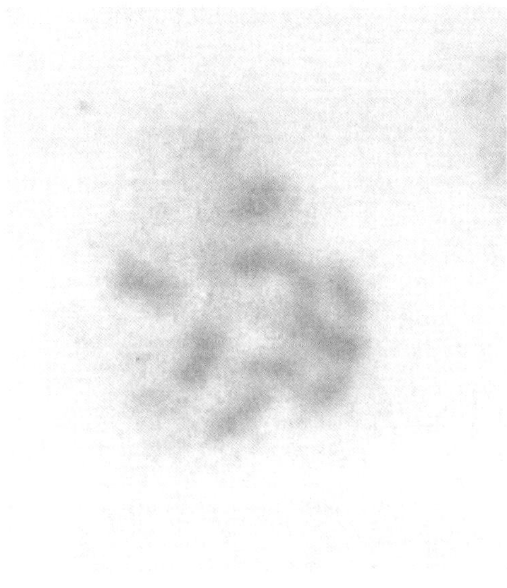

Fig. 7.21. A haploid set of chromosomes from a root tip cell of a regenerated plantlet ($n = x = 9$).

tems: phosphoglucoisomerase (PGI) and phosphoglucomutase (PGM), as reported by Grosser *et al.* (1988). Numbering for isozymes (PGI-1) and lettering for different allozymes are the same as used by Torres *et al.* (1978).

Citrus clementina is heterozygous for PGI-1 and PGM. According to Torres *et al.*

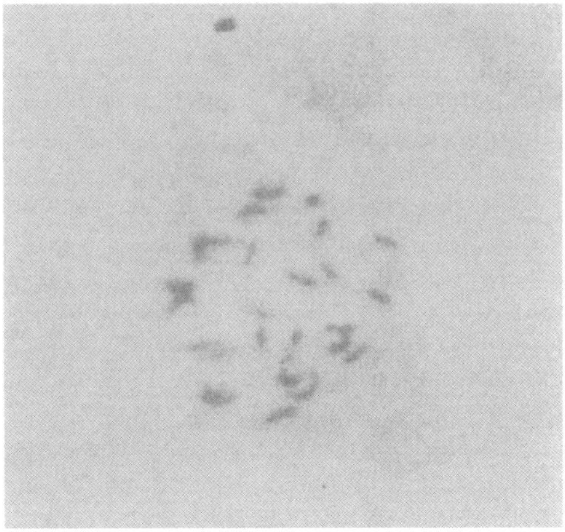

Fig. 7.22. A triploid set of chromosomes from a homozygous callus cell ($n = 3x = 27$).

(1978), the heterozygous clementine parent is FI (F = allele which specifies fast migration toward the anode enzyme; I = intermediate) in PGM, and WS (W = allele which specifies an enzyme migrating faster than F; S = allele which specifies a slowly migrating enzyme) in PGI. For analysis of calli and leaves obtained from anther culture, the presence of a single band was retained as the homozygous state (Fig. 7.23a and b). With one or two exceptions out of more than 100 samples analysed, both enzyme systems confirmed the androgenic nature of regenerants because of the contemporary lack of an allele.

X.X. Deng *et al.* (1992) used GOT (glutamate oxaloacetate transminase) isozyme analysis to show that the diploid plantlet obtained from anther culture of hybrid No. 14 of *C. ichangensis* × *C. reticulata* was homozygous and so pollen derived.

PGI, PGM2 and isocitrate dehydrogenase (IDH) have been used to characterize regenerants in gynogenesis by Ollitrault *et al.* (1996).

PGI and PGM have been employed by Germanà and Chiancone (2001) for the characterization of gynogenetic haploids in clementine.

RAPD analyses

A preliminary characterization of several doubled haploids of *C. clementina* has been carried out during a collaboration between our laboratory and the Istituto Sperimentale per l'Agrumicoltura of Acireale, using isozyme and random amplified polymophic DNA (RAPD) markers (Germanà *et al.*, 2000b). The aberrant transmission of RAPD markers due to the presence of a band found only in doubled haploids has been observed in homozygous clementine as well as in other genotypes (Pooler and Scorza, 1995). Further studies are in progress to determine the nature of these fragments.

Microsatellites

Microsatellites have also been employed to assess homozygosity and to characterize

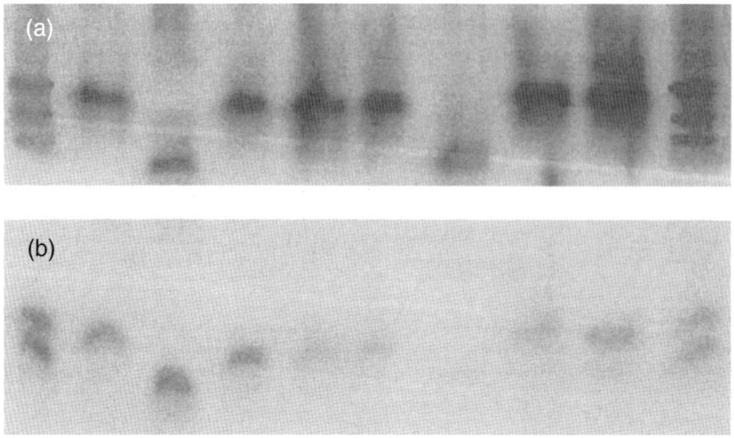

Figs. 7.23. Isozyme pattern of PGI (a) and PGM (b) of calli and leaves. The first lane on the left and the last lane on the right are the zymogram of the heterozygous Nules parent, the others are those of doubled haploid cultures.

regenerants obtained from citrus anther culture (Germanà and Chiancone, 2003; Germanà *et al.*, 2005b).

Applications of Haploids, Doubled Haploids and Gamete Biotechnology in Citrus Breeding

Production of triploids

The importance of triploids in *Citrus* improvement derives from the seedlessness of their fruits. This is a desirable trait of commercial importance and one of the main goals in *Citrus* breeding programmes.

Triploids can be conventionally produced by $2x \times 4x$ and $4x \times 2x$ crosses: two seedless triploid hybrids were produced from $2x$ pummelo and $4x$ grapefruit hybridization (Soost and Cameron, 1980, 1985).

Triploid plants can also be obtained through *in vitro* culture of endosperm, which, being the fusion of three haploid nuclei, is triploid. Triploid hybrid *Citrus* plants were recovered by *in vitro* embryogenesis from endosperm-derived calli (Gmitter *et al.*, 1990).

One of the most interesting applications of haploids in *Citrus* breeding is the possibility of obtaining triploid somatic hybrids by fusion between haploid and diploid protoplasts.

Protoplasts isolated from nucellar calli of 'Juman' Satsuma mandarin, 'Ohmishima' navel orange or 'Trovita' sweet orange were electrically fused with protoplasts of two haploid strains of clementine (Kobayashi *et al.*, 1997).

Triploid *Citrus* hybrids were obtained via somatic hybridization performed by electrofusion between a gynogenetic haploid cell line of clementine and nine diploid cultivars (Ollitrault *et al.*, 2000). The diploid protoplasts were isolated from mesophyll of Marumi kumquat and from embryogenic calli obtained from ovule culture of the following genotypes: Willow Leaf mandarin, Sunki mandarin, Murcott mandarin, Kinnow mandarin, Shamuti orange, Valencia late orange, Star Ruby grapefruit and Mexican lime. Triploid and tetraploid hybrids were obtained for each combination, along with a few pentaploid hybrids.

Ploidy analysis by flow cytometry of 94 regenerants from clementine anther culture showed that as many as 82% of them were tri-haploids, rather than haploids or doubled haploids as expected (Germanà *et al.*, 2005b). Regeneration from anther culture was therefore proposed as a rapid and attractive method of obtaining new triploid varieties in clementine.

Production of diploid somatic hybrids

Through haploid protoplast fusion, new interspecific and intergeneric hybrids can be obtained, by-passing the barriers that inhibit wide crosses and avoiding the tetraploid stage, which is usually characterized by undesirable traits such as thick rind.

Gametosomatic hybridization

The production of triploid interspecific hybrids in *Nicotiana* and intraspecific hybrids in *Petunia* by fusing tetrad protoplasts with callus protoplasts are examples of successful gametosomatic hybridization (Pirrie and Power, 1986; Lee and Power, 1988). A preliminary study on gametosomatic fusion between *P. trifoliata* tetrads and somatic protoplasts of *C. sinensis* cv. 'Jincheng' was reported by Z.A. Deng *et al.* (1992). One chimeric plantlet was regenerated.

Androclonal variation and its utilization

Generally, a large amount of variation is generated throughout pollen embryogenesis, including gene mutation and chromosome variation (Hu and Huang, 1987).

Inbreeding depression, due to complete homozygosity, may be associated with a reduced vigour observed in doubled haploid plants (Veilleux, 1994).

A higher incidence of genetic changes in plants derived from male cells has been observed in comparison with plants derived from female cells. However, this mechanism is poorly understood. It seems probable that since paternal organelles are not usually transmitted to the progeny, their mutations in nature are less important and male gamete cytoplasm is, for this reason, less stable (Huang, 1996).

Gametoclonal variation can be of nuclear origin or it can concern organelles as in albino (chlorophyll-deficient) plants, frequently found in cereals when microspores are induced to regenerate into haploid plants.

Haploid and albino embryoids have also been regenerated in *Citrus* from anther culture of 'Mapo' tangelo (*Citrus deliciosa* × *C. paradisi*) (Fig. 7.24) (Germanà and Reforgiato, 1997).

Male gametophytic selection (MGS)

The angiosperm pollen grains may easily be manipulated to improve the efficiency, the rapidity and the precision of plant breeding methods. The production of doubled haploids by culturing previously selected pollen could avoid the stress-buffering ability of the pistil during pollen germination and pollen tube growth which occurs when selective stress is applied to the whole plant. Gametophytic selection, based on the overlap (~70%) in genetic expression

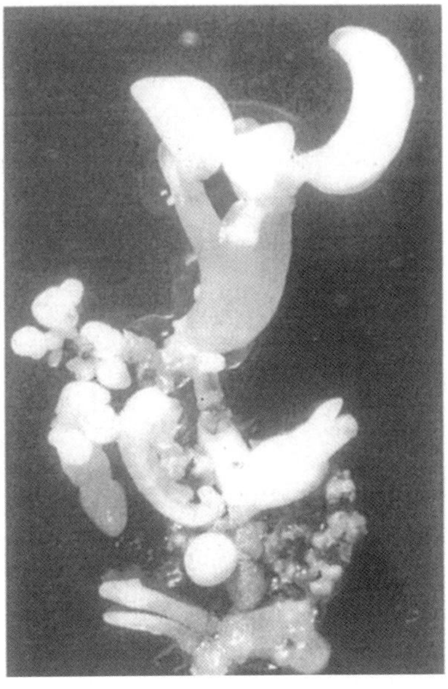

Fig. 7.24. Haploid and albino embryoids from tangelo 'Mapo' anther culture.

between the gametophytic and the sporophytic generation (Mascarenhas, 1990) and on the similarity in their response in the presence of several external factors (Hormaza and Herrero, 1992), could have several advantages deriving from the characteristics of the gametic phase: haploidy and large population size. Plant breeding programmes are costly and time consuming: large population sizes are needed to obtain the required combination of characters in the new genotypes. This is especially important in woody perennials due to the large size of the plants and their long juvenile phase. MGS can already be included in plant breeding programmes when the aim is to screen genotypes or to increase the vigour of the next sporophytic generation and, when knowledge regarding the reproductive processes is more advanced, it will be useful to transmit selected traits into the progeny.

Angiosperm pollen also has great potential by culturing the isolated microspores and exerting stress during culture. In this way, microspores, which result from recombination, are subjected to selection, and it is possible to recover, via gametic embryogenesis, mutants for physiological and biochemical traits (Evans *et al.*, 1990). For this aim, however, a well-defined procedure of regeneration through gametic embryogenesis is necessary.

Transformation

Microspores may also be considered attractive candidates for gene transfer by co-cultivation with *Agrobacterium tumefaciens* and eventual pollen culture, or by microinjection of transfer DNA into pollen embryos (Heberle-Bors *et al.*, 1990; Sangwan and Sangwan-Norrel, 1990). The delivery of DNA into embryogenic microspores further advances genetic improvement of crops by producing homozygous transgenic plants in a single step.

Embryogenic callus production in monoembryonic *Citrus*

Citrus clementina is characterized by monoembryony, and it is very difficult to obtain embryogenic callus from monoembryonic *Citrus* species. On the other hand, embryogenic calli are essential for the genetic improvement of crops through biotechnology.

In our experiments, embryogenic calli of different cultivars of *C. clementina* (Nules, SRA 63 and Monreal) have been obtained (Germanà *et al.*, 1994, 2000a, 2005b; Germanà and Chiancone, 2003).

Haploid and doubled haploid propagation

Haploid micropropagation by *in vitro* tissue culture is particulary important in cases, such as woody species, where the frequency of haploid induction by anther culture is low. It is possible to multiply haploid plants by culturing shoot meristems or axillary buds. Numerous haploid plants are easily obtained when highly embryogenic callus is produced by anther culture, as in *C. clementina.*

Protoplasts were isolated from the stem and leaf of a haploid golden delicious apple clone and protoplast-derived shoots were successfully *in vitro* propagated via organogenesis (Patat-Ochatt *et al.*, 1993). Similarly, haploid protoplasts could be used as a means of clonal propagation of rare haploids.

Conclusions

The great potential of employing haploidy, doubled haploidy and gametic embryogenesis in *Citrus* breeding is clearly evident. Haploids can improve the efficiency and the speed of the usually cumbersome, time-consuming, laborious and sometimes rather inefficient conventional breeding methods.

Although *in vitro* culture of gametes is more or less a standard tool for plant breeders in many crops, particularly *Brassi-*

caceae and cereals, this is yet to be achieved in *Citrus* breeding since deployment of gametic embryogenesis in *Citrus* improvement is still hampered by low frequencies of embryo induction, albinism, plant regeneration, plant survival and the genotype-dependent response.

A better understanding of the gametic embryogenesis process, the improvement of currently available techniques and the development of new technologies could make haploid production a powerful breeding tool in the near future for *Citrus*, allowing in this genus the effective exploitation of the potential of gamete biotechnology.

In order to make this possible, the fundamental goals are: to enlarge the number of respondent genotypes; to improve the induction rate (the frequency with which gametes form embryoids); and to increase the survival rate (the percentage of regenerated plants successfully transferred from *in vitro* to *in vivo* culture conditions).

Further goals are to characterize and to deploy haploids and doubled haploids in *Citrus* breeding (protoplast fusion, triploid production, transformation, etc.).

In conclusion, a better knowledge of the gametic embryogenesis process in *Citrus* is needed to transform this frontier of plant biotechnology into practical applications in this genus, as already achieved in other genera. Therefore, in our opinion, this area of research has enormous potential and merits a great deal of further support and attention.

Acknowledgements

The author wishes to thank Professor E. Rugini for critical reading of the manuscript.

References

Bajaj, Y.P.S. (1990) *In vitro* production of haploids and their use in cell genetics and plant breeding. In: Bajaj, Y.P.S. (ed.) *Biotechnology in Agriculture and Forestry, Vol.12: Haploids in Crop Improvement.* Springer-Verlag, Berlin, Part I, pp. 1–44.

Blakeslee, A.F. and Belling, J. (1924) Chromosomal mutations in the Jimson weed, *Datura stramonium. Journal of Heredity* 15, 195–206.

Blakeslee, A.F., Belling, J., Farnham, M.E. and Bergner, A.D. (1922) A haploid mutant in the Jimson weed, *Datura stramonium. Science* 55, 646–647.

Bouvier, L., Zhang Y.X. and Lespinasse, Y. (1993) Two methods of haploidization in pear, *Pyrus communis* L.: greenhouse seedling selection and *in situ* parthenogenesis induced by irradiated pollen. *Theoretical and Applied Genetics* 87, 229–232.

Button, J. and Kochba, J. (1977) Tissue culture in the Citrus industry. In: Reinert, J. and Bajaj, Y.P.S. (eds) *Applied and Fundamental Aspects of Plant Cell, Tissue and Organ Culture.* Springer-Verlag, Berlin, pp. 70–92.

Cappadocia, M., Cheng, D.S.K. and Ludlum-Simonette, R. (1984) Plant regeneration from *in vitro* culture of anthers of *Solanum chacoense* Bitt. and interspecific diploid hybrids *S. tuberosum* L. × *S. chacoense* Bitt. *Theoretical and Applied Genetics* 62, 155–159.

Chalak, L. and Legave, J.M. (1997) Effects of pollination by irradiated pollen in Hayward kiwifruit and spontaneous doubling of induced parthenogenetic trihaploids. *Scientia Horticulturae* 68, 83–93.

Chaturvedi, H.C. and Mitra, G.C. (1974) Clonal propagation of *Citrus* from somatic callus cultures. *HortScience* 9, 118–120.

Chaturvedi, H.C. and Sharma, A.K. (1985) Androgenesis in *Citrus aurantifolia* (Christm.) Swingle. *Planta* 165, 142–144.

Chen, Z. (1985) A study on induction of plants from *Citrus* pollen. *Fruit Varieties Journal* 39, 44–50.

Chen, Z., Wang, H. and Liao, H. (1980) The induction of *Citrus* pollen plants in artificial media. *Acta Genetica Sinica* 7, 189–192.

Chiancone, B., Tassoni, A., Bagni, N. and Germanà, M.A. (2006) Effect of polyamines on in vitro another culture of Citrus clementina Hort.ex Tan. *Plant Cell, Tissue and organ Culture* 87, 145-153.

Chu, C. (1978) The N6 medium and its applications to anther culture of cereal crops. In: *Proceedings of a*

Symposium on Plant Tissue Culture. Science Press, Peking. pp. 43–50.
D'Amato, F. (1977) Cytogenetics of differentiation in tissue and cell cultures. In: Reinert, J. and Bajaj, Y.P.S. (eds) *Applied and Fundamental Aspects of Plant Cell, Tissue, and Organ Culture.* Springer-Verlag, Berlin, pp. 343–356.
Deng, X.X., Deng, Z.A., Xiao S.Y. and Zhang, W.C. (1992) Pollen derived plantlets from anther culture of Ichang papeda hybrids No. 14 and Trifoliate orange. In: *Proceedings of the International Society of Citriculture.* Acireale, Italy, pp. 190–192.
Deng, Z.A., Deng, X.X., Zhang, W.C. and Wan S.Y. (1992) A preliminary report on gametosomatic fusion in Citrus. In: *Proceedings of the International Society of Citriculture.* Acireale, Italy, pp. 170–172.
Drira, N. and Benbadis, A. (1975) Analyse, par culture d'antheres *in vitro,* des potentialites androgenetiques de deux especes de Citrus (*Citrus medica* L. et *Citrus limon* L. Burm.). *Comptes Rendus Hebdomadaires de Sences de l'Academie des Sciences* 281, 1321–1324.
Dunwell, J.M. (1986) Pollen, ovule and embryo culture, as tools in plant breeding. In: Withers, L.A. and Alderson, P.G. (eds) *Plant Tissue Culture and its Agricultural Applications.* Butterworths, London, pp. 375–404.
Esen, A. and Soost, R.K. (1972) Unexpected triploids in citrus: their origin, identification, and possible use. *Journal of Heredity* 62, 329–333.
Evans, D.A. (1989) Somaclonal variation – genetic basis and breeding applications. *Trends in Genetics* 5, 46–50.
Evans, D.A., Sharp W.R. and Medina-Filho, H.P. (1984) Somaclonal and gametoclonal variation. *American Journal of Botany* 71, 759–774.
Evans, D.E., Singh, M.B. and Knox, R.B. (1990) Pollen development: applications in biotechnology. In: Blackmore, S. and Knox R.B. (eds) *Microspores: Evolution and Ontogeny.* Academic Press, New York, pp. 309–338.
Froelicher, Y. and Ollitrault P. (2000) Effects of the hormonal balance on *Clausena excavata* androgenesis. *Acta Horticulturae* 535, 139–146.
Frost, H.B. (1925) The chromosomes of *Citrus. Journal of Washington Academy Science* 15, 1–3.
Fusurato, K. (1957) Studies on polyembryony in *Citrus. Reports of the Kihara Institute of Biological Research* 8, 40–48.
Gamborg, O.L., Miller, R.A. and Ojima, K. (1968) Nutrient requirements of suspension cultures of soybean root cells. *Experimental Cellular Research* 50, 151–158.
Geraci, G. and Starrantino, A. (1990) Attempts to regenerate haploid plants from '*in vitro*' cultures of *Citrus* anthers. *Acta Horticulturae* 280, 315–320.
Germanà, M.A. (1992) Androgenesis in *Citrus*: a review. In: *Proceedings of the International Society of Citriculture.* Acireale, Italy, pp. 183–189.
Germanà, M.A. (1997) Haploidy in *Citrus.* In: Jain, S.M., Sopory S.K. and Veilleux, R.E. (eds) In Vitro *Haploid Production In Higher Plants,* Vol. 5. Kluwer Academic Publishers, Dordrecht, The Netherlands, pp, 195–217.
Germanà M.A. (2003a) Haploids and doubled haploids in Citrus spp. In: Maluszynsky, M., Kasha, K.J., Forster, B.P. and Szaejko, I. (eds) *Doubled Haploid Production in Crop Plants. A Manual.* FAO-IAEA. Kluwer Academic Publishers, Dordrecht, The Netherlands, pp. 303–308.
Germanà M.A. (2003b) Somatic embryogenesis and plant regeneration from anther culture of *Citrus aurantium* L and *Citrus reticulata* Blanco. *Biologia Bratislava* 58, 843–850.
Germanà, M.A. and Chiancone, B. (2001) Gynogenetic haploids of *Citrus* after *in vitro* pollination with triploid pollen grains. *Plant Cell, Tissue and Organ Culture* 66, 59–66.
Germanà, M.A. and Chiancone, B. (2003) Improvement of the anther culture protocol in *Citrus clementina* Hort. ex Tan. *Plant Cell Reports* 22, 181–187.
Germanà M.A., Chiancone B., Iaconia C. and Muleo R. (2005a) Preliminary results on the effect of light quality on anther culture of *Citrus clementina. Acta Physiologiae Plantarum* 27, 717–721.
Germanà, M.A., Chiancone, B., Lain, O. and Testolin, R. (2005b) Anther culture in *Citrus clementina*: a way to regenerate tri-haploids. *Australian Journal of Agricultural Research* 56, 839–845.
Germanà, M.A. and Reforgiato, G. (1997) Haploid embryos regeneration from anther culture of 'Mapo' tangelo (*Citrus deliciosa* × *C. paradisi*). *Advances in Horticultural Science* 11, 147–152.
Germanà, M.A., Crescimanno, F.G., De Pasquale, F. and Wang, Y.Y. (1991) Androgenesis in five cultivars of *Citrus limon* L. Burm. f. *Acta Horticulturae* 300, 315–324.
Germanà, M.A., Wang, Y.Y., Barbagallo, M.G., Iannolino, G., Crescimanno, F.G. (1994) Recovery of hap-

loid and diploid plantlets from anther culture of *Citrus clementina* Hort. ex Tan. and *Citrus reticulata* Blanco. *Journal of Horticultural Science* 69, 473–480.

Germanà, M.A., Scarano, M.T. and Crescimanno, F.G. (1996) First results on isolated microspore culture of *Citrus.* In: *Proceedings of the International Society of Citriculture.* South Africa, Vol. 2, pp. 882–885.

Germanà, M.A., Crescimanno, F.G. and Motisi A. (2000a) Factors affecting androgenesis in *Citrus clementina* Hort. ex Tan. *Advances in Horticultural Science* 14, 43–51.

Germanà, M.A., Crescimanno, F.G., Reforgiato, G., and Russo, M.P. (2000b) Preliminary characterization of several doubled haploids of *Citrus clementina* cv. Nules. *Acta Horticulturae* 535, 183–190.

Gmitter F.G., Jr, Ling, X.B. and Deng, X.X. (1990) Induction of triploid Citrus plants from endosperm calli *in vitro. Theoretical and Applied Genetics* 80, 785–790.

Grosser, J.W., Gmitter F.G., Jr and Chandler, J.L., Jr (1988) Intergeneric somatic hybrid plants from sexually incompatible woody species: *Citrus sinensis* and *Severina disticha. Theoretical and Applied Genetics* 75, 397–401.

Guha, S. and Maheshwari, S.C. (1964) *In vitro* production of embryos from anthers of *Datura. Nature* 204, 497–498.

Guha, S. and Maheshwari, S.C. (1966) Cell division and differentiation of embryos in the pollen grains of *Datura in vitro. Nature* 212, 97–98.

Heberle-Bors, E. (1980) Isolated pollen cultures and pollen dimorphism. *Naturwissenschaften* 67, 311–312.

Heberle-Bors, E. (1982) *In vitro* pollen embryogenesis in *Nicotiana tabacum* L. and its relation to pollen sterility, sex balance and floral induction of the pollen donor plants. *Planta* 156, 396–401.

Heberle-Bors, E. (1983) Induction of embryogenic pollen grains and subsequent embryogenesis in *Nicotiana tabacum* L. by treatments of the pollen donor plants with feminizing agents. *Physiologia Plantarum* 59, 67–72.

Heberle-Bors, E. (1985) *In vitro* haploid formation from pollen: a critical review. *Theoretical and Applied Genetics* 71, 361–374.

Heberle-Bors, E. (1989) Isolated pollen culture in tobacco: plant reproductive development in a nutshell. *Sexual Plant Reproduction* 2, 1–10.

Heberle-Bors, E., Benito Moreno, R.M., Alwen, A., Stòger, E. and Vincente, O. (1990) Transformation of pollen. In: Nijkamp, H. J.J., van der Plas, L.H.W. and van Hartrigik, J. (eds) *Progress in Plant Cellular and Molecular Biology.* IAPTC, Kluwer Academic Publishers, Dordrecht, The Netherlands, pp. 244–251.

Heslop-Harrison, J. (1979) The forgotten generation: some thoughts on the genetics and physiology of angiosperm gametophytes. In: Davies, D.R. and Hopwood, D.A. (eds) *The Plant Genome.* 4th J. Innes Symposium, Norwich, UK, pp. 1–14.

Hidaka, T. (1984a) Effects of sucrose concentration, pH of media, and culture temperature on anther culture of *Citrus. Japan Journal of Breeding* 34, 416–422.

Hidaka, T. (1984b) Induction of plantlets from anthers of 'Trovita' orange. *Journal of the Japanese Society for Horticultural Science* 53, 1–5.

Hidaka, T. and Omura, M. (1989a) Control of embryogenesis in *Citrus* cell culture. *Bulletin of the Fruit Tree Research Station* 16, 1–17.

Hidaka, T. and Omura, M. (1989b) Origin and development of embryoids from microspores in anther culture of Citrus. *Japan Journal of Breeding* 39, 169–178.

Hidaka, T., Yamada Y. and Shichijo, T. (1979) *In vitro* differentiation of haploid plants by anther culture in *Poncirus trifoliata* (L.) Raf. *Japan Journal of Breeding* 29, 248-254.

Hidaka, T., Yamada, Y, and Shichijo, T. (1981) Plantlet formation from anthers of *Citrus aurantium* L. In: *Proceedings of the International Society of Citriculture* pp. 153–155.

Hofer, M. (1995) *In vitro* androgenese bei Apfel. *Gartenbauwissenschaft* 60, 12–15.

Hofer, M. (1997) *In vitro* androgenesis in apple: optimization of the anther culture. *Acta Horticulturae* 447, 341–344.

Hofer, M. and Lespinasse, Y. (1996) Haploidy in apple. In: Jain, S.M., Sopory S.K. and Veilleux, R.E. (eds) *In Vitro Haploid Production In Higher Plants.* Kluwer Academic Publishers, Dordrecht, The Netherlands, Vol. 3, pp. 261–276.

Hormaza, J.I. and Herrero, M. (1992) Pollen selection. *Theoretical and Applied Genetics* 83, 663–672.

Hu, H. and Huang, B. (1987) Application of pollen-derived plants to crop improvement. In: Giles, K.L. and Prakash, J. (eds) *Pollen: Cytology and Development. International Review of Cytology.* Academic Press, New York, pp. 293–314.

Hu, H. and Yang, H.Y. (eds) (1986) *Haploids in Higher Plants* In Vitro. China Academic Publishers,

Beijing/Springer-Verlag, Berlin.
Huang, B. (1996) Gametoclonal variation in crop improvement. In: Jain, S.M., Sopory, S.K. and Veilleux, R.E. (eds) In Vitro *Haploid Production In Higher Plants,* Vol. 2, Kluwer Academic Publishers, Dordrecht, The Netherlands, pp. 73–91.
Huetteman, A, and Preece, J.E. (1993) Thidiazuron: a potent cytokinin for woody plant tissue culture. *Plant Cell, Tissue and Organ Culture* 33, 105–119.
Karasawa, K. (1971) On the occurrence of haploid seedlings in *Citrus natsudaidai* Hayata. *Bulletin of Sakushingakuin Junior College for Women* 1, 1–2.
Kasha, K.J. (ed.) (1974) *Haploids in Higher Plants. Advances and Potential.* University of Guelph Press, Canada.
Keller, W.A., Rajhathy, T. and Lacapra, J. (1975) *In vitro* production of plants from pollen in *Brassica campestris. Canadian Journal of Genetic Cytology* 17, 655–666.
Keller, W.A., Armstrong, K.C. and de La Roche, A.I. (1983) The production and utilization of microspore-derived haploids in *Brassica* crops. In: Sen, S.K. and Giles, K.L. (eds) *Plant Cell Culture in Plant Improvement.* Plenum, New York, pp. 169–183.
Khuroshvili, K.G., Goliadze, Sh.K. and Kashakashvili, Ts.S. (1982) Use of anther culture to obtain haploid plants of *Citrus* crops. *Subtropicheskie-Kul'tury* 5, 72–76.
Khush, G.S. and Virmani, S.S. (1996) Haploids in plant breeding. In: Jain, S.M., Sopory, S.K. and Veilleux, R.E. (eds) In Vitro *Haploid Production In Higher Plants,* Vol. 1, Kluwer Academic Publishers, Dordrecht, The Netherlands, pp. 11–33.
Kobayashi, S., Ohgawara, T., Saito, W., Nakamura, Y. and Omura, M. (1997) Production of triploid somatic hybrids in *Citrus. Journal Japanese Society Horticultural Science* 66, 453–458.
Kostoff, D. (1929) An androgenic *Nicotiana* haploid. *Zeitschrift für Zellforschung und Mikroskopische Anatomie* 9, 640–642.
Kumar, A., Altabella, T., Taylor, M.A. and Tiburcio A.F. (1997) Recent advances in polyamine research. *Trends in Plant Science* 2, 124–130.
Larkin, P.J. and Scowcroft, W.R. (1981) Somaclonal variation – a novel source of variability from cell culture for plant improvement. *Theoretical and Applied Genetics* 60, 197–214.
Lee, C.H. and Power, J.B. (1988) Intraspecific gametosomatic hybridisation in *Petunia* hybrids. *Plant Cell Reports* 7, 17–18.
Lillie, R.D. (1951) Simplification of the manufacture of Schiff reagent for use in histochemical procedures. *Stain Technology* 26, 163–165.
Ling, J., Iwamasa, M. and Nito, N. (1988) Plantlet regeneration by anther culture of Calamondin (*C. madurensis* Lour.). In: Goren, R. and Mendel, K. (eds) *Proceedings of the Sixth International Citrus Congress.* Balaban, Rehovot, Israel. Margraf, Germany, pp. 251–256.
Maheshwari, S.C., Tyagi, A.K., Malhotra, K. and Sopory, S.K. (1980) Induction of haploids from pollen grains in angiosperms – the current status. *Theoretical and Applied Genetics,* 58, 193–206.
Mascarenhas, J.P. (1990) Gene activity during pollen development. *Annual Review of Plant Physiology and Plant Molecular Biology* 41, 317–338.
Morrrison, R.A. and Evans, D.A. (1987) Gametoclonal variation. *Plant Breeding Reviews* 5, 359–391.
Mulchay, D.L. (1986) Preface in recognition of the forgotten generation. Biotechnology and ecology of pollen. In: *Proceedings of the International Conference on the Biotechnology and Ecology of Pollen 1985.* University of Massachusetts, Amherst, Massachusetts, pp. vi–xi.
Murashige, T. and Skoog, F. (1962) A revised medium for rapid growth and bioassays with tobacco tissue cultures. *Physiologia Plantarum* 15, 473–497.
Murashige, T. and Tucker, D.P.H. (1969) Growth factor requirement of *Citrus* tissue culture. In: *Proceedings of the First International Citrus Symposium* pp. 1155–1161.
Narayanaswamy, S. and George, L. (1982) Anther culture. In: Johri, B.M. (ed.) *Experimental Embryology of Vascular Plants.* Springer-Verlag, Berlin, pp. 79–103.
Nitsch, C. (1977) Culture of isolate microspore. In: Reinert, J. and Bajaj, Y.P.S. (eds) *Applied and Fundamental Aspects of Plant Cell, Tissue and Organ Culture.* Springer-Verlag, Berlin, pp. 268–278.
Nitsch, J.P. and Nitsch, C. (1969) Haploid plants from pollen grains. *Science* 163, 85–87.
Nitsch, C. and Norreel, B. (1973) Effet d'un choc thermique sur le pouvoir embryogène du pollen de *Datura innoxia* cultivé dans l'anthère ou isolé de l'anthère. *Comptes Rendu de l'Academie des Sciences Paris Serie D* 276, 303–306.
Ollitrault P., Allent, V. and Luro, F. (1996) Production of haploid plants and embryogenic calli of clementine

(*Citrus reticulata* Blanco) after *in situ* parthenogenesis induced by irradiated pollen. In: *Proceedings of the International Society of Citriculture*. South Africa, Vol. 2, pp. 913–917.

Ollitrault, P., Vanel, F., Froelicher, Y. and Dambier, D. (2000) Creation of triploid *Citrus* hybrids by electrofusion of haploid and diploid protoplasts. *Acta Horticulturae* 535, 191–198.

Oiyama, I.I. and Kobayashi, S. (1993) Haploid obtained from diploid × triploid crosses of *Citrus*. *Journal of the Japanese Society of Horticultural Science* 62, 89–93.

Osolnik, B., Bohanec, B. and Jelaska, S. (1993) Stimulation of androgenesis in white cabbage (*Brassica oleracea* var. *capatata*) anthers by low temperature and anther dissection. *Plant Cell, Tissue and Organ Culture* 32, 241–246.

Pandey K.K., Przywara L. and Sanders P.M. (1990) Induced parthenogenesis in kiwifruit (*Actinidia deliciosa*) through the use of lethally irradiated pollen. *Euphytica* 51, 1–9.

Patat-Ochatt, E.M., Boccon-Gibod, J., Duron, M. and Ochatt, S.J. (1993) Organogenesis of stem and leaf protoplast of a haploid golden delicious apple clone (*Malus* × *domestica* Borkh.). *Plant Cell Reports* 12, 118–120.

Pirrie, A. and Power, J.B. (1986) The production of fertile, triploid somatic hybrid plants (*Nicotiana glutinosa* (n) + *N. tabacum* (2n) via gametic: somatic protoplast fusion. *Theoretical and Applied Genetics* 72, 48–52.

Powell, W. (1988) The influence of genotype and temperature pre-treatment on anther culture response in barley (*Hordeum vulgare* L.). *Plant Cell, Tissue and Organ Culture* 12, 291–297.

Raghavan, V. (1990) From microspore to embryoid: faces of the angiosperm pollen grain. In: Nijkamp, H.J.J., van der Plas, L.H.W. and van Hartrigik, J. (eds) *Progress in Plant Cellular and Molecular Biology*. IAPTC. Kluwer Academic Publishers, Dordrecht, The Netherlands, pp. 213–221.

Ramirez, C., Chiancone, B., Testillano, P.S., Garcia-Fojeda, B., Germanà, M.A. and Risueno, M.C. (2003) First embryogenic stages of Citrus microspore-derived embryos. *Acta Biologica Cracoviensia Series Botanica* 45, 53–58.

Raquin, C. (1985) Induction of haploid plants by *in vitro* culture of *Petunia* ovaries pollinated with irradiated pollen. *Zeitschrift für Pflanzenzuchtung* 94, 166–169.

Reinert, J. and Bajaj, Y.P.S. (1977) Anther culture: haploid production and its significance. In: Reinert, J. and Bajaj, Y.P.S. (eds) *Applied and Fundamental Aspects of Plant Cell, Tissue and Organ Culture*. Springer-Verlag, Berlin, pp. 251–267.

Reynolds, T.L. (1997) Pollen embryogenesis. *Plant Molecular Biology* 33, 1–10.

Rode, J.C. and Dumas de Vaulx, R. (1987) Obtention de plantes haploides de carotte (*Daucus carota* L) issue de parthénogenèse *in situ* par du pollen irradié et culture *in vitro* de graines immatures. *Comptes Rendu de l'Academie des Sciences Paris Serie III* 305, 225–229.

Rotino, G.L. (1996) Haploidy in eggplant. In: Jain, S.M., Sopory, S.K., Veilleux, R.E. (eds) In Vitro *Haploid Production in Higher Plants*, Vol. 3. Kluwer Academic Publishers, Dordrecht, The Netherlands, pp. 115–142.

Sangwan, R.S. and Sangwan-Norrel, B.S. (1990) Anther and pollen culture. In: Bhajwari, S.S. (ed.) *Plant Tissue Culture. Applications and Limitations*. Elsevier, The Netherlands, pp. 220–241.

Sangwan-Norrel, B.S. (1983) Male gametophyte nuclear DNA content evolution during androgenic induction in *Datura innoxia*. *Zeitschrift für Pflanzenphysiologie* 111, 47–54.

Sangwan-Norrel, B.S., Sangwan R.S. and Pare, J. (1986) Haploidie et embryogenèse provoquée in vitro. *Bulletin de la Societe Botanique de France, Actualites Botaniques* 4, 7–39.

Sari N., K. Abak, K. Pitrat, M., Rode J.C. and Dumas de Vaulx, R. (1994) Induction of parthenogenetic haploid embryos after pollination by irradiated pollen in watermelon. *Hortscience* 29, 1189–1190.

Sauton, A. and Dumas de Vaulx, R. (1987) Obtention de plantes haploides chez le melon (*Cucumis melo* L.) par gynogénèse induite par du pollen irradié. *Agronomie* 7, 141–148.

Schenk, R.U. and Hildebrandt, A.C. (1972) Medium and techniques for induction and growth of monocotyledonous and dicotyledonous plant cell cultures. *Canadian Journal of Botany* 50, 199–204.

Shull, J.K. and Menzel, M.Y. (1977) A study of the reliability of synchrony in the development of pollen mother cells of *Lilium longiflorum* at the first meiotic prophase. *American Journal of Botany* 64, 670–679.

Snape, J.W., Sitch, L.A., Simpson, E. and Parker, B.B. (1988) Tests for the presence of gametoclonal variation in barley and wheat doubled haploids produced using the *Hordeum bulbosum* system. *Theoretical and Applied Genetics* 75, 509–513.

Sharma A.K. and Sharma A. (1972) *Chromosome Techniques*. Butterworths, London/University Park Press, Baltimore, Maryland.

Soost, R.K. and Cameron, J.W. (1980) 'Oroblanco', a triploid pummelo–grapefruit hybrid. *HortScience* 15, 667–669.

Soost, R.K. and Cameron, J.W. (1985) 'Melogold', a triploid pummelo–grapefruit hybrid. *HortScience* 29, 1134–1135.

Starrantino, A. (1986) Produzione di aploidi androgenetici di agrumi. In: *Il Recente Contributo della Ricerca allo Sviluppo dell'Agrumicoltura Italiana.* Delfino, Cagliari, pp. 31–37.

Stuart, D.A., Strickland, S.G. and Walker, K.A. (1987) Bioreactor production of alfalfa somatic embryos. *Hortscience* 22, 800–809.

Sunderland, N. (1974) Anther culture as means of haploid induction. In: Kasha, K.J. (ed.) *Haploids in Higher Plants. Advances and Potential.* University of Guelph Press, Canada, pp. 91–122.

Sunderland, N. (1978) Strategies on the improvement of yields in anther culture. In: *Proceedings of the Symposium on Plant Tissue Culture.* Science Press, Peking, pp. 65–86.

Sunderland, N. and Dunwell, J.M. (1977) Anther and pollen culture. In: Street, H.E. (ed.) *Plant Tissue and Cell Culture.* Blackwell, Oxford, pp. 223–265.

Torres, A.M., Soost, R.K. and Diedenhofen, U. (1978) Leaf isozymes as genetic markers in *Citrus. American Journal of Botany* 65, 869–881.

Touraev, A, Vincente, O. and Heberle-Bors, E. (1997) Initiation of microspore embryogenesis by stress. *Trends in Plant Science* 2, 297–301.

Vasil, I.K. (1967) Physiology and cytology of anther development. *Biological Reviews* 42, 327–373.

Vasil, I.K. (1980) Androgenic haploids. *International Review of Cytology* Suppl. 11 A, 195–223.

Veilleux, R.E. (1994) Development of new cultivars via anther culture. *Hortscience* 29, 1238–1241.

Wenzel, G., Hoffmann, F. and Thomas, E. (1977) Anther culture as a breeding tool in rape. I. Ploidy level and phenotype of androgenetic plants. *Zeitschrift für Pflanzenzuchtung* 78, 149–155.

Wenzel, G., Frei, U., Jahoor, A., Graner, A. and Foroughghi-Wehr, B. (1995) Haploids – an integral part of applied and basic research. In: Terzi, M., Cella R. And Falavigna, A. (eds) *Current Issues in Plant Molecular and Cellualar Biology.* Kluwer Academic Publishers, Dordrecht, The Netherlands, pp. 127–135.

Wernsman, E.A., Matzinger, D.F. and Rufty, R.C. (1989) Androgenetic vs. gynogenetic doubled haploids of tobacco. *Crop Science* 29, 1151–1155.

Zhang, Y.X. and Lespinasse, Y. (1991) Pollination with gamma-irradiated pollen and development of fruits, seeds and parthenogenetic plants in apple. *Euphytica* 54, 101–109.

Zhang, Y.X., Lespinasse, Y. and Chevreau, E. (1990) Induction of haploidy in fruit trees. *Acta Horticulturae* 280, 293–304.

8 Seedlessness and Ploidy Manipulation

Patrick Ollitrault[1], Yann Froelicher[1], Dominique Dambier[1], François Luro[2] and Masashi Yamamoto[3]

[1]CIRAD-BIOS, Avenue Agropolis – TA A-75/02, 34398 Montpellier Cedex 5, France; [2]SRA, INRA/CIRAD, San Giuliano, 20230 San Nicolao, France; [3]Faculty of Agriculture, Kagoshima University, 1-21-24, Korimoto, Kagoshima, 890-0032, Japan

Introduction

Seedlessness is a major characteristic for the fresh-fruit market and may have importance for the juice industry because the presence of seeds in the fruit could be associated with unfavourable aromatic compounds and bitterness. A lot of breeding work has been done to develop new seedless cultivars.

The seedlessness of citrus varies among accessions and, sometimes, environmental conditions. For instance, 'Mukaku kishiu' is completely seedless under any conditions. Navel orange and satsuma mandarin are usually seedless, but occasionally they produce seeds when pollinated. On the other hand, with cross-pollination, the fruit of some pummelos produce more than 100 seeds while they are seedless in self-pollination conditions. The number of seeds of almost all accessions is less than the number of ovules. Therefore, it is probable that most of the commonly cultivated accessions have some degree of ovule or pollen sterility (Frost, 1948).

Strong sterility coupled with parthenocarpy is necessary for the production of stable seedless fruits. Sterility could be divided into three types: female sterility, male sterility and self-incompatibility.

Yamamoto *et al.* (1995) proposed that the degree of female fertility/sterility should be rated on the basis of the average number of seeds per fruit obtained through hand pollination. High positive correlation ($r = 0.93$) was found between the number of seeds of open pollinated fruits and those of hand-pollinated fruits. This result indicated that female sterility is directly related to seediness.

The degree of male (pollen) sterility is variable in cultivated citrus, and usually pollen-sterile accessions produce seedless or few seedy fruits when cultivated in solid blocks. Male-sterile or self-incompatible accessions have the ability to produce seedless fruits when cross-pollination is prevented. Even in mixed planting with male-fertile accessions, male sterility and self-incompatibility reduce the seed production and increase the percentage of seedless fruits because those accessions have a smaller chance of fertilization than self-compatible accessions (Yamamoto *et al.*, 1995). However, those accessions may produce seedy fruits in mixed planting particularly if pollinating insects are frequent.

This implies that for diversification programmes, male and female sterilities must be selected, particularly for an area such as the Mediterranean Basin where the main seedless easy peel cultivar is the self-incompatible 'clementine'.

Some male-sterile and self-incompatible accessions cannot produce seedless fruits because of a lack of parthenocarpy. Thus, parthnenocarpy is an indispensable trait for seedless fruit production and this character seems to be widely present in *Citrus* germplasm. Some seedy accessions can also produce parthenocarpic fruit (Nagai and Tanikawa, 1928; Sykes and Possingham, 1992). Autonomic parthenocarpy, where seedless fruit is produced without any external stimulation (pollination), is the main type of parthenocarpy in citrus, such as in navel or satsuma mandarin. Stimulative parthenocarpy has also been reported (Vardi *et al.*, 1988). It was suggested that the autonomic parthenocarpy found in satsuma mandarin depends on three dominant complementary genes.

In *Citrus*, self-incompatibility seems to be of the gametophytic type, while male and female sterility may be due to different genetic factors such as sterility genes, chromosomal abnormalities (structural heterozygosity, inversion, translocation) and triploidy.

This chapter presents a review on the mechanism and genetic factors underlying seedlessness in *Citrus* and on their exploitation for new seedless cultivar breeding. It focuses mainly on the aspect of ploidy manipulation for triploid seedless breeding which has received much attention at the international level and for which new avenues have been opened up by biotechnology.

Seedlessness at the Diploid Level

Male and female sterility

Various levels of male sterility at the diploid level have been reported in citrus (Iwamasa, 1966) (Table 8.1). Chromosome aberration is one of the most important phenomena which cause pollen sterility. Asynapsis in 'Mukaku yuzu' is genetically controlled, while that in 'Eureka' lemon and 'Mexican' lime is induced by low temperature (Nakamura, 1943; Iwamasa and Iwasaki, 1962; Iwamasa, 1966). Reciprocal translocation is the main cause of sterility of 'Valencia' sweet orange, and this sterility is not found in other sweet oranges (Iwamasa, 1966). Inversion caused partial pollen sterility of 'Mexican' lime. Pollen sterility of 'Marsh' grapefruit is due to the failure of spindle formation (Raghuvanshi, 1962). Male sterility which is not caused by chromosome aberration is also well known. Anther abortion in satsuma hybrids is taken to be due to strict male sterility in citrus (Iwamasa, 1966). The sterile stamen appears only as a filament, and no pollen grains are produced (Fig. 8.1). Early degeneration of pollen mother cells (PMCs) was found in 'Washington' navel, 'Tahiti' lime and some other hybrids (Osawa, 1912; Uphof, 1931; Frost, 1948). Pollen sterility of satsuma mandarin is caused by plural sterility such as abnormal behaviour and degeneration of pollen grains (Nakamura, 1943; Yang and Nakagawa, 1970). A close relationship between the free proline content in citrus anther and pollen fertility has been found (Liu *et al.*, 1995), and the cultivars with normal fertility have a higher proline concentration than those with partial or complete sterility. To develop new seedless

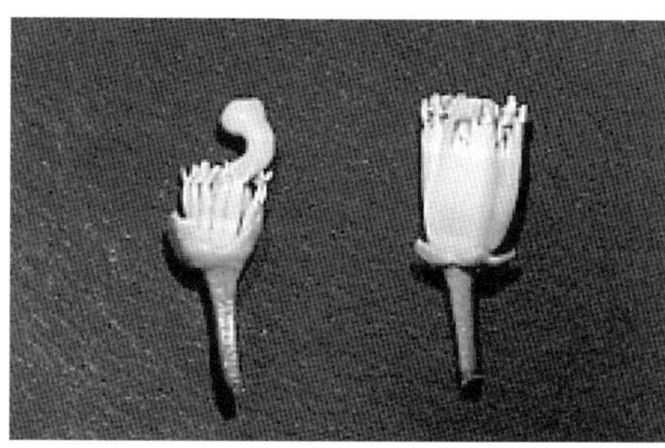

MS__MF

Fig. 8.1. Flowers of male sterile (aborted anthers, MS) and male fertile (normal anthers, MF).

Table 8.1. Diagramatic representation of various kinds of male sterility occurring in citrus, according to the sequential order of development (from Iwamasa, 1966).

Developmental stage	Nature of sterility	Cultivar or hybrid	Reference
Initiation of anther development	Anther abortion	Satsuma × Sweet orange, etc.	Iwamasa, 1966
↓ Archesporial stage			
↓ Resting stage	Degeneration of PMCs	Washington navel	Osawa, 1912
		Tahiti lime	Uphof, 1931
		Lemon × Valencia, etc.	Frost, 1948
		Satsuma × trifoliate orange	Iwamasa, 1966
↓ Meiosis			
Division I	Asynapsis (genic)	Mukaku yuzu	Iwamasa, 1966
	Asynapsis (by low temperature)	Eureka lemon	Nakamura, 1943
		Mexican lime	Iwamasa *et al.*, 1962
	Translocation	*C. assamensis*	Naithani *et al.*, 1958
		Valencia orange	Iwamasa, 1966
	Inversion	*C. assamensis*, etc.	Raghuvanshi, 1962a
		Mexican lime	Iwamasa, 1966
Division II	Failure of spindle	Marsh grapefruit	Raghuvanshi, 1962b
Liberation from tetrad			
	Degeneration	Jaffa orange	Oppenheim, 1929
↓ Mitotic division			
	Degeneration	Satsuma	Nakamura, 1943
↓ Mature pollen grain			

cultivars efficiently, genetic analysis of male sterility has been carried out. Among these studies, genetic analysis of anther abortion has progressed remarkably. This male sterility is due to gene–cytoplasmic interaction (satsuma mandarin and 'Encore' mandarin possess sterile cytoplasm) and is probably controlled by more than two major genes (Iwamasa, 1966; Yamamoto *et al.*, 1992, 1997; Nakano *et al.*, 2001). New seedless cultivars with aborted anthers have been released in Japan (Nishiura *et al.*, 1983; Matsumoto *et al.*, 1991). Anther male sterility and inheritance of pollen fertility/sterility were also studied (Ueno, 1986). Some pollen-sterile progeny had arisen from parents that were both pollen fertile. Vardi and Spiegel-Roy (1981) postulated that asynapsis was controlled by a single recessive gene because about a quarter of progeny arising from self-pollination of 'Wilking' showed asynapsis. Nakano *et al.* (2000) found DNA markers linked to male sterility for juvenile screening of male-sterile plants.

Female sterility is a very important trait closely related to seedlessness and is a heritable characteristic (Yamamoto *et al.*, 1995, 2001). 'Mukaku kishiu', a bud variation of the seedy kinokuni mandarin, is complete seedless, and is considered to have the strongest female sterility in citrus. This female sterility causes abortion of the zygote, and is controlled by two genes (Nesumi *et al.*, 2001). A new seedless cultivar and parental line with this sterility was bred in Japan. Navel orange and satsuma mandarin have strong female sterility. Only a few seeds developed when they were hand pollinated (Nishiura and Iwasaki, 1963). Osawa (1912) observed degeneration of the embryo sac in both navel orange and satsuma mandarin. Nesumi *et al.* (2000)

assumed that the female sterility of satsuma mandarin is controlled by two major genes, and they were genetically mapped (Omura *et al.*, 2000). Seediness of hand-pollinated fruits is low in both 'Valencia' orange and 'Marsh' grapefruit (Wong, 1939). Chromosome aberration, as mentioned with regard to male sterility, probably occurs in the embryo sac. In non-functional pistils of lemon, possible blocking of further stigma and style development is related to the presence or absence of receptive embryo sacs in the ovule (Wilms *et al.*, 1983).

Mutagenesis is efficient in developing seedless plants from seeded accessions because sterility is one of the most frequent effects of treatment with a mutagen. Hensz (1971) developed the seedless 'Star Ruby' grapefruit through irradiation of seed of the seedy 'Hudson' by thermal neutrons. After his success in developing a seedless cultivar using irradiation, this method was applied to several seedy cultivars. Hearn (1984) produced seedless strains of 'Pineapple' orange and 'Duncan' grapefruit from γ-ray irradiation of seeds. He also developed seedless strains of 'Foster' grapefruit through γ-irradiation of buds. Seedless strains of 'Eureka' lemon and 'Monreal' clementine were developed through γ-irradiation of budsticks (Spiegel-Roy *et al.*, 1985; Starrantino *et al.*, 1988). Chen *et al.* (1991) produced seedless strains of 'Jin Cheng' orange through γ-irradiation of seeds, and chromosome aberrations were observed in these strains. South Africa has developed an ambitious programme of mutagenesis by γ-irradiation of budwoods to obtain seedless cultivars (Froneman *et al.*, 1996). About 400 trees with 2400 branches exhibiting seedless fruits have been selected.

Biotechnology procedures such as transformation and genome analysis have been conducted to develop new seedless citrus (Koltunow *et al.*, 1998). Tobacco and *Arabidopsis* transformants have been regenerated containing chimeric genes of soybean conglycinin and storage protein gene promoters linked to the bacterial RNase gene, barnase. Reduction in seed size was only observed in *Arabidopsis* seeds (exalbuminous), and not in tobacco (albuminous). Some transformants of both species were male sterile and this correlated with the gene expression in anthers. Citrus forms exalbuminous seeds. The barnase constructions may be useful in eliciting a reduction in seed size. Transgenic West Indian lime plants containing genes for decreased seed set have already been produced from seedling hypocotyl and epicotyl segments by *Agrobacterium*-mediated gene transfer (Koltunow *et al.*, 2000). The shorter juvenile period of lime provides the opportunity to test the introduced genes for their ability to induce reduced seed set. Garcia *et al.* (2000) found a DNA marker for the number of seeds obtained from 69 DNA and isozyme markers, and this marker must be related to female sterility.

Self-incompatibility

Self-incompatibility is a genetically controlled phenomenon preventing seed set in self-pollinated plants producing functional gametes. In self-incompatible accessions, in conditions of self-pollination, no pollen tubes were found in the ovaries (Ton and Krezdorn, 1967). Nagai and Tanikawa (1928) found that some self-incompatible accessions produced seedless fruits when they were self-pollinated. Almost all pummelos, some mandarins and several natural or artificial hybrids are self-incompatible (Hearn, 1969). Some of the self-incompatible cultivars are seedy because of their female fertility and lack of parthenocarpy requiring cross-pollination to set fruits (Miwa, 1951; Mustard *et al.*, 1956; Krezdorn and Robinson, 1958). Some self-incompatible accessions can produce seedless fruits in a single planting, and clementine is probably the most famous of these. These fruits are sometimes smaller than seeded ones which tend to have reduced setting of fruit. Mixed cultivation of these accessions with male-fertile plants often yield seedy fruits unless they have female sterility (Soost, 1956; Reece and Register, 1961; Hearn, 1969; Iwamasa and

Oba, 1980; Li, 1980). However, seedless fruits can be produced in self-incompatible accessions without parthenocarpic ability by pollination with tetraploid pollen or application of growth regulators (Soost, 1961; Yamashita, 1976).

The incompatibility system of citrus is the gametophytic type, and Soost (1965, 1968) proposed 'S' genotypes in some accessions. At least three alleles are present in the pummelos tested. At least two of these differ from the alleles in 'Sukega'. At least four alleles are present in 'Dancy', clementine, 'Minneola', 'Orlando', 'Sukega' and 'Duncan' grapefruit. Incompatibility S alleles are distributed widely; not only in self-incompatible accessions but also in self-compatible ones such as satsuma mandarin, grapefruit and 'Dancy' (Soost, 1965, 1968; Vardi *et al.*, 2000). Thus, a self-incompatible individual can be produced from cross-combinations between both self-compatible parents, e.g. 'Orlando' and 'Minneola' arose from the combination 'Duncan' grapefruit and 'Dancy' (Swingle *et al.*, 1931). Wakana *et al.* (1998) demonstrated that glutamate oxaloacetate trans-minase (GOT) isozyme loci and the incompatibility locus are linked, which should be useful for early screening for self-incompatibility.

Ploidy Manipulation for Seedless Cultivar Breeding

The selection of triploid lines has been, and remains a very interesting way to develop seedless cultivars. Indeed triploidy is generally associated with both male and female sterility. Thus, most of the trees under field evaluation present these characters, and an efficient selection can be done for other traits. Moreover, the larger fruit size associated with triploidy should correct the reduction of fruit size generally observed in seedless mutants of seedy varieties. The more famous citrus triploid is 'Tahiti lime'. This natural triploid produces large seedless fruits while its diploid kin 'Mexican lime' produces small seedy fruits. Such programmes are being conducted in several countries (Starrantino, 1992; Khan *et al.*, 1996; Ollitrault *et al.*, 1998a; Grosser *et al.*, 2000; Guo *et al.*, 2000; Chandler *et al.*, 2001; Navarro *et al.*, 2004; Froelicher *et al.*, 2005; Handaji *et al.*, 2005; Wakana *et al.*, 2005; Wu *et al.*, 2005), and new triploid cultivars have been recently released in Italy (Starrantino, 1999; Russo *et al.*, 2004), in California (Anonymous, 2002) and in Japan (Yoshida *et al.*, 2003; Tokunaga *et al.*, 2005).

Natural polyploids

Polyploid germplasm

Diploidy is the general rule in *Citrus* and its related genera with the basic chromosome number $x = 9$ (Krug, 1943). However, some polyploid genotypes were detected early on in citrus germplasm. Longley (1925) was the first formally to identify a tetraploid wild form: the Hong Kong kumquat (*Fortunella Hindsii* Swing.). Triploid Tahiti lime (Jackson and Sherman, 1975), tetraploid strains of *Poncirus trifoliata* (Iwasaki, 1943), allotetraploid *Clausena excavata* Burm. F. (Froelicher *et al.*, 2000), tetraploid *Clausena harmandiana* and hexaploid *Glycosmis pentaphylla* are other examples of natural polyploidy found in the germplasm of the Aurantioideae subfamily (Froelicher, 1999).

Mechanisms of formation of polyploid citrus

TRIPLOIDS. The occurrence of spontaneous triploid seedlings was reported many years ago. Lapin (1937) mentioned about 4% triploidy in hybrid seedlings of *Citrus limon* crossed with eight other diploid species and varieties. Frost and Soost (1968) reported that more than 5% of 1200 hybrids from diploid parents grown at Riverside were probably triploids, of which 20 were proved by chromosome counts.

Studies have shown that most of the spontaneous triploids arising from diploid parents are found in small and abnormal seeds of monoembryonic female parents (Esen and Soost, 1971, 1973). The rates of spontaneous triploid formation vary among cultivars. Rates of 1 and 6–7% for clementine and King mandarin, respectively, have been observed in California and Sicily (Geraci *et al.*, 1975). 'Wilking' mandarin also shows a high rate (14.6%) of triploid seedlings (Soost, 1987).

Triploid embryos have also been observed in immature seeds of polyembryonic cultivars, with frequencies between 0 and 8% in *C. deliciosa* clones (cv 'Tardiva de Ciaculli' and 'Avana') and between 0 and 11.5% in different *C. limon* cultivars (Geraci, 1978). The rate of triploid embryos is very high in several sweet orange cultivars. Oiyama *et al.* (1980) found between 26 and 30% small seeds and between 8 and 33% triploid seedlings in sweet orange cultivars and intraspecific hybrids of sweet orange. A large percentage of triploid embryos has also been found in presumed interspecific hybrids of sweet orange such as 'Ortanique' tangor (25%) (Wakana *et al.*, 1981), Temple tangor (6.8%) and 'Sugeka' orangelo (23%) (Esen and Soost, 1971). Therefore, it appears that there is genetic control of triploid embryo formation as well as environmental factors because the same cultivar may produce different numbers of triploid hybrids in different years or different areas of production. Colder conditions seems to be favourable for triploid induction (our unpublished data). With the exception of the report concerning sweet orange (Oiyama *et al.*, 1980), it appears that the nucellar embryos present in the seeds are more developed than the triploid ones and considerably limit the probability of triploid seedling recovery from polyembryonic species (Wakana *et al.*, 1981).

Cytogenetic studies showed that triploid embryos are associated with pentaploid endosperm, which is a strong indication that triploid hybrids result from the fertilization of unreduced ovules by normal haploid pollen (Esen and Soost, 1977; Esen *et al.*, 1979; Wakana *et al.*, 1981). This ploidy ratio of 5/3 between endosperm and zygotic embryos is generally considered by citrus breeders as the origin of precocious abortion of endosperm development and subsequent overdevelopment of embryos found in small citrus seeds. It should also occur for monoembryonic and polyembryonic cultivars where the triploid embryo is associated with diploid nucellar embryos (Wakana *et al.*, 1981). However, another hypothesis should be considered: the endosperm balance number (EBN) theory proposed in the early 1980s (Johnston and Hanneman, 1980) to explain the basis of normal development of seeds in interspecific and interploidy crosses in the potato. According to this hypothesis, each species has a genome-specific effective ploidy (called the EBN) which must be in a 2/1 maternal to paternal ratio in the endosperm for normal development. The EBN, which can be different from the actual ploidy level, functions in an additive way during ploidy manipulation. This hypothesis, initially developed for potato, has also been applied in several other plants species such as *Trifolium*, *Lycopersicum*, *Avena*, *Datura* and *Impatiens* (Carputo *et al.*, 1999).

The frequency of occurrence of duplication in the female gametes varies between less than 1% and more than 20% (Soost, 1987; Iwamasa and Nito, 1988) and it is suggested to be due to the abortion of the second meiotic division in the megaspore (Esen *et al.*, 1979). This hypothesis is confirmed for Clementine by a molecular marker analysis showing that less than 50% of maternal heterozygosity is transmitted by the 2*n* ovules (Luro *et al.*, 2000, 2004). However, recently Chen *et al.* (2007) proposed that 2N gametes of sweet orange result from first division restitution. Very rare events of formation of triploid hybrids by fertilization of a haploid ovule by diploid pollen have also been demonstrated (Luro *et al.*, 2000).

TETRAPLOIDS. Tetraploidization seems to occur frequently in *Citrus* polyembryonic genotypes. Cameron and Frost (1968) mention from the Riverside (California) experiments, that 2.5% of 3600 nucellar progeny from a broad range of genotypes were tetraploid. In Russia, Lapin (1937) also found tetraploid seedlings among eight *Citrus* species (from <1% to 5.6%) and

Poncirus (4%). Russo and Torrisi (1951a) detected the tetraploid forms of *C. aurantium* and *C. limon* among nucellar seedlings. For the rootstocks citrange Troyer and Carrizo, respective frequencies of 3 and 2.5% of tetraploid seedlings were found from a three year experiment (Hutchison and Barrett, 1981).

Chromosome doubling of nucellar stock seems to be the general rule for the generation of such tetraploid seedlings (Cameron and Frost, 1968). Indeed these tetraploid seedlings are homogenous and do not display traits of the pollen parents in controlled crosses. Isoenzymic studies of tetraploid seedlings of *C. volkameriana* have proved this (Ollitrault and Jacquemond, 1995). It seems that this doubling occurs repeatedly in the ovule tissues because tetraploid seedlings are found in fruit where seeds are mainly diploids (Cameron and Frost, 1968). Moreover, diploid seedlings can arise from the same seeds as the tetraploid seedlings (Hutchison and Barrett, 1981). From a systematic search of autotetraploids in a wide range of taxa, Barrett and Hutchison (1978) postulated that the ability to produce such autotetraploid seedlings is a variable genetic trait present in polyembryonic *Citrus* and relatives. They also propose that the rates of tetraploid seedlings are affected by environmental conditions. This assertion is clearly demonstrated by Hutchison and Barrett (1981) who showed that the rates of tetraploid seedlings of Troyer and Carrizo citranges vary among years and position of the fruit on the tree. In a recent work, Luis Navarro from IVIA (Spain) also found differences in the numbers of tetraploids in seedlings of several rootstocks coming from different places throughout the world, suggesting that colder conditions are favourable for spontaneous tetraploidization in nucellar tissues (L. Navarro, personal comminication). Such an effect of cold on polyploidization events seems to be a general rule in both plants and animals (Ramsey and Schemske, 1998; Otto and Whitton, 2000).

Cases of chromosome doubling in somatic tissue have been reported by Raghuvanshi (1962). If it occurs in meristem, it should lead to chimeric shoots and branches. However, very few tetraploid budsports have been identified (Iwamasa and Nito, 1988) and these authors suggest that this is caused by unfavourable competition between diploid and tetraploid cells in the meristem.

The formation of fully developed tetraploid seeds from diploid female × tetraploid male hybridization has been described (Tachikawa *et al.*, 1961; Cameron and Soost, 1968; Esen and Soost, 1972). The authors have proposed that they originate from unreduced ovules fertilized by diploid pollen, leading to a good 3/2 ratio between endosperm and embryo. Considering the EBN theory, this also leads to an adequate 2/1 ratio of female EBN/male EBN in the endosperm and then normal endosperm and seed development.

In conclusion, two basic mechanisms are involved in spontaneous polyploidization in citrus:

- duplication of chromosome stocks in nucellar tissues that gave rise to autotetraploids
- $2n$ gametes arising mostly from second division restitution during meiosis of the megaspore that produce triploid hybrids in diploid × diploid hybridization and allotetraploids in diploid × tetraploid hybridization.

Even if the rates of these natural polyploidization events are very high in *Citrus* compared with those proposed as general rules for plants by Ramsey and Schemske (1998) (autotetraploid plants formed at a rate of $\sim 10^{-5}$ per individual per generation), it appears that polyploidy has played a minor role in citrus evolution. The lower vigour and fertility of most of the autotetraploids, and the lack of interest in them for human consumption due to thick peel and rough pulp, may have resulted in strong natural and human overselection of the autotetraploid nucellar plants. Moreover, the fact that autotetraploids are mostly polyembryonic strongly limits the potential for evolution of the natural tetraploid gene pool. Indeed they could principally act as male parents for monoembryonic diploids, with unfavourable competition with haploid pollen. Moreover, in the case of fertilization with diploid pollen, most of the seeds will present an abnormal development; thus a low probability of germination in natural conditions will result in triploids with very low fertility. For the same reason, very few triploids arising from $2n$ ovules have the chance to germinate and multiply in natural conditions.

Polyploid characteristics

Vegetative and fruit characteristics

An increase in cell size is the most common and universal effect of polyploidization. This change in cell volume may alter metabolic processes, especially those that involve membranes because of variation in the surface to volume ratio (Otto and Whitton, 2000). This could explain the slower developmental rate generally observed in some polyploids compared with diploids.

In citrus, the autotetraploid nucellar lines allow an accurate evaluation of the effect of polyploidization itself because of the homogeneity of the allelic constitution of these tetraploids and their parental diploids. Cameron and Frost (1968) have given a precise description of the autotetraploid plants cultivated at Riverside.

Tetraploid leaves are considerably broader relative to their length than diploids. They are also thicker and darker. Polyploids remain more thorny, thicker and darker than diploids. Growth is slower in tetraploids and the top is smaller and more compact. Tetraploids are slower than diploids to bloom and set fruit, and they generally produce less fruit. It appears that tetraploid grapefruits are much more vigorous and productive than most other autotetraploids. Tetraploid fruits are generally smaller and less elongate than diploid fruit. In most cases, the fruit shape is irregular and the rinds are thicker and rougher with more prominent oil glands. The proportion of juice relative to whole fruit weight is much lower than for diploids, but the flavour, acid and soluble contents are little affected. Seed number is generally lower as well as the number of embryos per seeds. An exception is a tetraploid 'Lisbon' lemon which produces more seeds and bigger fruit than the diploid parent. In conclusion, tree and fruit characteristics of autotetraploids are not valuable for production but they constitute an interesting germplasm for triploid breeding.

These observations are generally confirmed by more recent studies. Barrett (1992) described an autotetraploid of Key lime having darker leaves and larger fruit with the same flavour as the parental diploid. Autotetraploid lines of *P. trifoliata* used as rootstock confer a reduced vigour to the tree (Jacquemond and Blondel, 1986). Autotetraploid *C. volkameriana* display a very depressed growth compared with diploids (Ollitrault and Jacquemond, 1995), while autotetraploid 'Kinnow' mandarin has shown a moderately reduced growth and interesting morphological characters (Khan *et al.*, 1992). The authors suggested that the leaf thickness, size, breadth and the number and size of stomata should be good morphological markers to select spontaneous tetraploids in nucellar seedlings.

The identification of triploid plants by morphological traits appears much more difficult because they present a great variability due to their zygotic origin (Cameron and Frost, 1968). Starrantino (1992) observed that triploid hybrids arising from a diploid female × tetraploid male cross generally present a good vigour and display the fruit level similarities to the male parent providing the diploid gametes. For example, it appears that pigmentation of blood oranges used as male parent is highly heritable. According to Cameron and Frost (1968) and Starrantino (1992), most of the triploid citrus hybrids are nearly seedless.

Meiosis and fertility of polyploids

MEIOSIS OF TETRAPLOID CITRUS. Gametic viability is generally lower in autotetraploid genotypes having multivalent chromosome association during meiosis than in allotetraploids forming bivalents, leading to equilibrated disomic segregation. In citrus, it has been shown that the degeneration of PMCs is more frequent in autotetraploids than in their diploid parental genotypes (Frost and Soost, 1968). These authors also observed great variability in chromosome conjugation (tetravalents, trivalents, bivalents and univalents) during metaphase I and showed that one-third to one-half of sporads have more than the normal number of four microspores (generally six or seven). As a consequence, most autotetraploids generally produce few pollen grains with the normal chromosome complement and had a lower pollen viability than diploid parental lines. However, the remaining fertility is enough to allow controlled hand pollination.

MEIOSIS OF TRIPLOID CITRUS. Triploids have generally been considered to be an evolutionary dead-end because they have very low fertility and tend to produce aneuploid gametes, due to problems of chromosome pairing during meiosis. However triploids can produce haploid, diploid to triploid gametes at low rates that can lead to diploid, triploid and tetraploid progeny (Otto and Whitton, 2000). In the case of citrus, early cytogenetic studies have described triploid meiosis. Trivalent chromosomes and some univalents have been observed by Longley (1926). Frost and Soost (1968) described a predominance of trivalents but also the presence of numerous bivalents and univalents, as well as a great variation in the number of extra microspores in some genotypes. Abortion of megaspores has been observed for 'Oroblanco' and the LCNR46 *C. limon* × *C. sinenis* triploid hybrid, from the first division of the embryo sac to the stage of fertilized egg cell (Fatta Del Bosco *et al.*, 1992). In contrast, a low percentage of pollen abortion has been observed by the same authors. As a consequence of megasporogenesis abortion, most of the triploid hybrids obtained from diploid × autotetraploid crosses are seedless (Cameron and Frost, 1968; Starrantino, 1992). However, some triploid seeds have been found in triploid maternal progeny (Lapin, 1937; Russo and Torrisi, 1951b), while we found half triploids and half diploids among seeds of 'Oroblanco' (P. Ollitrault, unpublished data). In the same study, only diploids have been found in progeny of clementine fertilized by 'Oroblanco' pollen, suggesting an unfavourable competition of diploid pollen with the haploid pollen, or the absence of diploid pollen.

Ploidy manipulation for triploid breeding; practical and theorical aspects

Several strategies have been developed for triploid citrus breeding. Some of them exploit natural events such as $2n$ gametes, while the more recent strategies combine haplomethods and somatic hybridization for a direct synthesis of triploid hybrids. For each method, we will discuss briefly their biological limitations and practical aspects as well as their implications for gene segregation and recombination.

In the citrus literature, tetraploids arising from chromosome doubling (spontaneous in nucellar tissue, or induced by colchicine) are generally called autotetraploids while tetraploid somatic hybrids obtained by protoplast fusion are considered as allotetraploids. We adopt this terminology herein, keeping in mind that it is quite different from the general definition of auto- and allotetraploidy (Otto and Whitton, 2000) and that it may not be related to the disomic or tetrasomic mode of chromosomal segregation.

Selection of spontaneous triploid in 2x × 2x sexual crosses

BIOLOGICAL LIMITATIONS AND PRACTICAL ASPECTS. The selection of triploid hybrids arising from 2*n* megagametophytes was described in the 1970s (Esen and Soost, 1971, 1973; Geraci *et al.*, 1975). Triploids were sought on small seeds of monoembryonic cultivars with a high rate of diploid megagametophytes such as 'Temple', 'Wilking' or 'Fortune' (Esen and Soost, 1977; Wakana *et al.*, 1981; Soost, 1987). However, this approach was limited by a relatively low efficiency and the difficulty of screening large populations of seedlings by classical cytogenetic methods of chromosome counting. More recently the use of *in vitro* embryo rescue and ploidy evaluation by flow cytometry has provided much greater efficiency (Ollitrault *et al.*, 1996b). In this way, it is possible to exploit low rates of spontaneous diploid megagametophytes such as that of clementine (1%), and this strategy (Fig. 8.2) is actually routinely used by several teams in the Mediterranean Basin to select new easy peel cultivars (France, Spain and Morroco).

Concerning polyembryonic cultivars, Wakana *et al.* (1981) show that triploid zygotic embryos should be found with diploid nucellar embryos in small seeds. However, the practical possibility of selecting these triploid embryo is greatly limited by polyembryony (Wakana *et al.*, 1981). To avoid this problem, Geraci *et al.* (1977) have proposed carrying out a very early rescue of zygotic embryos from immature fruit, but it seems that selection of spontaneous triploids from polyembryonic seedlings has not found real applications in citrus breeding.

SEGREGATION AND RECOMBINATION. Second meiotic division restitution (SDR) has been proposed for diploid megagametophyte development in clementine (Luro *et al.*, 2004) while Chen *et al.* (2007) concluded first meiotic division (FDR) for sweet orange. In both cases only a part of the maternal heterozygosity is transmitted to the triploid hybrid but the structures of triploid populations are very different. The

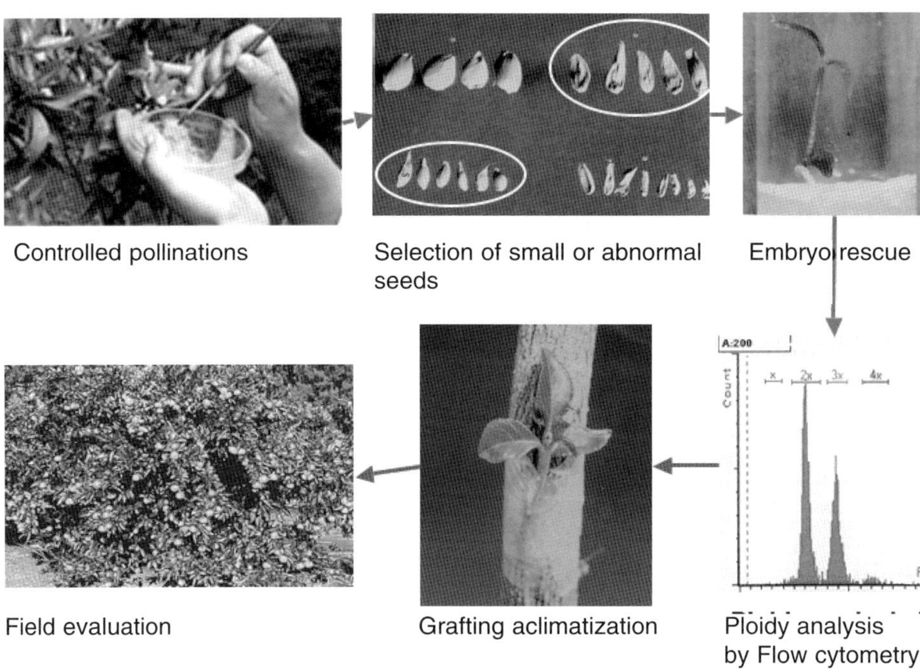

Fig. 8.2. Scheme for selection of triploid hybrids from dilpoid × diploid hybridization.

rate of maternal heterozygosity transmission (RHT) varies among the loci and is directly linked to the crossing over rate between the centromere and the locus (RHT = r for SDR and 1–1/2r for FRD, where r is the crossing over frequency between the locus and the centromere). For SDR and FDR respectively, RHT varies from the centromere to the telomere between 0 to 1 and 1 to 0.5 in the case of systematic single crossing over. Therefore, the homozygosity level of 2n gametes is higher for SDR than for FDR.

Segregations and recombinations occur for both the male and female parents. In the case of SDR for one locus, the maximum allelic combinations of the triploid hybrid vary between two in the case of systematic single crossing over, four for a locus very close to the centromere and six for others. In the case of FDR it is two for locus very close to the centromere and six for others.

Analysis of the origin of 2n gametes for several seed parents must be done to confirm that SDR observed in clementine is generalized in the mandarin group.

Sexual crosses between diploids and autotetraploids

BIOLOGICAL LIMITATION AND PRACTICAL ASPECTS. Most of the autotetraploids (doubled diploids) used for triploid breeding are spontaneous nucellar tetraploids from polyembryonic cultivars and are themselves polyembryonic. Obtaining such autotetraploid seedlings should be enhanced by selecting seeds from giga sectors of chimeric fruits (Grosser *et al.*, 1998). Relatively few studies have concerned chromosome stock duplication by colchicine treatments. Tachikawa (1971) and Barrett (1974) mentioned some autotetraploids as well as periclinal ploidy chimeras obtained by such treatments. To avoid chimera development, several authors have combined *in vitro* techniques and colchicine treatments. Gmitter and Ling (1991) have obtained a non-chimeric tetraploid of 'Valencia' sweet orange and 'Orlando' tangelo via somatic embryogenesis from culture of undeveloped ovules treated with colchicine, and Gmitter *et al.* (1991) selected 'Hamlin' and 'Ridhe Pineapple' tetraploid sweet oranges from embryogenic cultures treated with colchicine. Wakana *et al.* (2005) obtained tetraploid forms of acid citrus cultivars by top grafting of shoots with sprouting axial buds treated with colchicine.

Considering that autotetraploid selection from polyembryonic genotype seedlings is now very easy using flow cytometry (even for cultivars with low rates of chromosome duplication), colchicine treatment remains interesting mostly with the objective of monoembryonic tetraploid creation. It should be applied in budwoods (because *in vitro* techniques involving somatic embryogenesis are mostly useful for polyembryonic genotypes). Flow cytometry will help greatly to select non-chimeric tetraploid shoots. Tetraploid plants have also been regenerated by somaclonal variation from *C. limon* embryogenic callus obtained from ovule culture on medium containing 2,4-dichlorophenoxyacetic acid (2,4-D) (Vardi and Spiegel-Roy, 1982).

Crosses of $2x$ female × $4x$ male and $4x \times 2x$ have been used by breeders to obtain triploid hybrids (Cameron and Soost, 1968; Esen *et al.*, 1978; Oiyama *et al.*, 1981; Starrantino and Recupero, 1982). It appears that $4x \times 2x$ crosses have been more successful in producing triploid embryos (Cameron and Frost, 1968; Soost and Cameron, 1975). It was proposed that the 3/5 ratio between embryo and endosperm ploidy was more favourable than the 3/4 ratio obtained in the reciprocal crosses. However, most of the available tetraploid generations are polyembryonic. Thus, monoembryonic diploid female × tetraploid male hybridizations have been exploited much more than the reciprocal ones. Precocious embryo rescue, 3–4 months after anthesis has been proposed by Starrantino and Recupero (1982) to enhance the production of triploid hybrids from $2x \times 4x$ crosses. 'Oroblanco' and 'Melogold' triploid pummelo × grapefruit hybrids (Soost and Cameron, 1980, 1985) resulted from such crosses, as well as interesting tangor and mandarin triploid hybrids were selected in Italy (Starrantino, 1992).

Tetraploid hybrids are found in addition to triploids in $2x \times 4x$ hybridization (Cameron and Soost, 1968; Oiyama *et al.*, 1981), while only triploids are obtained in reciprocal crosses. These tetraploids are found in normal seeds, which is in agreement with the spontaneous production of diploid megagametophytes and the good embryo/endosperm ploidy ratio of EBN when a diploid pollen fertilized an unreduced megagametophyte. These tetraploid hybrids should be incorporated in the tetraploid gene pool for further triploid breeding. The occurrence of tetraploid hybrids in $2x \times 4x$ crosses implies that the ploidy of seedling hybrids should be checked before any other evaluations.

SEGREGATION AND RECOMBINATION. Segregation and recombination of tetraploids is complex (Wu *et al.*, 2001a, b). It involves preferential pairing between homeologous chromosomes that defines the proportion of bivalents and multivalents formed (Wu *et al.*, 2001a) and the double reduction frequency in the case of tetravalent formation (Mather, 1936). The latter parameter is constant for a specific locus depending on the distance to the centromere. A lot of economic *Citrus* species such as sweet orange and grapefruit are of interspecific origin. Moreover, differences in nuclear genome size exist among the three ancestral taxa of cultivated *Citrus* (Ollitrault *et al.*, 1995) certainly leading to structural heterozygosity in interspecific hybrids. It should explain the variability of chromosome association during meiosis observed in autotetraploids. Preferential pairing of duplicated chromosomes can be relatively frequent for doubled interspecific hybrids.

The restitution of the heterozygosity (ab) of the diploid line that generates the tetraploid (aabb) will be a function of the preferential pairing, of the rate of tetravalent formation and of the distance of the locus to the centromere. It will vary:

- from 100% in the case of total preferential pairing between duplicated chromosomes leading to systematic bivalent formation and disomic segregation. It will be 66% in case of random bivalent formation and tetrasonic segregation and less in case of tetravalent associations
- to 40% for loci independent of the centromere in the case of systematic tetravalent formation (Wu *et al.*, 2001a).

For 'autotetraploid' × diploid hybridization, segregations and recombinations occur for both the male and the female parents. For one locus, the maximum potential allelic combinations of triploid hybrids varies between:

- two (one heterozygous diploid gamete × two allelic haploid constitutions) in the case of total preferential pairing between duplicated chromosome (disomic segregation)
- and six ((two homozygous + one heterozygous diploid gamete constitution) × two allelic haploid constitutions).

Sexual hybridization between diploid females and tetraploid somatic hybrids

BIOLOGICAL LIMITATIONS AND PRACTICAL ASPECTS. The creation of tetraploid somatic hybrids for triploid breeding is described in Chapter 10; a review can also be found in Grosser *et al.* (2000). Additional combinations have been obtained recently (Calixto *et al.*, 2004; Grosser and Chandler, 2004; Wu *et al.*, 2005). Good levels of pollen fertility have been found in several allotetraploid somatic hybrids (Deng *et al.*, 1995) and some of them present an earlier flowering than autotetraploids. Due to an unbalanced ploidy ratio, embryo rescue is systematically used to recover triploid hybrids. The first interploid hybridizations with somatic hybrids were reported in 1991 (Oiyama, 1991) between diploid clementine and tetraploid Trovita sweet orange + *Poncirus trifoliata*. Embryos of underdeveloped seeds at fruit maturity were rescued *in vitro*. They produced mostly triploid hybrids and a few tetraploids (from normal seeds) resulting from the fertilization of unreduced ovules. Deng *et al.* (1996) observed a 74.5% pollen fertility in hybrid Hamlin Sweet orange + Rough lemon, and used it to fertilize diploid lines. Mostly abortive seeds were harvested three months after pollination for *in vitro* embryo rescue. Diploid, triploid and tetraploid plants have been obtained. This implies that the tetraploid somatic hybrid is able to produce haploid and diploid pollen. Similar observations were made by Tusa *et al.* (1996) in sexual crosses between diploid Feminello lemon and three allotetraploid somatic hybrids (Valencia sweet orange + Feminello lemon, Milam lemon + Feminello lemon and Key lime + Valencia sweet orange). An equivalent number of diploid, triploid and tetraploid hybrids was obtained. The use of diploid pollen from allotetraploid hybrids has been systematized in Florida, and several hundred triploids from interploid crosses with allotetraploids are already planted in the field (Grosser *et al.*, 2000).

SEGREGATION AND RECOMBINATION. As for doubled diploids, the gametic structures formed by tetraploid somatic hybrids depend on the mode of chromosome association at meiosis and on the position of the locus relative to the centromere. The main difference is that tetraploid somatic hybrids should have four different alleles (abcd) for the same locus. Therefore, a diploid gamete from an allotetraploid can display a very high heterozygosity. If the chromosomal differentiation between the two parents of the somatic hybrid is high at the interspecific level, it will lead to total preferential pairing of chromosomes from the same parent and disomic segregation. In this case, the diploid gametes will transmit interspecific heterozygosity. In other cases, diploid gametes should transmit either intraparental or interparental heterozygosity.

Recombination and segregation occur for both the female and the male parents, and this strategy leads to the greatest diversity of the triploid progeny. For each locus, the maximum number of allelic structure of the triploid varies between:

- eight (four allelic structures of the diploid gamete × two allelic structures of the haploid gamete) in the case of an allotetraploid with disomic segregation
- and 20 (ten allelic structures of the diploid gamete × two allelic structures of the haploid gamete) if there is the possibility of formation of tetravalents (Wu *et al.*, 2001a).

Endosperm culture of 2x × 2x sexual crosses

BIOLOGICAL LIMITATIONS AND PRACTICAL ASPECTS. This technique could be a tool to overcome the barriers to sexual hybridization that result from nucellar embryony and can theoretically be applied to all germplasm with female fertility. Successful regeneration of triploid plantlets has been reported by Wang and Chang (1978) and Gmitter *et al.* (1990). However, the step of shoot or embryo regeneration from endosperm calli appears to be critical (Jaskani *et al.*, 1996), which limits the practical application of this technique for breeding purposes. Indeed it appears difficult to obtain large recombining populations in order to apply efficient field selection.

SEGREGATION AND RECOMBINATION. The triploid structure of the endosperm results from the fertilization of two haploid polar nuclei of the embryo sac by a haploid vegetative nucleus of pollen. There is no restitution of maternal heterozygosity because the two polar embryos arise from the same haploid cell by mitotic division. This leads to a high level of homozygosity (in terms of a duplicated allele at each locus) in the triploid hybrids. Recombination and segregation occur for both the male and the female parent, and a maximum of four allelic constitutions should be found at each locus (two homozygous diploid ovule combinations × two allelic pollen constitutions).

Diploid + haploid protoplast fusion

BIOLOGICAL LIMITATIONS AND PRACTICAL ASPECTS. The first step is to establish haploid *Citrus* lines. These should be obtained by androgenesis (Germanà, 1992) or by induced gynogenesis (Ollitrault *et al.*, 1996a). Haplomethods in citrus are described in Chapter 7.

The technique of haploid + diploid somatic hybridization, developed simultaneously by Ollitrault *et al.* (1997) and Kobayashi *et al.* (1997), is described in Chapter 10. It should be applied to polyembryonic or monoembryonic diploid cultivars if haploid embryogenic callus lines are available. For combination with leaf protoplasts of haploid plantlets it is necessary to use diploid protoplasts from embryogenic callus lines arising mainly from polyembryonic cultivars. In the case of haploid callus lines, this method leads to triploid hybrids as well as tetraploid and pentaploid hybrids being obtained, due to the ploidy instability of haploid calli (Ollitrault *et al.*, 1998b).

The great limitation of this strategy is the scarcity of haploid lines available in *Citrus*. This could be overcome by the development of gametosomatic hybridization.

SEGREGATION AND RECOMBINATION. This strategy is the only one allowing a complete restitution of the diploid cultivar nuclear genome and its heterozygosity, without segregation and recombination. Therefore, it will be more interesting in terms of efficiency of multilocus selection at the diploid cultivar level to confer desirable traits to the triploid hybrid.

Segregation should occur at the level of cytoplasmic genomes when fusions are carried out with embryogenic callus protoplast for both diploid and haploid lines. This should lead to four different kinds of nucleocytoplasmic interaction.

For each nuclear locus, the maximum number of allelic combinations will be one in the case of combination with a haploid line, while it will be two with gametosomatic hybridization (segregation and recombination at the level of haploid cells).

Tools and protocols for ploidy manipulation in *Citrus*

A scheme for selection of triploids from diploids × diploids hybridization from using small seeds is presented in Fig. 8.2. *In vitro* techniques and flow cytometry have greatly modified the potential of ploidy manipulation for citrus breeding. Somatic hybridization methods are described in Chapter 10, so we will focus here on two technical aspects: embryo rescue and flow cytometry.

Embryo rescue

The difficulty of recovering triploid hybrids from interploid hybridization or diploid × diploid hybridization (due to an unbalanced EBN or ploidy ratio between embryo and endosperm as well as polyembryony) should be overcome by *in vitro* embryo rescue. *In vitro* culture of excised fully developed embryos was described 40 years ago to obtain zygotic plantlets from polyembryonic cultivars (Maheshwari and Ranga Swamy, 1958). Embryos should be rescued at different developmental stages, from the globular stage in undeveloped seeds of mature fruit (Starrantino and Russo, 1980) to heart-shaped embryos (Rangan *et al.*, 1968). Starrantino and Recupero (1981) obtained triploid hybrids from $2x \times 4x$ crosses by rescuing globular and heart-shaped embryos 3–4 months after anthesis.

Adenine sulphate (25 mg/l) and malt extract (500 mg/l) appear favourable for citrus embryo rescue. The addition of hormones should help embryo germination and shoot elongation. Many authors add 1 mg/l GA_3 (gibberellic acid), while a balance of 0.5 mg/l BA (6-benzylaminopurine), 0.5 mg/l kinetin and 0.11 mg/l NAA (1-naphthaleneacetic acid) are favourable when regeneration is difficult (Deng *et al.*, 1996).

As a standard culture medium, we can propose:

- Murashige and Tucker or Murashige and Skoog basic medium
- 30 g/l sucrose
- 25 mg/l adenine sulphate
- 500 mg/l malt extract
- 1 mg/l GA_3
- pH 5.7
- 8 g/l agar (agar is better than other gelling agents to avoid problems of hyperhydricity)

Seeds may be sterilized: 5 min in 70% alcohol + 10 min in sodium hypochorite solution (0.2% active chloride) and washed three times in sterilized water.

The culture may be incubated at 26/28°C, and a 16 h photoperiod is favourable.

After ploidy control, shoots can be grafted on greenhouse rootstocks. This will allow a better growth than classical acclimatization and prevents fungus problems, particularly for genotypes susceptible to *Phytophthora* spp. After grafting, it is necessary to maintain the plants for 2–3 weeks in saturated hygroscopy to prevent drying of the shoot.

Flow cytometry

Flow cytometry which was developed for medical research is now widely used for plant DNA content analysis (Arumuganathan and Earle, 1991a, b). This technique allows an estimation of the volume and fluorescence of isolated cells or nuclei. If the fluorescent probe used is DNA specific, nuclear DNA content can be quantified. A large number (10^4–10^6/min) of nuclei can be analysed rapidly, and the results are presented as a histogram of integral fluorescence. The position of the peak in the axis is proportional to the DNA content of the nuclei. Generally, if young leaves are analysed, two peaks are obtained. According to the cell cycle, the first peak is defined by nuclei of non-cycling cells (G_0) or in the pre-DNA synthesis stage (G_1), while the second peak represents nuclei in the post-DNA synthesis stage of mitosis. It is recommended to use an internal control (nuclei with already known DNA content or nuclei of genotypes of the same species with identified ploidy) for determination of absolute genome sizes or ploidy level of the sample.

Interest in flow cytometry for citrus ploidy analysis and then for polyploid breeding was first emphasized by Ollitrault and Michaux-Ferriere (1992). Since then, this technique has opened up new avenues for citrus ploidy manipulation. Indeed it allows the selection of rare events such as $2n$ gamete formation that are difficult to exploit for breeding by conventional cytogenetic techniques (Ollitrault *et al.*, 1996b; Luis Navarro *et al.*, 2004). Flow cytometry is now used by several teams around the world as a routine tool for citrus breeding for both sexual progeny (Tusa *et al.*, 1996; Ollitrault *et al.*, 1998a) and somatic hybrids (Ollitrault *et al.*, 1996c; Grosser *et al.*, 2000). It has also allowed genome size differentiation between cultivated diploid species (Ollitrault *et al.*, 1995). Larger genomes are observed for *C. medica* (0.81 pg/2C), while *C. reticulata* present the smallest ones (0.74 pg/2C). This differentiation does not allow prediction of the exact size of polyploid hybrids arising from sexual crosses because of chromosomic segregation.

The technique is very simple and allows the analysis of 150–200 genotypes per day. The procedure used is modified from Arumuganathan and Earle (1991a). Small pieces (50 µg) of leaf of the sample and the internal control are chopped with a razor blade in 300 µl of extraction buffer (phosphate-buffered saline (PBS) containing 1 mg/ml dithiothreitol and 0.3% Triton X-100). For exact nuclear DNA content evaluation, RNase (10^{-3} U/ml) must be added to the extraction buffer, while this is not necessary for routine ploidy analysis. The nucleus solution is filtered through a 30 µm nylon mesh. Then 150 µl of the filtered solution is mixed with 150 µl of extraction buffer with the addition of 200 µg/ml propidium iodide (DNA stain). After 10 min incubation, the stained nucleus solution is analysed in the flow cytometer.

Staining with 4',6-diamidino-2-phenylindole (DAPI) also allows analysis of the ploidy level.

The constitution of PBS is as follows (for 1 l): 8 g of NaCl; 0.2 g of KCl; 1.44 g of Na_2HPO_4; 0.24 g of KH_2PO_4. This is made up to 1 l with H_2O after the pH has been adjusted to 7.4 with HCl. The solution should be aliquoted and sterilized for long-term conservation.

Conclusion

Knowledge of the genetic and molecular factors affecting gametic fertility and seed development is growing rapidly and will enhance the potential of seedless breeding by sexual crosses at the diploid level (manipulation of male geno-cytoplasmic interaction, molecular markers for early selection). Mutagenesis programmes remain an interesting way for teams with important facilities for field evaluation to develop seedless cultivars from elite well known seedy ones. Transgenic plants with genes affecting meiosis or seed development should also have a use if and when consumers accept eating transgenic *Citrus*, which is far from the present situation in European countries. The ongoing programmes of ploidy manipulation will result in the near future in a broad range of triploid seedless cultivars and will be used to combine seedlessness and other traits such as resistances, quality traits or harvesting period. The different approaches for obtaining triploid hybrids can be applied to different kinds of cultivars (monoembryonic, polyembryonic, sterile) and lead to populations very different in terms of genetic segregation and recombination. The development of gametosomatic hybridization would considerably enhance the efficiency of this latter strategy. Overall, it is clear that field evaluation and market validation in close cooperation with different operators in the citrus industry is an absolute necessity to give a chance for the new seedless hybrids or mutants to reach the market as popular cultivars.

References

Anonymous (2002) Sweet Citrus Varieties with Deep Orange Rind Released by UC Riverside. http://www.newsroom.ucr.edu/cgi-bin/display.cgi?id=252

Arumuganathan, K. and Earle, E.D. (1991a) Estimation of nuclear DNA content of plants by flow cytometry. *Plant Molecular Biology Reporter* 9, 229–233.

Arumuganathan, K. and Earle, E.D. (1991b) Nuclear DNA content of some important plant species. *Plant Molecular Biology Reporter* 9, 210–220.

Barrett, H.C. (1974) Colchicine-induced polyploidy in citrus. *Botanical Gazette* 135, 29–41.

Barrett, H.C. (1992) An autotetraploid of the Key lime, *Citrus aurantifolia. Fruit Varieties Journal* 46, 166–170.

Barrett, H.C. and Hutchison, D.J. (1978) Spontaneous tetraploidy in apomictic seedlings of Citrus. *Economic Botany* 32, 27–45.

Calixto, M.C., Mourao Filho, F.A.A., Mendes, B.M.J. and Vieira, M.L.C. (2004) Somatic hybridization between *Citrus sinensis* (L.) Osbeck and *C. grandis* (L.) Osbeck. *Pesquisa Agropecuaria Brasileira*, 39, 721–724.

Cameron, J.W. and Frost, H.B. (1968) Genetics, breeding and nucellar embryony. In: Reuther W, Batchelor L.D. and Webber H. J. (eds) *The Citrus Industry*, Vol. 1. University of California, Riverside, California, pp. 325–370.

Cameron, J.W. and Soost, R.K. (1968) Characters of new populations of *Citrus* polyploids and the relation between tetraploidy in the pollen parent and hybrid tetraploid progeny. In: *Proceedings of the First International Citrus Symposium.* University of California, Riverside, California, pp. 199–205.

Carputo, D., Monti, L., Werner, J.E. and Frusciante, L. (1999) Use and usefulness of endosperm balance number. *Theoretical and Applied Genetics* 98, 478–484.

Chandler, J.L., Viloria, Z. and Grosser, J.W. (2001) Acid citrus fruit cultivar improvement via interploid hybridization. *Proceedings of the Florida State Horticultural Society* 113, 124–126.

Chen, S., Gao, F. and Zhang, J. (1991) Studies on the seedless character of citrus induced by irradiation. *Mutation Breeding Newsletter* 39, 8–9.
Chen, C., Lyon, M.T., O'Malley, D., Federici, C.T., Gmitter, J., Grosser, J.W., Chaparro, J.X., Roose, M.L. and Gmitter Jr., F.G. (2007) Origin and Frequency of 2n Gametes in *Citrus sinensis* × *Poncirus trifoliata* and their Reciprocal Crosses, *Plant, Plant Science* (in press).
Deng, X.X., Gmitter, F.G. and Grosser, J.W. (1995) Pollen fertility of auto- and allo-tetraploids of citrus. *Acta Horticulturae Sinica* 22, 16–20.
Deng, X.X., Yi, H.L., Li, F. and Guo, W. (1996) Triploid plants regenerated from crossing diploid pummelo and tangerine with allotetraploid somatic hybrid of citrus. In: *Proceedings of the Eighth International Citrus Congress.* International Society of Citriculture, Sun City, South Africa, Vol. 1, pp. 189–192.
Esen, A. and Soost, R.K. (1971) Unexpected triploids in Citrus: their origin, identification and possible use. *Journal of Heredity* 62, 329–333.
Esen, A. and Soost, R.K. (1972) Tetraploid progenies from $2x \times 4x$ crosses of Citrus and their origin. *Journal of the American Society for Horticultural Science* 97, 410–414.
Esen, A. and Soost, R.K. (1973) Precocious development and germination of spontaneous triploid seeds in Citrus. *Journal of Heredity* 64, 147–154.
Esen, A. and Soost, R.K. (1977) Relation of unexpected polyploids to diploid megagametophytes and embryo:endosperm ploidy ratio in Citrus. In: Carpena, O. (ed.) *Proceedings of the Congresso Mundial de Citricultura. 1973.* International Society of Citriculture, Vol. 2, pp. 53–63.
Esen, A., Soost, R.K. and Geraci, G. (1978) Seed set, size and development after $4x \times 2x$ and $4x \times 4x$ crosses in Citrus. *Euphytica* 27, 283–293.
Esen, A., Soost, R.K. and Geraci, G. (1979) Genetic evidence for the origin of diploid megagametophytes in Citrus. *Journal of Heredity* 70, 5–8.
Fatta Del Bosco, S., Matranga, G. and Geraci, G. (1992) Micro and macro-sporogenesis of two triploid hybrids of Citrus. In: *Proceedings of the Seventh International Citrus Congress.* International Society of Citriculture, Acireale, Italy, Vol. 1, pp. 122–124.
Froelicher, Y. (1999) Organisation de la diversité dans le genre Clausena Burm.F. (Rutaceae) Doctorat, Paris XI, Orsay, France.
Froelicher, Y., Luro, F. and Ollitrault, P. (2000) Analysis of meiotic behavior of the tetraploid *Clausena excavata* species by molecular marker segregation studies. In: *Proceedings of the Ninth International Citrus Congress.* International Society of Citriculture, Orlando, Florida, p. 210.
Froelicher, Y., Bouffin, J., Dambier, D., Bassene, J.B.,Vanh Tran, H. and Ollitrault P. (2005) The triploid breeding program at CIRAD. In: *Book of Abstracts of the Seventh International Congress of Citrus Nurserymen.* ISCN, Cairo, Egypt, pp. 65–66.
Froneman, I.J., Breedt, H.J., Koekemoer, P.J.J. and Van Rensburg, P.J.J. (1996) Producing seedless Citrus cultivars with gamma irradiation. In: *Proceedings of the Eighth International Citrus Congress.* International Society of Citriculture, Sun City, South Africa, Vol. 1, pp. 159–163.
Frost, H.B. (1948) Seed reproduction: development of gametes and embryos. In: Webber H.J. and Batchelor L.D. (eds) *The Citrus Industry*, Vol. 1. University of California Press, Berkley, pp. 767–816.
Frost, H.B. and Soost, R.K. (1968) Seed reproduction: development of gametes and embryos. In: Reuther W., Batchelor L.D.and Webber H.B. (eds) *The Citrus Industry*, Vol. 2. Univiversity of California Press, Berkley, pp. 290–324.
Garcia, M.R., Asins, M.J. and Carbonell, E.A. (2000) QTL analysis of yield and seed number in Citrus. *Theoretical and Applied Genetics* 101, 487–493.
Geraci, G. (1978) Percentage of triploid offspring of cross-pollinated diploid polyembryonic citrus. In: *Proceedings of the Third International Citrus Congress.* International Society of Citriculture, Sydney, Australia, Vol. 1, pp. 57–58.
Geraci, G., Esen, A. and Soost, R.K. (1975) Triploid progenies from $2x \times 2x$ crosses of Citrus cultivars. *Journal of Heredity* 66, 177–178.
Geraci, G., De Pasquale, F. and Tusa, N. (1977) Percentages of spontaneous triploids in progenies of diploid lemons and mandarins. In: *Proceedings of the Second International Citrus Congress.* International Society of Citriculture, Orlando, Florida, Vol. 2, pp. 596–597.
Germana, M.A. (1992) Androgenesis in Citrus: a review. In: *Proceedings of the Seventh International Citrus Congress.* International Society of Citriculture, Acireale, Italy, Vol. 1, pp. 183–189.
Gmitter, F.G. and Ling, X.B. (1991) Embryogenesis *in vitro* and nonchimeric tetraploid plant recovery from

undeveloped Citrus ovules treated with colchicine. *Journal of The American Society for Horticultural Science* 116, 317–321.
Gmitter, F.G., Ling, X.B. and Deng, X.X. (1990) Induction of triploid Citrus plants from endosperm calli *in vitro. Theoretical and Applied Genetics* 80, 785–790.
Gmitter, F.G., Ling, X., Cai, C. and Grosser, J.W. (1991) Colchicine induced polyploidy in Citrus embryogenic cultures, somatic embryos and regenerated plants. *Plant Science* 74, 135–141.
Grosser, J.W. and Chandler, J.L. (2004) Production of twelve new allotetraploid somatic hybrid citrus breeding parents with emphasis on late maturity and cold-hardiness. *Journal of the American Pomological Society* 58, 21–28.
Grosser, J.W., Gmitter, F.G., Fleming, G.H. and Chandler, J.L. (1998) Application of biotechnology to citrus cultivar improvment at the citrus research and education center. *Acta Horticulturae* 535, 213–220.
Grosser, J.W., Ollitrault, P. and Olivares, O. (2000) Somatic hybridization in Citrus: an effective tool to facilitate variety improvement. *In Vitro Cellular and Development Biology-Plant* 36, 434–449.
Guo, W.W., Deng, X.X. and Yi, H.L. (2000) Somatic hybrids between navel orange (*Citrus sinensis*) and grapefruit (*C. paradisi*) for seedless triploid breeding. *Euphytica* 116, 281–285.
Handaji, N., Dambier, D., Chahbar, A., Arsalane, N., Hmouni D., Benyahia, H. and Ollitrault P. (2005) Etude de facteurs influents sur la production de triploïdes par hybridations entre clémentiniers et mandariniers diploïdes. *Al Awamia* 113 (2–1): 29–46.
Hearn, C.J. (1969) Self-incompatibility and the effects of different pollen sources upon fruit characteristics of four Citrus hybrids. *Proceedings of the American Society for Horticultural Science* 87, 183–187.
Hearn, C.J. (1984) Development of seedless orange and grapefruit cultivars through seed irradiation. *Journal of the American Society for Horticultural Science* 109, 270–273.
Hensz, R.A. (1971) Star Ruby, a new deep-red-fleshed grapefruit variety with distinct tree characteristics. *Journal of the Rio Grande Valley Horticultural Society* 25, 54–58.
Hutchison, D.J. and Barrett, H.C. (1981) Tetraploid frequency in nucellar seedlings from single trees of Carrizo and Troyer *Citrus* hybrids. In: *Proceedings of the Fourth International Citrus Congress.* International Society of Citriculture, Tokyo, Japan, Vol. 1, pp. 27–29.
Iwamasa, M. (1966) Studies on the sterility in genus Citrus with special reference to the seedlessness. *Bulletin of the Horticulture Research Station* B6, 1–81.
Iwamasa, M. and Iwasaki, T. (1962) On the sterility phenomenon caused by low temperatures in the Mexican lime (*Citrus aurantifolia* Swingle). *Bulletin of the Horticulture Research Station* B2, 25–46.
Iwamasa, M. and Nito, N. (1988) Cytogenetics and the evolution of modern cultivated citrus. In: *Proceedings of the Sixth International Citrus Congress.* Margraf, Tel Aviv, Israel, Vol. 1, pp. 265–275.
Iwamasa, M. and Oba, Y. (1980) Seedlessness due to self-incompatibility in Egami-Buntan, a Japanese pummelo cultivar. *Bulletin of the Faculty of Agriculture, Saga University* 49, 39–45.
Iwasaki, T. (1943) On the big and small leaf strain of trifoliate orange (*Poncirus trifoliata* Raf.). *Journal of the Horticultural Association of Japan* 14, 302–305.
Jackson, L.K. and Sherman, W.B. (1975) Chromosome counts in 'Tahiti' lime. *Florida State Horticultural Society* pp. 458–459.
Jacquemond, C. and Blondel, L. (1986) Contribution à l'étude des porte-greffe des agrumes: le *Poncirus trifoliata* – 3eme partie: étude du comportement des porte-greffe après greffage. *Fruits* 41, 449–464.
Jaskani, M.J., Khan, I.A. and Yaqub, M.S. (1996) Regeneration of triploid from Citrus endosperm. In: *Proceedings of the Eighth International Citrus Congress.* International Society of Citriculture, Sun City, South Africa, Vol. 1, pp. 128–129.
Johnston, S.A. and Hanneman, R.E.J. (1980) Support of the endosperm balance number hypothesis utilising some tubearing Solanum species. *American Potato Journal* 57, 7–14.
Khan, I.A., Jaskani, M.J. and Haider Ali, S.N. (1996) Interploid hybridization for the improvement of Kinnow mandarin. In: *Proceedings of the Eighth International Citrus Congress.* International Society of Citriculture, Sun City, South Africa, Vol. 1, pp. 137–140.
Khan, M.M., Khan, I.A. and Mughal, A.H. (1992) Growth and morphological comparaison of diploid and tetraploid stains of Kinnow mandarin. In: Tribulato, E., Gentile, A. and Reforgiato, G. (eds) *Proceedings of the Seventh International Citrus Congress.* International Society of Citriculture, Acireale, Italy, Vol. 1, pp. 93–95.
Kobayashi, S., Ohgawara, T., Saito, W., Nakamura, Y. and Omura, M. (1997) Production of triploid somatic hybrids in Citrus. *Journal of the Japanese Society for Horticultural Science* 66, 453–458.

Koltunow, A.M., Brennan, P., Bond, J.E. and Barker, S.J. (1998) Evaluation of genes to reduced seed size in Arabidopsis and tobacco and their application to Citrus. *Molecular Breeding* 4, 235–251.

Koltunow, A.M., Brennan, P., Protopsaltis, S. and Nito, N. (2000) Regeneration of West Indian limes (*Citrus aurantifolia*) containing genes for decreased seed set. *Acta Horticulturae* 535, 81–91.

Krezdorn, A.H. and Robinson, F.A. (1958) Unfruitfulness in the Orlando tangelo. *Proceeding of the Florida State Horticultural Society* 71, 86–91.

Krug, C.A. (1943) Chromosomes numbers in the subfamily Aurantioideae with special reference to the genus *Citrus*. *Botanical Gazette* 48, 602–611.

Lapin, W.K. (1937) Investigations on polyploidy in Citrus. *USSR All-Union Science Research Institute Humid Subtropics Works* 1, 1–68.

Li, S.J. (1980) Self-incompatibility in 'Matou' wentan (*Citrus grandis* (L.) Osb.). *HortScience* 15, 298–300.

Liu, J., Qi, J., Dai, Z. and Chen, K. (1995) Study on relationship between free proline concentration of citrus anther and pollen sterility. *Acta Horticulturae* 403, 81–83.

Longley, A.E. (1925) Polycary, polyspory and polyploidy in Citrus and Citrus relatives. *Journal of the Washington Academy of Science* 15, 347–351.

Longley, A.E. (1926) Triploid Citrus. *Journal of the Washington Academy of Science* 16, 543–545.

Luro, F., Maddy, F., Ollitrault, P. and Rist, D. (2000) Identification of 2n gamete parental origin and mode of nuclear restitution of spontaneous triploid *Citrus* hybrids. In: *Proceedings of the Ninth International Citrus Congress*. Orlando, Florida, pp. 168–169.

Luro, F., Maddy, F., Jacquemond C., Froelicher, Y., Morillon, R., Rist, D. and Ollitrault, P. (2004) Identification and evaluation of diplogyny in clementine (*Citrus clementina*) for use in breeding. *Acta Horticulturae* 663, 841–847.

Maheshwari, P. and Ranga Swamy, N.S. (1958) Polyembryony and *in vitro* culture of embryos of Citrus and Mangifera. *Indian Journal of Horticulture* 15, 275–282.

Mather, K. (1936) Segregation and linkage in autotetraploids. *Journal of Genetics* 32, 287–314.

Matsumoto, R., Okudai, N., Oiyama, I., Takahara, T., Yamamoto, M., Asada, K., Ishiuchi, D. and Murata, H. (1991) New citrus cultivar 'Tsunokaori'. *Bulletin of the Fruit Tree Research Station* 21, 59–65.

Miwa, T. (1951) Pollination, fertilization and fruit drop in *Citrus tamurana*. *Horticultural Bulletin of the Miyazaki University (Natural Science)* 2, 1–67.

Mustard, J.M., Lynch, S.J. and Nelson, R.O. (1956) Pollination and floral studies of the Minneola tangelo. *Proceedings of the Florida State Horticultural Society* 69, 277–281.

Nagai, K. and Tanikawa, T. (1928) On citrus pollination. In: *Proceedings of the Third Pan-Pacific Science Congress* pp. 2023–2029.

Nakamura, M. (1943) Cytological and ecological studies on the genus Citrus with special reference to the occurrence of sterile pollen grain. *Memorial of the Faculty of Science and Agriculture, Tohoku University* 27, 53–159.

Nakano, M., Nesumi, H., Yoshioka, T., Omura, M. and Yoshida, T. (2000) Linkage analysis between male sterility of citrus and RAPD markers. In: *Proceedings of the Ninth International Citrus Congress.* International Society of Citriculture, Orlando, Florida, pp. 179–180.

Nakano, M., Nesumi, H., Yoshioka, T. and Yoshida, T. (2001) Segregation of plants with undeveloped anthers among hybrids derived from the seed parent 'Kiyomi'. *Journal of the Japanese Society for Horticultural Science* 70, 539–545.

Navarro, L., Olivares Fuster, O., Juarez, J., Aleza, P., Pina, J.A., Ballester Olmos, J.F., Cervera, M., Fagoaga, C., Duran Vila, N. and Pena, L. (2004) Applications of biotechnology to citrus improvement in Spain. *Acta Horticulturae* 632, 221–234.

Nesumi, H., Nakano, M. and Yoshida, T. (2001) Mode of inheritance on the abnormal development of impregenated ovules derived from Mukaku-kishu. *Journal of the Japanese Society for Horticultural Science* 70, 403.

Nesumi, H., Nakano, M., Yoshioka, T., Omura, M. and Yoshida, T. (2000) Genetic analysis of female sterility from satsuma. *Journal of the Japanese Society for Horticultural Science* 69, 192.

Nishiura, M. and Iwasaki, T. (1963) Studies on the citrus breeding I. Variation of seed formation in citrus crossing. *Bulletin of the Horticulture Research Station* 2, 1–13.

Nishiura, M., Shichijo, T., Ueno, I., Iwamasa, M., Kihara, T., Yamada, Y., Yoshida, T. and Iwasaki, T. (1983) New citrus cultivar 'Kiyomi' tangor. *Bulletin of the Fruit Tree Research Station* 10, 1–9.

Oiyama, I., Okudai, N. and Takahara, T. (1980) Studies on polyploid breeding in Citrus (3). Triploid progenies from small seeds of diploid sweet orange cultivars. In: *Meeting of the Japanese Society of Horicultural Science* pp. 52–53.

Oiyama, I., Okuday, N. and Takarahara, T. (1981) Ploidy levels of seedlings obtained from $2x \times 4x$ crosses in citrus. In: *Proceedings of the Fourth International Citrus Congress.* International Society of Citriculture, Tokyo, Japan, Vol. 1, pp. 32–34.

Oiyama, I., Kobayashi, S., Yoshinaga, K., Ohgawara, T. and Ishii, S. (1991) Use of pollen from a somatic hybrid between *Citrus* and *Poncirus* in the production of triploids. *HortScience* 26, 1082.

Ollitrault, P. and Jacquemond, C. (1995) Facultative apomixis, spontaneous polyploidization and inbreeding in *Citrus volkameriana* seedlings. *Fruits* 49, 398–400.

Ollitrault, P. and Michaux-Ferriere, N. (1992) Application of flow cytometry for citrus genetic and breeding. In: *Proceedings of the Seventh International Citrus Congress.* International Society of Citriculture, Acireale, Italy, Vol. 1, pp. 193–198.

Ollitrault, P., Dambier, D., Luro, F. and Duperray, C. (1995) Nuclear genome size variations in Citrus. *Fruits* 49, 390–393.

Ollitrault, P., Allent, V. and Luro, F. (1996a) Production of haploid plants and embryogenic calli of clementine (*Citrus reticulata* Blanco) after *in situ* parthenogenesis induced by irradiated pollen. In: *Proceedings of the Eighth International Citrus Congress.* International Society of Citriculture, Sun City, South Africa, Vol. 2, pp. 913–917.

Ollitrault, P., Dambier, D., Allent, V., Luro, F. and Jacquemond, C. (1996b) *In vitro* rescue and selection of spontaneous triploids by flow cytometry for easy peeler citrus breeding. In: *Proceedings of the Eighth International Citrus Congress.* International Society of Citriculture, Sun City, South Africa, Vol. 1, pp. 254–258.

Ollitrault, P., Dambier, D., Sudahono S. and Luro, F. (1996c) Somatic hybridisation in Citrus: some new hybrid and alloplasmic plants. In: *Proceedings of the Eighth International Citrus Congress.* International Society of Citriculture, Sun City, South Africa, Vol. 2, pp. 907–912.

Ollitrault, P., Dambier, D., Sudahono, S., Mademba-Sy, F., Vanel, F., Luro, F. and Aubert, B. (1997) Biotechnology for triploid mandarin breeding. In: Aubert, B. (ed.) *Fifth International Congress of Citrus Nurserymen.* ISCN, Montpellier, France, p. 337.

Ollitrault, P., Dambier, D., Sudahono, S., Mademba-Sy, F., Vanel, F., Luro, F. and Aubert, B. (1998a) Biotechnology for triploid mandarin breeding. *Fruits* 53, 307–317.

Ollitrault, P., Dambier, D., Vanel, F. and Froelicher, Y. (1998b) Creation of triploid Citrus hybrids by electrofusion of haploid and diploid protoplasts. *Acta Horticulturae* 535, 191–197.

Omura, M., Ueda T., Kita M., Komatsu A., Takanokura Y., Shimada T., Endo-Inagaki T., Nesumi H. and. Yoshida T. (2000) EST mapping in *Citrus.* In: *Proceeding of the Ninth International Citrus Congress.* International Society of Citriculture, Orlando, Florida, pp. 71–74.

Osawa, I. (1912) Cytological and experimental studies in Citrus. *Journal of the College of Agriculture Tokyo University* 4, 83–116.

Otto, S.P. and Whitton, J. (2000) Polyploid incidence and evolution. *Annual Review of Genetics* 34, 401–437.

Raghuvanshi, S.S. (1962) Cytologenetical studies in genus Citrus IV. Evolution in genus Citrus. *Cytologia* 27, 172–188.

Ramsey, J. and Schemske, D.W. (1998) Pathways, mechanisms and rates of polyploid formation in flowering plants. *Annual Review of Ecology and Systematics* 29, 467–501.

Rangan, T.S., Murashige T. and Bitters W.P. (1968) *In vitro* studies of zygotic and nucellar embryogenesis in Citrus. In: *Proceedings of the First International Citrus Symposium.* University of California, Riverside, California, Vol. 1, pp. 225–229.

Reece, P.C. and Register, R.O. (1961) Influence of pollinators on fruit set on Robinson and Oceola tangerine hybrids. *Proceedings of the Florida State Horticultural Society* 74, 104–106.

Russo, F. and Torrisi, M. (1951a) Forme tetraploidi in *Citrus limon* L., e *C. aurantium* L. *Annali Sperimentazione Agraria* 5, 13–20.

Russo, F. and Torrisi, M. (1951b) Il poliploidismo nei Citrus autopoliploidi e allopoliploidi. *Annali Sperimentazione Agraria* 5, 1041–1062.

Russo, G., Recupero, S., Puglisi, A. and Recupero, G.R. (2004) New triploid citrus hybrids by Italian genetic improvement. Nuovi hibridi triploidi di agrumi dal miglioramento genetico italiano. *Rivista di Frutticoltura e di Ortofloricoltura* 66 (3), 14–18.

Soost, R.K. (1956) Unfruitfulness in the clementine mandarin. *Proceedings of the American Society for Horticultural Science* 67, 171–175.

Soost, R.K. (1961) Effects of gibberellin on yield and fruit characteristics of clementine mandarin. *Proceedings of the American Society for Horticultural Science* 77, 194–201.

Soost, R.K. (1965) Incompatibility allele in the genus Citrus. *Proceedings of the American Society for Horticultural Science* 87, 176–180.

Soost, R.K. (1968) The incompatibility gene system in citrus. In: *Proceeding of the First International Citrus Symposium.* University of California, Riverside, California, Vol. 1, pp. 189–190.

Soost, R.K. (1987) Breeding citrus – genetics and nucellar embryony. In: Abbott A.J. and Atkin R.K. (eds) *Improving Vegetatively Propagated Crops.* Academic Press, London, pp. 83–110.

Soost, R.K. and Cameron, J.W. (1975) Citrus. In: Janick. J. and Moore. J.N. (eds) *Advance in Fruit Breeding.* Purdue University Press, West Lafayette, Indiana, pp. 507–540.

Soost, R.K. and Cameron, J.W. (1980) 'Oroblanco' a triploid pummelo–grapefruit hybrid. *HortScience* 15, 667–669.

Soost, R.K. and Cameron, J.W. (1985) 'Melogold' a triploid pummelo–grapefruit hybrid. *HortScience* 20, 1134–1135.

Spiegel-Roy, S., Vardi, A. and Elhanati, A. (1985) Seedless induced mutant in lemon (*Citrus limon*). *Mutation Breeding Newsletter* 26, 1–2.

Starrantino, A. (1992) Use of triploids for production of seedless cultivars in citrus improvment programs. In: *Proceedings of the Seventh International Citrus Congress.* International Society of Citriculture, Acireale, Italy, Vol. 1, pp. 117–121.

Starrantino, A. (1999) Tacle, a new triploid clementine × Tarocco hybrid. 'Tacle', nuovo hibrido triploide di clementine × 'Tarocco'. *Rivista di Frutticoltura e di Ortofloricoltura* 61 (1), 45–47.

Starrantino, A. and Recupero, G.R. (1981) Citrus hybrids obtained *in vitro* from 2*x* females × 4*x* males. In: *Proceedings of the Fourth International Citrus congress.* International Society of Citriculture, Tokyo, Japan, Vol. 1, pp. 31–32.

Starrantino, A. and Russo, F. (1980) Seedlings from undeveloped ovules of ripes fruits of polyembryonic citrus cultivars. *Hortscience* 15, 296–297.

Starrantino, A., Mannino, P. and Russo, F. (1988) The genetic stability of three seedless clonal selections obtained by gamma rays of the seedy 'Monreal' clementine. In: *Proceeding of the Sixth International Citrus Congress.* Margraf, Tel Aviv, Israel, Vol. 1, 175–182.

Swingle, W.T., Robinson, T.R. and Savage, E.M. (1931) New citrus hybrids. *United States Department of Agricultural Circular* 181, 1–19.

Sykes, S.R. and Possingham, J.V. (1992) The effect of excluding insect pollinators on seediness of Imperial mandarin fruits. *Australian Journal of Experimental Agriculture* 32, 409–411.

Tachikawa, T. (1971) Investigation in the breeding of Citrus trees. IV The chromosome number in citrus. *Bulletin of the Shizuoka Prefectural Citrus Experiment Station* 9, 11–25.

Tachikawa, T., Tanaka, Y. and Hara, S. (1961) Investigation on the breeding of Citrus trees. I Study on the breeding of triploid Citrus varieties. *Bulletin of the Shizuoka Prefectural Citrus Experiment Station* 4, 33–44.

Tokunaga, T., Nii, M., Tsumura, T. and Yamao, M. (2005) Production of triploids and breeding seedless cultivar 'Tokushima 3X No. 1 from tetraploid × diploid crosses in sudachi (*Citrus sudachi* Shirai). *Horticultural Research Japan* 4 (1), 11–15.

Ton, L.D. and Krezdorn, A.H. (1967) Growth of pollen tubes in three incompatible varieties of citrus. *Proceedings of the American Society for Horticultural Science* 89, 211–215.

Tusa, N., Fatta Del Bosco, S., Nardi, L. and Lucretti, S. (1996) Obtaining triploid plants by crossing *Citrus limon* cv 'Feminello' 2*n* × 4*n* allotetraploid somatic hybrids. In: *Proceeding of the Eighth International Citrus Congress.* International Society of Citriculture, Sun City, South Africa, Vol. 1, pp. 133–136.

Ueno, I. (1986) Studies on the inheritance of citrus flower characteristics. I. Segregation of viable pollen production in hybrid seedlings. *Bulletin of the Fruit Tree Research Station* 13, 1–9.

Uphof, J.C.T. (1931) Wissenscaftliche Beobachtungen und Versuche an Agrumen. I. Ueber die Bluetenverhaeltnisse der Tahtilimonella. *Gartenbauwissenschaft* 4, 513–520.

Vardi, A. and Spiegel-Roy, P. (1981) Gene-controlled meiosis in *Citrus reticulata.* In: *Proceedings of the Fourth International Citrus Congress.* The International Society of Citriculture, Tokyo, Japan, Vol. 1, pp. 26–27.

Vardi, A. and Spiegel-Roy, P. (1982) Plant regeneration from citrus protoplast: variability in methodological requirements among cultivated species. *Theoretical and Applied Genetics* 62, 171–176.

Vardi, A., Frydman Shani, A. and Weinbaum, S.A. (1988) Assessment of parthenocarpic tendency in citrus using irradiated marker pollen. In: *Proceedings of the Sixth International Citrus Congress.* Margraf, Tel Aviv. Israel, Vol. 1, pp. 225–230.

Vardi, A., Neuman, H., Frydman-Shani, A., Yaniv, Y. and Spiegel-Roy, P. (2000) Tentative model on the inheritance of juvenility, self-incompatibility and parthenocarpy. *Acta Horticulturae* 535, 199–205.

Wakana, A., Ngo, X.B. and Isshiki, S. (1998) Self-incompatibility in Citrus: linkage between GOT isozyme loci and the incompatibility loci. In: Omura M., Hayashi T. and Scott N.S. (eds) *Breeding and Biotechnology for Fruit Trees; Second Japan–Australia International Workshop.* NIFTS/CSIRO, pp. 90–93.

Wakana, A., Hanada, N., Park SungMin, Fukudome, I. and Kajiwara, K. (2005) Production of tetraploid forms of acid citrus cultivars by top grafting of shoots with sprouting axially buds treated with colchicine. *Journal of the Faculty of Agriculture, Kyushu University* 50 (1), 93–102.

Wakana, S., Iwamasa, M. and Uemoto, S. (1981) Seed development in relation to ploidy of zygotic embryo and endosperm in polyembryonic citrus. In: *Proceedings of the Fourth International Citrus Congress.* International Society of Citriculture, Tokyo, Japan, Vol. 1, pp. 35–39.

Wang, D. and Chang, C.J. (1978) Triploid citrus plantlet from endosperm culture. *Scientia Sinica* 21, 822–827.

Wilms, H.J., Van Went, J.L., Cresti, M. and Ciampolini, F. (1983) Structural aspects of female sterility in *Citrus limon. Acta Botanica Neerlandica* 32, 87–96.

Wong, C.Y. (1939) The influence of pollination on seed development in certain varieties of citrus. *Proceedings of the American Society for Horticultural Science.* 37, 161–164.

Wu, J.H., Ferguson, A.R., Mooney, P.A. (2005) Allotetraploid hybrids produced by protoplast fusion for seedless triploid Citrus breeding. *Euphytica* 141, 229–235.

Wu, R., Gallo Meagher, M., Littell, R.C. and Zhao-Bang, Z. (2001a) A general polyploid model for analyzing gene segregation in outcrossing tetraploid species. *Genetics* 159, 869–882.

Wu, S.S., Wu, R., Ma, C.X., Zeng, Z.B., Yang, C.K. and Casella, G. (2001b) A multivalent pairing model of linkage analysis in autotetraploids. *Genetics* 159, 1339–1350.

Yamamoto, M., Okudai, N. and Matsumoto, R. (1992) Segregation for aborted anthers in hybrid seedlings using *Citrus nobilis* × *C. deliciosa* cv. Encore as the seed parent. *Journal of the Japanese Society for Horticultural Science* 60, 785–789.

Yamamoto, M., Matsumoto, R. and Yamada, Y. (1995) Relationship between sterility and seedlessness in citrus. *Journal of the Japanese Society for Horticultural Science* 64, 23–29.

Yamamoto, M., Matsumoto, R., Okudai, N. and Yamada, Y. (1997) Aborted anthers of Citrus result from gene-cytoplasmic male sterility. *Scientia Horticulturae* 70, 9–14.

Yamamoto, M., Matsumoto, R. and Kuniga, T. (2001) Inheritance of female sterility in Citrus. *Breeding Research* 3, 49–51.

Yamashita, K. (1976) Production of seedless fruits in Hyuganatsu, *Citrus tamurana* Hort. ex Tanaka and Hassaku, *Citrus hassaku* Hayata through pollination with pollen grains from the 4*x* Natsudaidai, *Citrus natsudaidai* Hayata. *Journal of the Japanese Society for Horticultural Science* 45, 225–230.

Yang, H.J. and Nakagawa, S. (1970) Cyto-histological studies on the male sterility of satsuma orange (*Citrus unshiu* Marc.) *Journal of the Japanese Society for Horticultural Science* 58, 239–245.

Yoshida, T., Nesumi, H., Yoshioka, T., Ieki, H., Ito, Y., Nakano, M., Ueno, I., Yamada,Y., Murase, S. and Takishita, F. (2003) New kumquat cultivar 'Puchimaru'. *Bulletin of the National Institute of Fruit Tree Science* 2, 9–16.

9 Somaclonal Variation in Sweet Orange: Practical Applications for Variety Improvement and Possible Causes

J. W. Grosser[1], **X. X. Deng**[2] **and R. M. Goodrich**[1]

[1]*University of Florida, Citrus Research and Education Center, Lake Alfred, FL 33850, USA;* [2]*Citrus Research Institute, Huazhong Agricultural University, Wuhan, Hubei, 430070, PR China*

Introduction

Sweet orange (*Citrus sinensis* L. Osbeck) is the most horticulturally important and widely grown *Citrus* species in Florida and worldwide. Approximately 90% of the Florida crop and more than 30% of the world crop are used for processing. Frozen concentrate orange juice has historically been the primary product of the Florida industry, but more recently there has been a strong shift to fresh pasteurized not-from-concentrate (NFC) product. Strong international competition in the concentrate market and a demand for increased quality in juice products has fuelled this change. The development of higher quality oranges with expanded maturity dates will facilitate this change and increase the competitive ability of the Florida industry. No true sweet orange cultivars have been produced by conventional breeding techniques due to its complex biology including large plant size, extended juvenility and nucellar polyembryony. Commonly grown sweet orange cultivars probably originated by the selection of chance seedlings well adapted to a particular area or from a mutation in a particular cultivar or seedling (Nishiura, 1965; Hodgson, 1968). Alternative methods are therefore required to develop improved sweet oranges, and we have been investigating somaclonal variation for improving sweet orange.

The term 'somaclonal variation' was introduced by Larkin and Skowcroft (1981) to define genetic variation that is present in plants regenerated from tissue cultures, either uncovered or induced by a tissue culture process. Reports of significant somaclonal variation have been primarily from solanaceous or cereal crops, affecting a wide range of traits including plant height, overall growth habit, flower, fruit and leaf morphology, juvenility, maturity date, disease resistance, yield and biochemical characteristics. Long-term studies in woody perennial fruit crops until now have been lacking. For the past 17 years, we have been investigating the use of tissue culture methods to produce genetic variation in 'Hamlin' and 'Valencia' sweet oranges, the two most important cultivars in Florida.

Sweet orange is well suited for studies of somaclonal variation due to its biology and excellent performance in tissue culture. Sweet orange has a natural tendency to form adventive somatic embryos from nucellar tissue, and many studies of somatic embryogenesis have been reported (Rangan *et al.*, 1968; Button and Bornman, 1971; Kochba *et al.*, 1972). Since the first report by Vardi *et al.* (1982), numerous sweet orange plants have been regenerated via somatic embryogenesis from protoplasts isolated from embryogenic callus or suspension cultures. Sweet orange also responds quite well organogenically, as adventitious buds can be induced efficiently from non-meristematic juvenile nucellar seedling explants (Grinblat, 1972; Barlass and Skene, 1982). This chapter will discuss ongoing research that shows how somaclonal variation can be used effectively as a tool to develop improved sweet oranges, with focus on the development of improved 'Hamlin' and 'Valencia' clones for processing (Grosser *et al.*, 1997).

Production of Sweet Orange Somaclones

A summary of tissue culture-derived plants of 'Hamlin' and 'Valencia' sweet oranges from various sources is provided in Table 9.1. Smaller populations of nucellar seedlings of each cultivar were also included in the study as a control and to provide an assessment of the natural variation possible in these cultivars. Regenerated sweet orange plants were budded to either Carrizo citrange, Swingle citrumelo or the precocious dwarfing 'Hamlin' + Flying Dragon somatic hybrid rootstock (Grosser *et al.*, 1988) to expedite fruiting.

Embryogenesis-derived plants were regenerated from secondary embryogenic callus cultures according to the method of Gmitter and Moore (1986). Organogenic-derived plants were regenerated primarily from nucellar seedling internodes via adventitious bud induction on DBA3 medium (Deng *et al.*, 1992), and shoots were rooted on RMAN medium (Grosser and Gmitter, 1990). Plants were regenerated from protoplasts isolated from embryogenic callus or suspension cultures, or nucellar seedling leaves from somatic hybridization experiments requiring cybridization (Grosser *et al.*, 1996), according to Grosser and Gmitter (1990). All grafted trees were planted in the field between 1988 and 1990, split among four locations. Some trees began producing fruit during the 1993–94 season, and fruit quality data collection was initiated at this time. All clones were evaluated by comparing Brix, acid and juice colour data measured by standard methods. Data from small representative fruit samples (6–8 fruit per tree) taken over several years were used to make primary selections for more in-depth evaluation. For the past few years, larger samples (40–60 fruit per tree) from promising clones were run through the state-certified FMC 091B State Test juice extractor located in the Citrus Research and Education Center (CREC) pilot plant to obtain more meaningful data. Pilot plant data from two or more seasons are provided for selected superior clones with consistent performance for altered traits. Recently, sensory analyses have been added as additional screening for clone selection. Properly conducted, this type of analysis can provide quantitative data with

Table 9.1. Summary of 'Valencia' and 'Hamlin' sweet orange somaclone trees in the field.

	Cultivar	
Regeneration pathway	'Valencia'	'Hamlin'
Organogenesis	242	229
Embryogenesis	261	219
Protoplasts	352	500
Nucellar (control)	118	20
Totals	973	968

Rootstocks: Carrizo citrange, Swingle citrumelo and 'Hamlin' + Flying Dragon somatic hybrid. Trees planted from 1988 to 1991. Most of the 'Hamlin' embryogenic plants were regenerated by F. G. Gmitter, Jr.

respect to taste preference and acceptance, and introduces objectivity to discussions of flavour (Lawless and Heymann, 1998; Resurreccion, 1998). Clear clonal differences in flavour have been observed. All clones showing promise have been propagated for further evaluation, including collection of yield data.

Results of Fruit/Juice Analyses and Discussion

Although the majority of the sweet orange somaclones appear to be normal, significant stable variation has been observed in tree and fruit characteristics. Altered tree characteristics include: canopy size/shape; leaf size/shape; ploidy level; and juvenility/thorniness/vigour. Altered fruit characteristics include: Brix:acid ratio; colour (fruit/juice); maturity date; size; rind thickness; rag; juice content; and seediness. Two organogenic-derived tetraploid clones of 'Hamlin' and an embyogenic-derived tetraploid clone of 'Valencia' were recovered and are being used as pollen parents in our interploid breeding programme to generate seedless triploids. The remainder of this chapter will focus on selected superior somaclones.

Early Maturing 'Valencia' Clones for Processing

Probably the most important clones being developed in this project are early 'Valencia' clones, and CREC pilot plant data over the past six years from some of the more interesting selections are provided in Table 9.2. Presently, there is a significant gap between optimum harvest dates for our midseason juice oranges ('Pineapple' and 'Midsweet' – usually harvested during January) and the highest quality late maturing standard 'Valencia' (harvest usually runs from March to June). This creates a gap in the processing season (usually lasting the entire month of February), resulting in inefficient operation of some processing plants during this time. An earlier maturing

Table 9.2. Selected 'Valencia' somaclones with a maturity date 2–4 weeks earlier than standard 'Valencia': 1997–2002 data (ratio/soluble solids[a]/colour).

	Harvest date					
Clone	14 January 1997	15 January 1998	28 January 1999	10 February 2000	4 February 2001	28 January 2002
B4-70	15.30	ND	16.95	14.88	ND	ND
	5.72		6.50	6.70		
	36.4		36.8	36.9		
B6-66	13.46	15.72	16.08	13.36	12.48	14.40
	7.00	6.15	6.93	7.23	6.44	6.20
	36.6	37.2	36.4	36.9	36.1	37.6
B6-68	13.94	15.34	13.84	13.94	10.92	14.06
	6.60	6.38	7.15	6.47	6.10	6.19
	37.9	38.4	38.2	38.0	38.1	37.7
B10-81	ND	14.58	14.98	13.51	12.54	14.06
		6.34	7.01	6.70	6.31	6.48
		38.3	38.7	38.5	37.7	38.3
Control	10.53	12.61	12.50	12.20	10.34	11.61
	7.15	6.96	6.89	6.65	6.11	6.43
	36.3	37.6	38.1	37.6	37.5	37.8

[a]Pounds of solids per 90 lb box.
ND = not determined.

'Valencia' would of course solve this problem. Other advantages to having a superior early 'Valencia' include: (i) the possibility of harvesting grade A fruit in the event that there is a fruit-destructive freeze any time after mid-January; (ii) the availability of superior quality and highly coloured juice to blend with 'Hamlin' as needed to achieve grade A juice during the early–middle part of the season; (iii) expanding the NFC season; and (iv) the possibility of harvesting fruit prior to fruit set of the following season's crop. Standard 'Valencia' has the new and old fruit on the tree at the same time, making it difficult to use abscission chemicals to facilitate mechanical harvesting.

The most promising early maturing 'Valencia' clones for processing from the first group studied (Table 9.2) are B6-68 and B10-81, and the data suggest that they should be amenable to a February harvest (in Florida) during most years. The original trees of these clones are of normal size and vigour, and both appear to be yielding well. Fruits are less seedy than typical 'Valencia' and consistently have a high juice content. Juice of both clones has excellent colour and a typical 'Valencia' flavour. This is supported by sensory data that indicate that B10-81 is as acceptable as the control Midsweet as early as mid-January, well before 'Valencia' fruit can usually be utilized. The two earliest clones of this group, B4-70 and B4-83, can usually be harvested in mid–late January, but often their juice content is lower than average (data not shown), and their juice colour is also less than average. The original trees of these two clones are smaller and less thrifty than typical 'Valencia'. Juice from somaclone B6-66 has an excellent rich flavour and the fruit is totally seedless. The superior flavour of B6-66 is confirmed by preliminary sensory testing that shows it is significantly preferred to the control 'Valencia' at both mid-February and mid-March harvests. Unfortunately, this clone also has lower than average juice colour scores and juice content. The original tree is also smaller than average. The five clones described in Table 9.2 were all regenerated from embryogenic callus.

More recently, a protoplast-derived clone SF14-62W was identified that is even earlier maturing than the clones mentioned above (Table 9.3). This clone has a maturity date very comparable with that of 'Midsweet', the currently preferred mid-season cultivar in Florida. Somaclone SF14-62W has significantly better juice colour than 'Midsweet', and was favoured over 'Midsweet' in preliminary sensory testing. Brix:acid ratio data also indicate that SF14-62W is significantly earlier than either B10-81 or B6-68. SF14-62W, therefore, has excellent potential as a mid-season

Table 9.3. Recently selected very early maturing 'Valencia' somaclone (4–8 weeks earlier than standard 'Valencia'):(Brix/ratio/colour/soluble solids[a]).

	Harvest date		
Clone	28 January 2000	4 February 2001	15 January 2002
SF14-62W	11.0	12.8	11.8
	14.7	15.8	13.8
	38	38	37.6
	4.9	7.0	ND
Control	Valencia 10.1	Valencia 11.7	Midsweet 12.0
	10.8	10.3	13.4
	37.6	37.5	36.4
	5.6	6.1	ND

[a]Pounds of solids per 90 lb box.
ND = not determined.

Table 9.4. Late maturing 'Valencia' somaclones: 1999–2000 CREC pilot plant data.

Clone		Harvest date 24 May 1999	24 May 2000
T3-56	Juice content (%)	59.8	52.4
	Brix:acid ratio	19.1	14.6
	Pounds of SS[a]/90 lb box	8.25	6.86
	Juice colour score	38.5	39.3
T3-58	Juice content (%)	59.8	58.2
	Brix:acid ratio	19.1	16.8
	Pounds of SS/90 lb box	7.83	7.28
	Juice colour score	38.2	39.7
T3-53	Juice content (%)	56.9	59.7
	Brix:acid ratio	19.7	17.9
	Pounds of SS/90 lb box	7.74	7.22
	Juice colour score	39.0	39.7
T3-51	Juice content (%)	58.6	51.9
	Brix:acid ratio	19.6	17.4
	Pounds of SS/90 lb box	8.45	6.41
	Juice colour score	39.3	39.6

Control fruit were not available.
[a]SS = soluble solids.

clone, which could be especially useful for blending with low-coloured 'Hamlin' juice.

Late Maturing 'Valencia' Clones for Processing

The shift from a frozen concentrate industry to an NFC industry requires more fruit to go directly from the tree into the carton, which minimizes opportunities for juice storage and blending. As a result, there is an increased demand for clones that maintain a low sugar/acid ratio into the late spring and early summer (May and June). Acid levels generally decline rapidly during this time of year, causing juice ratios to elevate into the mid-twenties. Processors are seeking clones that can maintain a brix:acid ratio below or near 18. We have identified a few clones showing potential to achieve this, and data from two seasons are presented in Table 9.4. Maintaining an adequate juice content is also critical during this time of the year (>55% being desirable), and this must be considered when making selections. It is interesting to note that all four of these clones are organogenic derived. Several other clones are showing promise for late maturity (data not shown) and have been included in sensory analyses.

'Valencia' Somaclones with Fresh Market Potential

Although a small amount of 'Valencia' fruit in Florida goes to the fresh market, the value of this cultivar to the fresh market is much greater for other citrus-producing regions, including South Africa, California and Brazil. Improved 'Valencia' clones for the fresh market would therefore make a significant contribution to citriculture worldwide. We have identified eight seedless somaclones covering a range of maturity dates (three early maturing, three standard maturity and two late maturing). We have identified other low-seeded clones (averaging ~1 seed/fruit) that repeatedly produce larger fruit than average and are easier to eat due to a softer rag (more of a melting flesh) than standard 'Valencia'.

Table 9.5. Promising 'Valencia' somaclones for the fresh fruit market.

Clone	Traits
T2-21	Seedless
	Early maturity (February)
SF15-62W	Seedless
	Large fruit size
	Normal maturity
N12-11	Soft rag and candy flavour
	Large fruit size
	Low seed content
	Normal maturity
N11-18	Seedless
	Large fruit size
	Late maturity (May)

Some of the more promising clones are described in Table 9.5. These clones should be tested in a Mediterranean environment to determine their potential as improved fresh fruit clones. Such clones should be more attractive to the consumer and also expand the fresh 'Valencia' season.

'Hamlin' Somaclones with Improved Colour and Higher Soluble Solids

'Hamlin' has been a popular processing orange in Florida because it matures early and produces excellent yields. However, the juice quality of 'Hamlin' is poor, lacking colour and flavour. As a result, 'Hamlin' juice is usually blended with mandarin or 'Valencia' juice to produce a grade A product. Selection of somaclones with improved colour and soluble solids would increase the value of 'Hamlin' as a processing orange, particularly for NFC production. Our studies indicate that 'Hamlin' is much more stable than 'Valencia', and we have observed much less useful somaclonal variation. Three-year CREC Pilot Plant data from selected clones are provided in Table 9.6. We have identified a few clones that consistently produce better colour and higher solids (although not grade A colour), and a clone with higher solids that appears to mature a bit earlier than standard 'Hamlin' (clone N14-10). These and other promising clones have been propagated for further evaluation. Note that the scores from the control trees of 'Hamlin' were very poor during 1998 and 1999, but we are not sure why. We have also noticed that typical 'Hamlin' has a colour score that generally reaches a maximum of approximately 35.5 if fruit is harvested into late December and January. It will be several years before we know if selected improved clones yield as well or better than standard 'Hamlin'. Interest in improved 'Hamlin' clones has diminished since the recent release of highly coloured and flavoured early maturing clones including 'Early Gold' and 'Itaborai'.

Studies to Determine the Cause of Citrus Somaclonal Variation

Variation occurring from the *in vitro* culture of citrus, namely somaclonal variation, is of importance not only to the genetic conservation of germplasm *in vitro*, but also to genetic improvement. Theoretically, somaclonal variation is nearly the same thing as the somatic variation (namely budsport mutations), in the field. The difference is that the former is referred to the *in vitro* system, and the latter to the *in vivo* system. With the *in vitro* system, it is much easier to examine the variation at the cytological and molecular levels than it is with what occurs in the field, due to the much higher frequency of variation. For

Table 9.6. 'Hamlin' sweet orange somaclones selected for improved juice colour and soluble solids. CREC pilot plant data.

Clone		Harvest date		
		12 December 1997	15 December 1998	1 November 1999
C2-46	Ratio	16.96	19.41	11.29
	SS	7.32	6.29	6.01
	Colour score	34.6	33.7	32.8
C1-45	Ratio	17.54	17.38	11.21
	SS	6.65	6.29	5.76
	Colour score	34.6	33.4	33.2
N14-10	Ratio	21.14	23.7	12.59
	SS	7.00	6.45	5.41
	Colour score	33.6	32.5	33.2
Control	Ratio	18.94	14.92	9.92
	SS	6.97	4.71	4.18
	Colour score	33.8	33.5	31.0

SS = pounds of soluble solids/90 lb box.

germplasm conservation, unwanted somaclonal variation is a key issue that affects the system's efficiency for maintaining the genetic stability of the conserved material; on the other hand, for genetic improvement as previously discussed, somaclonal variation is an important source of novel variation, particularly for species that are difficult to improve by conventional breeding such as sweet orange and grapefruit.

As described earlier, the reported variations in sweet orange include maturity date, morphology, ploidy level and others. Changes in maturity date and seed content may have the most practical value. Autotetraploid sweet orange plants have been obtained via organogenesis from non-embryogenic callus and nucellar seedling internodes, and from embryogenic callus. Generally, somaclonal variation can be classified into two categories, one is stable genetic variation that can be transmitted to the next generation, which includes ploidy variation, single gene mutation, movement of retroelements/transposons, etc. (Grandbastien, 1998; Kubis *et al.*, 2003). The other is the epigenetic variation, such as DNA methylation (Fig. 9.1; Finnegan *et al.*, 1996; Hao and Deng, 2002b), which leads to variable gene transcription and translation. Epigenetic variation in general cannot be transmitted to the next generation.

For the origin of stable somaclonal variation, there are two possible primary pathways; one is that the variation comes from pre-existing variation in the explant. Deng *et al.* (1985) examined the chromosome number of the meristems of citrus cultivars, and verified the existence of variation in ploidy level in 2–7% of cells examined. Such pre-existing variation that occurs in somatic cells is rarely expressed or recovered, but may be recovered if such cells have an opportunity to regenerate *in vitro*, and as a result variable plants will be found among the regenerates. Another pathway is that the variation originated during *in vitro* culture. Potential induction factors include the medium, temperature and the growth regulators used during the culture. However, how these factors may cause variation is poorly understood. Previous experiments have verified that 2,4-dichlorophenoxyacetic acid (2,4-D) enhanced ploidy variation in citrus non-embryogenic callus (Deng *et al.*, 1985), but kinetin did not. Generally, the longer the time in culture (culture age), the higher the

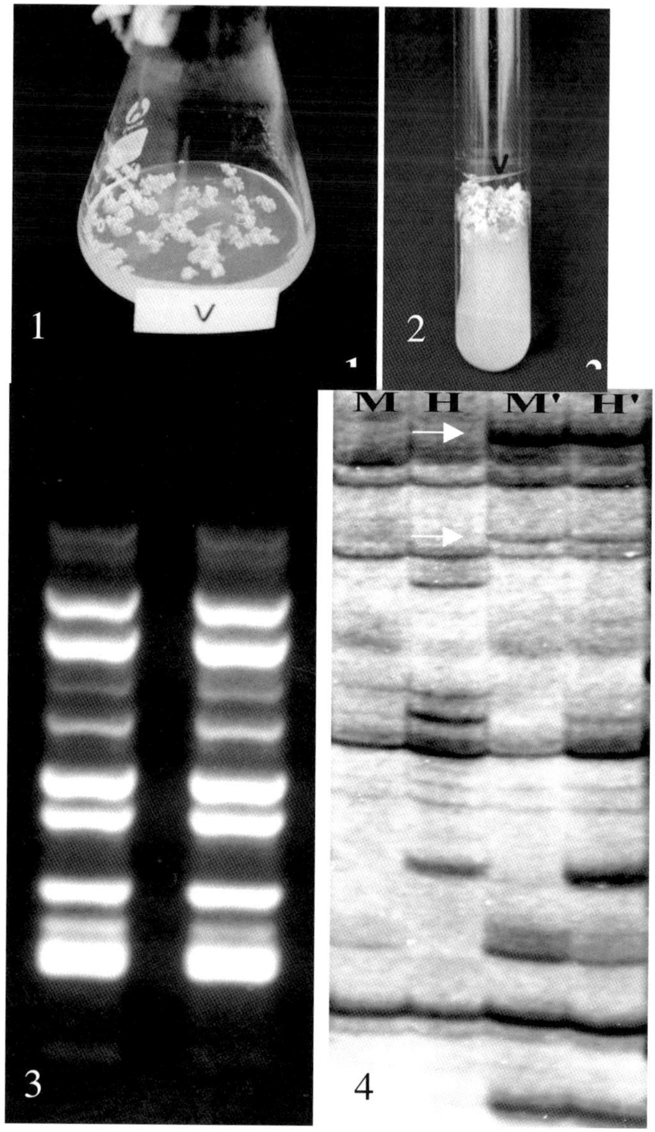

Fig. 9.1. An embryogenic callus incapable (1) and capable (2) of embryogenesis of Newhall navel orange did not show any differences in their random amplified polymorphic DNA (RAPD) profile (3) with primer OPV-12. However, the methylation state of DNA was different; some of the sites were demethylated. (4) Methylation-sensitive amplification polymorphism (MSAP) analysis with the primer pair HM-TCAA/E-ATG indicated the difference. Lanes M and H were calli incapable of embryoid regeneration, while 'M' and H' were calli capable of regeneration. Arrowheads point to the demethylated sites.

rate of associated somaclonal variation. The most easily observed phenomenon is that the regeneration capacity of citrus embryogenic callus decreases with culture age, probably due to the accumulation of mutations. This phenomenon has been exploited to enhance the selection of somatic hybrids in protoplast fusion experiments by reduc-

Table 9.7. Chromosomal ploidy of different genotypes of citrus calluses.

Varieties	Chromosome ploidy levels (%)				
	Subculture years	Cells counted	Diploid	Aneuploid	Tetraploid
Newhall navel orange	6	45	93.3	2.1	4.6
Murcott tangor	10	123	93.2	1.4	4.4
Guoqing No. 1 satsuma mandarin	1	143	95.1	2.1	2.8
Marsh grapefruit	4	155	94.8	1.9	3.3
Valencia sweet orange	3	127	90.9	2.7	6.2
Jincheng sweet orange	15	65	92.3	2.9	4.8
Ponkan mandarin	1	22	93.5	1.9	3.9

ing or eliminating the regeneration of diploid plants from the embryogenic parent (Grosser *et al.*, 2000). To get rid of the interruption of pre-existing variation, single protoplast sibling lines of Newhall sweet orange were first established by two rounds of low-density protoplast culture in agar-solidified medium. These diploid lines originated from a single protoplast and have the same genetic background. During the regeneration process, chromosome counts were conducted, and the results showed that polyploid cells were found as early as 40 days after culture initiation (Hao and Deng, 2002a).

Chromosome number variation in citrus embryogenic callus is a common phenomenon. Embryogenic callus from eight citrus cultivars was examined, and the percentage of cells with variable chromosome numbers varied from 4.9 to 9.1% of the total cells (Table 9.7; Fig. 9.2). Most of the cells with variable chromosome numbers were tetraploid. Ten microcalli, all of single protoplast descent, were examined for ploidy level. Data presented in Table 9.8 showed that eight out of the ten microcalli had clear chromosomal images and four out of the eight calli had cells with variable chromosome numbers. Out of the total observed cells of the eight calli, 91.7% of the cells were diploid, and 3.7% were variable. Interestingly, the percentage of cells with variable chromosome numbers did not increase over time (based on examining cells from subsequent subcultures). From

Table 9.8. Chromosomal variations of citrus calluses derived from protoplasts 40 days after culture initiation.

Sibling line series number	Total cells observed	Diploid cells	Tetraploid cells	Cells with chromosomes between 2*x* and 4*x*
1	20	18	1	1
2	0	0	0	0
3	12	11	0	1
4	9	8	0	1
5	17	15	1	1
6	7	7	0	0
7	0	0	0	0
8	19	17	1	1
9	10	10	0	0
10	14	13	1	0
Total (%)	108 (100)	99 (91.7)	4 (3.7)	5 (4.6)

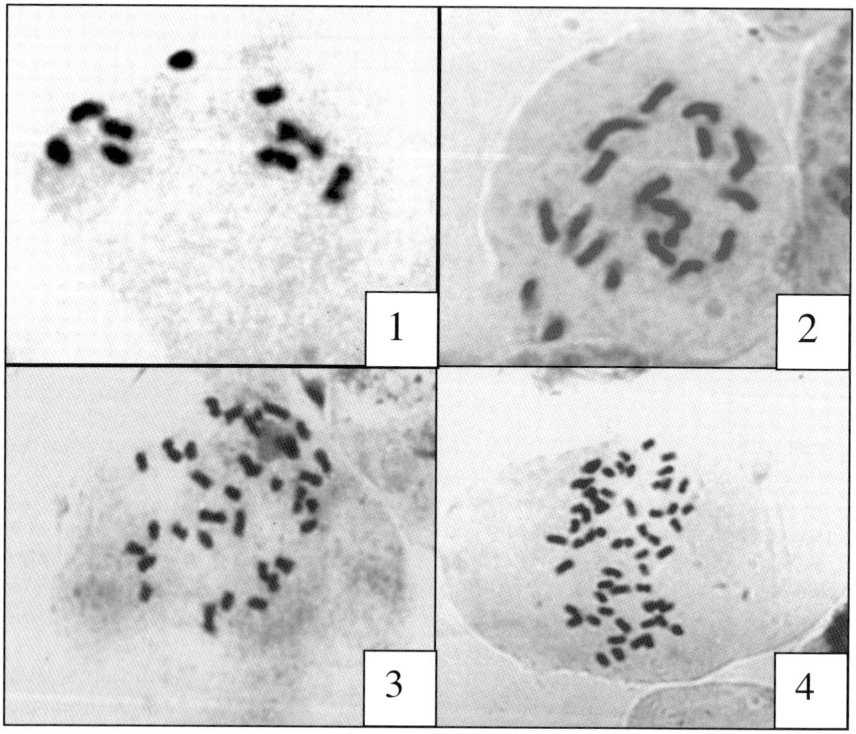

Fig. 9.2. Different ploidy cells found in the embryogenic callus of Valencia sweet orange: (1) Haploid $2n = x = 9$; (2) diploid $2n = 2x = 18$; (3) tetraploid $2n = 4x = 36$; (4) hexaploid $2n = 6x = 54$.

Table 9.7, data indicate that the subculture year was not positively correlated with the variation.

Sibling lines of Newhall sweet orange were examined for ploidy stability. Data from Table 9.9 showed that the ploidy level was relatively stable after one year in culture. Ploidy level changes apparently reach an equilibrium, in contrast to other types of variation that appear to increase over time in culture. In this case, the Newhall embryogenic callus has been subcultured for more than six years, much longer than an annual crop, yet the callus still has capacity for somatic embryogenesis. The chromosomal stability is probably one of the reasons. Another reason for this stability may be due to the culture medium used, which contains no variation-inducing growth regulators.

To understand better the mechanism of ploidy level stability in citrus callus, we tested the mitotic index of cells with different ploidy levels (Table 9.10). The results showed that tetraploid cells have a lower mitotic index, and a higher percentage of apoptosis (Fig. 9.3; Hao *et al.*, 2002), which can explain the ploidy stability phenomenon, i.e. diploid cell cultures continuously generate a small percentage of cells with variable chromosome numbers, but these cells with variable chromosome numbers generally have a lower competitive ability and, therefore, remain at a low level in the mixed population.

We also examined the chromosome numbers of callus, embryoids and plantlets derived from the same callus line of ‘Anliucheng’ sweet orange. We observed more cells with variable chromosome numbers in the callus than in the embryoids, which had more variable cells than the resulting plantlets. This result indicates that

Table 9.9. The relative stability of ploidy constitution of Newhall sweet orange calluses during subculture.

	Before subculture			After subculturing for 1 year		
Sibling line series number	Diploid (%)	Aneuploid (%)	Tetraploid (%)	Diploid (%)	Aneuploid (%)	Tetraploid (%)
1	93.4	2.2	4.4	94.5	2.0	3.5
2	93.7	2.3	4.0	93.0	3.6	3.4
3	92.5	2.6	4.9	93.1	2.4	4.5
4	92.8	2.9	4.3	92.8	3.3	3.9
5	93.9	2.3	3.8	92.7	3.0	4.3
6	92.9	2.9	4.2	92.1	2.4	4.6
7	93.1	3.1	3.8	93.5	2.0	4.5
8	94.8	1.8	3.4	95.1	1.6	3.2
Average	93.26	2.79	4.08	93.22	2.6	3.99

the regeneration pathway screens for normal diploid cells. Normal cells seem to have a higher capacity for plant regeneration.

Another point worthy of further study is the hypothesis that the *in vitro* system can generate more variation than *in vivo* culture over the same length of time. If such a comparison is based on the timing of cell division, maybe both systems would have the same variation rate. However, cell division in the *in vitro* system is continuous, whereas over the same period of time, cell division in the *in vivo* system is interrupted. Therefore, the *in vitro* system offers more opportunities for variation to occur.

Protocols

Protocol for producing citrus somaclones via organogenesis

Table 9.10. The mitotic index of different ploidy cells in the same cell population.

	Mitotic index (%)	
	Diploid	Tetraploid
Cultivars	cells	cells
Newhall sweet orange	41.2	4.8
Valencia sweet orange	39.8	4.9

Nucellar seedlings of polyembryonic citrus cultivars grown under aseptic conditions are the most commonly used and most efficient explant source. Root, epicotyl, hypocotyl and stem internodes are all amenable to adventitious bud induction, but stem internodes are most often utilized. Sections approximately 1 cm in length are cultured on DBA3 shoot induction medium (MT basal medium (Murashige and Tucker, 1969) containing 13.3 μM 6-benzylaminopurine, 0.045 μM 2,4-D and 20 ml/l coconut water; Deng *et al.*, 1992). Two to three 1-month passages on DBA3 may be required to obtain multiple shoots. Shoots are commonly rooted on RMAN rooting medium (half-strength MT basal medium containing 0.11 μM naphthalene acetic acid (NAA) and 0.5 g/l activated charcoal; Grosser and Gmitter, 1990). Rooted plantlets are acclimated under low-light humid conditions prior to greenhouse culture. This technique can also be utilized to regenerate plants from enlarged, malformed somatic embryos that fail to germinate.

Protocol for producing citrus somaclones via somatic embryogenesis

Embryogenic callus from many polyembryonic citrus genotypes (including sweet oranges, mandarins and lemons) can be initiated by culturing unfertilized or undevel-

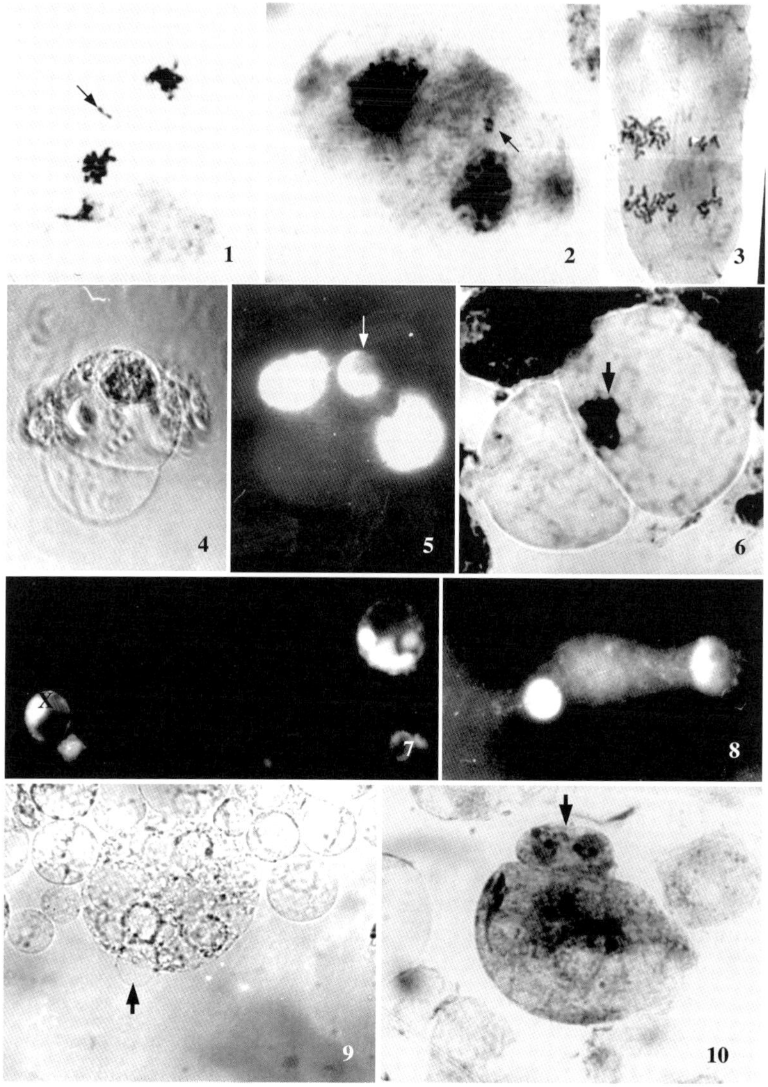

Fig. 9.3. The cytogenetic mechanisms and control of chromosome variations in the embryogenic callus of Newhall navel orange. (1) Chromosome lagging at anaphase (1060×). The arrow points to a lagging chromosome. (2) Lagged chromosomes formed a small nuclear body beside the normal nuclei (850×). The arrow points to a mininucleus adjacent to the normal nucleus. (3) Meiosis-like division (920×). (4) Asymmetrical distribution of duplicated chromatins to one protoplast at telophase (520×). (5) Fluorescent observation of a protoplast stained with Hoechst 33258 (520×). The arrow points to the duplicated chromatins distributed to one protoplast. (6) Asymmetrical distribution of duplicated chromatins to one cell at telophase (850×). The arrow points to the duplicated chromatins distributed to one cell. (7) Apoptotic tetraploid cell and normal diploid cell of Newhall navel stained with Hoechst 33258 (320×). (8) Comet electrophoresis of an apoptotic tetraploid cell and a normal diploid cell of Newhall navel (260×). (9) An apoptotic giant protoplast (320×). The arrow points to an apoptotic body budded from the giant protoplast. (10) An apoptotic giant cell (320×). The arrow points to an apoptotic body budded from the giant cell.

oped ovules (from either developing or mature fruit) on an MT basal medium (Murashige and Tucker, 1969) supplemented with 5 mg/l kinetin and 0.5 g/l malt extract (V. Sapp, personal communication). Large numbers of ovules should be plated, and 1–3 passages on this medium are generally required to obtain callus induction. Although it is difficult and time-consuming to establish friable embryogenic citrus callus cultures, habituated culture lines can be established and maintained indefinitely on either EME or H+H liquid or solid culture medium (Grosser and Gmitter, 1990). Solid medium cultures require maintenance transfers every 4–6 weeks, whereas suspension cultures are generally maintained on a 2-week cycle.

Citrus somatic embryos are routinely matured on 1500 medium (MT basal medium supplemented with 1.5 g/l malt extract; Grosser and Gmitter, 1990). However, abnormal embryo development occurs frequently. Niedz *et al.* (2002) recommend the use of cellulose acetate semi-permeable membranes over semi-solid medium to affect the normalizing of citrus embryogenesis positively, thereby improving the efficiency of plant recovery.

Mature citrus embryos are routinely germinated on MT basal medium supplemented with 1–3 g/l gibberellic acid (Grosser and Gmitter, 1990). Germinated embryos generally exhibit long, thin and often spindly roots that are inferior for acclimatization. Re-rooting of shoots dissected from germinated embryos on RMAN rooting medium (half-strengh MT basal medium containing 0.02 mg/l NAA and 0.5 g/l neutralized activated charcoal) improves acclimatization efficiency (Grosser and Gmitter, 1990).

Somaclones can also be regenerated via somatic embryogenesis from protoplasts (protoclones) isolated from solid or liquid embryogenic citrus cultures. Protoplast-derived populations appear to have the most useful variation. Protocols for producing protoclones and protoplast-derived cybrids can be found in Chapter 10.

Concluding Remarks

Significant somaclonal variation in whole plant characteristics can be found in both 'Valencia' and 'Hamlin' sweet oranges, with 'Valencia' showing more profound and useful variation. Genetically stable variation for useful traits can be obtained. We have identified early and late maturing somaclones of 'Valencia' with superior quality that should facilitate the processing industry, particularly NFC production. Sensory testing has been added as another tool for evaluating commercial potential of specific clones, and clones with preferred flavour are being identified. Improved somaclones of 'Valencia' with fresh market potential have been identified, including seedless clones with different maturity dates, and large-fruited clones with a more melting flesh. These clones should be evaluated in a Mediterranean environment. Improved somaclones of 'Hamlin' have been selected for better juice colour and increased soluble solids. The next step is to determine if the improved traits are expressed in trees propagated from the original selections, and if these selected somaclones yield adequately. Second-generation trees of selected clones have been propagated on various rootstocks and planted in the three representative citriculture areas of Florida. Results obtained so far indicate that the generation and evaluation of somaclonal variation in sweet orange is a viable method for developing improved sweet oranges both for processing and for the fresh market. New cultivars from this should facilitate NFC production by providing juice of higher quality throughout an expanded season. Somaclonal variation can certainly be exploited as a tool to expand the maturity dates (and associated market window) and reduce the seed content of commercially important citrus cultivars.

Studies to determine the potential causes of somaclonal variation revealed that polyploid and aneuploid cells are produced consistently in cultured citrus cells. However, their percentages remain relatively constant over increased time in culture. Also, the percentages of cells with altered chromosome numbers decrease as cultures proceed through the plant regeneration process. These facts indicate that more normal diploid cells are more competitive in culture, and that the plant regeneration process selects against cells with radical genetic changes. This suggests that long-term cultures may accumulate minor genetic changes (i.e. mutations/movement of transposons (Kubis *et al.*, 2003) or cytological aberrations such as inversions or translocations) that do not significantly affect the mitotic index or interfere with the plant regeneration process, thereby resulting in regenerated plants with the observed minor but significant variation.

Acknowledgements

The authors thank Becker Holding Company and the CREC Pilot Plant crew for their valuable assistance in growing somaclone trees and collecting data, respectively. Technical assistance from Patricia Brickman is also appreciated. This research was supported by the Florida Agricultural Experiment Station and a grant from Florida Citrus Production Advisory Council (#991-981; entitled 'New Scion Variety Development for Florida Citrus Growers'), and approved for publication as Journal Series No. R-9132.

References

Barlass, M. and Skene, K.G.M. (1982) *In-vitro* plantlet formation from *Citrus* species and hybrids. *Scientia Horticulturae* 17, 333–341.

Button, J. and Bornman, C.H. (1971) Development of nucellar plants from unpollinated and unfertilized ovules of the Washington navel orange. *Journal of South African Botany* 37, 127–134.

Deng X.X., Liu, G.B. and Zhang, W.C. (1985) Study on the chromosomal variation of citrus callus. *China Citrus* 1985 (3), 4–7.

Deng, X.X., Grosser, J.W. and Gmitter, F.G., Jr (1992) Intergeneric somatic hybrid plants from protoplast fusion of *Fortunella crassifolia* cultivar 'Meiwa' with *Citrus sinensis* cultivar 'Valencia'. *Scientia Horticulturae* 49, 55–62.

Finnegan, E.J., Peacock, W.J. and Dennis, E.S. (1996). Reduced DNA methylation in *Arabidopsis thaliana* results in abnormal plant development. *Proceedings of the National Academy of Sciences of the USA* 93, 8449–8454.

Gmitter, F.G., Jr and Moore, G.A. (1986) Plant regeneration from undeveloped ovules and embryogenic calli of citrus: embryo production, germination, and plant survival. *Plant Cell Tissue and Organ Culture* 6, 139–147.

Grandbastien, M.A. (1998). Activation of plant retrotransposons under stress conditions. *Trends in Plant Science* 3, 181–187.

Grinblat, U. (1972) Differentiation of *Citrus* stems *in-vitro*. *Journal of the American Society of Horticultural Science* 97, 559–603.

Grosser, J.W. and Gmitter, F.G., Jr (1990) Protoplast fusion and citrus improvement. *Plant Breeding Reviews* 8, 339–374.

Grosser, J.W., Gmitter, F.G., Jr and Chandler, J.L. (1988) Intergeneric somatic hybrid plants of *Citrus sinensis* cv. Hamlin and *Poncirus trifoliata* cv. Flying Dragon. *Plant Cell Reports* 7, 5–8.

Grosser, J.W., Gmitter, F.G., Jr, Tusa, N., Reforgiato-Recupero, G. and Cucinnota, P. (1996) Further evidence of a cybridization requirement for plant regeneration from citrus leaf protoplasts following somatic fusion. *Plant Cell Reports* 15, 672–676.

Grosser, J.W., Gmitter, F.G., Jr and Chandler, J.L. (1997) Development of improved sweet orange cultivars using tissue culture methods. *Proceedings of the Florida State Horticultural Society* 110, 13–16.

Hao, Y.J. and Deng, X.X. (2002a) Occurrence of chromosomal variation and regeneration of plant from long term cultured citrus callus. *In Vitro Cell and Developmental Biology-Plant* 38, 472–476.

Hao, Y.J. and Deng, X.X. (2002b) Stress treatments and DNA methylation affect somatic embryogenesis from citrus callus. *Acta Botanica Sinica*, 44, 673–677.

Hao, Y.J., You, C.X. and Deng, X.X. (2002) Cell size is used as a morphological marker to calculate mitotic index and ploidy level of citrus callus. *Plant Cell Reports* 20, 1123–1127.

Hodgson, R.W. (1968) Horticultural varieties of citrus. In: Reuther, W., Webber, H.J. and Batchelor, L.D. (eds) *The Citrus Industry*, Vol. 1. University of California, Berkeley, California, pp. 431–591.

Kochba, J., Spiegel-Roy, P. and Safran, H. (1972) Adventive plants from ovules and nucelli in citrus. *Planta* 106, 237–245.

Kubis, S.E., Castilho, A.M., Vershinin, A.V. and Heslop-Harrison, J.S. (2003) Retroelements, transposons and methylation status in the genome of oil palm (*Elaeis guineenis*) and the relationship to somaclonal variation. *Plant Molecular Biology* 52, 69–79.

Larkin, P.J. and Skowcroft, W.R. (1981) Somaclonal variation – a novel source of variability from cell cultures for plant improvement. *Theoretical and Applied Genetics* 60, 197–214.

Lawless, H.T. and Heymann, H. (1998) Acceptance and preference testing. In: Lawless, H.T. and Heymann, H. (eds) *Sensory Evaluation of Food: Principles and Practices*. Chapman & Hall, New York, pp. 430–479.

Murashige, T. and Tucker, D.P.H. (1969) Growth factor requirements of *Citrus* tissue culture. *Proceedings of the First International Citrus Symposium* 3, 1155–1161.

Niedz, R.P., Hyndman, S.E., Wynn, E.T. and Bauscher, M.G. (2002) Normalizing sweet orange (*C. sinensis* (L.) Osbeck) somatic embryogenesis with semi-permeable membranes. *In Vitro Cell and Developmental Biology-Plant* 38, 552–557.

Nishiura, M. (1965) Natural mutation and its utilization in the selection of citrus fruits. In: *Use of Spontaneous Mutations in Plant Breeding, Gamma Field Symposia No. 4*. Institute of Radiation Breeding. Ministry of Agriculture and Forestry, Japan, pp. 27–38.

Rangan, T.S., Murashige, T. and Bitters, W.P. (1968) *In-vitro* initiation of nucellar embryos in monoembryonic *Citrus*. *HortScience* 3, 226–227.

Resurreccion, A.V.A. (1998) Applications of the electronic nose in product research and testing. *Center for Food Safety and Quality Enhancement Annual Meeting*, Atlanta, Georgia, March 10–11.

Vardi, A., Spiegel-Roy, P. and Galun, E. (1982) Plant regeneration from Citrus protoplasts: variability in methodological requirements among cultivars and species. *Theoretical and Applied Genetics* 62, 171–176.

10 Somatic Hybridization

Patrick Ollitrault[1], Wenwu Guo[2] and Jude W. Grosser[3]

[1]Center for International Cooperation in Agricultural Research for Development (CIRAD), Avenue Agropolis – TA 71/09, 34398 Montpellier, Cedex 5, France; [2]National Key Laboratory of Crop Genetic Improvement, Huazhong Agricultural University, Wuhan 430070, PR China; [3]University of Florida, Citrus Research and Education Center (CREC), 700 Experiment Station Road, Lake Alfred, FL 33850, USA

Introduction

Although somatic hybridization techniques are being ignored by variety improvement programmes for most commodities, their contribution to citrus variety improvement continues to expand with increasing complexity. During the last decade, a primary application has been for citrus rootstock breeding. Somatic hybridization allows the addition of all dominant traits, irrespective of the heterozygosity level of the breeding material. At the international level, agronomic evaluations are ongoing for several interspecific and intergeneric hybrids with some very interesting results. Somatic hybrids have been obtained between sexually incompatible species, but their interest for breeding seems to be limited by unfavourable traits and problems of nuclear genome instability for the intersubtribal and intertribal combinations. A second primary application is the exploitation of polyembryonic (apomictic) and sterile cultivars for the synthesis of fertile tetraploid hybrids. The final objective of such ploidy manipulation is the synthesis of seedless triploid cultivars. Somatic hybridization has resulted in a great diversification of the tetraploid gene pool available for sexual hybridization with diploid cultivars, and thousands of triploids have been produced from interploid crosses using somatic hybrid parents. Somatic hybridization has also allowed the direct synthesis of triploid hybrids by protoplast fusion between diploid and haploid lines. Several alloplasts and cybrids have also been obtained by symmetric somatic hybridization. This material provides an opportunity for nucleo-cytoplasmic interaction studies and should open new avenues for citrus germplasm valorization. An extensive review that describes the fundamental roles of somatic hybridization in broad-based citrus improvement programmes, along with comprehensive lists of somatic hybrids and cybrids produced around the world, was recently provided by Grosser *et al.* (2000). The primary goal of this chapter is to complement this previous review with evolving molecular techniques that facilitate somatic hybrid and cybrid genome characterization, and new applications of somatic hybridization and cybridization technologies to facilitate citrus scion and rootstock development. These new applications are possible due to feedback from field trials of scion and rootstock citrus somatic hybrids produced previously, along with molecular marker information that identifies the progenitors of important citrus cultivars. These new applications have great potential to facilitate the development of high quality seedless fresh market citrus and superior widely adapted rootstocks. Extensive protocols for the production and characterization of citrus somatic hybrids and cybrids are also provided.

Genomic Characterization of Somatic Hybrids

Selection of somatic hybrids

After fusion, protoplast suspensions contain parental, homofused, heterofused and multifused protoplasts. Regeneration is done without any selection pressure. It is therefore necessary to select the somatic hybrids and alloplasmic plants among the regenerated plants. Morphological characters can be useful for somatic hybrid identification (Grosser and Gmitter, 1990). However, a much more efficient selection is generally done by the combination of molecular or isozyme marker analysis and ploidy evaluation.

Restriction fragment length polymorphism (RFLP) analysis with rDNA probes was one of the first molecular techniques for somatic hybrid identification (Ohgawara *et al.*, 1985; Takayanagi *et al.*, 1992; Miranda *et al.*, 1997). However, isozyme analysis is still a very easy and powerful tool for such applications. Several isozyme systems are routinely used, e.g. peroxidase (Tusa *et al.*, 1990; X.X. Deng *et al.*, 1992; Grosser *et al.*, 1992a; Ye *et al.*, 1992), phosphoglucomutase (Grosser *et al.*, 1992a; Tusa *et al.*, 1992; Ye *et al.*, 1992; Ollitrault *et al.*, 1996b) and phosphoglucose isomerases (PGIs; Tusa *et al.*, 1990, 1992; Ollitrault *et al.*, 1996b). Relative band intensity is generally efficient for identification of somatic hybrids when a homozygous genotype is combined with a heterozygous one sharing a common allele with diploid plants. It is particularly true with dimeric enzymes such as PGI or isocitrate dehydrogenase (IDH). Moreover, these enzymes generally allow distinction between triploid and tetraploid hybrids arising from diploid + haploid protoplast fusion (P. Ollitrault, 2002, France, unpublished data). Random amplified polymorphic DNA (RAPD; Grosser *et al.*, 1996b; Kobayashi *et al.*, 1997; Shi *et al.*, 1998b; Guo *et al.*, 2000) has also been used and should be particularly useful when co-dominant markers such as isozymes or RFLP do not exhibit any polymorphism between parents. STMS (sequence-tagged microsatellites) displaying high polymorphism between mandarins (Luro *et al.*, 2001) should also be interesting for combinations inside this cultivar group. Chromosome counting can now be favourably by-passed by flow cytometry for quicker ploidy analysis (Ollitrault *et al.*, 1996b; Grosser *et al.*, 2000). By using flow cytometry and isozymes or polymerase chain reaction (PCR) markers, it is possible to select somatic hybrids at the *in vitro* stage of plant regeneration.

Accurate nuclear genome characterization and somatic hybrid meiotic analysis

Nuclear genome instability has been observed for several crops, particularly for wide somatic hybridization (Téoulé, 1992; Kisaka *et al.*, 1997; Oberwalder *et al.*, 1998). Correlations between genetic distance and chromosome elimination have been described in *Brassica* (Sundberg and Glimelius, 1991). The multiple subcultures generally necessary to regenerate wide hybrids should be favourable to chromosome elimination or recombination (Oberwalder *et al.*, 1998; Guo and Deng,

1999). Moreover, the systematic use of embryogenic callus lines as one of the parents in citrus somatic hybridization could be another source of nuclear genome instability. In the case of asymmetric hybridization, it is clearly necessary to manage tools to evaluate the contribution of the donor parent in the hybrid genome. It is therefore important to be able to analyse precisely the nuclear genome constitution for both symmetric and asymmetric hybridization. Tetraploid somatic hybrids are used as parents for sexual crosses for triploid scion or tetraploid rootstock breeding (Grosser *et al.*, 2000). Thus, the knowledge of meiotic behaviour of allotetraploid somatic hybrids appears essential to establish efficient breeding schemes.

Molecular tools currently used for somatic hybrids nuclear genome characterization

Several tools are currently used for somatic hybrid nuclear genome analysis. Schematically, two main classes of markers can be distinguished. Markers allowing a large random coverage of the genome such as RAPD (Forsberg *et al.*, 1998b), single sequence repeats (SSRs; Harding and Millam, 2000; Cheng *et al.* 2002), inter-simple sequence repeats (ISSRs; Scarano *et al.*, 2002) and amplified fragment length polymorphism (AFLP; Tian and Rose, 1999; Guo *et al.*, 2002) are very useful. Specific locus analysis allowing marking of each individual chromosome will be the better approach when addition or deletion lines are studied. For this purpose, RFLP (Rutgers *et al.*, 1997) or STMS analysis with mapped markers will be preferred.

None of these methods will allow the display of evidence of interspecific chromosome recombination. For such structural studies and global visualization of complex genomes, genomic *in situ* hybridization (GISH) has proved to be efficient for several crops including woody persimmon (Choi *et al.*, 2002). For example, this approach has been used to demonstrate that hexaploid hybrids arising from somatic hybridization between *Lycopersicon esculentum* (2*x*) and *Solanum lycopersicoïdes* (2*x*) contain two sets of *L. esculentum* with several chromosome rearrangements between the two genera (Escalante *et al.*, 1998). Specific elimination of *Allium cepa* in somatic hybrids between *A. cepa* and *Allium ampeloprasum* has also been demonstrated (Buiteveld *et al.*, 1998). GISH is also very powerful for studying somatic hybrid meiosis (Garriga-Calderé *et al.*, 1999; Gavrilenko *et al.*, 2001).

Potential of GISH for analysis of citrus somatic hybrid genomes

To determine the potential of GISH for citrus, Ollitraut *et al.* (2000d) have analysed three diploid species of *Citrus* (*C. medica* cv 'Poncire', *C. reticulata* cv 'Willow leaf' and *C. maxima* cv 'Pink'), *Poncirus trifoliata* cv 'Pomeroy' and *Fortunella japonica* cv 'Marumi'. GISH has also been applied on diploid sexual and tetraploid somatic hybrids between these genotypes. GISH did not allow colouring of all the set of chromosomes. In most cases, only five pairs from nine were coloured. Moreover, the staining appeared limited to the extremity of chromosomes and some central rDNA sites. The double staining of the *C. reticulata* + *P. trifoliata* intergeneric somatic hybrid, with genomic DNA of each parent marked with different colours, showed that genomic differentiation between the two parental species was sufficient to identify the stained chromosomes of each species (Fig. 10.1). In the case of *C. reticulata* + *F. japonica* as well as *C. reticulata* × *C. maxima* and *C. maxima* × *C. medica*, the differentiation of chromosomes appeared much more difficult and many spots displayed similar affinity with the genomic DNA probes of both parents. A better specificity of *in situ* hybridization at the interspecific level was found for the C. *reticulata* × *C. medica* hybrid.

In situ hybridization with 18S–25S rDNA has also been tested by Ollitrault *et al.* (2000d). They found that the number of chromosomes marked with the rDNA probe

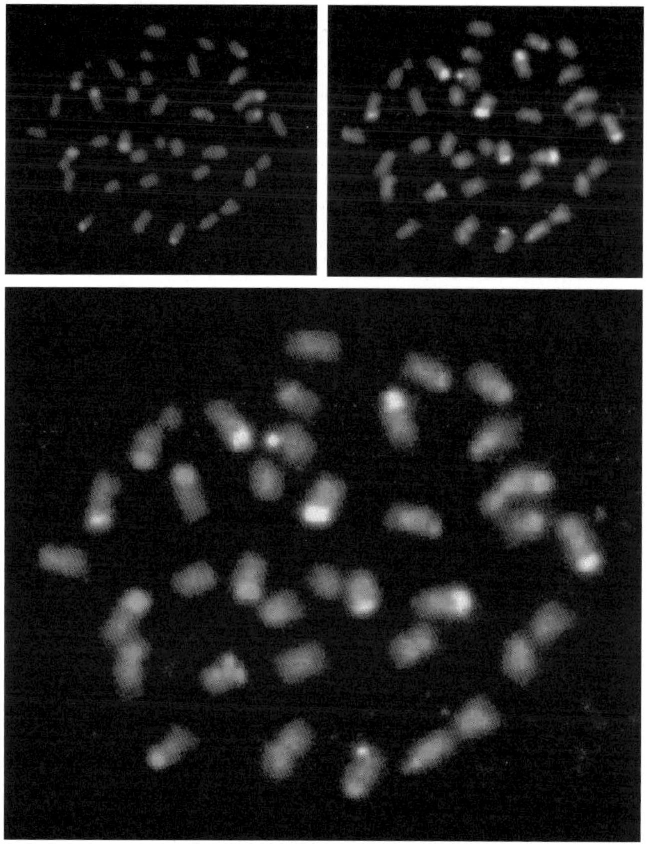

Fig. 10.1. Double *in situ* hybridization of an allotetraploid hybrid *C. deliciosa* + *P. trifoliata* with total genomic DNA of *C. reticulata* (green) *and P. trifoliata* (red), from Ollitrault *et al.* (2000d).

(pTA71) varied between two for *C. medica* and *C. limon* to six for *P. trifoliata* (Fig. 10.2). Three sites were found for *C. sinensis* and *C. aurantium*, and four for *F. japonica*. These results are similar to those of Roose *et al.* (1998), who found six major sites and one occasional minor site for *P. trifoliata*, and three strong and two minor sites for *C. sinensis*.

From these results, it appears that the incomplete staining of the chromosome set and the very partial coloration of each chromosome will limit the application of GISH for citrus somatic hybrid genome analysis. It could be used to identify the relative contribution of each parental species to multiploid intergeneric genomes. Such an application should also be found at the inter- and intrageneric levels, with rDNA probes if parental species present a different number of rDNA sites. The use of bacterial artificial chromosome (BAC) probes for *in situ* hybridization could perhaps be an interesting complementation of rDNA for chromosome distinction and then for the analysis of specific introgression or elimination in breeding schemes.

Nuclear genomic structure of citrus somatic hybrids

The majority of publications report on the symmetric addition of nuclear parental

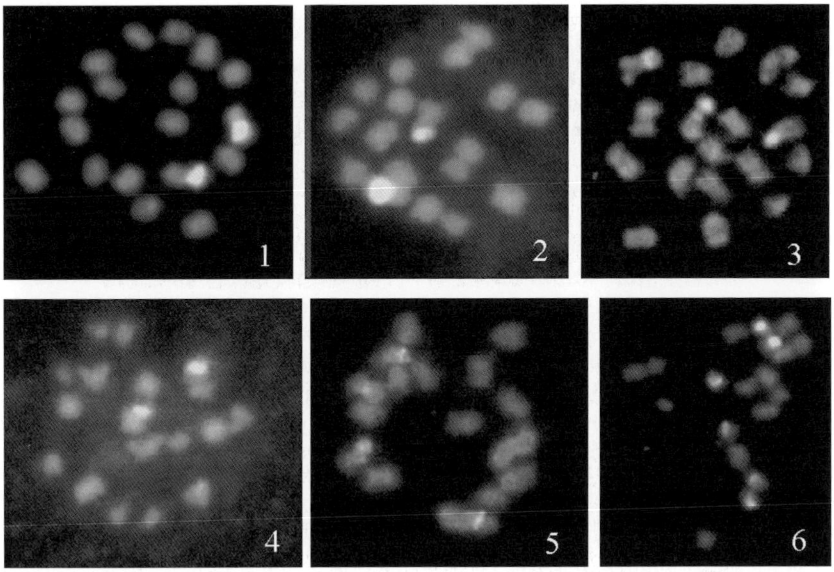

Fig. 10.2. *In situ* rDNA hybridization on chromosomes of six species of citrus (1, *C. medica*; 2, *C. lemon*; 3, *C. sinensis*; 4, *C. aurantium*; 5, *F. japonica*; 6, *P. trifoliata*); from Ollitrault *et al.* (2000d).

genomes in citrus somatic hybrids arising from polyethylene glycol (PEG) or electrically mediated protoplast fusions at the intrasubtribal level (Citrinae) (Grosser *et al.*, 1996b; Guo and Deng, 2001). However, unexpected ploidy levels are observed among somatic hybrids for some combinations. Triploid plants were obtained from *Severinia buxifolia* (2*x*) + *C. sinensis* (2*x*) (Grosser *et al.*, 1992b) as well as pentaploid hybrids from tetraploid *Fortunella hindsii* + diploid *P. trifoliate* (Miranda *et al.*, 1997). A somatic hybrid between *Fortunella crassifolia* cv 'Meiwa' and *C. sinensis* cv 'Valencia' with abnormal growth was proven to be a chimera containing non-tetraploid cells along with amphidiploids (Shi *et al.*, 1998a; Guo and Deng, 2001). Aneuploid cells and chromosomal variations were also observed in embryoids arising from *C. sinensis* + *C. reticulata* Blanco hybridization (Ye *et al.* 1992). These results suggest that chromosome elimination could occur in citrus somatic hybrids as has been observed for other crops (Sundberg and Glimelius, 1991; Kisaka *et al.*, 1997; Oberwalder *et al.*, 1998). At the intertribal level, Guo and Deng (1998) obtained symmetric tetraploid hybrids between *Citrus* and *Murraya*, while all the plants regenerated from *C. sinensis* (2*x*) and *Clausena lansium* (2*x*) were hexaploid (Guo and Deng, 1999). Froelicher (1999) also obtained unexpected polyploid hybrid embryoids from intersubtribal and intertribal combinations: hexaploids from *Triphasia trifolia* (2*x*) + *C. aurantifolia* (2*x*), and 10*x* and 11*x* from the 2*x* *C. aurantifolia* + 6*x* *Clausena excavata* androgenetic line. As discussed by Guo and Deng (2001), these high ploidy levels observed in wide somatic hybrids could be due to: (i) multifusion leading to more favourable genomic structures; (ii) ploidy variation at the level of parental embryogenic callus lines; and (iii) chromosome doubling of either parent or global doubling of the hybrid genome followed by specific chromosome eliminations.

Ollitrault *et al.* (1998) observed a greater ploidy level diversity on regenerated material from diploid + haploid somatic hybridization than that generally described for diploid + diploid combinations. Indeed, triploid, tetraploid and pentaploid hybrids were obtained from most of the $2x + x$ combinations. This result could be due to the ploidy instability in the 'haploid callus line'. Indeed, the presence of diploid and triploid cells in this callus has been demonstrated by flow cytometry analysis (Ollitrault *et al.*, 1998). Furthermore, the fact that Kobayashi *et al.* (1997) have obtained only diploids (with one of the nuclear parental genomes) or triploid somatic hybrids by combining protoplasts from diploid calli and haploid leaves confirms that triploid cells arising from haploid + diploid combinations do not present a specific nuclear genome instability during mitosis.

Cytoplasmic genome analysis

Somatic hybridization allows combining nuclear, chloroplastic and mitochondrial genomes in new patterns with no predefined rule, unlike sexual hybridization. Moreover, organelle genome recombination (mainly mitochondria) has been observed in several crops following somatic hybridization (Belliard *et al.*, 1979; Rothenberg *et al.*, 1985; Galun *et al.*, 1987; Kanno *et al.*, 1997). Thus, the organelle genomic constitution must be characterized for each regenerated plant. For citrus plants, cytoplasmic genome analysis is classically done by RFLP (Kobayashi *et al.*, 1991; Saito *et al.*, 1993, 1994; Yamamoto and Kobayashi, 1995; Grosser *et al.*, 1996a; Ollitrault *et al.*, 1996b; Moriguchi *et al.*, 1997; Moreira *et al.*, 2000a, b; Cabasson *et al.*, 2001). This method is powerful but not applicable to plants *in vitro* because it requires too much fresh leaf material. Thus PCR methods for cytoplasmic analysis are required for easier and earlier organelle analysis. Direct sequencing or restriction analysis of fragments amplified with universal primers defined by Demesure *et al.* (1995) for organelle DNA has been used for different plants (Gielly and Taberlet, 1994), and was applied to somatic hybrids of *Brassica* by Bastia *et al.* (2001).

CAPS analysis of citrus organelle genomes

Following Luro and Ollitrault (1996), Ollitrault *et al.* (2000c) have analysed the potential of cleaved amplified polymorphic sequences (CAPS) for differentiation of organelle genomes at the intrageneric level (*Citrus*) and intergeneric levels within the Aurantioideae subfamily. Fourty-four genotypes from 31 species were studied combining tagged PCR amplification with two mitochondrial and five chloroplast universal primers (Demesure *et al.*, 1995) revised for citrus by Lotfy *et al.* (2002) and restricted with 12 enzymes.

From this study, the CAPS technique appears more powerful for displaying polymorphisms for chloroplasts than for mitochondria. Polymorphisms have only been found at the intergeneric level for mitochondria, while intrageneric diversity was revealed for chloroplasts. The chloroplast genomes of all cultivated *Citrus* species can be distinguished, except *C. maxima*, *C. sinensis* and *C. paradisi* for one group, and *C. aurantium* and *C. limon* for a second group. This is in agreement with the generally accepted hypothesis that sweet orange and grapefruit originated from female pummelo interspecific hybridization. One combination of primer pair/enzyme (trnT3/trnD2/*Dra*I) allows the distinction of *C. lemon*/*C. aurantium* plastones from those of *C. maxima*/*C. sinensis*/*C. paradisi*. Such differentiation was not observed in the study of plastone restriction carried out by Green and Vardi (1986). Nicolosi *et al.* (2000) hypothesized, from chloroplastic CAPS and nuclear genome analysis, that *C. limon* was derived from hybridization between *C. aurantium* (female) and *C. medica*. In the same study, including a broad range of citrus species, a very close relationship was found between *C. micrantha* and *C. aurantifolia* chloroplasts, and they are very different from the other cultivated species.

CAPS analysis has been successfully applied on interspecific and intergeneric somatic hybrids for mitochondrial and chloropast genomes (Fig. 10.3, Lofty *et al.* (2002) and Fig.10.4). This method is much more simple, rapid and less expensive than traditional methods, and can be applied to small *in vitro* plants. Recently, new PCR markers for cytoplasmic genomes have been successfully developed: microsatellite markers for chloroplast (Lofty *et al.*, 2003; Cheng *et al.*, 2005) and SSCP markers for mitochondria (Olivares *et al.*, in press).

Cytoplamic genomes of citrus somatic hybrids

Genetic studies on the regenerated citrus somatic hybrids and cybrids after fusion of callus-derived protoplasts with leaf-derived protoplasts demonstrate the non-segregation of mitochondrial genomes (the mitochondrial genome from the embryogenic parent always prevails in cybrids, as well as complete somatic hybrids) and segregation of chloroplastic genomes (random segregation of the chloroplast genome in both cybrids and hybrids) (Kobayashi *et al.*, 1991; Saito *et al.*, 1993, 1994; Yamamoto and Kobayashi, 1995; Grosser *et al.*, 1996a; Ollitrault *et al.*, 1996b; Moriguchi *et al.*, 1997; Moreira *et al.*, 2000a, b; Cabasson *et al.*, 2001; Guo *et al.*, 2002). In the case of fusion between callus-derived protoplasts of the two parents, random segregation of the chloroplast and mitochondria is observed, allowing four different cytoplasmic constitutions with the same nuclear genome (Grosser *et al.*, 2000). Moreover, rearrangements of the cytoplasmic genomes that have often been observed in somatic hybrids and cybrids of many plant species were also reported in some citrus studies following asymmetric (Vardi *et al.*, 1987, 1989) and standard (Motomura and Hidaka, 1995; Moriguchi *et al.*, 1997; Cheng *et al.*, 2002) somatic hybridizations. Moreira *et al.* (2000a, b) observed non-parental mitochondrial fragments in several somatic hybrids and cybrids. Moreira *et al.* (2000b) also made a unique observation of the addition of choroplastic parental genomes on all the 14 'Succari' + *Citropsis gilletiana* somatic hybrid plants analysed.

The sole presence of the mitochondrial genome from the embryogenic parent in all regenerated cybrids and somatic hybrids suggests a critical role for these organelles in plant regeneration via somatic embryogenesis (Kobayashi *et al.*, 1991; Saito *et al.*, 1993; Grosser *et al.*, 1996a; Moriguchi *et al.*, 1996; Ollitrault *et al.*, 1996b; Moreira *et al.*, 2000a, b; Cabasson *et al.*, 2001). Moreira *et al.*

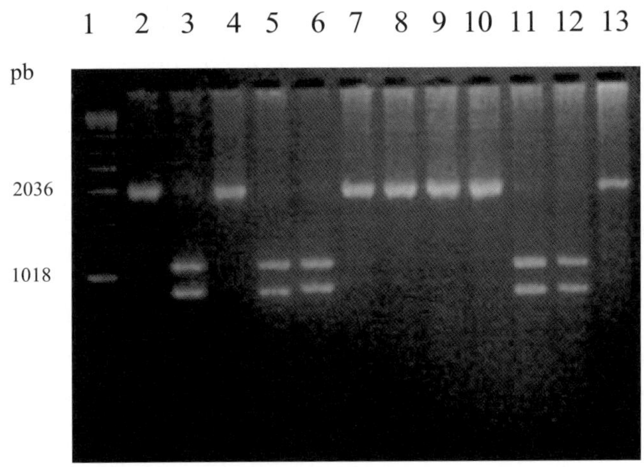

Fig. 10.3. CAPS analysis of chloroplast segregation in a population of somatic hybrids between 'Star Ruby' grapefruit and 'Willow Leaf' mandarin (from Ollitrault *et al.*, 2000a). Products generated after amplification and restriction using trnT3/trnD2 primers and *Dra*I restriction enzyme are from somatic hybrids (lanes 2–11), 'Willow Leaf' mandarin (lane 12) and 'Star Ruby' grapefruit (lane 13). Lane 1 = molecular size marker.

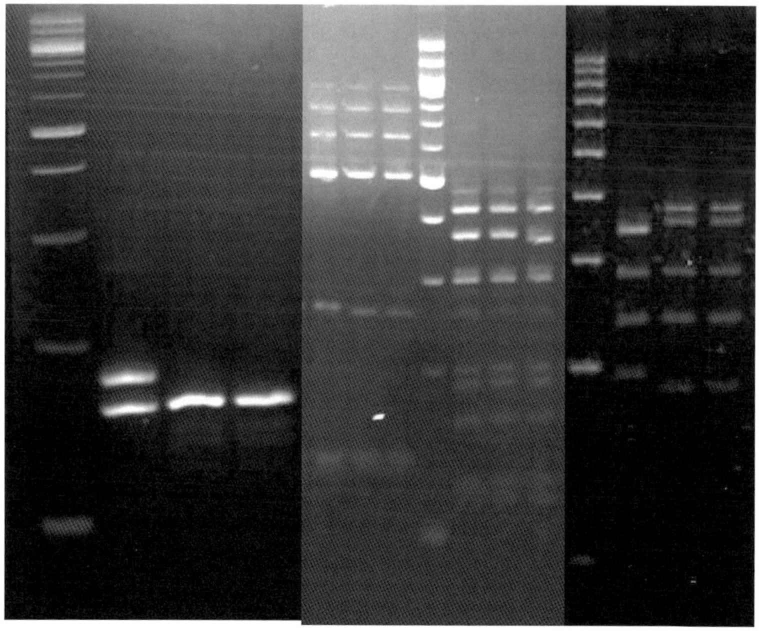

Fig. 10.4. Molecular analysis of G1 satsuma + HB pummelo diploid cybrid plant. Genotype identification per lane: 1, 8 and 12, 100 bp DNA ladder; 2, 7, 9 and 13: satsuma mandarin cv. Guoqing No 1 (G1); 3, 6, 10 and 14: G1 + HB pummelo diploid regenerant; 4, 5, 11 and 15: HB pummelo. Lanes 1–4: SSR analysis by primer pair TAA15 confirmed that the nuclear genome was derived from HB pummelo. Lanes 5–11: CAPS analysis of the mitochondria genome by primer pairs 18S rRNA/5S rRNA and Nad4exon1/Nad4exon2, respectively, and cut by the enzyme *Taq*I, which showed that the mtDNA was derived from G1. Lanes 12–15: CAPS analysis of the chloroplast genome by primer pair TrnD/TrnT and cut by the enzyme *Taq*I, which showed that the chloroplast DNA was derived from HB pummelo. Note: a 3% Metaphore agarose gel was used to separate the small size DNA fragments.

(2000a, b) hypothesized that this may be a quantitative effect and that only cultured cells have adequate quantities of mitochondria to provide the necessary energy for somatic embryogenesis. They also verified that the number of mitochondria per embryogenic culture cell was significantly higher than per leaf cells.

New Avenues for Somatic Hybridization in Citrus

Haploid + diploid protoplast fusions and gametosomatic hybridizations

Triploid breeding is an important method in citrus variety improvement, mainly for small citrus fruit and lemons. For all the strategies involving $2n$ gametes, interploid sexual crosses or endosperm culture, both of the parental genomes are submitted to meiotic recombination. Maximum heterozygosity in triploid progeny will be obtained from interploid crosses involving an allotetraploid parent. Concerning triploids arising in diploid sexual crosses, it appears that only a part of parental heterozygosity is present in the $2n$ gametes producing the spontaneous triploids (Ollitrault *et al.*, 1998). Considering that cultivars are generally highly heterozygous, it is clear that in most cases the selected genetic balances of the diploid parental cultivars will be lost in triploid hybrids. So, the effectiveness of selection made at the diploid level for complex characters is low, and it is necessary to make final selections from a large number of triploid hybrids. The production of allotriploids from somatic hybridizations between haploid potato and diploid tomato has been previously described (Schoenmakers *et al.*, 1991), while triploid somatic hybrid plants of *Nicotiana* and *Petunia* have been obtained by gametic + somatic ($n + 2n$) protoplast fusions (Pirrie and Power, 1986; Lee and Power, 1988). These methods allow the synthesis of triploids in one cycle of hybridization, and provide an interesting alternative to diploid × tetraploid sexual crosses or spontaneous triploid selection. Indeed it is the only method allowing the addition of a haploid genome to the whole genome of a diploid cultivar without recombination.

Triploid citrus hybrids have been synthesized by somatic hybridization between diploid cultivars and haploid lines (Kobayashi *et al.*, 1997; Ollitrault *et al.*, 1997, 1998). Indeed, the regeneration of haploid plants and cell lines by anther culture (Germaná, 1992) or by induced gynogenesis (Oiyama and Kobayashi, 1993; Ollitrault *et al.*, 1996a) has opened up avenues for this new breeding scheme. Ollitrault *et al.* (1998, 2000b) have combined two haploid cell lines of clementine obtained by induced gynogenesis with 11 diploid cultivars. Triploid and tetraploid hybrids have been obtained for each combination as well as a few pentaploid hybrids. Triploid hybrids should be exploited directly and will soon be evaluated in tropical, subtropical and Mediterranean areas. Tetraploid hybrids obtained by this approach will join the pool of allotetraploids for further diploid × tetraploid sexual crosses.

Citrus is the first example of triploid hybrids obtained by somatic hybridization in fruit crops. The main limitation of this strategy is the lack of haploid lines. This could be overcome by the application of gameto-somatic hybridization mentioned by Z. A. Deng *et al.* (1992) and X.X. Deng *et al.* (1995), who have regenerated only chimeric plants with 18 and 19 chromosomes. This technique is currently being developed by CIRAD (France), and will hopefully allow the production of polymorph triploid progeny recombining only for the haploid source. If successful, this would provide a new avenue for citrus genetics and breeding research.

Asymmetric nuclear hybridization

For numerous crops, somatic hybridization is a way to by-pass sexual incompatibility between species. In this case, it is of interest to transfer only a limited amount of genetic material from a foreign species to a crop rather than to combine two complete genomes (Forsberg *et al.*, 1998b). The difficulty in obtaining viable somatic hybrids between *Citrus* and very distant species suggests that this strategy should be interesting for introgression of resistance characters from wild germplasm, such as resistance of *Murraya paniculata* against Huanglongbing. Moreover the diversification and breeding of species such as sweet orange or grapefruit (that are not amenable to conventional breeding) would probably benefit greatly from such a strategy, even for partial transfer of the genome of compatible species. Two main methods can be distinguished for partial nuclear genome transfer leading to the development of asymmetric hybrids: chromosome fragmentation before somatic hybridization and microprotoplast fusion.

Transfer of fragmented chromosomes

The most common method used for asymmetric hybridization has been to expose the donor protoplasts to γ irradiation (Bates *et al.*, 1987) or X irradiation (Dudits *et al.*, 1980). Correlations between irradiation dose and the degree of chromosome elimination have been demonstrated in some studies (Melzer and O'Connell, 1992; Schoenmakers *et al.*, 1994). More recently,

UV irradiation has also been proven to be efficient to fragment and eliminate chromosomes (Jazdzewska *et al.*, 1995; Forsberg *et al.*, 1998a, b). The transfer of fragmented chromosomes demonstrates its efficiency when it was combined with *in vitro* selection pressure. For instance, it has allowed transfer of resistances to methotrexate and 5-methyltryptophan from carrot to tobacco (Dudits *et al.*, 1987), resistance to *Phoma linguam* from *Brassica juncea*, *Brassica carinata* and *Brassica nigra* to *Brassica napus* (Sjödin and Glimenius, 1989) or resistance to tobacco mosaic virus from *Nicotiana repanda* to *Nicotiana tabacum* (Bates, 1990). More recently, the use of fluorescence-activated cell sorting (FACS) has been proven to be efficient to select asymmetric hybrids just after fusion (Rasmussen *et al.*, 1997). The proportion of foreign genome transfer is usually evaluated with RFLP (Forsberg *et al.*, 1998b), RAPD (Rasmussen *et al.*, 1997) or AFLP (Tian and Rose, 1999).

This technique has not been explored in citrus. It should find application to breed tolerant cultivars when *in vitro* selection is possible both for pathogens such as Mal Secco by application of toxin in cell cultures (Gentile, 1992) and abiotic stresses such as salinity (Spiegel Roy and Ben Hayvin, 1985). This should also be a way to induce genetic diversity in very monomorphic species such as sweet orange and grapefruits. However, the uncontrolled transfer of multiple chromosome fragments would require a very strong effort of selection in the field as well as the development of molecular markers to select specific traits in large populations.

Chromosome transfer by microprotoplast technique

The microprotoplast-mediated chromosome transfer (MMCT) method was originally developed for mammalian cells. Micronuclear induction in plant by herbicides (amiprosphosmethyl, oryzalin) was demonstrated in *Solanum tuberosum*, *Daucus carota*, *Nicotiana plumbaginifolia* and *Helianthus* sp. (Morejohn *et al.*, 1987; Verhoeven *et al.*, 1990; Ramulu *et al.*, 1994; Binsfeld *et al.*, 2000). Successful induction of micronuclei and microprotoplast isolations for asymmetric hybridization have allowed the transfer of a single potato chromosome to tomato and tobacco (Ramulu *et al.*, 1996a, b), and addition lines in sunflower (Binsfeld *et al.*, 2000).

For citrus, Louzada *et al.* (2002) have developed a method to produce citrus microprotoplasts containing a limited number of chromosomes, and used them in protoplast fusions with diploid lines. If successful, this method should have very interesting applications both for genetic study with monosomic addition line and for breeding. Indeed, it is possible that the addition of a single chromosome from wild species may be more efficient to obtain viable and interesting cultivars than symmetric hybridization. Moreover, the preservation of pomological and organoleptic specific traits of species such as sweet orange and grapefruit should be more efficient by adding single chromosomes of the *Citrus* gene pool for improving disease tolerance/resistance.

Somatic cybridization

Production of diploid somatic hybrid plants containing the nuclear genome of one parent and either the cytoplasmic genome of the other parent or a combination of both parents (cybrids) has been a common approach in plant improvement. Characterization of such cybrids can determine cytoplasm inheritance and cytoplasm-coded agronomic traits, and may lead to improved selections (Kumar and Cocking, 1987). Male sterility and tolerance traits with cytoplasmic determinism have been the more common objective of such strategies (Pelletier *et al.*, 1983; Barsby *et al.*, 1987; Thomzik and Hain, 1988; Varotto *et al.*, 2001). Classically, cybrid and alloplasmic plants are obtained by asymmetric protoplast fusions between irradiated donor protoplasts (with destroyed nuclei) and recipient protoplasts whose mitochondria and chloroplasts have been inactived by iodoacetate treatments (Sidorov *et al.*, 1981; Galun *et al.*, 1987; Vardi *et al.*, 1987; Li *et al.*, 1993). Cytoplasmic genome recombination should occur and concern principally the mitochondrial genome (Belliard *et al.*, 1979; Rothenberg *et al.*, 1985; D'Hont *et al.*, 1987) while reports of new chloroplast profiles are very rare (Medgyest *et al.*, 1985).

Besides the success achieved in symmetric citrus somatic hybridization worldwide, only a few reports can be found pursuing the production of citrus cybrid plants by asymmetric somatic hybridization (Vardi *et al.*, 1987, 1989; Li and Deng, 1997; Liu and Deng, 2002). The challenging methodology involved with the original donor–recipient method and the lack of information concerning cytoplasmic traits initially slowed the evolution of cybridization in citrus. However, chance and nature have played an important role in this field, allowing the regeneration of several citrus alloplasmic plants as a by-product from the application of standard somatic hybridization procedures (Kobayashi *et al.*, 1988; Ohgawara *et al.*, 1989, 1991; Tusa *et al.*, 1990; Saito *et al.*, 1993, 1994; Yamamoto and Kobayashi, 1995; Grosser *et al.*, 1996a; Moreira *et al.*, 2000a; Moriguchi *et al.*, 1996, 1997; Ollitrault *et al.*, 1996b, 2000b; Cabasson *et al.*, 2001).

Potential seedlessness via cybridization

As mentioned, seedlessness is a prerequisite for new fresh market citrus cultivars. Seedlessness in diploid citrus generally relates to male and/or female sterility. The seedless satsuma mandarin is typically male sterile, and its male sterility has been identified to be a cytoplasmic male-sterile (CMS) type (Yamamoto *et al.*, 1997). CMS in higher plants is known to be controlled by mitochondrial DNA (mtDNA). Navel orange is also male sterile, accompanied by partial ovule sterility, and it has not been determined whether its sterility is the CMS type. Due to complex citrus biology, it is not easy to transfer the sterility character from satsuma mandarin to other seedy citrus cultivars by conventional breeding. However, it may be possible to transfer the CMS trait of satsuma (and possibly navel orange) into commercially important diploid cultivars via cybridization, and efforts are underway to achieve this.

For citrus somatic hybridization, the fusion model of 'diploid embryogenic protoplasts + diploid leaf-derived protoplasts' has been used extensively. Normally, unfused leaf protoplasts do not divide and regenerate into plants. However, diploid plants resembling the leaf parent morphologically have been recovered unexpectedly from more than 30 symmetrical fusion combinations (Deng *et al.*, 2000; Grosser *et al.*, 2000). In all such cases examined, RFLP analysis indicated that these plants were not directly regenerated from unfused leaf protoplasts, but were cybrids with the nuclear DNA from the leaf parent and the mtDNA from the corresponding embryogenic parent (Saito *et al.*, 1993; Yamamoto and Kobayashi, 1995; Grosser *et al.*, 1996a; Moriguchi *et al.*, 1996; Moreira *et al.*, 2000a; Cabasson *et al.*, 2001; Guo *et al.*, 2002, Wuhan, PR China, unpublished data). Chloroplast DNA was randomly inherited in these plants. Moreira *et al.* (2000a) theorized that leaf protoplasts do not have an adequate quantity of mitochondria as needed to undergo somatic embryogenesis, and that cybridization fulfils this need. These results suggest that it should be possible to transfer the mtDNA from male-sterile cultivars to seedy diploid fresh fruit cultivars by simply conducting symmetric fusion experiments, and the research groups directed by X.X. Deng and J.W. Grosser are collaborating to achieve this. Using an embryogenic suspension culture of 'Guoqing No. 1' satsuma mandarin provided by X.X. Deng, the Grosser laboratory has produced diploid cybrid plants of 'Hirado Buntan Pink' pummelo, 'Sunburst' mandarin and an unnamed 'Clementine' × 'Murcott' hybrid (Guo *et al.*, 2004a).

Confirmation of the cybridity of the latter two combinations has been difficult due to the close relatedness of the parental mitochondrial genomes. These plants will be fruited and flowered as soon as possible to determine if the substitution of satsuma mtDNA can result in a new mitochondria–nucleus interaction that could result in making these cultivars seedless without otherwise altering their cultivar integrity. If successful, this strategy could be applied to remove seed from many superior diploid cultivars. With the same objective, Xu *et al.* (2006) have developed a very promising new technique of cytoplast isolation and fusion to generate citrus hybrids.

Breeding agronomic traits via cybridization

With the exception of potential seedlessness, the agronomic value of citrus cybrids is currently unknown. The evaluation of citrus cybrids in the field will allow characterization of agronomic traits encoded by the cytoplasmic genome. Tusa *et al.* (2000) suggested from cybrid evaluation that specific mechanisms of resistance against Mal secco could be activated in these genotypes. Mandarin and sweet orange cybrids in the field at the CREC (Florida) are showing significant variation in agronomically important traits including fruit maturity date and seed content. In the same way, Fanciullino *et al.* (2005) found significant quantitative variation of aromatic compounds in cybrids indicating that cybridization is a potential source of genetic variation for citrus cultivar improvement. As an additional interest, the development of citrus cybrid callus should increase the range of somatic hybridization parents (Saito *et al.*, 1994; Grosser *et al.*, 1996a).

Scion improvement: potential tetraploid somatic hybrid cultivars

Several allotetraploid somatic citrus hybrids produced for the purpose of serving as tetraploid breeding parents in interploid crosses to generate seedless triploids have flowered and produced fruit. Although still emerging through juvenility, a few somatic hybrids are producing fruit with cultivar potential. A somatic hybrid of 'Succari' sweet orange + 'Page' tangelo produces seedless fruit with a shape similar to that of sweet orange. Fruits of this hybrid are very early maturing and have had sugar/acid ratios over 15 by the beginning of October of the past two seasons. These fruits also have adequate juice content and excellent flavour. This hybrid therefore has potential as an early season fresh fruit cultivar. Another somatic hybrid of 'Valencia' sweet orange + 'Murcott' tangor produces nearly seedless fruits (about one seed every three fruit). Fruits of this hybrid are very much intermediate to that of the parents, possessing excellent flavour, and are peelable. This hybrid has potential as a mid–late season fresh fruit cultivar. It is possible that some somatic hybrids are seedless because of mutations that accumulate in the embryogenic callus lines used as parents. These results suggest potential for designing fusion experiments to generate hybrids with cultivar potential at the tetraploid level, and CREC have conducted numerous fusion experiments to achieve this. CREC has focused on using parents that have been successful in previous work, or complementary parents that provide low acidity, high juice content and/or excellent fruit quality (i.e. colour and flavour). Of course any resulting hybrids would also have great potential as tetraploid breeding parents. Recent successes of CREC in the endeavour are summarized in Table 10.1. (Guo *et al.*, 2004b). As with many other recent somatic fusion experiments, flow cytometry (using a Partec table-top model) was used to identify tetraploid embryos early on in the plant regeneration process to save time and money. These hybrids will all be top-worked to mature field trees to expedite flowering and fruiting as needed to accelerate the evaluation process.

Table 10.1. New allotetraploid citrus somatic hybrids with direct cultivar potential produced at the CREC.

Embryogenic callus parent	+	Leaf parent
'Guoquing' Satsuma No. 1	+	'Murcott' tangor
'Page' tangelo	+	LB8-9 (Clementine × Minneola)
'Page' tangelo	+	Lee hybrid (Clementine × Murcott)
'Page' tangelo	+	unnamed (Clementine × Satsuma)
'Page' tangelo	+	'Ortanique' tangor
'Page' tangelo	+	'Murcott' tangor
'Murcott' tangor	+	'Dancy' tangerine

Rootstock improvement: building a better sour orange

Sour orange was formerly the most important rootstock worldwide due to its wide adaptation, tolerance to citrus blight and ability to produce good yields of high quality fruit. However, due to its susceptibility to citrus tristeza virus (CTV)-induced quick decline disease, it can no longer be used in most situations. A suitable replacement rootstock has yet to be developed. Molecular marker analyses indicate that sour orange is a hybrid of pummelo and mandarin (Nicolosi *et al.*, 2000), but it is unlikely that it was produced from the best pummelo or mandarin. This report also demonstrates that the vigorous rootstocks including rough lemon, rangpur, Palestine sweet lime and Volkamer lemon have a significant genetic contribution from citron. Such vigorous rootstocks are also highly susceptible to citrus blight, and generally produce fruit of poor quality. Perhaps avoiding citron-based material in building new rootstocks could minimize these problems. CREC therefore use somatic hybridization to combine widely adapted mandarins (Amblycarpa and Shekwasha mandarins) with tristeza-resistant pummelos or superior pummelo seedlings selected following germination of seed in flats of winder soil (calcareous soil, pH = 5.8) inoculated with both *Phytophthora nicotianae* and *P. palmivora*. (Note: a successful replacement for sour orange must also be able to handle challenging soils and *Phytophthora*.) To date, ten such mandarin + pummelo somatic hybrids have been produced, and several are showing excellent nursery vigour (Table 10.2) (Grosser *et al.*, 2004). These promising hybrids and others produced subsequently will be immediately tested for resistance to tristeza-induced quick decline disease and entered into field trials. It should be possible to develop a quick decline-resistant replacement for sour orange that also provides some level of tree size control.

Rootstock breeding at the tetraploid level: production of 'tetrazygs'

The breeding of somatic hybrids at the tetraploid level provides an opportunity to mix the genes from three to four proven rootstocks (or other valuable contributing parents) simultaneously, thereby maximizing genetic diversity in progeny. CREC has used two somatic hybrids, Nova mandarin + Hirado buntan pummelo (zygotic) and sour orange + rangpur, that are performing well in field trials and produce high percentages of zygotic seed, as females in such crosses at the tetraploid level. Selected high performance somatic hybrids including sour orange + Carrizo, Cleopatra + trifoliate orange and sour orange + Palestine sweet lime are being used as pollen parents. Grosser *et al.* (2003) have coined the term 'tetrazygs' to identify tetraploid progeny from such crosses. They routinely screen progeny from such crosses by germinating

Table 10.2. New mandarin + pummelo somatic hybrids produced recently at the CREC – efforts to rebuild a better 'sour orange' rootstock.

Embryogenic parent	+	Leaf parent
'Murcott' tangor	+	*Citrus grandis* 'Hirado Buntan Pink' (HBP)[a]
'Murcott'	+	HBP sdl-JL1[a]
Amblycarpa mandarin (*C. amblycarpa*)	+	HBP[a]
Amblycarpa	+	*C. grandis* 'Chandler'[a]
Amblycarpa	+	HBP sdl-JL1
Amblycarpa	+	HBP sdl-5-1-99-1B[a]
Amblycarpa	+	HBP sdl-JL2B[a]
Amblycarpa	+	*C. grandis* 'LingPingYau' sdl-8-1-99-4A[a]
Amblycarpa	+	HBP sdl-JL4
Amblycarpa	+	Chandler sdl-A1-11
Shekwasha mandarin	+	HBP[a]
Shekwasha	+	'Chandler'[a]
Amblycarpa	+	HBP sdl-JL12

[a]Hybrid confirmed by RAPD analyses

seed directly in a high pH, calcareous 'winder' soil inoculated with *Phytophthora*. Superior 'tetrazyg' hybrids are then grafted with sweet orange infected with a quick decline isolate of CTV to determine their resistance to CTV-induced quick decline disease. Simultaneously, the selected hybrids can be propagated by topworking and/or rooted cuttings to provide clonal material for further evaluation. This approach is expected to shorten the time required to develop a new rootstock by 8–10 years. This programme began at the CREC in 1999, and so far more than 300 genetically diverse 'tetrazyg' hybrids have been selected for further evaluation. 'Tetrazygs' are also being screened for tolerance to salinity and to the *Diaprepes* root weevil–*Phytophthora* complex (Grosser *et al.*, 2003).

Protocols

Protocol for PEG-induced fusion and plant regeneration (CREC)

Citrus somatic hybrids are most often produced from the fusion of protoplasts isolated from embryogenic callus or suspension cultures of one parent with leaf-derived protoplasts of the second parent. At least one parent in any fusion combination must be embryogenic to provide the capacity for plant regeneration in the somatic hybrid progeny. This protocol has been used successfully to produce somatic hybrid plants from more than 120 parental combinations at the University of Florida's Citrus Research and Education Center (CREC), Lake Alfred, Florida, USA.

Protoplast isolation: embryogenic parent

Embryogenic callus or suspension cultures are generally maintained on either EME (MT basal medium containing 0.5 g/l malt extract (Murashige and Tucker, 1969)) or H+H (modified MT basal medium containing 1.55 g/l glutamine, 0.5 g/l malt extract and 50% less KNO_3 and NH_4NO_3). Cultured cells used for protoplast isolation should be in the log phase of growth. For protoplast isolation, transfer 1–2 g of friable callus tissue into a 60 × 15 mm Petri dish (for suspension, transfer ~2 ml of suspension with a wide-mouth pipette and drain off the liquid using a Pasteur pipette). Resuspend the cells in 2.5 ml of 0.7 M BH3 medium (Grosser and Gmitter, 1990) (BH3 is a nitrate-free MT basal medium containing

3.1 g/l glutamine, 20 ml/l coconut water and the Kao and Michayluk (1974) organic addenda), and then add 1.5 ml of filter-sterilized enzyme solution containing 0.7 M mannitol, 12.0 mM $CaCl_2$, 6.0 mM MES buffer, 1.4 mM NaH_2PO_4, 1% Onozuka RS cellulase, 1% macerase, 0.2% pectolyase Y-23, pH 5.6. Seal dishes with Nescofilm (or Parafilm) and incubate overnight on a rotary shaker at 20 r.p.m., in either low light or darkness.

Leaf parent

The best results are generally obtained using fully expanded leaves of new flush that have not fully hardened, taken from seedlings or recently budded plants maintained in a growth chamber or heavily shaded greenhouse. Use of *in vitro* grown leaf material precludes the need for decontamination prior to isolation. Leaf material is decontaminated by immersion in 1 N HCl for a few seconds followed by a 12–15 min immersion in 10–15% commercial bleach containing three drops of Liquinox soap or other surfactant, followed by a 5 min rinse and two 10 min rinses in double-distilled H_2O. Damaged vascular tissue and the midvein region are removed with a sharp scalpel. Remaining leaf material is feathered or cut into thin strips with a sharp scalpel and incubated in 3 ml of enzyme solution combined with 8 ml of 0.7 M BH3 medium in a 125 ml side-armed Erlenmeyer flask (with the side arm covered with Miracloth to prevent contamination). Leaf material in the enzyme cocktail is then evacuated for 15 min at 50 kPa to facilitate enzyme infiltration. Preparations are then sealed and incubated as above.

Protoplast purification (using a sucrose/mannitol gradient)

Following incubation, preparations are passed through a 45 μm stainless steel (or nylon) mesh screen to remove undigested cell clumps and debris. Protoplast-containing filtrates are then centrifuged for 4–10 min at 100 *g* in 15 ml calibrated screw-top centrifuge tubes. The supernatant is removed with a Pasteur pipette, and the pellet containing the protoplasts is gently resuspended in 5 ml of a 25% sucrose solution containing CPW nutrients (27.2 mg/l KH_2PO_4, 100 mg/l KNO_3, 150 mg/l $CaCl_2$, 250 mg/l $MgSO_4$, 2.5 mg/l $Fe_2(SO_4)_3 \cdot 6H_2O$, 0.16 mg/l KI, 0.00025 $CuSO_4$, pH 5.8) (Frearson *et al.*, 1973). This is followed by slowly pipetting 2 ml of a 13% mannitol solution (containing CPW salts) directly on top of the sucrose layer (avoid mixing). Centrifuge for 6 min at 100 *g*. Viable protoplasts form a band at the interface between the sucrose and the mannitol. Carefully remove the protoplasts from this interface and resuspend them in the appropriate volume of BH3 medium as necessary for further manipulation.

PEG-induced protoplast fusion

This protocol is simple, efficient, inexpensive and non-toxic to citrus protoplasts. Mix approximately equal volumes of purified protoplasts from each parental source in 0.6 M BH3 medium and centrifuge for 4 min at 100 *g*. Resuspend the pellet of mixed protoplasts in a volume of BH3 medium equal to 4–20 times the volume of the original pellet (10× is recommended for initial experiments, followed by fine tuning). Pipette two drops of the resuspended mixture into 60 × 15 plastic Petri dishes (the number of dishes determined by the volume of mixed protoplasts). Immediately add two drops of PEG solution (40% polyethylene glycol 8000, 0.3 M glucose and 66 mM $CaCl_2$ at pH 6.0) to each fusion dish and incubate for 8 min. Note that the PEG solution rapidly acidifies over time – the pH should be checked prior to use. Add two drops of A + B solution (9:1 (v/v), A = 0.4 M glucose, 66 mM $CaCl_2$ and 10% dimethylsulphoxide at pH 6.0; and B = 0.3 M glycine at pH 0.5 using KOH pellets) to each fusion dish. To avoid precipitation, the A + B solution should be mixed just prior to use. Following another incubation of 12 min, add 12–15 drops of BH3 medium around the periphery of the fusing protoplasts. After incubating for 5 min, carefully remove the PEG plus [A + B] solution with a Pasteur pipette and replace it with 15 drops of BH3 medium. After incubating for another 10 min, remove the BH3 medium with a Pasteur pipette and replace it with 12–15 drops of fresh BH3 medium. Repeat this washing step twice more, carefully avoiding the loss of protoplasts. After the final wash, protoplasts are cultured directly in the fusion Petri dish in either a shallow pool (8–12 drops of medium) or thin-layer culture (1.5 ml of medium) in either BH3 medium, EMEP medium or a 1:1 (v/v) mixture of BH3 and EMEP (Grosser and Gmitter, 1990). Seal plates with Nescofilm, and culture in either darkness or low light.

This protocol is not only useful for generating somatic hybrids, but also frequently produces cybrids as a by-product (as does the electofusion protocol below). The cybrids generally contain the nucleus of the leaf parent, the mitochondrial genome of the callus parent, and the chloroplast genome is randomly inherited.

Protoplast culture and plant regeneration

Following incubation for 4–6 weeks, cultures can be supplemented with medium containing reduced osmoticum, which is accomplished by adding 10–12 drops of a 1:2 (v/v) mixture of 0.6 M BH3 medium and 0.38 M EME medium (MT basal containing 125 g/l sucrose and 0.5 g/l malt extract). After another two weeks, cultures can be transferred to solid medium in 100 × 20 Petri dishes with further osmoticum reduction as follows: add 2 ml of a 1:2 (v/v) mixture of BH3 medium and 0.15 M EME medium (MT basal containing 50 g/l sucrose and 0.5 g/l malt extract) to each fusion dish and pour the entire contents on to solid medium plates containing standard agar-solidified EME medium. The liquid medium containing the protoplast-derived colonies should be spread evenly over the entire plate. Vigorously growing cultures may require dilution in order to achieve somatic embryo induction. Cultures recalcitrant to embryo induction can also be transferred to EME medium containing 50 g/l maltose instead of sucrose as the carbohydrate source. Recovered somatic embryos are enlarged and germinated using any standard citrus somatic embryogenesis media sequence (Grosser and Gmitter, 1990). Niedz *et al.* (2002) have recently shown that the use of certain semi-permeable membranes (i.e. cellulose acetate) can positively affect the normalizing of citrus embryogenesis, thereby improving the efficiency of plant recovery. Abnormal embryos that fail to germinate can be dissected into large sections and cultured on DBA3 medium (X.X. Deng *et al.*, 1992) for shoot induction. Resulting shoots can be rooted on RMAN medium (Grosser and Gmitter, 1990). Rooted plants can be transferred to any suitable commercial potting mixture and maintained under cover at high humidity for 2–3 weeks for acclimatization.

Protocol for protoplast electrofusion and plant regeneration

Suspension and leaf protoplasts were isolated and purified as described by Grosser and Gmitter (1990). The purified protoplast bands are transferred to new centrifugation tubes, and then washed twice by centrifugation at 100 *g* for 10 min in electrofusion solution containing 0.6 M mannitol and 0.25 mM $CaCl_2$, pH 5.8. Fusions are conducted using an SSH-2 instrument (Shimadzu Somatic Hybridizer-2, Japan). The electrofusion chamber is the FTC-03 with 0.8 ml volume. A protoplast mixture (0.8 ml) containing 3–5 × 10^5 callus protoplasts and 10–15 × 10^5 mesophyll protoplasts was transferred to the FTC-03 chamber, sealed by parafilm and kept still for 5 min before fusion treatment. The electrical fusion parameters were carefully determined prior to fusion (Guo *et al.*, 1998). For most fusion combinations, the following parameters can work well: AC (alternate current) field, 60 s, 125 V/cm; DC (direct current) field, 1250 V/cm, 30 μs in duration, five times at 0.5 s intervals; final time, 5 s. For some fusion combination such as red tangerine + trifoliate orange, where the protoplast size of the latter was only 1/3–1/5 the size of the former, altered parameters (i.e. AC field, 5 s, 200 V/cm; then AC field, 30 s, 100 V/cm; DC field, 1250 V/cm, 30 μs in duration, five times at 0.5 s intervals; final time, 5 s) are required, which not only resulted in good formation of protoplast pearl chains, and a 5–6% binuclear heterokaryon rate, but also circumvented the negative effect of the long duration time of high AC field strength on the viability of protoplasts (Guo *et al.*, 2002). After fusion treatment, the protoplast mixtures are incubated for at least 10 min before being transferred to 10 ml centrifuge tubes, and then centrifuged at 100 *g* for 4 min. This results in an optimum rate of binuclear heterokaryons. The supernatant is then discarded, and the fusion products are resuspended at a density of $1–2 \times 10^5$ cells/ml in BH3 medium (Grosser and Gmitter, 1990) by liquid thin-layer culture. The subsequent protoplast culture and plant regeneration procedures are the same as mentioned above.

Protocol for *in situ* hybridization in citrus (CIRAD, from D'Hont *et al.*, 1996, 1998)

Small fragments (1 mm^2) of leaf or root mitotic areas are digested in pectinase + cellulase solution before performing chromosome squashes. For GISH, total genomic DNA of each species is used as a probe. DNA probes of approximately 300 bp are marked by nick translation of dNTPs linked with two kinds of molecule (biotin or digoxigenin) that should be revealed by two florochromes with different emission colours. For rDNA study, the probe pTa71 (18S–25S rDNA) marked with digoxigenin is used. The chromosome squashes are hybridized with probes after a step of denaturation. Then the biotin probes are revealed by avidin marked with Texas red (red coloration), while digoxigenin probes are revealed with anti-digoxigenin antibodies marked with fluorescein isothiocyanate (FITC; green coloration). After an amplification step, a counterstaining of chromosomes with 4',6-diamino-2-phenylindole (DAPI) is performed. See D'Hont *et al.* (1996, 1998) for detailed protocols.

Protocol for CAPS analysis of citrus mitochondrial and chloroplastic genomes (CIRAD, from Lotfy *et al.*, 2002)

DNA amplification

The 13 pairs of universal cytoplasmic primers are described in Demesure *et al.* (1995). Some of them have been modified to obtain better amplification with citrus (Lotfy *et al.*, 2002). The primers and PCR conditions selected are described in Table 10.3. The PCR mixture (25 μl) contains 67 mM Tris–HCl (pH 8.8), 16 mM $(NH_4)_2SO_4$, 0.01% Tween-20, 0.2 μM of each primer and 300 μM of dNTP (Eurobio), 1.5–2.5 mM $MgCl_2$, 0 or 4% glycerol (depending on the primer pair used, Table 10.3), 0.5 U of *Taq* DNA polymerase (Eurobio) and 50 ng of citrus DNA sample. The mixture is covered with a drop of mineral oil, and the reaction is performed in a DNA thermal cycler (model PTC-100 MJ Research), programmed for an initial denaturing cycle of 4 min at 94°C then 30 cycles of 45 s denaturation at 92°C, 45 s annealing at 55 or 58°C (depending on the primer pair used, Table 10.3), 3 min elongation at 72°C and a final step of 10 min at 72°C to complete the synthesis of DNA strands.

DNA restriction

Amplified DNA fragments are digested using 4–6 base recognition restriction endonucleases (*Dra*I, *Alu*I, *Bsp*143-I, *Hae*III, *Rsa*I, *Eco*RI, *Mva*I, *Hin*fI, *Hin*dIII, *Ava*II and *Ama* 87I) (Eurogentec or Amersham), in a final volume of 25 μl containing 1× specific buffer (Eurogentec or Amersham) for each restriction enzyme, 5 U of endonuclease and 15 μl of amplification product. The reaction medium is incubated for 3 h at 37°C.

Table 10.3. Description of seven pairs of universal primers of chloroplast and mitochondrial DNA used for *Citrus* CAPS development by Lotfy *et al.* (2002)[a].

		PCR conditions for *Citrus*			
Primer 1	Primer 2	Observed length in citrus	Annealing temperature	$MgCl_2$ (mM)	Glycerol (%)
Chloroplast primers					
psaA (PSI (P700 apoprotein A1)) 5'-ACTTCTGGTTCCGGCGAACGAA-3'	trnS (tRNA-Ser(GGA)) 5'-AACCACTCGGCCATCTCTCCTA-3'	3054 bp	58.0°C	2	0
trnM (tRNA-Met (CAU)) 5'-TGCTTTCATACGGCGGGAGT-3'	rbcL (RuBisCo large subunit) 5'-GCTTTAGTCTCTGTTTGTGG-3'	3000 bp	58.0°C	2	0
trnH (tRNA-His (GUG))[b] 5'-ACGGGAATTGAACCCGCGCA-3'	trnK (tRNA-Lys (UUU) exon 3) 5'-CAACGGTAGAGTACTCGGCTTTTA-3'	4072 bp	55.0°C	1.5	4
trnC (tRNA-Cys (GCA) exon 2)[c] 5'-CCAGTTCAAATC**C**GGGTG**C**C-3'	trnD [tRNA-Asp (GUC) exon 1) 5'-GGGATTGTAGTTCAATTGGT-3'	3500 bp	55.0°C	2	0
trnT (tRNA-Thr (GGU) exon 3)[d] 5'-CCCTTTTAACTCAGTGGTAG-3'	trnD (tRNA-Asp (GUC) exon 2) 5'-ACCAATTGAACTACAATCCC-3'	1850 bp	55.0°C	2	0
Mitochondrial primers					
nad4 exon 1 5'-CAGTGGGTTGGTCTGGTATG-3'	nad4 exon 2 5'-TCATATGGGCTACTGAGGAG-3'	2036 bp	58.0°C	1.5	4
nad4 exon 3 5'-TGTTTCCCGAAGCGACACTT-3'	nad4 exon 4 5'-GGAACACTTTGGGGTGAACA-3'	3054 bp	55.0°C	2	4

[a]Universal primers previously described by Demesure *et al.* (1995) are used but some modifications are introduced.
[b]New combinations.
[c]The trnC2 sequence contains two different bases compared with Demesure *et al.* (1995).
[d]The trnT3 sequence is inverted compared with Demesure *et al.* (1995).

DNA analysis

Native and digested amplification products are separated by electrophoresis in a 1.8% agarose gel with 1× TBE during 5 h, and then visualized by UV fluorescence after staining with ethidium bromide (3 μg/ml). The size of the separated fragments is estimated by comparison with the 1 kb DNA ladder (0.5–10 kb) (Sigma).

Propidium iodide staining for flow cytometry (Yu *et al.*, 1993)

Ploidy analysis and cell cycle distribution analysis

PROTOCOL

1. Pellet 2–3 × 10^6 cells in a 15 ml centrifuge tube and wash once with cold phosphate-buffered saline (PBS)/azide solution.
2. Resuspend cells in 1 ml of low salt stain by gently vortexing. Cover the tube with foil and incubate in a 37°C water bath for 20 min with gentle mixing every 5 min.
3. Add 1 ml of high salt stain by gently vortexing and store at 4°C for at least 1 h, and *preferably overnight.*

The Bauer DNA staining technique uses a hypotonic solution of Triton X-100 detergent to strip cell membranes and produce a nuclear suspension. RNase digestion destroys RNA which also stains with propidium iodide. The high salt buffer is added to re-establish an isotonic solution.

SOLUTIONS

1. Sterile PBS + 0.1% sodium azide. Store a 4°C.
2. Low salt stain (100 ml).

- 3.0 g of PEG 8000.
- 5.0 ml of propidium iodide solution (1 mg/ml). Dissolve 0.01 g of propidium iodide in 10 ml of autoclaved distilled water; use 5 ml for low salt stain and 5 ml for high salt stain. Cover the cylinder with foil.
- 18,000 units of RNase A (Worthington Biochemicals supplied as DNase-free liquid) add 180 units/ml to staining solution just before use.
- 1 ml of 10% (v/v) Triton X-100 with 9 ml of PBS/azide, vortex. Use 1 ml for low salt stain and 1 ml for high salt stain.
- Add 4 mM sodium citrate to bring the volume to 100 ml (~94 ml).
- Add the above reagents, bring the volume to 100 ml with citrate and cover the cylinder with foil.

3. High salt stain (100 ml).

- 3.0 g of PEG 8000.
- 5 ml of propidium iodide solution (1 mg/ml).
- 1 ml of 10% (v/v) Triton X-100 (see above).
- Add 400 mM sodium chloride to bring the volume to 100 ml (~94 ml).
- Pipette 4 ml of each stain into sterile pop-top tubes and store at –20°C covered with foil.

The stain is stable for 1 year.

Concluding Remarks

Many laboratories around the world are now engaged in citrus improvement research that utilizes somatic hybridization and cybridization techniques. In this regard, *Citrus* is certainly the model crop for applying such techniques to variety improvement efforts. Verification of targeted somatic hybrids and cybrids, and characterization of their genomes have been enhanced by emerging plant molecular biology methods in combination with flow cytometry. Feedback from commercial field trials and/or molecular marker-generated progenitor species information regarding important scions and rootstocks is guiding current somatic hybridization research and suggests new applications to produce improved scions and rootstocks. At a time when most plant improvement programmes are focused on production of transgenics, which tends to narrow the germplasm base and may have consumer acceptance problems, successful somatic hybridization offers opportunities to expand the germplasm base greatly. The most successful citrus improvement programmes are integrating somatic hybridization and cybridization techniques along with conventional breeding, standard tissue culture and evolving plant molecular biology methods – whatever it takes to get the job done efficiently.

Acknowledgements

Part of this research was supported by the Florida Agricultural Experiment Station and grants from Florida Citrus Production Advisory Council, and approved for publication as Journal Series No. R-9133. J.W.G. thanks the Florida Citrus Production Advisory Council for ongoing grants supporting portions of this research conducted at the CREC.

References

Barsby, T.L., Chuong, P.V., Yarrow, S.A., Wu, S.C., Coumans, M., Kemble, R.J., Powel, A.D., Beversdorf, W.D. and Pauls, K.P. (1987) The combination of polima cms and cytoplasmic triazine resistance in *Brassica napus*. *Theoretical and Applied Genetics* 73, 809–814.

Bastia, T., Scotti, N. and Cardi, T. (2001) Organelle DNA analysis of *Solanum* and *Brassica* somatic hybrids by PCR with 'universal primers'. *Theoretical and Applied Genetics* 102, 1265–1272.

Bates, W.G. (1990) Transfer of tobacco mosaic virus resistance by asymmetric protoplast fusion. In: Nijkamp, H.J.J., Van der Plas, L.H.W. and Van Aartrijk, J. (eds) *Progress in Plant Cellular and Molecular Biology*. Kluwer Academic, Dordrecht, pp. 293–298.

Bates, G.W., Hasenkampf, C.A., Contolini, C.L. and Piastuch, W.C. (1987) Asymmetric hybridization in *Nicotiana* by fusion of irradiated protoplasts. *Theoretical and Applied Genetics* 74, 718–726.

Belliard, G., Vedel, F. and Quetier, F. (1979) Mitochondrial recombination in cytoplasmic hybrids of *Nicotiana tabacum* by protoplast fusion. *Nature* 281, 401–403.

Binsfeld, P.C., Wingender, R. and Schnabl, H. (2000) Characterization and molecular analysis of transgenic plants obtained by microprotoplast fusion in sunflower. *Theoretical and Applied Genetics* 101, 1250–1258.

Buiteveld, J., Suo, Y., Campagne, M.M.V. and Creemers-Molenaar, J. (1998) Production and characterization of somatic hybrid plants between leek (*Allium ampeloprasum* L.) and onion (*Allium cepa* L.). *Theoretical and Applied Genetics* 96, 765–775.

Cabasson, C., Luro, F., Ollitrault, P. and Grosser, J.W. (2001) Non random inheritance of mitochondriale genomes in Citrus hybrids produced by protoplast fusion. *Plant Cell Reports* 20, 604–609.

Cheng, Y.J., Guo, W.W. and Deng, X.X. (2002) Molecular characterization of cytoplasmic and nuclear genomes in phenotypically abnormal Valencia orange (*Citrus sinensis*) + Meiwa kumquat (*Fortunella crassifolia*) intergeneric somatic hybrids. *Plant Cell Reports* 21, 445–451.

Cheng, Y., de Vicente, M.C., Meng, H., Guo, W., Tao, N. and Deng, X. (2005) A set of primers for analyzing chloroplast DNA diversity in Citrus and related genera. *Tree Physiology* 25(6), 661–672.

Choi, Y.A., Tao, R., Yonemori, K. and Sugiura, A. (2002) Multi-color genomic *in situ* hybridization identifies parental chromosomes in somatic hybrids of *Diospyros kaki* and *D. glandulosa*. *HortScience* 37, 184–186.

Demesure, B., Sodzi, N. and Petit, R.J. (1995) A set of universal primers for amplification of polymorphic non coding regions of mitochondrial and chloroplast DNA in plants. *Molecular Ecology* 4, 129–131.

Deng, X.X., Grosser, J.W. and Gmitter, F.G. (1992) Intergeneric somatic hybrid plants from protoplast fusion of *Fortunella crassifolia* cultivar 'Meiwa' with *Citrus sinensis* cultivar 'Valencia'. *Scientia Horticulturae* 49, 55–62.

Deng, X.X., Grosser, J.W. and Gmitter, F.G. (1995) Interspecific gameto-somatic fusion and culture in *Citrus*. *Acta Genetica Sinica* 22, 316–321.

Deng, X.X., Guo, W.W. and Yu, G.H. (2000) Diploid mesophyll parent type plants regenerated from nine symmetrical protoplast fusion combination in *Citrus*. *Acta Horticulturae Sinica* 27, 1–5. (In Chinese.)

Deng, Z.A., Deng, X.X., Zhang, W.C. and Wan, S.Y. (1992) A preliminary report on gametosomatic fusion in citrus. *Proceedings of the International Society of Citriculture* 1, 170–172.

D'Hont, A., Quetier, F., Teoule, E. and Dattee, Y. (1987) Mitochondrial and chloroplast DNA analysis in interspecific somatic hybrids *of Medicago. Plant Science* 53, 287–288.

D'Hont, A., Grivet, L., Feldmann, P., Rao, S., Berding, N. and Glaszmann, J.C. (1996) Characterisation of the double genome structure of modern sugarcane cultivars (*Saccharum* sp) by molecular cytogenetic. *Molecular and General Genetics* 250, 405–413.

D'Hont, A., Ison, D., Alix, K., Roux, C. and Glaszmann, J.C. (1998) Determination of basic chromosome numbers in the genus *Saccharum* by physical mapping of rDNA genes. *Genome* 41, 221–225.

Dudits, D., Fejer, O., Hadlaczky Koncz, G.C., Lazar, G.B. and Horvath, G. (1980) Intergeneric gene transfer mediated by plant protoplast fusion. *Molecular and General Genetics* 179, 283–288.

Dudits, D., Maroy, E., Praznovsky, T., Olah, Z., Gyorgyey, J. and Cella, R. (1987) Transfer of resistance traits from carrot into tobacco by assymmetric somatic hybridization: regeneration of fertile plants. *Proceedings of the National Academy of Sciences of the USA* 84, 8434–8438.

Escalante, A., Imanishi, S., Hossain, M., Ohmido, N. and Fukui, K. (1998) RFLP analysis and genomic *in situ* hybridization (GISH) in somatic hybrids and their progeny between *Lycopersicon esculentum* and *Solanum lycopersicoides. Theoretical and Applied Genetics* 96, 719–726.

Fanciullino, A.L., Gancel, A.L., Froelicher, Y., Luro, F., Ollitrault, P. and Brillouet, J.M. (2005) Effects of nucleo-cytoplasmic interactions on leaf volatile compounds from citrus somatic diploid hybrids. *Journal of Agricultural and Food Chemistry* 53, 4517–4523.

Forsberg, J., Dixelius, C., Lagercrantz, U. and Glimelius, K. (1998a) UV dose-dependent DNA elimination in asymmetric somatic hybrids between *Brassica napus* and *Arabidopsis thaliana. Plant Science* 131, 65–76.

Forsberg, J., Lagercrantz, U. and Glimelius, K. (1998b) Comparison of UV light, X-ray and restriction enzyme treatment as tools in production of asymmetric somatic hybrids between *Brassica napus* and *Arabidopsis thaliana. Theoretical and Applied Genetics* 96, 1178–1185.

Froelicher, Y. (1999) Organisation de la Diversite dans le Genre *Clausena* Burm.F. (Rutaceae). Doctorat, Paris XI, Orsay.

Frearson, E.M. Power, J.B. and Cocking, E.C. (1973) The isolation, culture and regeneration of petunia leaf protoplasts. *Developmental Biology* 33, 130–137.

Galun, E., Aviv, D., Breiman, A., Fromm, H., Perl, A. and Vardi, A. (1987) Cybrids in *Nicotiana, Solanum* and *Citrus:* isolation and characterization of plastone mutants, pre-fusion treatments, selection and analysis of cybrids. In: Wettstein, D.V. and Chua, N.H. (eds) *Plant Molecular Biology*. Plenum Press, New York, pp. 199–207.

Garriga-Caldere, F., Huigen, D.J., Jacobsen, E. and Ramanna, M.S. (1999) Prospect for introgressing tomato chromosomes into potato genome: an assessment through GISH analysis. *Genome* 42, 282–288.

Gavrilenko, T., Thieme, R. and Rokka, V.M. (2001) Cytogenetic analysis of *Lycopersicon esculentum* (+) *Solanum etuberosum* somatic hybrids and their androgenetic regenerants. *Theoretical and Applied Genetics* 103, 231–239.

Gentile, A. (1992) *In vitro* selection of nucellar lemon callus and regeneration of plants tolerant to *Phoma tracheiphila* toxin. In: *Proceedings of the International Society of Citriculture* pp. 150–153.

Germanà, M.A. (1992) Androgenesis in *Citrus:* a review. In: *Proceedings of the International Society of Citriculture* pp. 183–189.

Gielly, L. and Taberlet, P. (1994) The use of chloroplast DNA to resolve plant phylogenies: noncoding versus rbcL sequences. *Molecular Biology and Evolution* 11, 769–777.

Green, R.M. and Vardi, A. (1986) The plastome of *Citrus*: physical map, variation among *Citrus* cultivars and species and comparison with related genera. *Theoretical and Applied Genetics* 72, 170–177.

Grosser, J.W. and Gmitter, F.G. (1990) Protoplast fusion and *Citrus* improvement. *Plant Breeding Review* 8, 339–374.

Grosser, J.W., Gmitter, F.G., Louzada, E.S. and Chandler, J.L. (1992a) Production of somatic hybrid and autotetraploid breeding parents for seedless *Citrus* development. *HortScience* 27, 1125–1127.

Grosser, J.W., Gmitter, F.G., Sesto, F., Deng, X.X. and Chandler, J.L. (1992b) Six new somatic citrus hybrids and their potential for cultivar improvement. *Journal of the American Society of Horticultual Science* 117, 169–173.

Grosser, J.W., Gmitter, F.G., Tusa, N., Reforgiatio Recupero, G. and Cucinotta, P. (1996a) Further evidence of a cybridization requirement for plant regeration from citrus leaf protoplasts following somatic fusion.

Plant Cell Reports 15, 672–676.

Grosser, J.W., Mourao-Fo, F.A.A., Gmitter, F.G., Louzada, E.S., Jiang, J., Baergen, K., Quiros, A., Cabasson, C., Schell, J.L. and Chandler, J.L. (1996b) Allotetraploid hybrids between *Citrus* and seven related genera produced by somatic hybridization. *Theoretical and Applied Genetics* 92, 577–582.

Grosser, J.W., Graham, J.H., McCoy, C.W., Hoyte, A., Rubio, H.M., Bright, D.B. and Chandler, J.L. (2003) Development of 'Tetrazyg' rootstocks tolerant of the Diaprepes/Phytophthora complex under greenhouse conditions. *Proceedings of the Florida State Horticultural Society* 116, 263–267.

Grosser, J.W., Medina-Urrutia, V., Ananthakrishnan, G. and Serrano, P. (2004) Building a replacement sour orange rootstock: somatic hybridization of selected mandarin + pummelo combinations. *Journal of the American Society of Horticultural Science* 129, 540–534.

Grosser, J.W., Ollitrault, P. and Olivares, O. (2000) Somatic hybridization in *Citrus*: an effective tool to facilitate variety improvement. *In Vitro Cell Developmental Biology-Plant* 36, 434–449.

Guo, W.W. and Deng, X.X. (1998) Somatic hybrid plantlets regeneration between *Citrus* and its wild relative, *Murraya paniculata* via protoplast electrofusion. *Plant Cell Reports* 18, 297–300.

Guo, W.W. and Deng, X.X. (1999) Intertribal hexaploid somatic hybrid plants regeneration from electrofusion between diploids of *Citrus sinensis* and its sexually incompatible relative, *Clausena lansium. Theoretical and Applied Genetics* 98, 581–585.

Guo, W.W. and Deng, X.X. (2001) Wide somatic hybrids of *Citrus* with its related genera and their potential in genetic improvement. *Euphytica* 118, 175–183.

Guo, W.W., Deng, X.X. and Shi, Y.Z. (1998) Optimization of electrofusion parameters and interspecific somatic hybrid regeneration in *Citrus. Acta Botanica Sinica* 40, 417–424.

Guo, W.W., Deng, X.X. and Yi, H.L. (2000) Somatic hybrids between navel orange (*Citrus sinensis*) and grapefruit (*C. paradisi*) for seedless triploid breeding. *Euphytica* 116, 281–285.

Guo, W.W., Cheng, Y.J. and Deng, X.X. (2002) Regeneration and molecular characterization of intergeneric somatic hybrids between *Citrus reticulata* and *Poncirus trifoliata. Plant Cell Reports* 20, 829–834.

Guo, W.W., Prasad, D., Cheng, Y.J., Serrano, P., Deng, X.X. and Grosser, J.W. (2004a) Targeted cybridization in citrus: transfer of satsuma cytoplasm to seedy cultivars for potential seedlessness. *Plant Cell Reports* 22, 752–758.

Guo, W.W., Prasad, D., Serrano, P., Gmitter, F.G. and Grosser, J.W. (2004b) Citrus somatic hybridization with potential for direct tetraploid scion cultivar development. *Journal of Horticultural Science and Biotechnology* 79, 400–405.

Harding, K. and Millam, S. (2000) Analysis of chromatin, nuclear DNA and organelle composition in somatic hybrids between *Solanum tuberosum* and *Solanum sanctaerosae. Theoretical and Applied Genetics* 101, 939–947.

Jazdzewska, E., Niklas, A. and Majewska-Sawka, A. (1995) Progress towards sugar beet improvement through somatic hybridization. *Acta Societatus Botanicorum Poloniae* 64, 341–347.

Kanno, A., Kanzaki, H. and Kameya, T. (1997) Detailed analyses of chloroplast and mitochondrial DNAs from the hybrid plant generated by asymmetric protoplast fusion between radish and cabbage. *Plant Cell Reports* 16, 479–484.

Kao, K.N. and Michayluk, M.R. (1974) Method for high frequency intergeneric fusion of plant protoplasts. *Planta* 115, 355–367.

Kisaka, H., Kisaka, M., Kanno, A. and Kameya, T. (1997) Production and analysis of plants that are somatic hybrids of barley (*Hordeum vulgare L.*) and carrot (*Daucus carota L.*). *Theoretical and Applied Genetics* 94, 221–226.

Kobayashi, S., Ohgawara, T., Ohgawara, E., Oiyama, I. and Ishii, S. (1988) A somatic hybrid plant obtained by protoplast fusion between navel orange (*C. sinensis*) and satsuma mandarin. *Plant Cell and Tissue Organ Culture* 14, 63–69.

Kobayashi, S., Ohgawara, T., Fujiwara, K. and Oiyama, 1. (1991) Analysis of cytoplasmic genome in somatic hybrids between navel orange (*Citrus sinensis* Osb.) and 'Murcott' tangor. *Theoretical and Applied Genetics* 82, 6–10.

Kobayashi, S., Ohgawara, T., Saito, W., Nakamura, Y. and Omura, M. (1997) Production of triploid somatic hybrids in *Citrus. Journal of the Japanese Society of Horticultural Science* 66, 453–458.

Kumar, A. and Cocking, E.D. (1987) Protoplast fusion: a novel approach to organelle genetics in higher plants. *American Journal of Botany* 74, 1289–1303.

Lee, C.H. and Power, J.B. (1988) Intraspecific gametosomatic hybridization in *Petunia* hybrids. *Plant Cell Reports* 7, 17–18.

Li, F. and Deng, X.X. (1997) Preliminary study of asymmetric fusion between citrus protoplasts. *Journal of Huazhong Agricultural University* 16, 87–90.

Li, Y.G., Tanner, G.J., Delves, A.C. and Larkin, P.J. (1993) Assymetric somatic hybrid plants between *Medicago sativa* L. (alfalfa, lucerne) and *Onobrychis viciifolia* Scop. (sainfoin). *Theoretical and Applied Genetics* 87, 455–463.

Liu, J.H. and Deng, X.X. (2002) Regeneration and analysis of citrus interspecific mixoploid hybrid plants from asymmetric somatic hybridization. *Euphytica* 125, 13–20.

Lotfy, S., Carreel, F., Froelicher, Y., Rist, D. and Ollitrault, P. (2002) Application of cleaved amplified polymorphic sequence (CAPS) method for analysis of cytoplasmic genome among Aurantioideae intergeneric somatic hybrids. *Journal of the American Society of Horticultural Science* 128, 225–230.

Lofty, S., Luro, F., Froelicher, Y., Risterucci, A.M., Filloux, D. and Ollitrault, P. (2003) Development of microsatellite markers of chloroplastic genome studies in Citrus. In: Seventh International Congress of Plant Molecular Biology, ISPMB 2003, Barcelona (Spain), June 23–28. Book of abstracts, Barcelona, ISPMB Office, 2003, p. 408.

Louzada, E.S., del Rio, H.S., Xia, D. and Moran-Mirabal, J.M. (2002) Preparation and fusion of *Citrus* sp. microprotoplasts. *Journal of the American Society of Horticultural Science* 127, 484–488.

Luro, F. and Ollitrault, P. (1996) A new set of cytoplasmic markers based on amplification reaction: application for taxonomy and study of somatic hybridization. *Proceedings of the International Society of Citriculture* 2, 899–902.

Luro, F., Rist, D. and Ollitrault, P. (2001) Evaluation of genetic relationships in *Citrus* genus by means of sequence tagged microsatellites. *Acta Horticultarae* 546, 237–242.

Medgyest, P., Fejes, E. and Maliga, P. (1985) Interspecific chloroplst DNA in *Nicotiana* somatic hybrids. *Proceedings of the National Academy of Sciences of the USA* 82, 6960–6964.

Melzer, J.M. and O'Connell, M.A. (1992) Effect of radiation dose on the production of and the extent of asymmetry in tomato asymmetric somatic hybrids. *Theoretical and Applied Genetics* 83, 337–344.

Miranda, M., Motomura, T., Ikeda, F., Ohgawara, T., Saito, W., Endo, T., Omura, M. and Moriguchi, T. (1997) Somatic hybrids obtained by fusion between *Poncirus trifoliata* (2*x*) and *Fortunnella hindsii* (4*x*) protoplasts. *Plant Cell Reports* 16, 401–405.

Moreira, C.D., Chase, C.D., Gmitter, F.G. and Grosser, J.W. (2000a) Inheritance of organelle genomes in citrus somatic cybrids. *Molecular Breeding* 6, 401–405.

Moreira, C.D., Chase, C.D., Gmitter, F.G. and Grosser, J.W. (2000b) Transmission of organelle genomes in citrus somatic hybrids. *Plant Cell Tissue and Organ Culture* 61, 165–168.

Morejohn, L.C., Bureau, T.E., Mole-Bajer, J., Bajer, A. and Fosket, D.E. (1987) Oryzalin, a dinitroaniline herbicide, binds to plant tubulin and inhibits microtubule polymerization. *Planta* 172, 141–147.

Moriguchi, T., Hidaka, T., Omura, M., Motomura, T. and Akihama, T. (1996) Genotype and parental combination influence efficiency of cybrid induction in *Citrus* by electrofusion. *HortScience* 31, 275–278.

Moriguchi, T., Motomura, T., Hidaka, T., Akihama, T. and Omura, M. (1997) Analysis of mitochondrial genomes among *Citrus* plants produced by the interspecific somatic fusion of 'Seminole' tangelo with rough lemon. *Plant Cell Reports* 16, 397–400.

Motomura, T. and Hidaka, T. (1995) Intergeneric somatic hybrids between *Citrus* and *Atalantia* or *Severinia* by electrofusion and recombination of mitochondrial genomes. *Breeding Science* 45, 309–314.

Murashige, T. and Tucker, D.P.H. (1969) Growth factor requirement of *Citrus* tissue culture. *Proceedings of the First International Citrus Symposium* 3, 1155–1161.

Nicolosi, E., Deng, Z.N., Gentile, A., La Malfa, S., Continella, G. and Tribulato, E. (2000) *Citrus* phylogeny and genetic origin of important species as investigated by molecular markers. *Theoretical and Applied Genetics* 100, 1155–1166.

Niedz, R.P., Hyndman, S.E., Wynn, E.T. and Bauscher, M.G. (2002) Normalizing sweet orange (*C. sinensis* (L.) Osbeck) somatic embryogenesis with semi-permeable membranes. *In Vitro Cell Developmental Biology-Plant* 38, 552–557.

Oberwalder, B., Schilde-Rentschler, L., Ruo, B., Wittemann, S. and Ninnemann, H. (1998) Asymmetric protoplast fusions between wild species and breeding lines of potato effect of recipients and genome stability. *Theoretical and Applied Genetics* 97, 1347–1354.

Ohgawara, T., Kobayashi, S., Ohgawara, E., Uchimiya, H. and Ishii, S. (1985) Somatic hybrid plants obtained by protoplast fusion between *Citrus sinensis* and *Poncirus trifolata*. *Theoretical and Applied Genetics* 71, 1–4.

Ohgawara, T., Kobayashi, S., Ishii, S., Yoshinaga, K. and Oiyama, 1. (1989) Somatic hybridization in *Citrus*: navel orange (*C. sinensis* Osb.) and grapefruit (*C. paradisi* Macf.). *Theoretical and Applied Genetics* 78, 609–612.

Ohgawara, T., Kobayashi, S., Ishii, S., Yoshinaga, K. and Oiyama, I. (1991) Fertile fruit trees obtained by somatic hybridization: navel orange (*Citrus sinensis*) and Troyer citrange (*C. sinensis* × *Poncirus trifoliata*). *Theoretical and Applied Genetics* 81, 141–143.

Oiyama, I.I. and Kobayashi, S. (1993) Haploid obtained from diploid × triploid crosses of *Citrus*. *Journal of the Japanese Society of Horticultural Science* 62, 89–93.

Olivares-Fuster, O., Hernandez-Garrido, M., Guerri, J. and Navarro, L. (In press) Plant somatic hybrid cytoplasm characterization by single strand conformation polymorphysm. *Tree Physiology* (in press).

Ollitrault, P., Allent, V. and Luro, F. (1996a) Production of haploid plants and embryogenic calli of clementine (*Citrus reticulata* Blanco) after *in situ* parthenogenesis induced by irradiated pollen. *Proceedings of the International Society of Citriculture* 2, 913–917.

Ollitrault, P., Dambier, D., Sudahono and Luro, F. (1996b) Somatic hybridization in *Citrus*: some new hybrid and alloplasmic plants. *Proceedings of the International Society of Citriculture* 2, 907–912.

Ollitrault, P., Dambier, D., Sudahono, Mademba-Sy, F., Vanel, F., Luro, F. and Aubert, B. (1997) Biotechnology for triploid mandarin breeding. In: Aubert, B. (ed.) *Fifth International Congress of Citrus Nurserymen*. ISCN, Montpellier, p. 337.

Ollitrault, P., Dambier, D., Vanel, F. and Froelicher, Y. (1998) Creation of triploid *Citrus* hybrids by electrofusion of haploid and diploid protoplasts. *Acta Horticulturae* 535, 191–197.

Ollitrault, P., Dambier, D., Froelicher, Y., Carceel, F., D'Hont, A., Luro, F., Cabasson, C., Lotfy, S., Joumaa, A., Vanel, F., Maddi, F., Treanton, K. and Grisoni, M. (2000a) Apport de l'hybridation somatique pour l'exploitation des resources genetiques des agrumes. *Cahier de Pagriculture* 9, 223–236.

Ollitrault, P., Dambier, D., Froelicher, Y., Luro, F. and Cottin, R. (2000b) La diversite des agrumes: structuration et exploitation par hybridation somatique. *Comptes Rendus de l'Academie d'Agriculture* 86, 197–221.

Ollitrault, P., Lotfy, S., Rist, D., Froelicher, Y., Luro, F. and Carreel, F. (2000c) Development of CAPS (cleaved amplified polymorphic sequences) for cytoplasmic genome studies in *Aurantioideae* sub-family. *Proceedings of the International Society of Citriculture* 1, 207.

Ollitrault, P., Treanton, K., Dambier, D. and D'Hont, A. (2000d) *In situ* hybridization for polyploid *Citrus* genome analysis. *Proceedings of the International Society of Citriculture* (in press).

Pelletier, G., Primard, C., Vedel, F., Chetrit, P., Remy, R., Rousselle, P. and Renard, M. (1983) Intergeneric cytoplasmic hybridization in *Cruciferae* by protoplast fusion. *Molecular and General Genetics* 191, 244–250.

Pirrie, A. and Power J.B. (1986) The production of fertile, triploid somatic hybrid plants (*Nicotiana glutinosa* (*n*) + tabacum (2*n*)) via gametic–somatic protoplast fusion. *Theoretical and Applied Genetics* 72, 48–52.

Ramulu, K.S., Dijkhuis, P., Famelaer I., Cardi, T., and Verhoeven, H.A. (1994) Cremart: a new chemical for efficient induction of micronuclei in cells and protoplasts for partial genome transfer. *Plant Cell Reports* 13, 687–691.

Ramulu, K.S., Dijkhuis, P.A., Rutgers, E., Blaas, J., Krens, F.A., Don, J.J.M., Colijn-Hooymans, C.M. and Verhoeven, H. (1996a) Microprotoplast mediated transfer of single chromosomes between sexually-incompatible plants. *Genome* 39, 921–933.

Ramulu, K.S., Dijkhuis, P.A., Rutgers, E., Blaas, J., Krens, F.A., Verbeek, W.H.J., Colijn-Hooymans, C.M. and Verhoeven, H. (1996b) Intergeneric transfer of a partial genome and direct production of monosomic addition plants by microprotoplast fusion. *Theoretical and Applied Genetics* 92, 316–325.

Rasmussen, J.O., Waara, S. and Rasmissen, O.S. (1997) Regeneration and analysis of interspecific asymmetric potato–*Solanum* ssp. hybrid plants selected by micromanipulation or fluorescence-activated cell sorting (FACS). *Theoretical and Applied Genetics* 95, 41–49.

Roose, M.L., Schwarzacher, T. and Heslop-Harrison, J.S. (1998) The chromosomes of *Citrus* and *Poncirus* species and hybrids: identification of characteristic chromosomes and physical mapping of rDNA loci using *in situ* hybridization and fluorochrome banding. *Journal of Heredity* 89, 83–86.

Rothenberg, M., Boeshore, M.L., Hanson, M.R. and Izhar, S. (1985) Intergenomic recombination of mitochondrial genome in somatic hybrid plant. *Current Genetics* 9, 615–618.

Rutgers, E., Ramulu, K.S., Diikhuis, P., Blaas, J., Krens, F.A. and Verhoeven, H.A. (1997) Identification and molecular analysis of transgenic potato chromosomes transferred to tomato through microprotoplast fusion. *Theoretical and Applied Genetics* 94, 1053–1059.

Saito, W., Ohgawara, T., Shimizu, J., Ishii, S. and Kobayashi, S. (1993) *Citrus* cybrid regeneration following cell fusion between nucellar cells and mesophyll cells. *Plant Science* 88, 195–201.
Saito, W., Ohgawara, T., Shimizu, J. and Kobayashi, S. (1994) Somatic hybridization in *Citrus* using embryogenic cybrid callus. *Plant Science* 99, 89–95.
Scarano, M.T., Abbate, L., Ferrante, S., Lucretti, S. and Tusa, N. (2002) ISSR-PCR technique: a useful method for characterizing new allotetraploid somatic hybrids of mandarin. *Plant Cell Reports* 20, 1162–1166.
Schoenmakers, H.C.H., Wolters, A.M.A., Nobel, E.M. and Koornneef, M. (1991) The production of triploid somatic hybrids via protoplast fusion of diploid tomato (*Lycopersicon esculentum*) and haploid potato (*Solanum tuberosum*). *Physiologia Plantarum* 82, 24.
Schoenmakers, H.C.H., Van der Meulen-Muisers, J.J.M. and Koorneef, M. (1994) Assymetric fusion between protoplast of tomato (*Lycopersicon esculentum* Mill.) and gamma irradiated potato (*Solanum tuberosum* L.): the effect of gamma irradiation. *Molecular General Genetics* 242, 313–320.
Shi, Y.Z., Deng, X.X. and Yi, H.L. (1998a) Variation of intergeneric somatic hybrids between *Citrus sinensis* cv Valencia and *Fortunella crassifolia* cv. Meiwa. *Acta Botanica Sinica* 40, 1060–1066. (In Chinese.)
Shi, Y.Z., Guo, W.W. and Deng, X.X. (1998b) The establishment of RAPD techniques and identification of somatic hybrids in *Citrus. Acta Horticulturae Sinica* 25, 105–110. (In Chinese.)
Sidorov, V., Menczel, L., Nagy, F. and Maliga, P. (1981) Chloroplast transfer in *Nicotiana* based on metabolic complementation between irradiated and iodoacetate treated protoplasts. *Planta* 152, 341–345.
Sjödin, C. and Glimenius, K. (1989) Transfer of resistance against *Phoma lingam* to *Brassica napus* by asymmetric somatic hybridization combined with toxin selection. *Theoretical and Applied Genetics* 78, 513–520.
Spiegel Roy, P. and Ben Hayvin, G. (1985) Selection and breeding for salt tolerance *in vitro. Plant and Soil* 89, 243–252.
Sundberg, E. and Glimelius, K. (1991) Effects of parental ploidy level and genetic divergence on chromosome elimination and chloroplast segregation in somatic hybrids within *Brassicaceae. Theoretical and Applied Genetics* 83, 81–88.
Takayanagi, R., Hidaka, T. and Omura, M. (1992) Regeneration of intergeneric somatic hybrids by electrical fusion between *Citrus* and its wild relatives: Mexican lime (*Citrus aurantifolia*) and Java feroniella (*Feroniella lucida*) or Tabog (*Swinglea glutinosa*). *Journal of the Japanese Society of Horticultural Science* 60, 799–804.
Téoulé, Y. (1992) Fusions de protoplastes et variabilite genetique. In: Bureau des Resources Génétique (ed.). *Complexes d'Especes, Flux de Genes et Ressources Genetiques des Plantes.* Paris, pp. 269–277.
Thomzik, J.E. and Hain, R. (1988) Transfer and segregation of triazine tolerant chloroplasts in *Brassica napus. Theoretical and Applied Genetics* 76, 165–171.
Tian, D. and Rose, R.J. (1999) Asymmetric somatic hybridisation between the annual legumes *Medicago truncatula* and *Medicago scutellata. Plant Cell Reports* 18, 989–996.
Tusa, N., Grosser, J.W. and Gmitter, F.G. (1990) Plant regeneration of 'Valencia' sweet orange, 'Femminello' lemon and the interspecific somatic hybrid following protoplast fusion. *Journal of the American Society of Horticultural Science* 116, 1043–1046.
Tusa, N., Grosser, J.W., Gmitter, F.G. and Louzada, E.S. (1992) Production of tetraploid somatic hybrid breeding parents for use in lemon cultivar improvement. *HortScience* 27, 445–447.
Tusa, N., Fatta del Bosco, S., Nigro, F. and Ippolito, A. (2000) Response of cybrids and somatic hybrids of lemon to *Phoma tracheiphila* infections. *HortScience* 35, 125–127.
Vardi, A., Breiman, A. and Galun, E. (1987) Citrus cybrids: production by donor–recipient protoplast fusion and verification by mitochondrial-DNA restriction profiles. *Theoretical and Applied Genetics* 75, 51–58.
Vardi, A., Arzee-Gonen, P., Frydman-Shani, A., Bleichman, S. and Galun, E. (1989) Protoplast-fusion-mediated transfer of organelles from *Microcitrus* into Citrus and regeneration of novel alloplasmic trees. *Theoretical and Applied Genetics* 78, 741–747.
Varotto, S., Nenz, E., Lucchin, M. and Parrini, P. (2001) Production of asymmetric somatic hybrid plants between *Cichorium intybus* L. and *Heliantus annuus* L. *Theoretical and Applied Genetics* 102, 950–956.
Verhoeven, H.A., Ramulu, K.S. and Dijkhuis, P.A. (1990) Comparison of the effect of various spindle toxins on metaphase arrest and formation of micronuclei in cell suspension culture of *Nicotiana plumbaginifolia. Planta* 182, 408–411.
Xu, X.Y., Liu, J.H. and Deng, X.X. (2006) Isolation of cytoplasts from satsuma mandarin (Citrus unshiu Marc.)

and production of alloplasmic hybrid calluses via cytoplast-protoplast fusion. *Plant Cell Rep.* 2006, 25(6), 533–539.

Yamamoto, M. and Kobayashi, S. (1995) A cybrid plant produced by electrofusion between *Citrus unshiu* (satsuma mandarin) and *C. sinensis* (sweet orange). *Plant Tissue Culture Letters* 12, 131–137.

Yamamoto, M., Matsumoto, R., Okudai, N. and Yamada, Y. (1997) Aborted anthers of *Citrus* result from gene-cytoplasmic male sterility. *Scientia Horticulturae* 70, 9–14.

Ye, X.R., Deng, X.X., Xiao, S.Y. and Zhang, W.C. (1992) *Citrus* plant regeneration and chromosomal variation following protoplast fusion. *Proceedings of the International Society of Citriculture* 1, 173–176.

Yu, H., Bauer, B., Lipke, G.K., Phillips, R.L. and Van Zant, G. (1993) Apoptosis and hematopoiesis in murine fetal liver. *Blood* 81, 373–384.

11 Microprotoplast-mediated Chromosome Transfer and its Potential for Citrus Breeding

Eliezer S. Louzada

Texas A&M University, Kingsville Citrus Center, 312 N. International Blvd, Weslaco, TX 78596, USA

Introduction

The development of new improved cultivars is the key for the economic success of the agricultural marketing of any crop. Cultivars with unique traits facilitate expansion of crop areas and attract small growers interested in developing niche markets to avoid the competition with large and well established growers. Conventional breeding has been the most used method to create cultivars in most crops; however, for citrus this is not very practicable. Conventional breeding has been irrelevant for the development of improved cultivars for the most economically important citrus species, specifically sweet orange, grapefruit and lemon. Barriers such as sterility (Soost and Cameron, 1975) and self- and cross-incompatibility (Soost, 1969) prevent important crosses from being successful. Additionally, the widespread nucellar embryony (Frost and Soost, 1968; Soost and Cameron, 1975) in conjunction with high heterozygosity makes the production of a large segregating population for the selection of a specific trait an almost impossible task.

The development in the last decade of molecular methods for the introduction of genes into several citrus species (Gutierrez *et al.*, 1997; Bond and Rose, 1998; Dominguez *et al.*, 2000; Yang *et al.*, 2000) opened up new opportunities for the creation of improved citrus varieties in the near future. Many horticulturally important traits, however, are polygenically encoded and, therefore, are not amenable to be transferred by the currently available genetic transformation procedures. Additionally, for the gene to be transferred by genetic transformation, it needs to be isolated, and currently the availability of genes of horticultural importance for citrus is still very limited. Somatic hybridization by protoplast fusion is a method that allows combination of the entire genome of two different species. This has been an excellent method in citrus for combining closely (Louzada *et al.*, 1992; Grosser *et al.*, 1998) and distantly related genotypes (Grosser *et al.*, 1988, 1990; Louzada *et al.*, 1993), overcoming incompatibility barriers. Fertility seems to be no problem among several somatic hybrids produced to date (J.W. Grosser, Florida, 2002, personal communication). This method has been very successful in producing tetraploid breeding parents to be used in interploid crosses targeting seedless triploid cultivars. In addition, it has been of great importance for the production of new genotypes to be tested directly as rootstocks. For direct use as scion varieties, however, the tetraploid nature of the somatic hybrids seems to be, in most cases, less beneficial probably due to the overexpression of some genes. The production of citrus somatic hybrids containing the complete genome of a recipient species, and only one or a few chromosomes, or chromosome segments from a donor species would be of great interest for the citrus industry; so far there are no reports on the production of such hybrids.

To achieve partial genome transfer, usually asymmetric somatic hybridization is performed by irradiating donor protoplasts with high doses of UV, X- or [γ]-rays, before fusion, to induce donor chromosome elimination or recombination in fusion products. Sometimes, most donor chromosomes are eliminated in the asymmetric hybrids (Vlahova *et al.*, 1997; Yemets *et al.*, 2000), and at other times only a few are eliminated (McCabe *et al.*, 1993), and, further, donor chromosomes may undergo fragmentation and recombine with the recipient genome (Lie *et al.*, 1999; Tian and Rose, 1999; Tian *et al.*, 2002). Highly asymmetric hybridization, however, is more common when distantly related species are used as parents, being more difficult to accomplish when the parents are from the same genus (Dudits *et al.*, 1987). Asymmetric hybridization involving distantly related species is important to introgress genes from incongruent species into cultivated species. For citrus, it would also be of great interest to have a method that would facilitate the transfer of one or a few chromosomes from one species to a closely related one, for the production of new cultivars in one step without further sexual hybridization. Additionally, it would probably be more profitable if partial genome transfer could be performed without the use of irradiation to minimize the damage to the donor genome. Partial nuclear genome transfer in citrus is still to be explored since hybrids with very limited chromosome numbers from a donor species have never been produced. In this chapter, we will discuss the methodology of microprotoplast-mediated chromosome transfer (MMCT) as a new approach for partial genome transfer in citrus.

Partial Genome Transfer in Mammals

In mammals, a method for partial transfer called chromosome-mediated gene transfer (CMGT) was established more than three decades ago, which allowed the transfer of subchromosome fragments from Chinese hamster to mouse (McBride and Ozer, 1973). The method consists of isolation of metaphase chromosomes from cells arrested in mitosis by colchicines or colcemid. The donor chromosome enters recipient cells by phagocytosis, which is facilitated by the addition of poly-L-ornithine. The donor genotype carries complementary genes that, when transferred to the recipient cells, allow their proliferation in a selective medium. Isozyme analysis of transformant cell lines demonstrate expression of the complementary gene derived from the donor cells (McBride and Ozer, 1973). This method was later improved by Miller and Ruddle (1978) to increase the frequency of donor chromosome transfer. Since CMGT transfers chromosome fragments, the breakage event disrupts syntenic relationships of genes on a chromosome and, therefore, this method can be used to produce mapping data (Klobutcher and Ruddle, 1981). CMGT mapping was successfully used to map genes in human chromosome 17 (Klobutcher and Ruddle, 1979). The CMGT procedure may be stable or unstable, and no donor chromosome fragments can be detected cytologically (McBride and Ozer, 1973) even though the transfer is confirmed by gene complementarity.

Another technique called microcell-mediated chromosome transfer was later developed by Fournier and Ruddle (1977), which enabled them to transfer 1–5 murine chromosomes to mouse, hamster or human recipient cells. Unlike CMGT, microcell-mediated chromosome transfer allows intact chromosomes to be transferred from one cell to another, resulting in karyotypically simple somatic cell hybrids for cytogenetic and genetic analysis (Killary and Lott, 1996). To date, microcell-mediated chromosome transfer is one of the most important tools in mammalian cells and especially for humans for gene mapping, analysis of gene function and molecular cloning of defined chromosomal regions (Jacob *et al.*, 1999). Most of the 22 human chromosomes are currently available in monochromosomal cell hybrids, produced using microcell-mediated chromosome transfer (Cuthbert *et al.*, 1995; Murakami *et al.*, 2000; Inoue *et al.*, 2001). The procedure, as initially set up (Fournier and Ruddle, 1977), consists of exposure of donor cells to colcemid, a microtubule polymerization inhibitor, which prevents formation of a functional mitotic spindle, and thereby arrests the cells in metaphase. This treatment will induce chromosomes from donor cells to be scattered throughout the cytoplasm. When the cells eventually exit mitosis, a nuclear membrane reforms around a single chromosome or small clusters of chromosomes to produce micronuclei. Individual micronuclei may be isolated from the micronucleated cells by the action of cytochalasin B (CB) together with gradient centrifugation to form microcells. CB induces nuclear extrusion which, when combined with centrifugation, causes extruded micronuclei to break out on a thin rind of cytoplasm (McNeill and Brown, 1980). Purer microcells can be produced by incorporating a microfiltration procedure after the centrifugation step (Stubblefield and Pershouse, 1992). The microcells are then fused with recipient cells to produce microcell hybrids.

Chromosome Transfer in Plants

Partial nuclear genome transfer in plants has generally been performed by asymmetric hybridization; however, in most cases, the amount of donor chromosomes eliminated in hybrids tends to be low. It is even more difficult to eliminate chromosomes if the species involved are closely related. In plants, little progress has been made in the development of new methods for the transfer of a limited number of chromosomes from one species to another. Szabados *et al.* (1981) established a procedure for mass isolation of chromosomes from wheat (*Triticum monococcum*) and parsley (*Petroselinum hortense*), and showed evidence of introduction into parsley, maize or wheat protoplasts. Unfortunately, the culture period for hybrid protoplast was too short to verify the feasibility of this procedure as a partial genome transfer method. Later, de Laat and Blaas (1987) microinjected metaphase chromosomes isolated from kanamycin-resistant *Nicotiana plumbaginifolia* suspension cells into protoplasts of wild-type *N. plumbaginifolia*; however, only visual observation was made on the transfer process. Griesbach (1987) microinjected *Petunia alpicola* chromosomes into protoplasts of *P. hybrida* and demonstrated biochemically that enzymes from the donor could be detected in the calli formed; however, no cytological analysis was performed, and no plants were obtained. One of the most promising methodologies, which proved to be an excellent tool in mammalian cells is microcell-mediated chromosome transfer, which for plants is called microprotoplast-mediated chromosome transfer (MMCT). This technique is a combination of several important steps, some of which were developed more than a decade ago. For example, the first step in MMCT is to induce donor chromosomes to be scattered in small numbers throughout the cytoplasm to form micronucleated cells. This phenomenon is induced by microtubule (MT)-depolymerizing agents, which prevent formation of a mitotic spindle, arresting the cells in metaphase. Micronucleation has been known in mammalian cells for many years (Phillips and Phillips, 1969); however, for a long time the lack of an efficient and reversible anti-MT drug in plants hampered the development of MMCT. In 1987, de Laat *et al.* (1987) demonstrated that Amiprophos-methyl (APM) induced a high degree of chromosome metaphase arrest in cells of *N. plumbaginifolia* and, after a prolonged exposure, the chromosome decondensed and formed micronuclei. Falconer and Seagull (1987) observed that APM is a rapid and reversible anti-MT agent for plant cells. APM was reported to be a highly efficient mitosis-arresting agent that induced formation of a high frequency of micronuclei in *N. plumbaginifolia* (Ramulu *et al.*, 1988a) and in *Solanum tuberosum*, *Daucus carota* and *Haplopappus gracilis* cells (Ramulu *et al.*, 1988b). They also observed that a large percentage of the micronuclei contained 1–3 chromosomes and suggested the application of their data to initiate a new method to produce microcell hybrids in plants, which would be suitable for genetic manipulation and gene mapping. Ramulu *et al.* (1994) reported that the herbicide Cremart® (butamiphos) was also very efficient as an anti-MT agent to induce mitosis arrest and the formation of micronuclei in plants.

Micronucleation is only the first step in the development of a chromosome transfer procedure, because individual micronuclei must to be isolated from the cells by an efficient enucleation process. Lorz *et al.* (1981) enucleated protoplasts of *Hyoscyamus muticus*, *Nicotiana tabacum* and *Zea mays* using a density gradient of percoll, calcium chloride and mannitol to produce cytoplasts and miniprotoplasts. Later, Lesney *et al.* (1986) enucleated *Solanum nigrum* L. protoplasts using a mannitol/sucrose gradient. In mammals, the most efficient enucleation process, which has been used for MMCT since the first establishment as a partial genome transfer procedure (Fournier and Ruddle, 1977), has been CB, which is a fungal metabolite that interferes with the microfilament attachment to the cell membrane (Carter, 1967) inducing the extrusion of the nuclei from the cell. Combination of this treatment with centrifugation proved to be very efficient for enucleation in mammalian cells (Fournier and Ruddle, 1977). Wallin *et al.* (1978) observed that CB plus centrifugation was efficient for enucleation of plant protoplasts; however, Lorz *et al.* (1981) reported reduction of viability and plating efficiency of miniprotoplasts produced using this chemical in concentrations from 1 to 200 μg/ml and incubation times of up to 24 h. Verhoeven and Ramulu (1991) demonstrated that a continuous iso-osmotic gradient of mannitol and percoll in association with CB was very efficient to produce of microprotoplasts of *N. plumbaginifolia*. Additionally, the negative effect of CB on regeneration of subprotoplasts was compensated by a better fractionation and by higher yields of evacuolated, intact subprotoplasts. Ramulu *et al.* (1993) observed that the most important parameters for production of a large number of small microprotoplasts were the synchronization of the suspension cell cycle with hydroxyurea or aphidicolin, the presence of CB during protoplast isolation and ultracentrifugation, and passage of microprotoplasts through sieves of 48, 20, 15 and 5 μm. With these parameters, these authors were able to isolate a fraction with approximately 80% of microprotoplasts containing 1–4 chromosomes. The culmination in the establishment of MMCT for plants occurred when Ramulu *et al.* (1995) reported the application of this method for transgenic *S. tuberosum* and *N. plumbaginifolia* as donors, and *Lycopersicon peruvianum* or *N. tabacum* as recipients. They successfully produced hybrid plants containing 24 or 48 *L. peruvianum* chromosomes and one *S. tuberosum* chromosome. Additionally, they observed several plants with phenotypes between that of wild tomato and potato; one of these plants had 71 tomato chromosomes, five potato chromosomes and two chromosomes with interchanged or reciprocally translocated parts of the wild tomato and potato chromosomes. For the *S. tuberosum* + *N. tabacum* combination, all plants resembled the recipient parent, *N. tabacum*; however, they expressed donor characteristics such as kanamycin resistance and β-glucuronidase (GUS) activity.

Later, Ramulu *et al.* (1996a) using the anti-MT agent cremart and transgenic potato suspension cells as chromosome donor, were able to produce several hybrid plants containing one chromosome of potato, carrying a single copy of the *nptII* and *gus* genes, and a complete set of tobacco or wild tomato. In the first backcross progeny of these hybrids, they recovered monosomic and disomic additions, and introgression plants showing integration of *gus* and *nptII* genes. This demonstrates that the transferred potato chromosomes had a normal function in the wild tomato genome background. In all the above microprotoplast hybrids, the chromosome donor parents were transgenic and the hybrids were regenerated under selection pressure. If there is a need for a donor parent to carry a selectable marker for MMCT to be applied, the broad use of this procedure would be limited. Binsfeld *et al.* (2000) using non-transgenic donors obtained microprotoplast hybrid plants of *Helianthus annuus* containing 2–8 chromosomes of *H. giganteus* or *H. maximiliani*. Further analysis of 12 microprotoplast hybrids indicated that despite some meiotic abnormality (bridges, laggard chromosomes, univalent or multivalent pairing), most of the hybrids presented regular chromosome pairing (Binsfeld *et al.*, 2001). In addition, the hybrids produced highly viable pollen permitting sexual transmission of the transferred chromosomes to their progeny. Chromosome elimination or rearrangement also seems to occur in the progeny. These results clearly indicate that there is no need for selection pressure to maintain an alien chromosome in the recipient parent.

Establishment of MMCT in Citrus

MMCT is still an emerging technology for partial genome transfer in plants, with the procedure being completely established only for a few members of the Solanaceae and Compositae families. This technology has a great potential for partial genome transfer in citrus and especially for the highly polyembryonic species, which are completely dependent on natural or induced mutation for the development of new improved cultivars, as previously discussed. The most important parameter to establish MMCT in citrus is to know the growth rates of suspension cells of several species. According to Verhoeven *et al.* (1991), a high division activity is very important for the formation of a large number of micronucleated cells, since micronuclei are formed during cell division. Ramulu *et al.* (1993), in order to obtain a sustained division activity of suspension cells of *N. plumbaginifolia*, subcultured the cells at 3–4 day cycles. A day after subculture, at an early log phase, the cell cycle was synchronized by applying hydroxyurea (HU) or aphidicolin for 24 h, after which, APM was applied. Similar conditions were applied to *S. tuberosum* suspension cells (Ramulu *et al.*, 1996b). Binsfeld *et al.* (2000) observed that on the fourth day of cultivation, suspension cells of *H. giganteus* and *H. maximiliani* had their highest mitotic activity and, therefore, cells were treated with APM on the third day, during early log-phase growth without cell cycle synchronization. The conditions used for the induction of high micronucleation rates of *N. plumbaginifolia* and *S. tuberosum* suspension cells were very suitable for all citrus species tested and therefore we have been using a 3–4 day subculture cycle, with cell cycle synchronization performed a day after subculture, followed by APM treatment (Louzada, 2001, unpublished data). Cell cycle synchronization, 24 h before APM treatment was critical to maintain high micronucleation rates in citrus. Concentrations of APM ranging from 24 to 48 μM for 24 h, preceded by treatment with 10 mM HU for 24 h, were very effective in inducing high micronucleation rates in 'Ruby Red' grapefruit (*Citrus paradisi*), 'Valencia' sweet orange (*C. sinensis*), 'Changsha' mandarin (*C. reticulata*), 'Murcott' tangor (*C. sinensis* × *C. reticulata*), and the citrus relatives *Swinglea glutinosa* and *Microcitrus papuana*. The best concentration, however, for mass production of microprotoplasts was 32 μM APM preceded by 10 mM HU (Louzada, 2002, unpublished data). After this initial APM treatment, protoplasts need to be isolated and maintained highly micronucleated. In *N. plumbaginifolia* (Ramulu *et al.*, 1993) and *S. tuberosum* (Ramulu *et al.*, 1996b), the presence of APM or cremart plus CB during the entire process of donor protoplast manipulation was required for efficient fractionation of micronucleated protoplast by ultracentrifugation. APM prevented the reformation of MTs in micronucleated protoplasts, and CB disrupted the microfilaments while maintaining the integrity of the plasma membrane. Louzada *et al.* (2002) produced large quantities of microprotoplasts from 'Ruby Red' grapefruit and from the citrus relative *S. glutinosa* using the procedure described by Ramulu *et al.* (1993) for *N. plumbaginifolia*, with a few modifications as briefly described. Early log phase suspension cells (1.0 g fresh weight drained) of 'Ruby Red' grapefruit and *S. glutinosa* were harvested 1 day after subculture (3–4 day cycle) and treated with a freshly prepared solution of 10 mM HU for 24 h. Non-treated suspension cells were used as a control. Treated and control cells were washed four times with H+H medium (Grosser and Gmitter, 1990) and incubated with 32 μM APM (Bayer Corp., Agricultural Division, Kansas City, Missouri) for 24 h. During this time, cells with scattered chromosomes (Fig. 11.1) of with several micronuclei (Fig. 11.2) could be visualized. After the initial APM treatment, suspension cells were incubated for 24 h in a cell wall digesting mixture containing equal parts of enzyme solution and 0.6 M BH3 medium (Grosser and Gmitter, 1990) supplemented with 32 μM APM, and 20 μM CB (Sigma, St Louis, Missouri). Protoplasts were then filtered through a 45 μm stainless steel mesh screen and pelleted at 100.*g* to remove enzyme. Protoplasts were purified by centrifugation using a 25% sucrose–13% mannitol gradient and washed once with 0.4 M mannitol. During the protoplast manipulations, all solutions contained 32 μM APM and 40 μM CB. A continuous iso-osmotic gradient of percoll was prepared by adding 7.2% (w/v) mannitol to a percoll solution (Amersham Pharmacia Biotech, Piscataway, New Jersey) followed by centrifugation for 30 min at 100,000.*g* in a swinging bucket rotor (SW 41 Ti-Beckman Instruments, Inc., Fullerton, California) and 6 × 13.2 ml tubes. The protoplast suspension was layered in the pre-formed gradient and centrifuged for 2 h using the above conditions. After cen-

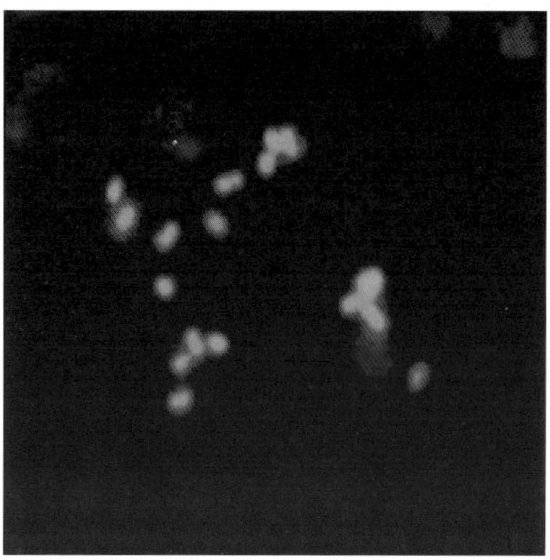

Fig.11.1. A cell with scattered chromosomes.

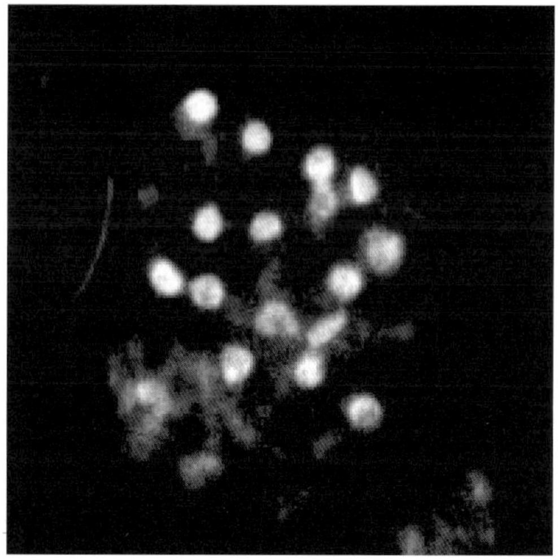

Fig.11.2. A cell with several micronuclei.

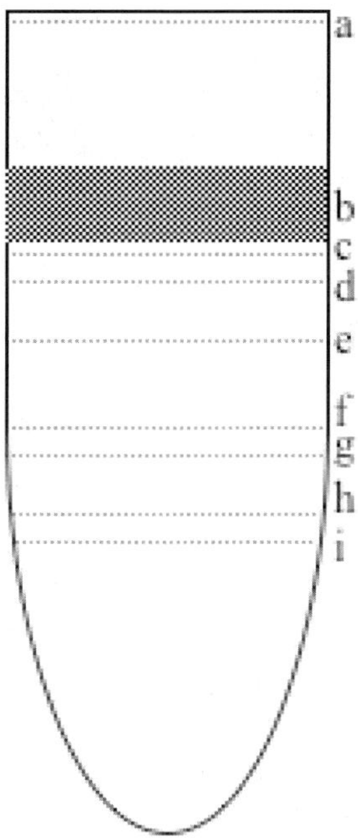

Fig.11.3. A percoll gradient.

trifugation, bands 'b' and 'c' (Fig. 11.3) were collected together in 15 ml of 0.6 M BH3 (Grosser and Gmitter, 1990) and filtered sequentially through a 20 μm nylon mesh screen (Small Parts Inc., Miami Lakes, Florida), 14 μm nucleopore membrane (Corning, Action, Massa-chusetts) and 10 and 5 μm nylon mesh screens (Small Parts, Inc.). When bands 'd' and 'e' were dense enough, they were collected together with bands 'a' and 'b'. Filtration was performed by gravity flow, and a light pressure was applied, if necessary. Microprotoplasts were collected by two rounds of centrifugation at 80.*g* and 100.*g* using a table top centrifuge, and resuspended in an appropriate volume of BH3 medium (Grosser and Gmitter, 1990). Microprotoplasts were stained with one drop of acridine orange solution at 10 μg/ml and checked for integrity using fluorescence microscopy. To determine yield, microprotoplasts were counted under a light microscope using a haemacytometer. In the original protocol described by Ramulu *et al.* (1993) and others for plants of the *Solanaceae* and *Compositae* families, the microprotoplast bands collected from the percoll gradient were diluted in 0.4 M mannitol, and also used to wash the sieves during the microfiltration. For citrus, if too much mannitol was collected from the gradient, or if the bands were diluted in mannitol, the microprotoplasts could not be precipitated after filtration. Therefore, special attention must be paid during the removal of bands from the percoll gradient. In cases where more mannitol was collected together with the bands, it was very important to dilute further with 0.6 M BH3 medium.

The procedure has been very consistent, always producing a large amount of microprotoplasts, of the order of 2×10^6 microprotoplasts/g of drained suspension cells, and almost 80% of the microprotoplasts have only one chromosome. A size comparison between protoplast (P) and microprotoplast (M) is shown in Fig 11.4. 'Ruby Red' microprotoplasts produced were used in fusion with 'Succari' sweet orange protoplasts isolated from embryogenic suspension cells. No plants were regenerated; however, two embryos were produced. One of the embryos produced roots, enabling us to determine cytologically the chromosome number, which was 22.

The *swinglea glutinosa* microproto-

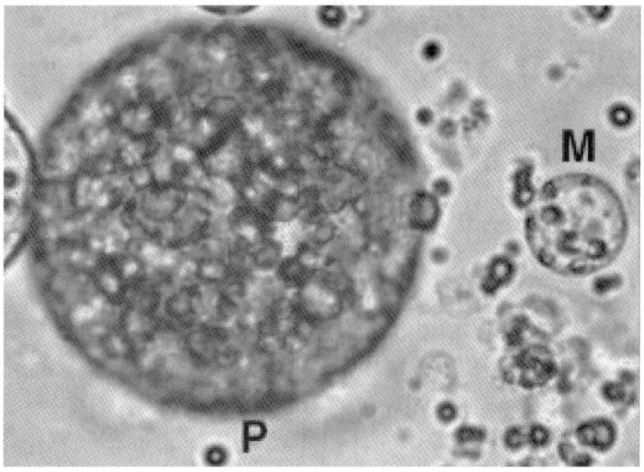

Fig.11.4. Size comparison between a protoplast (P) and a microprotoplast (M).

plasts were fused with sour orange (*C. aurantium*) mesophyll protoplasts, but no embryos were obtained, even though the fusion products grew very rapidly and produced microcallus. Suspension cells were produced from the fusion products and many cells containing 2–6 extra chromosomes were observed. Since mesophyll protoplasts from sour orange do not regenerate by themselves, neither do the *S. glutinosa* microprotoplasts (Louzada, 2001, unpublished data), it was expected that many cells would have extra chromosomes. Several other combinations were attempted but, even though callus were obtained, no embryos were produced. Later, we found that the reduction or inhibition of embryogenesis of the fusion products was directly related to the concentration of CB used during the MMCT process (Louzada, 2001, unpublished data). Lorz *et al.* (1981) reported a reduction of viability and plating efficiency of miniprotoplasts using concentrations of CB from 1 to 200 μg/ml and incubation times of up to 24 h. By reducing the concentration of CB from 20 to 10 μM during the enzyme incubation, and from 40 to 10 μM during the percoll gradient, we were able to produce a large amount of embryos from several fusion combinations. The majority of the embryos were bipolar and produced roots and shoots; however, the roots were initially very vigorous, but when the shoots were approximately 1.5 cm, they reduced their growth rates dramatically. Attempts to produce adventitious roots failed in all cases, therefore, the shoots were micro-grafted *in vivo* (Skaria, 2000). Cytology performed in root tips of several plantlets from the fusion 'Succari' sweet orange (recipient) + 'Ruby Red' grapefruit (donor) revealed chromosome numbers ranging from $2n + 1$ to $2n + 6$. This fusion combination was cultured in EME medium (Grosser and Gmitter, 1990) which contains 0.6 M sucrose. It is well known that the embryogenesis capacity of sweet orange is inhibited in medium with a sucrose concentration higher than 0.3 M (Ohgawara *et al.*, 1985), and the microprotoplasts do not divide when cultured alone (Louzada, 2001, unpublished data). We therefore, expected that only those with extra chromosomes would be able to regenerate into plants. All cytological analyses performed so far in root tips revealed additional chromosomes. A micro-grafted plant is shown in Fig. 11.5, and a root tip cell containing 24 chromosomes ($2n + 6$) is shown in Fig. 11.6. We are currently micro-grafting plantlets from several other fusion combinations. Molecular characterization of all the plants is underway.

Fig.11.5. A micro-grafted microprotoplast hybrid.

MMCT in Citrus – Detailed Procedure

The following is the complete protocol for microprotoplast isolation. It is important to emphasize that the suspension cells to be used as donor must be fast growing ones for the isolation of a large quantity of microprotoplasts, since the micronucleation occurs during cell division.

1. Grow suspension cells in a 3-4 day subculture cycle in H+H medium (Grosser and Gmitter, 1990) containing 50% of macronutrients, under appropriated lighting conditions.

2. One day after subculturing, add 10 mM of Hu for 24 h.

3. Wash out the HU four times with H+H medium.

4. Add 32 μM APM for 24 h. Cells with chromosomes scattered (Fig. 11.1) or with several micronuclei (Fig. 11.2) can be visualized by staining fixed cells with 4′,6-diamidino-2-phenyindole (DAPI) at 0.4 μg/ml.

5. Drain and divide the cells into approximately 300 mg amounts in 60 × 15 mm Petri dishes and add 32 μM APM + 10 μM CB + one part of enzyme solution to three parts of 0.6 M BH3 medium (enzyme and 0.6 M BH3, as described by Grosser and Gmitter, 1990). Incubate in the dark with gentle agitation for 15–18 h.

6. Filter the micronucleated protoplasts through a 45 μm stainless steel screen (Small Parts, Inc.,) and pellet them at 100.*g* to remove enzyme. Purify the protoplasts by centrifugation through a 25% sucrose – 13% mannitol gradient and wash once with 0.4 M mannitol. Remove most of the mannitol and maintain the protoplasts as a dense solution. All solutions must contain 32 μM APM + 10 μM CB during the whole process of protoplast isolation and purification.

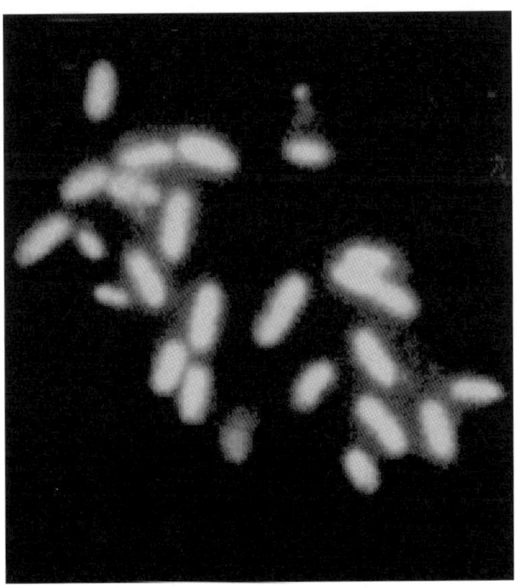

Fig.11.6. A root tip cell containing 24 chromosomes, $2n + 6$.

7. Prepare an iso-osmotic gradient of percoll by adding 7.2% mannitol to a percoll solution (Amersham Pharmacia Biotech), and spin at 100,000.*g* at 20°C for 30 min in a swinging bucket rotor.

8. To the top of the pre-formed percoll gradient add the dense solution of micronucleated protoplasts and spin at 100,000.*g* for 2 h at 20°C in a swinging bucket rotor.

9. After centrifugation, usually nine bands are obtained (Fig. 11.3). The number of bands will depend on the amount of protoplasts loaded and on the efficiency of micronucleation. Band 'a', forms at the surface of the gradient and is usually composed of a sticky material. This band should not be collected since it will contaminate the microprotoplast suspension. A glass rod or the tip of a Pasteur pipette can be used to remove the central part of this band to facilitate the collection of the others. Band 'b' is normally very thick and contains a large amount of microprotoplasts, and is usually collected together with band 'c'. In all microprotoplast isolations performed so far, for citrus, the amount and quality of microprotoplasts originated from these two bands has been more than can be handled by one person per day in fusion experiments. The other bands below 'c' contain very clean and good quality small size microprotoplasts; however, the amount of mannitol collected together with the bands makes it very difficult to precipitate the microprotoplasts, even when diluted with large amounts of 0.6 M BH3.

10. Collect bands 'b' and 'c' (Fig. 11.3) and dilute it to at least 15 ml with 0.6 M BH3.

11. Filter the microprotoplasts suspension sequentially through a 20 μm nylon mesh screen, 14 μm nucleopore membrane, and 10 and 5 μm nylon mesh screens. The nylon screens are manufactured by Small parts

Inc., and the Nucleopore membranes by Corning. If necessary, a light pressure can be applied to facilitate the flow; if high pressure is applied, the microprotoplasts may be damaged and larger microprotoplasts will flow through.
12. The microprotoplasts can be visualized under fluorescence after staining with 3 μl of an acridine orange solution at 10 μg/ml. A size comparison of a protoplast (P) and microprotoplasts (M) is shown in Fig. 11.4.
13. Protoplasts from a recipient species should by this time already be isolated and purified as described by Grosser and Gmitter (1990).
14. Protoplasts and microprotoplasts are mixed in a proportion of 1:2 or 1:3 and fused by polyethylene glycol (PEG) method as described by Grosser and Gmitter (1990).
15. The fusion products are cultured as described by Louzada *et al.* (2002) and plants are regenerated using B germination medium (Grosser and Gmitter, 1990) with half strength of macronutrients.
16. Regenerated plants are rooted or micro-grafted on suitable rootstock.
17. Chromosome counting is performed according to Louzada *et al.* (2002), or by any other cytological method. Molecular characterization can be carried out by polymerase chain reaction (PCR)-based methods or amplified fragment length polymorphism (AFLP). If the species are distantly related, probably GISH (genome *in situ* hybridization) will be feasible. Fluorescence *in situ* hybridization (FISH) or (GISH) for MMCT between closely related citrus species is very difficult to accomplish.

Potential and Prospects of MMCT in Citrus

MMCT in citrus is still in its developmental stages; however, it is possible to envisage its potential for citrus breeding. The transfer of a limited number of chromosomes from one citrus species or relative to another citrus could greatly impact the development of new cultivars. For sweet orange, grapefruit, lemon and limes, which are highly polyembryonic, this would be especially important, since so far natural or induced mutation has been the primary method for cultivar development for these groups. In mammalian cells, as previously discussed, microcell-mediated chromosome transfer is still one of the most powerful tools for gene mapping, analysis of gene function and the molecular cloning of defined chromosome regions (Jakobs *et al.*, 1999). Most of the 22 human chromosomes are currently available in monochromosomal cell hybrids, produced using microcell-mediated chromosome transfer (Cuthbert *et al.*, 1955; Murakami *et al.*, 2000; Inoue *et al.*, 2001). For citrus, MMCT could also become an important tool for cultivar improvement and for others genetic and molecular analysis. Theoretically, several different individuals can be produced in a single microprotoplast fusion experiment. Considering that the haploid chromosome number of citrus is nine and that a single chromosome is transferred from a donor to a recipient species, at least nine different hybrids may be produced. Since the microprotoplast suspension contains a mixture of donor chromosomes, the fusion with the recipient genome will occur in a completely random way. Therefore, if two chromosomes are transferred, the number of different hybrids would theoretically be the combination of nine chromosomes two by two. More than two chromosomes can be transferred, which further increases the probability of different individuals per fusion. We have been regenerating plants with $2n + 1$ to $2n + 6$ chromosomes so far.

Additionally, more than one donor can be used per single recipient species. Recently we fused microprotoplasts from grapefruit plus mandarin as donor, with sweet orange as recipient. We have so far obtained four plantlets from this combination, which have been micro-grafted into sour orange and will later be analysed cytologically and molecularly. Additionally, if delay of the DNA replication of the donor chromosomes in the immediate cell cycle after fusion occurs, the progression of the recipient genome to metaphase may induce pre-chromosome condensation (PCC) of the S-phase donor chromosome, leading to donor DNA or chromosome breakage. The released DNA may be integrated into the recipient genome through the transformation and repair process (Ramulu *et al.* 1955), further increasing the possibilities for new hybrid genotypes.

References

Binsfeld, P.C., Wingender, R. and Schnabl, H. (2000) Characterization and molecular analysis of transgenic plants obtained by microprotoplast fusion in sunflower. *Theoretical and Applied Genetics* 101, 1250–1258.

Binsfeld, P.C., Wingender, R. and Schnabl, H. (2001) Cytogenetic analysis of interspecific sunflowers hybrids and molecular evaluation of their progeny. *Theoretical and Applied Genetics* 102, 1280–1285.

Bond, J.E. and Roose, M.L. (1998) *Agrobacterium*-mediated transformation of the commercially important citrus cultivar Washington navel orange. *Plant Cell Reports* 18, 229–234.

Carter, S.B. (1967) Effect of cytochalasins on mammalian cells. *Nature* 218, 261–264.

Cuthbert, A.P., Trott, D.A., Ekong, R.M., Jezzard, S., England, N.L., Themis, M., Todd, C.M. and Newbold, R.F. (1995) Construction and characterization of a highly stable human:rodent monochromosomal hybrid panel for genetic complementation and genome mapping studies. *Cytogenetics and Cell Genetics* 71, 68–76.

De Laat, A.M.M. and Blaas, J. (1987) An improved method for protoplast micro-injection suitable for transfer of entire plant chromosomes. *Plant Science* 50, 1761–1769.

De Laat, A.M.M., Verhoeven, H.A., Ramulu, K.S. and Dijkhuis, P. (1987) Efficient induction by amiprophos-methyl and flow-cytometric sorting of micronuclei in *Nicotiana plumbaginifolia. Planta* 172, 473–478.

Dominguez, A., Guerri, J., Cambra, M., Navarro, L., Moreno, P. and Peña, L. (2000) Efficient production of transgenic citrus plants expressing the coat protein gene of citrus tristeza virus. *Plant Cell Reports* 19, 427–433.

Dudits, D., Maroy, E., Praznovszky, T., Olah, Z. and Gyorgyey, J. (1987) Transfer of resistance traits from carrot into tobacco by asymmetric somatic hybridization: regeneration of fertile plants. *Proceedings of the National Academy of Sciences of the USA* 84, 8434–8438.

Falconer, M.M. and Seagull, R.W. (1987) Amiprophos-methyl (APM): a rapid reversible, anti-microtubule agent for plant cell cultures. *Protoplasma* 136, 118–124.

Fournier, R.E.K. and Ruddle, F.H. (1977) Microcell mediated transfer of murine chromosome into mouse, Chinese hamster, and human somatic cells. *Proceedings of the National Academy of Sciences of the USA* 74, 319–323.

Frost, H.B. and Soost, R.K. (1968) Seed reproduction:development of gametes and embryos. In Reuther, W., Batchelor, L.D. and Webber, H.J. (eds) *The Citrus Industry*, Vol. 2. University of California Press, Berkeley, pp. 292–294.

Griesbach, R.J. (1987) Chromosome-mediated transformation via micro-injection. *Plant Science* 50, 69–77.

Grosser, J.W. and Gmitter, F.G. Jr (1990) Protoplast fusion and citrus improvement. *Plant Breeding Review* 8, 339–374.

Grosser, J.W., Gmitter, F.G. Jr and Chandler, J.L. (1988) Intergeneric somatic hybrid plants from sexually incompatible woody species: *Citrus sinensis and Severinia disticha. Theoretical and Applied Genetics* 75, 397–401.

Grosser, J.W., Gmitter, F.G. Jr, Tusa, N. and Chandler, J.L. (1990) Somatic hybrid plants from sexually incompatible wood species: *Citrus reticulata* and *Citropsis gilletiana. Plant Cell Reports* 8, 656–659.

Grosser, J.W., Jiang, J., Mourao-Fo, F.A.A., Louzada, E.S., Baergen, K., Chandler, J.L. and Gmitter, F.G. Jr (1998) Somatic hybridization, an integral component of citrus cultivar improvement: I. Scion improvement. *HortScience* 33, 1057–1059.

Guitierrez, E.A.A., Luth, D. and Moore, A.G. (1997) Factors affecting *Agrobacterium* mediated transformation in citrus and production of sour orange (*Citrus aurantium* L.) plants expressing the coat protein gene of citrus tristeza virus. *Plant Cell Reports* 16, 745–753.

Inoue, J., Mitsuya, K., Maegawa, S., Kugoh, H., Kadota, M., Okamura, D., Shinohara, T., Nishihara, S., Takehara, S., Yamauchi, K., Schulz, T.C. and Oshimura, M. (2001) Construction of 700 human/mouse

A9 monochromosomal hybrids and analysis of imprinted genes of human chromosome 6. *Journal of Human Genetics* 46, 137–145.

Jakobs, P.M., Smith, L., Thayer, M. and Grompe, M. (1999) Embryonic stem cells can be used to construct hybrid cell lines containing a single, selectable murine chromosome. *Mammalian Genome* 10, 381–384.

Killary, A.M. and Lott, S.T. (1996) production of microcell hybrids. *Methods in Enzymology* 9, 3–11.

Klobutcher, L.A. and Ruddle, F.H. (1979) Phenotype stabilization and integration of transferred material in chromosome mediated gene transfer. *Nature* 280, 657–660.

Klobutcher, L.A. and Ruddle, F.H. (1981) Chromosome mediated gene transfer. *Annual Review of Biochemistry* 50, 533–554.

Lesney, M.S., Callow, P.W. and Sink, K.C. (1986) A technique for bulk production of cytoplasts and miniprotoplasts from suspension culture-derived protoplasts. *Plant Cell Reports* 3, 115–118.

Liu, B., Liu, Z.L. and Li, X.W. (1999) Production of a highly asymmetric somatic hybrid between rice and *Zizania latifolia* (Griseb): evidence for inter-genomic exchange. *Theoretical and Applied Genetics* 98, 1099–1103.

Lorz, H., Paszkowski, J., Diersks-Ventling, C. and Potrykus, I. (1981) Isolation and characterization of cytoplasts and miniprotoplasts derived from protoplasts of cultured cells. *Physiologia Plantarum* 53, 385–391.

Louzada, E.A., Grosser, J.W., Gmitter, F.G. Jr, Nielsen, B., Chandler, J.L., Deng, X.X. and Tusa, N. (1992) Eight new somatic hybrid citrus rootstocks with potential for improved disease resistance. *HortScience* 27, 1033–1036.

Louzada, E.A., Grosser, J.W., and Gmitter, F.G. Jr (1993) Somatic hybrid plants from sexually incompatible parents: *Citrus sinensis* cv. Hamlin and *Atalantia ceylanica. Plant Cell Reports* 12, 687–690.

Louzada, E.S., del Rio, H.S., Xia, D. and Moran-Mirabal, J.M. (2002) Preparation and fusion of microprotoplasts from *Citrus* spp. *Journal of The American Society for Horticultural Science* 127, 484–488.

McBride, O.W. and Ozer, H.L. (1973) Transfer of gentic information by purified metaphase chromosomes. *Proceedings of the National Academy of Sciences of the USA* 70, 1258–1262.

McCabe, P.F., Dunbar, L.J., Guri, A. and Simk, K.C. (1993) T-DNA-tagged chromosome 12 in donor *Lycopersicon esculentum* [x] *L. pennellii* is retained in asymmetric hybrids with recipient *Solanum lycopersicoides. Theoretical and applied Genetics* 86, 377–382.

McNeill, C.A. and Brown, R.L. (1980) Genetic manipulation by mean of microcell-mediated transfer of normal human chromosomes into recipient mouse cells. *Proceedings of the National Academy of Sciences of the USA* 77, 5394–5398.

Miller, C.L. and Ruddle, F.H. (1978) Co-transfer of human X-linked markers into murine somatic cells via isolated metaphase chromosomes. *Proceedings of the National Academy of Sciences of the USA* 75, 3346–3350.

Murakami, Y., Uejima, H., Fukuhara, H., Maruyama, T., Oshimura, M. and Sekiya, T. (2000) Construction of human-rodent hybrid cells containing single transferable fragments of human chromosome 10p. *Journal of Human Genetics* 45, 370–373.

Ohgawara, T., Kobayasshi, S., Ohgawara, E., Uchimiya, H. and Ishii, S. (1985) Somatic hybrid plants obtained by protoplast fusion between *Citrus sinensis* and *Poncirus trifoliate. Theoretical and Applied Genetics* 71, 1–4.

Phillips, S.G. and Phillips, D.M. (1969) Site of nucleolus production in cultured Chinese hamster cells. *Journal of Cell Biology* 40, 248–268.

Ramulu, K.S., Verhoeven, H.A. and Dijkhuis, P. (1988a) Mitotic dynamics of micronuclei induced by amiprophos-methyl and prospects for chromosome-mediated gene transfer in plants. *Theoretical and Applied Genetics* 75, 575–584.

Ramulu, K.S., Verhoeven, H.A., Dijkhuis, P. and Gilissen, L.J.W. (1988b) Chromosome behavior and formation of micronuclei after treatment of cell suspension cultures with amiprophos-methyl in various plant species. *Plant Science* 56, 227–237.

Ramulu, K.S., Dijkhuis, P., Famelaer, I., Cardi, T. and Verhoeven, H.H. (1993) Isolation of sub-diploid microprotoplasts for partial genome transfer in plants: enhancement of Micronucleation and enrichment of microprotoplasts with one or few chromosomes. *Planta* 190, 190–198.

Ramulu, K.S., Dijkhuis, P., Famelaer, I., Cardi, T. and Verhoeven, H.H. (1994) Cremart: a new chemical for efficient induction of micronuclei in cells and protoplasts for partial genome transfer. *Plant Cell Reports* 13, 687–691.

Ramulu, K.S., Dijkhuis, P., Rutgers, E., Blaas, J., Verbeek, W.H.J., Verhoeven, H.A. and Colijn-Hooymans, C.M. (1995) Microprotoplast fusion technique: a new tool for gene transfer between sexually-incongruent plant species. *Euphytica* 85, 255–268.

Ramulu, K.S., Dijkhuis, P., Rutgers, E., Blaas, J., Krens, F.A., Dons, J.J.M., Colijn-Hooymans, C.M. and Verhoeven, H.A. (1996a) Microprotoplast-mediated chromosome transfer of single specific chromosomes between sexually incompatible plants. *Genome* 39, 921–933.

Ramulu, K.S., Dijkhuis, P., Rutgers, E., Blaas, J., Krens, F.A., Verbeek, W.H.J., Colijn-Hooymans, C.M. and Verhoeven, H.A. (1996b) Intergeneric transfer of partial genome and direct production of monosomic addition plants by microprotoplast fusion. *Theoretical and Applied Genetics* 92, 316–325.

Skaria, M (2000) A microbudding technique for biological indexing and ultra high density planting of citrus. In: da Graca, J.V., Lee, R.F. and Yokomi, R.K. (eds) *Proceedings of the 14th Conference of the International Organization of Citrus Virologists.* IOCV, Riverside, California, p. 411.

Soost, R.K. (1969) The incompatibility gene system in citrus. In: Chapman, H.D. (ed.) *Proceedings of the First International Citrus Symposium.* University of California, Riverside, pp. 189–190.

Soost, R.K. and Cameron, J.W. (1975) Citrus. In: Janick, J. and Moore, J.N. (eds) *Advances in Fruit Breeding.* Purdue University, West Lafayette, Indiana, pp. 507–540.

Stubblefield, E. and Pershouse, M. (1992) Direct formation of microcells from mitotic cells for use in chromosome transfer. *Somatic Cell and Molecular Genetics* 18, 485–491.

Sazbados, L., Hadlaczky, G. and Dudits, D. (1981) Uptake of isolated chromosomes by plant protoplasts. *Planta* 151, 141–145.

Tian, D. and Rose, R.J. (1999) Asymmetric somatic hybridization between the annual legumes *Medicago truncatula* and *Medicago scutellata. Plant Cell Reports* 18, 989–996.

Tian, D., Niu, C. and Rose, R.J. (2002) DNA transfer by highly somatic asymmetric hybridization in *Medicago truncatula* (+) *Medicago rugosa,* and *Medicago truncatula* (+) *Medicago scutellata. Theoretical and Applied Genetics* 104, 9–16.

Verhoeven, H.A. and Ramulu, K.S. (1991) Isolation and characterization of microprotoplasts from APM-treated suspension cells of *Nicotiana plumbaginifolia. Theoretical and Applied Genetics* 82, 346–352.

Verhoeven, H.A., Ramulu, K.S., Gilissen, L.J.W., Famelaer, I., Dijkhuis, P. and Blaas, J. (1991) Partial genome transfer through micronuclei in plants. *Acta Botanica* 40, 97–113.

Vlahova, M., Hinnisdaels, S., Frulleux, F., Claeys, M., Atanassov, A. and Jacobs, M. (1997) UV irridation as a tool for obtaining asymmetric somatic hybrids between *Nicotiana plumbaginifolia* and *Lycopersicon esculentum. Theoretical an Applied Genetics* 94, 184–191.

Wallin, A., Glimelius, K. and Eriksson, T. (1978) Enucleation of plant protoplasts by cytochalasin B. *Zeitschrift für Pflanzenphysiologie* 87, 333–340.

Yang, Z.N., Ingelbrecht, I.L., Louzada, E.S., Skaria, M. and Mirkov, T.E. (2000) *Agrobacterium*-mediated transformation of the commercially important grapefruit cultivar Rio Red (*Citrus paradisi* Macf.). *Plant Cell Reports* 19, 1203–1211.

Yemets, A.I., Kundel'chuk, O.P., Smertenko, A.P., Solodushko, V.G., Rudas, V.A., Gleba, Y.Y. and Blume, Y.B. (2000) Transfer of amiprophosmethyl resistance from *Nicotiana plumbaginifolia* mutant by somatic hybridization. *Theoretical and Applied Genetics* 100, 847–857.

12 Mapping and Marker-assisted Selection

M. L. Roose

Department of Botany and Plant Sciences, University of California, Riverside, CA 92521, USA

Introduction and Objectives of Mapping

Genetic linkage maps are a statistical description of the organization of DNA segments into chromosomes and the order and recombination frequency of DNA segments within chromosomes. Maps can include both DNA markers and trait genes that are detected by affecting the phenotype of an individual. A good map should include one linkage group that corresponds to each chromosome of the genome (nine in *Citrus* and closely related genera). Each linkage group will be comprised of a linearly ordered set of markers and/or trait genes, with distances between adjacent markers determined by the proportion of gametes that have a crossover event between the markers. A map is created by studying segregation of heterozygous markers and genes in progeny of a specific genotype or cross, and is therefore specific to that genotype or cross. Maps created for different genotypes may differ in marker order because genotypes actually differ in the order of DNA segments (due to various DNA rearrangements). Distances between the same marker pair may differ in different maps because genotypes differ in the rate of recombination. Since maps are statistical descriptions, such differences can also arise by chance sampling errors, particularly if the number of progeny tested is small.

For citrus breeders, the objective of mapping is to identify markers that can be used in marker-assisted selection (MAS). MAS involves identification of DNA markers that are near a variable trait gene, and selection of progeny carrying markers that predict which progeny will carry the desired trait gene allele. This process is described later in this chapter.

Mapping Populations

The first step in developing a genetic linkage map is to choose a population in which to make the map. Since the map is based on segregation of heterozygous markers in one or both parents, at least one parent should be heterozygous. If trait genes are to be mapped in the same population, then these also must be heterozygous in the same parent(s). Generally, mapping populations are formed by crossing a specific female with a specific male and using the progeny as the mapping population. The population should be large enough to allow mapping to the desired marker density. If the breeder wishes to develop a dense map, with very closely spaced markers, then the popula-

 Citrus Genetics, Breeding and Biotechnology (ed. I.A. Khan) tion size must be large in order for there to be an adequate number of progeny with crossovers between closely spaced markers. For example, with a population size of 100, if one parent is heterozygous for markers having 2% recombination, then we expect only about two recombinant progeny, and there is a fair chance that we will observe none, in which case the two markers cannot be ordered. In the early days of mapping, citrus geneticists often used relatively small mapping populations (50–60 progeny) because resources were limited and development of dense maps seemed unlikely. The development of new high-throughput methods for marker analysis makes it possible to develop much denser maps and, if this is anticipated, then the mapping population should be larger (100–300 progeny). A discussion of mapping population sizes is given by Maliepaard *et al.* (1997).

A second choice that the mapper must face is the type of population. In annual crops where homozygous lines are available, mapping is much simpler because segregation of markers and traits is predictable. In citrus, virtually all genotypes are heterozygous. An advantage of this situation is that elite cultivars can be used as parents, but a disadvantage is that markers will show various types of segregation which complicates mapping. This can be illustrated by considering the mapping population used by Jarrell *et al.* (1992). This population is a cross of two citrus rootstocks, Sacaton citrumelo × Troyer citrange ((*Citrus paradisi* × *Poncirus trifoliata*) × (*C. sinensis* × *P. trifoliata*)). For each linkage group, each parent contains one chromosome derived from a *Citrus* parent and one derived from a *Poncirus* parent. Depending on the specific allele carried by each chromosome, three types of segregation can be observed for each marker locus (see Table 12.1). These different segregation types lead to different mapping strategies and several different maps. The different map types are listed in the last column of Table 12.1. Markers that segregate in only one parent, say Sacaton, can be used to construct a map of that parent that reflects the frequency of crossovers between homologous chromosomes in that genotype. A parallel map can be prepared for those markers that segregate only in the other parent. Markers showing the 1:1:1:1 segregation can also be combined with those segregating 1:1 by 'renaming' the two alleles in the homozygous parent. For example, segregation type $ab_1 \times ab_2$ can be combined with ab × aa by renaming all b_2 alleles as a alleles. Markers segregating in 3:1 and 1:2:1 ratios can be combined with each other, but cannot be accurately mapped with 1:1 markers (Maliepaard *et al.*, 1997), essentially because recombinant gametes from one parent are identical to non-recombinant gametes from the other parent. Similarly, for dominant markers in an F_2 configuration, only markers in coupling phase (dom-

Table 12.1. Types of marker segregation that can be expected in a cross between two heterozygous parents (Sacaton and Troyer).

Sacaton	Troyer	Progeny	Expected segregation ratio	Map
aa	ab	aa, ab	1:1	Troyer
bb	ab	bb, ab	1:1	Troyer
ab	aa	aa, ab	1:1	Sacaton
ab	bb	aa, ab	1:1	Sacaton
ab	ab	aa, ab, bb	1:2:1	Sacaton*, Troyer*, F_2
Aa	Aa	A_, aa	3:1	F_2
a_1b	a_2b	a_1a_2, a_1b, ba_2, bb	1:1:1:1	Sacaton, Troyer, F_2
a b_1	ab_2	aa, ab_2, b_1a, b_1b_2	1:1:1:1	Sacaton, Troyer, F_2
a_1b_1	a_2b_2	a_1a_2, a_1b_2, b_1a_2, b_1b_2	1:1:1:1	Sacaton, Troyer, F_2

inant alleles on one chromosome and recessive alleles on the homologue) can be accurately mapped (Knapp *et al.*, 1995).

One mapping strategy is the 'pseudo-testcross' in which two maps are prepared, one for each parent, using only those loci that segregate 1:1 or 1:1:1:1. An 'F_2-type' map can also be prepared using loci that segregate 1:2:1, 3:1 or 1:1:1:1. The F_2 map can then be merged with the single parent maps, using the 1:1:1:1 loci that occur on all maps as anchors. The F_2 population type is more efficient because recombination frequency from two meioses (one in each parent) per progeny plant is determined. However, if recombination frequencies are quite different in the two parents, then this map is a compromise that does not reflect the biology of either parent.

Another issue that must be considered in choosing a mapping population is inbreeding in the parents. Whether the cross considered is a backcross type, or F_2, if the heterozygous parent(s) are somewhat inbred, then there may be regions of the genome that are homozygous (identical-by-descent) in that parent. Such regions cannot be mapped because there will be no heterozygous markers. It should be noted that such regions may occur in parents not known to be inbred. For example, *P. trifoliata* genotypes are capable of self-pollination, and many zygotic seedlings apparently originate by selfing (Khan and Roose, 1988). It is possible that standard cultivars originated through selfing or mating between relatives and, if so, then we would expect to find such homozygous regions. Comparison between maps of *Poncirus* and *Poncirus* × *Citrus* hybrids, which should not contain homozygous segments, should reveal homozygous segments provided that the maps share a sufficient density of common markers.

Map Construction

Maps are prepared using computer programs to identify loci in linkage groups and then determine the order of the loci and distances between them. Linkage groups are generally identified by computing pairwise tests for evidence of linkage at a range of criteria, usually the LOD score. The LOD score is a 'log of odds' to the base 10, so that a LOD score of 2 means that it is 100 times more likely that the data are explained by linkage at the most probable recombination fraction, than that they are explained by no linkage. A locus is considered a member of a linkage group if it shows significant linkage (at the current LOD value) to any other member of the group. When the LOD score is low (say 1–2), all loci may fall into a single or a few linkage groups. Generally, with error-free data that meet the assumptions of the analysis, a LOD score of 3–4 will identify a number of linkage groups equal to the number of chromosomes. The number of larger linkage groups will be fairly stable at this value, gradually increasing as the LOD score is increased. Linkage groups identified at scores of 3 or 4 are generally used for mapping, but higher values may be appropriate for some data sets.

Once linkage groups are identified, the loci within each group are ordered and distances between loci are computed. These calculations, and locus grouping, are performed by computer programs. Commonly used programs for construction of linkage maps are MAPMAKER (Lander *et al.*, 1987) and JOINMAP (Stam and Van Ooijen, 1995). MAPMAKER performs multipoint linkage analysis, an approach with great power. The major limitation of this program for analysis of citrus maps is that it can analyse only a backcross or F_2-type segregation. This means that 1:1:1:1 loci must be recoded with a loss of information, and that maps that combine backcross and F_2-type segregation patterns cannot be prepared. The MAPMAKER interface is command line (DOS or Unix), making it somewhat difficult to use. JOINMAP uses multiple two-point linkage estimates, an intrinsically less powerful approach, but it allows analysis of all types of segregation patterns. JOINMAP also allows maps developed in different populations with shared markers to be merged. JOINMAP 3.0 has a modern

Windows interface and is considerably easier to use than MAPMAKER, but it is a commercial product that requires purchase of a licence, whereas MAPMAKER is free. It is probable that additional mapping programs will become available, so those conducting such projects should carry out an Internet search for linkage mapping programs to identify suitable programs.

Another issue that needs to be addressed in citrus mapping is marker phase. This refers to the actual physical chromosome on which allelic markers are located. For example, dominant markers can be in coupling (A B/a b) or repulsion (A b/a B) phase. If the grandparents of the mapping population are known, then the marker phase is also generally known. However, in many citrus crosses, the parents are not hybrids of known ancestry. In such situations, the phase must be inferred from the data. JOINMAP performs this analysis, while with MAPMAKER data of unknown phase must be entered in both possible phases, linkage groups identified and one member of each pair of linkage groups discarded. If known, the phase can be specified by the allele code: for a/b-type allele codes, all a alleles are assumed to be on one homologue, and b alleles on the other. For some programs, allele position in the genotype code indicates phase. For example, for two loci if the parents are specified as ab × cd and ab × ab, then the chromosomes of the two parents are a a/b b and c a/d b.

Marker Types

Mapping can be conducted with a variety of marker types (Table 12.2). Two general classes can be identified: markers that generally are expressed in a co-dominant fashion so that the heterozygote is distinguishable from both homozygotes, and dominant markers in which the heterozygote is identical to one homozygote. Isozymes, RFLPs (restriction fragment length polymorphisms), CAPS (cleaved amplified polymorphisms) and SSRs (simple sequence repeats) are usually co-dominant markers, while many polymerase chain reaction (PCR)-based markers such as RAPDs (random amplified polymorphic DNAs), ISSRs (inter-simple sequence repeats) and AFLPs (amplified fragment

Table 12.2. Characteristics of some common types of DNA markers.

AFLP	Amplified fragment length polymorphism – PCR based, gel analysis, multilocus, sequence anonymous
CAPS	Cleaved amplified polymorphic sequence – PCR followed by restriction enzyme digestion, gel analysis, single locus per gel, sequence known, can target genes
ISSR	Inter-simple sequence repeat – PCR based, gel analysis, multilocus, sequence anonymous
RAPD	Random amplified polymorphic DNA – PCR based, gel analysis, single to a few loci per gel, sequence anonymous
RFLP	Restriction fragment length polymorphism – restriction enzyme and gel based, single to a few loci per gel, can target genes
SNP	Single nucleotide polymorphism – PCR based, gel, plate or chip analysis; single locus, but multiplexing possible, sequence known, can target genes
STS	Sequence-tagged site – PCR based, gel analysis, single locus, sequence known
SSR (STMS, STR)	Simple sequence repeat – PCR based, gel analysis, single locus, sequence known, can target some genes

The method of detection of the DNA sequence (e.g. PCR) is listed, the method for distinguishing alleles (e.g. gel analysis), whether the method generally detects one or several to many marker loci in a single analysis, whether the DNA sequences detected are known or anonymous, and whether the method can target specific genes.

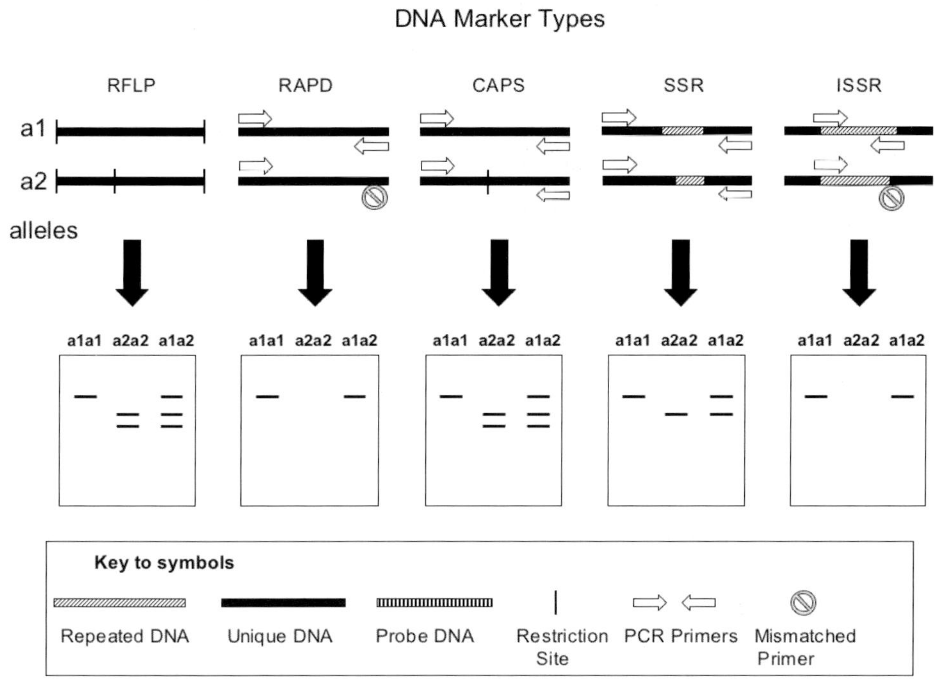

Fig. 12.1. Relationship between DNA sequence differences between two alleles and banding patterns observed for five types of gel-based marker systems.

length polymorphisms) are dominant. For a cross in which most markers are expected to have 1:1 segregation, dominance does not affect mapping. However, for an F_2-type population, or one in which a variety of segregation types occurs (as is usual in citrus), co-dominant markers are more informative, meaning that, for a given population size, the accuracy of a linkage estimate from co-dominant markers is greater than that for dominant markers. Figure 12.1 illustrates the relationship between DNA sequence variation and banding patterns for several gel-based marker systems. Note how dominant markers arise when sequence variation in the PCR primer-binding site prevents amplification of one allele. Figure 12.2 shows examples of each marker system in citrus.

A second classification of markers is based on the number of loci detected per analysis. Maps can be developed much more rapidly and inexpensively using high-throughput markers where several to many markers can be detected in a single analysis. This is often called multiplexing. Marker systems in this class include AFLPs, ISSRs and RAPDs, all PCR-based techniques in which anonymous DNA sequences are analysed. In contrast, RFLPs, CAPS and SSR markers are based on detection of known DNA sequences through hybridization or PCR amplification. These markers are typically analysed 'one at a time'.

An advantage of the sequence-based markers such as RFLPs and CAPS is that the primers or probes can be designed to target genes, while most anonymous DNA sequences are not likely to be genes. Mapping genes allows comparison of maps with those of other plant species, a process known as synteny analysis. Such markers are also relatively useful for quantitative trait locus (QTL) analysis because markers may themselves be candidate genes to control the trait (see below).

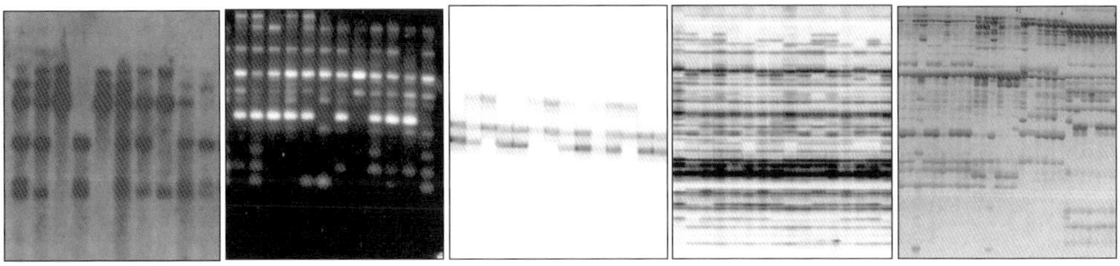

Fig. 12.2. Examples of marker variation for five types of DNA marker systems. The RFLP and RAPD markers were separated on agarose gels. The SSR and AFLP patterns were analysed on an automated DNA analyser. The ISSRs were analysed on a silver-stained polyacrylamide gel and show patterns in seedlings from different source trees, not a single segregating population.

At present, an efficient mapping approach is to use a moderate number (100) of co-dominant markers such as SSRs as anchors for the map, and a larger number (200) of AFLP or other high-throughput markers to increase marker density. SSRs are particularly informative because they have many alleles and are heterozygous in many individuals. These characteristics allow them to be used to combine and compare maps developed in different populations. CAPS or single nucleotide polymorphism (SNP) markers could also be used for this purpose if they are sufficiently polymorphic.

On the horizon, but not yet developed in citrus, are very high-throughput mapping methods that involve parallel analysis of many markers (e.g. SNPs) per individual, or many individuals per marker. These approaches are being developed in humans, and in model organisms such as *Arabidopsis* and rice. These methods could revolutionize citrus mapping by facilitating analysis of much larger populations and/or development of much denser maps.

Marker Methods

As indicated above, a variety of different marker methods are available for citrus mapping and marker-aided selection. Methods involving PCR amplification are generally preferred over older methods such as RFLP because they require much less DNA from each individual to be tested, and PCR-based methods require less labour per marker. Most marker methods require separation of PCR products on agarose or polyacrylamide gels. Polyacrylamide gels have higher resolution than agarose gels so that more allelic variants can be distinguished. Polyacrylamide gels can also reveal more loci per gel than agarose if they are stained with silver or very sensitive DNA-binding dyes such as SYBR-Gold, which reveal more fragments than the ethidium bromide stains traditionally used with agarose gels. However, polyacrylamide gels require more labour for gel preparation and generally have lower throughput than agarose gels, so choosing between these systems is not an easy decision.

Citrus Maps

Six groups have published linkage maps of citrus (Table 12.3), and additional maps are being developed. It is difficult to compare these maps because they have been developed with different mapping programs and generally share few markers (Roose *et al.*, 2000), but see Ruiz and Asins (2003) for a comparison of *Citrus* and *Poncirus* maps. These maps are incomplete in that the

Table 12.3. Summary of linkage maps of citrus (modified from Roose et al., 2000).

Cross	Total Markers[z]	Marker types[y]	Pop. size	Map length cM	Major linkage groups[x]	Percent distorted	Reference
C. maxima x (*C. maxima* x *P. trifoliata*)	310	I, F, R, S, O	60	874	9	26	Sankar and Moore, 2001
C. maxima x (*C. reshni* x *P. trifoliata*)	34, 97	I, F, R, O	52	600, 1500	4, 7	6, 37	Luro *et al.*, 1995
C. sunki x *P. trifoliata*	63, 62	R	80	732, 867	5, 5	nd	Cristofani *et al.*, 1999
C. volkameriana x *P. trifoliata*	97, 73	I, R, F, S, O	80	460, 342	5, 4	29, 40	Ruiz and Asins., 2003
C. latipes x *C. aurantium*	92, 247	A, R, F	120?	433, 964	5, 8	nd	Recupero *et al.*, 2000
(*C. paradisi* x *P. trifoliata*) x (*C. sinensis* x *P. trifoliata*)	153	I, F, R, S, O	57	701	14	17	Roose *et al.*, 2000

[z] Where two values are given, these refer to maps for the female and male parent respectively
[y] I=isozymes, F=RFLP, R=RAPD, A=AFLP, S=SSR, O=Other
[x]Major linkage groups is the number of groups with more than 4 markers.

number of major linkage groups is often less than the number of chromosomes (nine), and in most cases the total number of markers is too few to give complete coverage of the genome. These criticisms do not indicate that the maps are not useful, but rather that development of effective linkage maps in citrus is incomplete. A second criticism of some of the maps is that the mapping populations are too small. Resolution of a map is determined by the number of recombination events on each chromosome. When too few meioses are studied, then few or no recombinants will be found between some loci and these cannot then be ordered accurately. Mapping populations should consist of at least 100 progeny to minimize these problems.

Markers for Specific Genes

Markers for several genes have been identified in citrus. In many cases, the approach used was bulked segregant analysis (Michelmore *et al.*, 1991) rather than complete mapping. In this strategy, a population in which the trait segregates is identified, and equal amounts of DNA from 6–8 individuals with contrasting phenotypes for the trait are combined to form two bulks. For example, with a disease resistance gene, one bulk is composed of resistant individuals, and one of susceptible individuals. These two samples are then screened for a large number of PCR-based markers. Only if a marker is closely linked to the trait gene is it likely that one bulk will have a specific marker band that is absent in all individuals of the other bulk. The basis of this approach is that a 1/8 dilution of template DNA still allows detection of a band. Candidate markers are then tested on individual plant samples to determine whether the marker–trait association is robust. It is efficient to analyse two bulks of each phenotypic class to reduce the frequency of false positives. Often many candidate markers are discarded based on analysis of these single-individual samples. For the remaining markers, the analysis is then extended to a larger population of 100 or more individuals to confirm the association and estimate a map distance between the trait and marker. For traits determined by major genes, bulked segregant analysis is quite effective in citrus. However, because the individual plants are not homozygous, only some linked markers are informative for dominant traits. For example, in an F_2-type segregation, a dominant marker allele (*M*) linked in repulsion phase to a trait allele (*t*) will not be discovered because the dominant phenotypic class (genotypes *TT* and *Tt*) will have marker genotypes *mm* and *Mm*. The recessive phenotypic class (*tt*) will have marker genotype *MM* and so the marker band will be observed in both trait classes.

Cheng and Roose (1995) located markers for dwarfing by the rootstock Flying Dragon trifoliate orange, but these are not needed in breeding because the dwarfing trait appears associated with visible markers, curved thorns and 'zigzag' stem growth. Additional research (M. L. Roose, unpublished) indicates that the dwarfing trait is not transmitted to progeny of citrus × Flying Dragon crosses. Markers for *Ctv*, the citrus tristeza virus resistance gene, in trifoliate orange have been developed by several groups (Gmitter *et al.*, 1996; Mestre *et al.*, 1997, Fang *et al.*, 1998b), and markers very close to the resistance are available. Another gene for which markers have been identified is *acitric*, a pummelo gene that reduces citric acid in fruit when heterozygous and nearly eliminates it when homozygous (Fang *et al.*, 1998a). Markers for genes involved in nucellar embryony (Garcia *et al.*, 1999), citrus nematode tolerance (Ling *et al.*, 2000) and salinity tolerance (Tozlu *et al.*, 1999b) have also been located. There is considerable ongoing work in this area, and it is likely that markers for many additional traits will be reported in the future. Such markers are useful to the breeder in that they allow MAS as described below.

QTL Analysis

QTL analysis refers to a genetic analysis of a trait in which genes that influence the trait are located on a genetic map. The quantitative value of a trait is normally considered to be influenced by one to many genes and the environment. Analytical approaches are now quite sophisticated, but all depend upon particular marker alleles being associated with particular trait alleles (linkage disequilibrium). If we consider a genetic linkage map, each progeny individual has a particular genotype for each locus. If a marker locus is near a gene that influences the trait, then there will have been little recombination between the marker alleles and the trait alleles. Various statistical tests are applied to identify those marker loci or positions for which values of the trait differ significantly. These are putative QTLs. Many different computer packages are available to conduct QTL analysis, but not all are easy to use for citrus crosses because some require populations (such as F_2, backcross or recombinant inbred lines) derived from homozygous parents. QTL CARTOGRAPHER (Basten *et al.*, 1998) and MAPQTL (Van Ooijen and Maliepaard, 1996) have been used in previous QTL mapping studies. See Liu (1998) for a through discussion of QTL analysis.

The conventional approach to QTL analysis is to study a measurable trait on progeny (replicated if possible) in a segregating population for which a linkage map has been developed. Marker and trait data are then jointly analysed to identify QTLs. However, citrus crosses are more complex than generally assumed by programs designed to analyse populations derived from inbred lines. Consider for example a 'backcross' between pummelo, as the recurrent parent, and trifoliate orange. Both pummelo and trifoliate orange are somewhat heterozygous, so their potential genotypes are A_1A_2 and A_3A_4, although for any locus A_1 may be identical to A_2 and/or A_3 to A_4. This means that the 'backcross' parents could be $A_1A_2 \times A_1A_3$, $A_1A_2 \times A_1A_4$, $A_1A_2 \times A_2A_3$, $A_1A_2 \times A_2A_4$, etc. instead of the simple $A_1A_1 \times A_1A_3$ configuration assumed by models developed for a true backcross. Thus this design does not detect effects of all alleles in trifoliate orange, only the particular alleles that are passed to the F_1 parent. Furthermore, the effects of trifoliate alleles are not cleanly separated from effects of pummelo alleles if the pummelo parent is heterozygous for a QTL.

An alternative approach, linkage disequilibrium mapping, is now being developed; it depends only on natural linkage disequilibrium between traits and markers. These methods do not require a mapping population derived from specific parents. Instead, existing individuals (such as a variety collection) are measured for the trait and genotyped for previously mapped markers. In this approach, it is essential that marker genotypes reflect the specific DNA sequence (haplotype) present at the locus because the marker information is used to infer that individuals of unknown ancestry but having the marker allele also share alleles for genes in the surrounding region. Statistically significant associations between marker alleles and trait values are then identified. The extent of linkage disequilibrium varies among plants depending on their history, mating system and other factors. If disequilibrium extends for large distances (hundreds of kilobases), then it is fairly easy to locate regions containing QTLs, but fine-scale mapping of the QTL is not possible. Conversely, if disequilibrium covers only short distances, fine mapping is easier if an initial QTL–marker association can be found. We know little about the extent of disequilibrium in citrus, but it is likely to vary greatly with the population. With a species group having sexual reproduction, such as pummelo or mandarin, disequilibrium is likely to cover smaller regions than if interspecific hybrids such as orange are included. In general, current citrus maps are not sufficiently dense to apply these methods, but this is likely to improve in the future. See Flint Garcia *et al.* (2003) for a review of linkage disequilibrium mapping in plants.

Several QTL studies have been reported in citrus. Tozlu *et al.* (1999a, b) studied QTLs for many traits related to growth and ion accumulation under normal and saline conditions in 47 progeny from a backcross of pummelo and trifoliate orange. Despite the rather small population size, they identified 70 putative QTLs that influenced various traits. Some QTLs influence growth under both normal and saline conditions, while others were specific to only one environment. In some cases, QTLs for ion accumulation localized to the same region as QTLs for growth under saline conditions.

A second example is provided by Garcia *et al.*, (1999) who studied QTLs associated with nucellar embryony (apomixis) in 50 progeny of a cross of *C. volkameriana* × *P. trifoliata*. They found QTLs affecting polyembryony in both parents, indicating that genes influencing the trait segregate in both parents. No QTLs of major effect were found. However, the single-parent maps on which this analysis was based included only 45 and 38 markers, and it is not clear what portion of the genome is covered by these maps. Therefore, it is likely that only some of the heterozygous genes influencing nucellar embryony in these parents were discovered. Any QTL analysis is limited to discovery of QTLs in the portion of the genome covered by segregating markers, so map coverage is an important issue.

Marker-assisted Selection

MAS can increase the efficiency of citrus breeding, and may speed release of new cultivars. Essentially, the procedure involves identification of a marker that is polymorphic in the target population and closely linked to a gene that is segregating and the target of selection. In such situations, selection for the marker allows the breeder to identify seedlings that carry the trait gene with a low error rate. Only seedlings with desirable genotypes are carried forward for additional testing. Conditions in which MAS is more efficient that conventional breeding have been discussed by a number of authors (Luby and Shaw, 2001). Traits that are costly to score by phenotype, such as traits that require fruiting trees, or those that have low heritability and can be accurately scored only with a progeny test (percentage of nucellar seedlings in a rootstock) are likely to be good candidates for MAS.

Markers must be informative about the trait, i.e. the marker genotype must predict the progeny genotype. A dominant marker (*M*) that is in repulsion phase with a dominant trait gene (*T*) is not efficient in an F_2 configuration: *Mt*/*mT* × *Mt*/*mT* will give 25% *Mt*/*Mt*, 50% *Mt*/*mT* and 25% *mT*/*mT* in the progeny if there is no recombination between the marker and the trait gene. All *mm* progeny will be *TT*, but 75% of the population will be discarded, including 50% of the *T*-carrying progeny. If the marker and trait gene are in coupling phase, *MT*/*mt* × *MT*/*mt*, the progeny will contain 25% *MT*/*MT*, 50% *MT*/*mt* and 25% *mt*/*mt*; all progeny with the *M* phenotype will carry the T gene and therefore selection will only require discarding 25% of the population, and no progeny with the desirable allele *T* will be discarded. Since it is not easy to generate very large populations in citrus, discarding progeny with desirable genes reduces the overall efficiency of the program. Co-dominant, multiallelic markers are more informative in general, but still can be inefficient in certain crosses.

A common situation where MAS is useful is in transferring a gene from one species to another. It is easy to find informative markers in such crosses.

The breeder should always consider the economic efficiency of MAS before implementing it. Such a program may require identification of markers, and this may be costly and take at least 1–2 years. The cost of DNA extraction and marker analysis should also be considered. If the same degree of improvement can be achieved by phenotypic or family selection, then MAS is not justified (Luby and Shaw, 2001).

As markers for more traits become available, the costs of DNA extraction and marker testing will effectively decrease because they will be spread over multiple traits. Citrus biology limits what can be achieved with these methods. If we start with a population of 1000 hybrids, which is larger than usually obtained in citrus breeding, and impose MAS for three independent traits that segregate 1:1, we will have 1000 $(1-0.5^3) = 125$ selected hybrids to evaluate in the field for other traits. If the target traits segregated 3:1 and the recessive was selected, then we would expect only 1000 $(1-0.75^3) = 15.6$ progeny to have the desired marker genotype. This is too few to have much chance of finding hybrids with additional desired traits. Therefore, it is not realistic to expect that we can select for a large number of traits by MAS in a single generation. Similarly, it may often be difficult to select for a complex QTL and still have sufficient plants to select for other traits. A two-generation strategy involving MAS for different sets of traits in two populations, then intercrossing selections from these populations, might be more effective, but the long generation time of citrus reduces the value of this approach.

References

Asins, M.J. (2002) Present and future of quantitative trait locus analysis in plant breeding. *Plant Breeding* 121, 281–291.

Basten, C.J., Weir, B.S. and Zeng, Z.B. (1998) *QTL Cartographer: a Reference Manual and Tutorial for QTL Mapping*. Department of Statistics, North Carolina State University, Raleigh, pp. 1–32.

Cheng, F.S. and Roose, M.L. (1995) Origin and inheritance of dwarfing by the citrus rootstock *Poncirus trifoliata* 'Flying Dragon'. *Journal of the American Society for Horticultural Science* 120, 286–291.

Cristofani, M., Machado, M.A. and Grattapaglia, D. (1999) Genetic linkage maps of *Citrus sunki* Hort. Ex. Tan. and *Poncirus trifoliata* (L.) Raf. and mapping of citrus tristeza virus resistance gene. *Euphytica* 109, 25–32.

Fang, D.Q., Federici, C.T. and Roose, M.L. (1998a) Development of molecular markers linked to a gene controlling fruit acidity in Citrus. *Genome* 40, 841–849.

Fang, D.Q., Federici, C.T. and Roose, M.L. (1998b) A high-resolution linkage map of the citrus tristeza virus resistance gene region in *Poncirus trifoliata* (L.) Raf. *Genetics* 150, 883–890.

Flint Garcia, S.A., Thornsberry, J.M. and Buckler, E.S., IV (2003) Structure of linkage disequilibrium in plants. *Annual Review of Plant Biology* 54, 357–374.

Garcia, R., Asins, M.J., Forner, J. and Carbonell, E.A. (1999) Genetic analysis of apomixis in *Citrus* and *Poncirus* by molecular markers. *Theoretical and Applied Genetics* 99, 511–518.

Gmitter, F.G., Xiao, S.Y., Huang, S., Hu, X.L., Garnsey, S.M. and Deng, Z. (1996) A localized linkage map of the citrus tristeza virus resistance gene region. *Theoretical and Applied Genetics* 92, 688–695.

Jarrell, D.C., Roose, M.L., Traugh, S.N. and Kupper, R.S. (1992). A genetic map of citrus based on the segregation of isozymes and RFLPs in an intergeneric cross. *Theoretical and Applied Genetics* 84, 49–56.

Khan, I.A. and Roose, M.L. (1988) Frequency and characteristics of nucellar and zygotic seedlings in three cultivars of trifoliate orange. *Journal of the American Society for Horticultural Science* 113, 105–110.

Knapp, S.J., Holloway, J.L., Bridges, W.C. and Liu B.H. (1995) Mapping dominant markers using F2 matings. *Theoretical and Applied Genetics* 91, 74–81.

Lander, E.S., Green, P., Abrahamson, J., Barlow, A., Daly, M.J., Lincoln, S.E. and Newburg, L. (1987) MAPMAKER: an interactive computer package for constructing primary genetic linkage maps of experimental and natural populations. *Genomics* 1, 174–181.

Ling, P., Duncan, L.W., Deng, Z., Dunn, D., Hu, X., Huang, S. and Gmitter, F.G., Jr (2000) Inheritance of citrus nematode resistance and its linkage with molecular markers. *Theoretical and Applied Genetics* 100, 1010–1017.

Liu, B.H. (1998) *Statistical Genomics: Linkage, Mapping and QTL Analysis.* CRC Press, Boca Raton, Florida, pp. 611.

Luby, J.J. and Shaw, D.V. (2001) Does marker-assisted selection make dollars and sense in a fruit breeding program? *HortScience* 36, 872–879.

Luro, F., Lorieux, M., Laigret, F., Bové, J.M. and Ollitraut, P. (1995) Cartographie du génome des agrumes á l'aide des marquers moléculaires et distorsions de ségrégation. In: INRA (ed.) *Techniques et Utilisations des Marqueurs Moléculaires.* INRA, Paris, pp. 69–82.

Maliepaard C., Jansen J. and Van Ooijen J.W. (1997) Linkage analysis in a full-sib family of an outbreeding plant species: overview and consequences for applications. *Genetical Research* 70, 236–250.

Mestre, P.F., Asins, M.J., Pina, J.A., Carbonell, E.A. and Navarro, L. (1997) Molecular markers flanking citrus tristeza virus resistance gene from *Poncirus trifoliata* (L.) Raf. *Theoretical and Applied Genetics* 94, 458–464.

Michelmore, R.W., Paran, I. and Kesseli, R.V. (1991) Identification of markers linked to disease-resistance genes by bulked segregant analysis: a rapid method to detect markers in specific genomic regions by using segregating populations. *Proceedings of the National Academy of Sciences of the USA* 88, 9828–9832.

Roose, M.L., Fang, D., Cheng, F.S., Tayyar, R.A., Federici, C.T. and Kupper, R.S. (2000) Mapping the *Citrus* genome. *Acta Horticulturae* 535, 25–32.

Ruiz, C. and Asins, M.J. (2003) Comparison between *Poncirus* and *Citrus* genetic linkage maps. *Theoretical and Applied Genetics* 106, 826–836.

Sankar, A.A. and Moore, G.A. (2001) Evaluation of inter-simple sequence repeat analysis for mapping in *Citrus* and extension of the genetic linkage map. *Theoretical and Applied Genetics* 102, 206–214.

Stam, P. and Van Ooijen, J.W. (1995) *JoinMap (tm) Version 2.0: Software for the Calculation of Genetic Linkage Maps.* CPRO-DLO Wageningen, The Netherlands.

Tozlu, I., Guy, C.L. and Moore, G.A. (1999a) QTL analysis of morphological traits in an intergeneric BC_1 progeny of *Citrus* and *Poncirus* under saline and nonsaline environments. *Genome* 42, 1020–1029.

Tozlu, I., Guy, C.L. and Moore, G.A. (1999b) QTL analysis of Na^+ and Cl^- accumulation related traits in an intergeneric BC_1 progeny of *Citrus* and *Poncirus* under saline and nonsaline environments. *Genome* 42, 692–705.

Van Ooijen, J.W. and Maliepaard, C. (1996) *MAPQTL™ Version 3: Software for the Calculation of QTL Positions on Genetic Maps.* CPRO-DLO, Wageningen, The Netherlands.

13 Cloning and Characterization of Disease Resistance Genes

Fred G. Gmitter Jr[1], Zhanao Deng[2] and Chunxian Chen[1]

[1]University of Florida, Institute of Food and Agricultural Sciences, Citrus Research and Education Center, Lake Alfred, FL 33850, USA; [2]University of Florida, Institute of Food and Agricultural Sciences, Gulf Coast Research and Education Center, Bradenton, FL 34203, USA

Introduction

The establishment of the new science of genomics in the 1980s has revolutionized biological research into a completely new era. This revolution is enabling scientists through the use of powerful new tools to know the sequence of the entire genome, to understand the structure and function of every single gene, to study the organization and evolution of any sequence and to determine the molecular bases of genetic variation. Undoubtedly, the Human Genome Plan (HGP) and other similar genome programmes on model organisms (e.g. *Arabidopsis* and rice as model plant organisms) have been playing the leading role in the incredible developments in this area. Among the special focuses and major objectives in these genomic programmes are the characterization, isolation and manipulation of the agriculturally important disease resistance (*R*) genes responsible for defending against diseases caused by various pathogens (Takken and Joosten, 2000).

Dozens of plant *R* genes have been isolated from various plant species or crops since the first four were published (Martin *et al.*, 1993; Bent *et al.*, 1994; Jones *et al.*, 1994; Whitham *et al.*, 1994): *Pto* from tomato resistant to *Pseudomonas syringae* pv. tomato, *N* from tobacco resistant to tobacco mosaic virus (TMV), *RPS2* from *Arabidopsis* resistant to *Pseudomonas syringae* pv. tomato and *Cf-9* from tomato resistant to *Cladosporium fulvum*. A few excellent reviews have been written on plant *R* genes regarding their structure, function, evolution, organization, classification, defence mechanisms, and so on (Ellis *et al.*, 2000; Takken and Joosten, 2000; Young, 2000; Lehmann, 2002). Briefly, there are six major classes of *R* genes (Table 13.1), five classified earlier plus a new class from *RPW8* in *Arabidopsis* (Xiao *et al.*, 2001). Most *R* gene products have the LRR (leucine-rich repeat) domain, and also largely are in the NBS (nucleotide-binding site)–LRR form. Two groups of NBS–LRR *R* genes were clearly distinguished by the conserved amino acid motifs in the NBS domain. One group, which was comprised of sequences encoding an N-terminal domain with homology to Toll/inter-

Table 13.1. Classes of resistance genes, classified by their structural domains.

Class	Example of R genes	Structural domain description	First reference
PK	***Pto***	Serine/threonine protein kinase with myristoylation site	Martin *et al.*, 1993
TIR–NBS–LRR	***N***, *RPP1*, *RPP10*, *RPP14*, *L6*, *L1-12*, *M*, *RPP5*, *RPS4*	Cytoplasmic protein with homologies to Toll cytoplasmic domain, apoptotic ATPases CED4 and Apaf1, and C-terminal LRRs	Whitham *et al.*, 1994
CC–NBS–LRR	***RPS2***, *Prf*, *RPM1*, *RPS5*, *RPP8*, *Mi*, *I2*, *Dm3*, *Pi-B*, *Xa1*	Cytoplasmic protein with homologies to CC, apoptotic ATPases CED4 and Apaf1, and C-terminal LRRs (also called LZ–NBS–LRR)	Bent *et al.*, 1994
LRR–TM	***Cf-9***, *Cf-2*, *Cf-4*, *Cf-5*, *Hcr9-4E*, *HSI*$^{pro\text{-}1}$	Transmembrane protein with extracellular LRRs (eLRRs)	Jones *et al.*, 1994
LRR–TM–PK	***Xa21***	Transmembrane protein; extracellular LRRs, cytoplasmic (LRR kinase)	Song *et al.*, 1995
SA–CC	***RPW8.1*, *RPW8.2***	Putative signal anchor (SA) for membrane insertion, and putative CC domain	Xiao *et al.*, 2001

Genes in bold are the first reported in each class.
LRR = leucine-rich repeat; NBS = nucleotide-binding site; TIR = Toll/interleukin-1 receptor; CC = coiled-coil; LZ = leucine zipper; PK = protein kinase.

leukin-1 receptor (TIR), surprisingly was entirely absent from monocot plants. The other group was found in both monocot and dicot plants. The conservation of *R* gene structural domains provides not only an informative guide for isolation of other *R* genes, but also a broad potential to enable use of natural genetic resistance more efficiently by rapid transfer of genes among crops for specific resistance breeding objectives.

Citrus is one of the most important fruit crops widely planted in the tropical and subtropical zones of the world. It is a major source of revenue in many regions of the developed world, as well as an important nutrient source in some less industrialized areas. However, in every growing region of the world, cultivated citrus species are susceptible to various diseases or pests, including but not limited to citrus tristeza virus (CTV), citrus nematode (*Tylenchulus semipenetrans*), *Phytophthora* species, citrus canker (*Xanthomonas axonopodis*), viroids, phytoplasmas, etc. that can result in tremendous economic losses of citrus production. Traditional genetic manipulation methods and conventional hybridization breeding programmes, which are time consuming and labour intensive, have been proven ineffective or irrelevant for many citrus breeding objectives. Such goals as breeding disease-resistant cultivars of most species are nearly impossible to achieve because of high genetic heterozygosity, long periods of juvenility and little genetic knowledge of the inheritance of most traits. Most scion cultivar groups, specifically sweet oranges, grapefruit, most lemons and many mandarins, have diversified by somatic mutations and not through sexual recombination and segregation; these groups, by virtue of their limited genetic base in addition to the other breeding impediments, are not amenable to genetic manipulation through hybridization strategies. Other characters, such as nucellar embryony, self-incompatibility or large tree size in most citrus varieties, will significantly increase the difficulties because these characters generally reduce the possibilities of obtaining genetic information on the transmission and heritability of useful agronomic traits in citrus. The rapidly developed molecular marker technology and MBC (map-based cloning) approach in the past several years have greatly accelerated understanding of the citrus genome (Durham *et al.*, 1992; Cai *et al.*, 1996; Liou *et al.*, 1996) and the genetics of several agriculturally important single gene or quantitative trait loci (QTLs) related to disease, pest and stress resistances (Cai *et al.*, 1994; Gmitter *et al.*, 1996, 1997; Deng *et al.*, 1997, 2000; Fang *et al.*, 1997, 1998; Ling *et al.*, 2000). The polymorphic DNA markers closely linked to an important specific trait can greatly facilitate early selection (called marker-assisted selection; MAS) and minimize costs associated with plant size and juvenility. Genetic transformation methods allow trait-specific modification of commercial cultivars, including those listed above for which sexual hybridization is precluded as a strategy for improvement. We will summarize the advances in MBC of the single dominant CTV resistance gene, designated *Ctv* (Gmitter *et al.*, 1996), and in characterization of a major citrus nematode resistance QTL, named *Tyr1* (Ling *et al.*, 2000). We will discuss the cloning and characterization of other *R* genes within the citrus genome, through a strategy based on sequence homologies among plant *R* genes in general, and provide information on the distribution of *R* genes and resistance gene candidates (RGCs). Finally, the chapter will conclude with consideration of the future prospects for manipulating disease resistance through genomic approaches.

Map-based Cloning Approach – A Powerful Tool for Citrus Genetic Improvement

MBC, also called positional cloning, is a very powerful approach used to isolate genes without prior knowledge of the gene product or even the mechanism of function, as long as the gene is associated with a segregating natural or mutant phenotype (Jander *et al.*, 2002). MBC consists of four basic steps; Fig. 13.1 shows the schematic diagram of the entire process. The first step is to develop a local genetic map where at least two closely linked or co-segregating

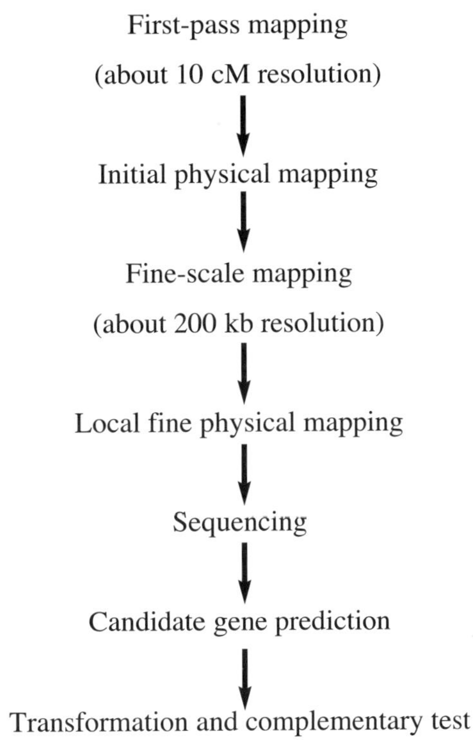

Fig. 13.1. A schematic diagram of the map-based cloning process.

molecular markers flanking the gene have been identified. The genetically closer to the gene of interest the markers are, the less difficult the subsequent screening will be. A larger mapping population is used after the initial marker screening of a smaller population so that the recombination between those markers and the gene is evaluated more precisely. The second step is to screen a large insert genomic library such as BAC (bacterial artificial chromosome) with the markers and conduct chromosome walking to construct a physical contig covering the flanked region. This step also includes the essential task of finding additional co-segregating (no recombination) markers to narrow the gene further into a short physical region, facilitating the selection of the BAC clones to sequence from the contig. The third step is to sequence the selected large insert clones and to predict the candidate gene(s). This step relies on the application of various programmes for gene prediction. The last step generally is to perform genetic complementation tests through transformation of the candidate gene(s) into a selection without the known phenotype to confirm the gene function. If the transgenic selection expresses the expected phenotype such as a disease resistance, the gene of interest should be enclosed within that transformed fragment. Once the gene is successfully isolated, an in-depth molecular and biochemical analysis can follow.

The tremendous advantage of the MBC process is the ability to tag and clone any gene associated with a distinct and scorable phenotype. Such resources, either natural or mutant, appear nearly unlimited. One of the most interesting applications of MBC is the isolation of resistance genes to the diseases and pests that cause tremendous production losses in food species, as well as resulting in environmental pollution due to use of pesticides for control. The process is likely to become more routine as the so-called post-genome era of most model organisms is coming. Already for some organisms there exist powerful and publically available resources, including high density genetic and physical maps, and even full genome sequences, which greatly speeds up the whole process of MBC. For example, in *Arabidopsis*, it took a total effort of 3–5 person-years to isolate a gene in 1995, while now less than 1 person-year is required (Jander *et al.*, 2002). For citrus, an intensive effort in genomic tool development and MBC will enable greater understanding of citrus genetics and cloning of important genes. Application of the improved tools for genetic improvement is essential to keep the world's citrus industries functioning as profitable enterprises into the future.

Citrus Tristeza Virus Resistance Gene (*Ctv*) – A Single Dominant Allele

CTV is the most significant viral pathogen of citrus worldwide, inducing two major serious disease syndromes: 'quick decline' and 'stem-pitting' (Garnsey and Lee 1988). The former may cause tree death resulting from phloem necrosis below the graft union of the infected scions on sour orange (*C. aurantium* L.) rootstock; the latter will decrease tree vigour, fruit size and quality, and productivity regardless of rootstock (Bar-Joseph *et al.*, 1989). There is no known source of broad resistance to CTV among *Citrus* species. The CTV resistance gene was characterized as a single dominant gene (Yoshida 1985, 1993) and designated *Ctv* (Gmitter *et al.*, 1996); it is found only in *Poncirus trifoliata* (the trifoliate orange), a sexually compatible species related to *Citrus*.

Several populations were used initially to evaluate the inheritance of CTV resistance and susceptibility by enzyme-linked immunosorbent assay (ELISA) to detect the virus in rootstocks and in the grafted hybrids, using either a monoclonal antibody or polyclonal antibodies (Garnsey *et al.*, 1981, 1987; Gmitter *et al.*, 1996). A minimum of four repeated ELISAs over time were run independently at two laboratories to ensure identical results for individuals from a larger population, consisting of 65

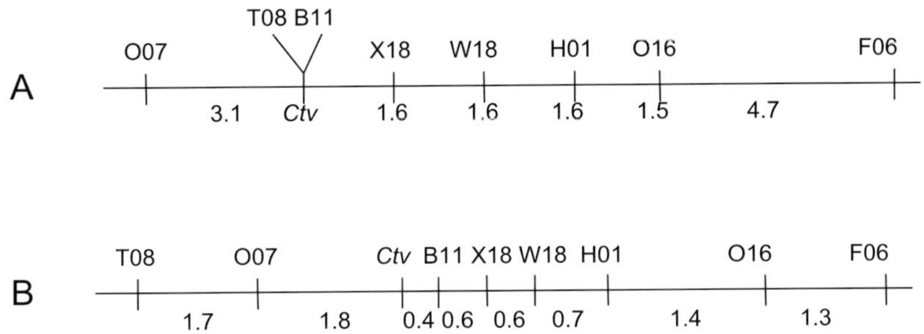

Fig. 13.2. The first genetic linkage map of the *Ctv* region, based on the R family (*C. grandis* and *P. trifoliata* backcross to *C. grandis*), containing eight linked RAPD markers (a). The consensus map of the *Ctv* region based on the R family and the other four families (b). MAPMAKER 3.0 and JOINMAP 1.3 were used for mapping, respectively. All RAPD markers were identified initially using bulked segregant analysis (BSA; Michelmore *et al.*, 1991).

BC_1 hybrids of 'Thong Dee' pummelo (*C. grandis* L.) × USDA 17-40 ('Thong Dee' pummelo × Pomeroy trifoliate orange (*P. trifoliata*)). It was designated the R family and used as the primary marker screening and genetic mapping population. The first localized map derived from the R family, as well as the consensus map based on all populations used, is presented in Fig. 13.2 (Gmitter *et al.*, 1996). This map started the powerful engine to move the exploration of the gene forward.

Because of limitations in using quickly identified random amplified polymorphic DNA (RAPD) markers in genomic library screening and other applications, the amplified fragments were cloned, sequenced, and used to develop the specific sequence characterized amplified region (SCAR) markers linking the *Ctv* gene (Deng *et al.*, 1997). In this stage, a preliminary screening of a BAC library using those specific markers resulted in several individual BAC clones in this region (Gmitter *et al.*, 1997; Chen *et al.*, 1999). Continued marker development and mapping in a larger segregating population led to an increase of marker density on the genetic map, and development of additional BAC libraries harbouring larger inserts were screened so that both the genetic and physical maps in this region were integrated (Deng *et al.*, 2001a, b; Fig. 13.3). Finally, BAC contigs that spanned both the resistance and susceptibility allelic regions were constructed, additional fine mapping was conducted to delimit the gene region, and full sequences of both allelic chromosome regions were determined.

Similar work has been conducted at the University of California at Riverside, and Texas A & M University (Fang *et al.*, 1998), and a full-length sequence of the resistant chromosome region has been published by Yang *et al.* (2003). This sequence reveals numerous RGCs, as well as other gene sequences and mobile genetic elements, based on gene prediction software. Many of these candidate genes have been cloned into transformation vectors, and several are already confirmed to have been integrated into the genomes of CTV-susceptible citrus lines for genetic complementation assays (F. G. Gmitter, T. E. Mirkov and M. L. Roose, unpublished data). To date, however, the gene or genes ultimately responsible for the broad-spectrum resistance of the Mendelian gene, *Ctv*, have not yet been confirmed.

Citrus Nematode Resistance Gene (*Tyr1*) – A Major QTL

Citrus nematode (*T. semipenetrans* (Cobb 1914)) is an important rhizospheric para-

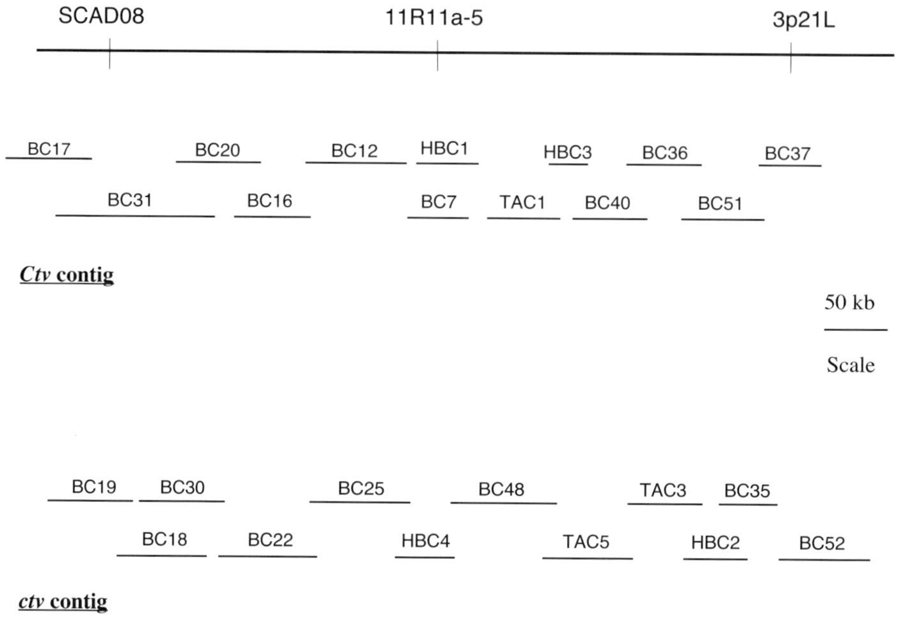

Fig. 13.3. BAC contigs of the CTV resistance gene region (*Ctv* contig) and its allelic susceptibility gene region (*ctv* contig).The insert sizes of these clones vary between 70 and 175 kb. Clones from the *Bam*HI, *Hin*dIII and transformation-competent artificial chromosome (TAC) libraries were named with the pre-fix BC, HBC and TAC, respectively. The genetic map of the *Ctv* locus region was integrated with BAC contigs and is shown above the contigs.

sitic pest that causes severe damage to citrus roots and results in decreased production (Duncan and Cohn 1990). Selecting citrus nematode-resistant rootstocks is an important objective for citrus breeding. The traditional screening and selection approach is very labour intensive and inefficient. Molecular linkage markers have made it possible to understand the quantitative inheritance of citrus nematode resistance. Once their linkage with the nematode resistance was identified and mapped, it became possible to perform early selection in citrus-resistant rootstock breeding programmes, through the MAS process. It was found that most biotypes of *Poncirus* are resistant to citrus nematode (Cameron *et al.*, 1954) and that the resistance is inherited in some fertile intergeneric hybrids from the trifoliate orange (Swingle and Reece 1967). An intergeneric backcross family from (Clementine mandarin (*C. reticulata*) × Hamlin orange (*C. sinensis*), designated LB6-2) × (Swingle citrumelo (*C. paradisi* × *P. trifoliata*)) was used to identify associated molecular markers and to evaluate the mode of inheritance of citrus nematode resistance.

It was critical to determine the phenotype of citrus nematode resistance and susceptibility segregating among individuals of this family. The individuals, including parents and hybrids, were inoculated at the same time to monitor the nematode population development with live citrus nematodes directly collected from the heavily infected roots of field citrus trees. The numbers of citrus nematode females, eggs and juveniles were estimated by counting from nematode extractions prepared from root samples (Baines *et al.*, 1968). Nematode extracts were taken three times from each replicate, and the mean number of female nematodes per gram of fresh root (females/g

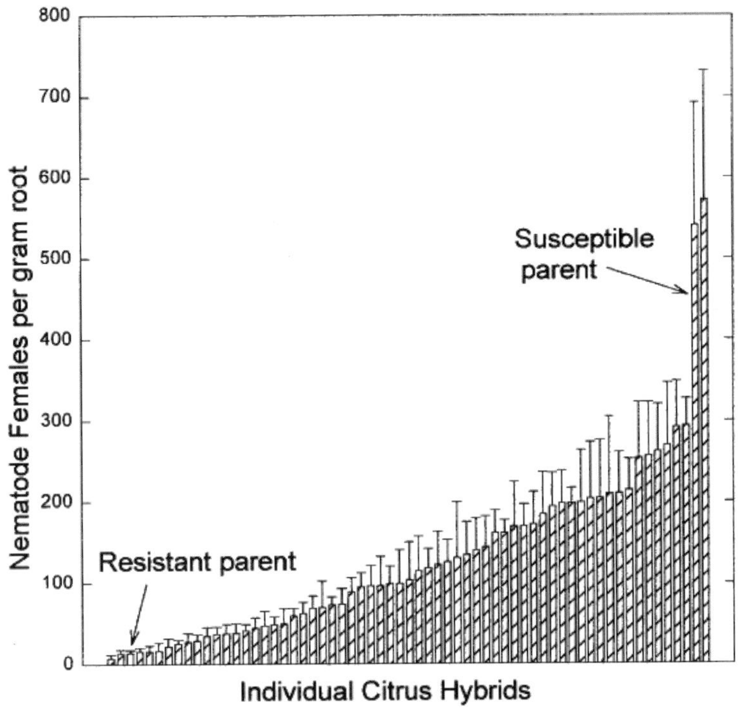

Fig. 13.4. The phenotypic distribution (mean and standard deviation from six replicates) of 62 hybrids in response to citrus nematode inoculation. The infestation levels of the resistant (Swingle citrumelo) and susceptible parent (LB6-2) are indicated (Ling *et al.*, 2000).

of root) was calculated for each hybrid. All the hybrids and parental plants were evaluated for their response to citrus nematode inoculation by counting the female nematodes in each of the root samples. Data were subjected to analysis of variance using the MSTATC computer program.

Analysis of variance revealed that there were significant differences among the hybrids for the numbers of females/g of root ($P < 0.001$). The mean values among the hybrid individuals were continuously distributed in a wide range, indicating the quantitative nature of the resistant trait (Fig. 13.4; Ling *et al.*, 2000). The most resistant and susceptible individuals were assorted into groups by mean number of females/g of root to perform bulked segregant analysis (BSA) and identify the associated DNA markers, which resulted in a local map with 14 RAPD, SCAR and RGC markers (Fig. 13.5; Ling *et al.*, 2000).

Some 15 specific markers were developed in the *Tyr1* region and used to evaluate their potential in selecting for nematode resistance in a citrus rootstock breeding programme, by applying the markers to several populations and other individuals with known phenotype (unpublished data). This was done by screening the BAC library described above with NBS–LRR class RGC sequences; over 200 positive clones were identified by two RGCs that mapped in the *Tyr1* QTL region. A few of the BAC clones were found to be closely linked with *Tyr1*, when the primers based on the clones' insert sequences yielded polymorphic fragments that were associated with the phenotypes in the population previously phenotyped by Ling *et al.* (2000). By applying the primer walking approach, three

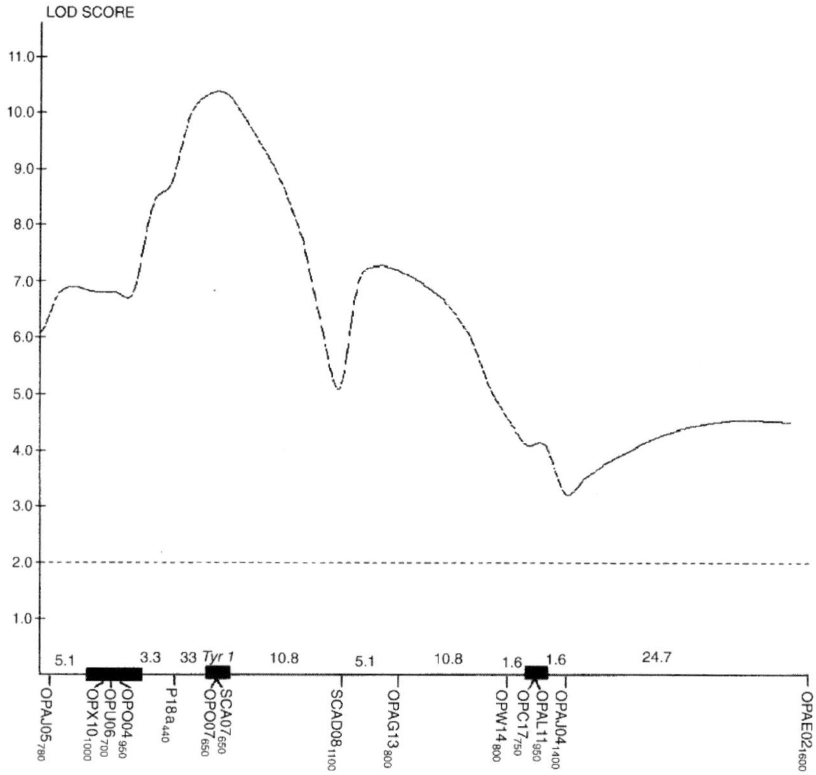

Fig. 13.5. The localized linkage map of the citrus nematode resistance gene region, generated by MAPMAKER/QTL. The map distances (cM) were calculated using the Kosambi function and are indicated between markers. The major citrus nematode resistance gene (*Tyr1*) region corresponded to the highest QTL peak. The LOD score value indicated that the *Tyr1* region was most closely associated with marker loci OPO07650 and SCA07650. The co-segregating marker loci are indicated by solid bars.

complete NBS–LRR class resistance gene sequences were tagged and identified separately in three BAC clones. More specific markers were developed from these tagged sequences and relatively high density genetic maps were constructed by incorporating the newly developed markers and previously developed markers in the CN (citrus nematode) family. These markers are being applied in a larger mapping population to enrich the marker linkage map between the *Ctv* and *Tyr1* region. In addition, the utility for MAS for citrus nematode resistance is being explored (X. Xiang, Q. Zheng, S. Huang, C. Chen, L. W. Duncan, Z. Deng, K. D. Bowman and F. G. Gmitter, Jr, unpublished results).

Homology-based Cloning of RGCs – A Tool for Gene Discovery

Technological primers

Homology-based cloning of RGCs has become feasible in the last several years after a number of plant *R* genes had been isolated through transposon tagging or MBC (Kanazin *et al.*, 1996; Leister *et al.*, 1996; Yu *et al.*, 1996). Characterization of the predicted protein products from these cloned *R* genes indicates that different plant species may use just a few classes of proteins in conferring resistance to various bacterial, fungal and viral pathogens, and nematodes (Hammond-Kosack and Jones

1997; Staskawicz *et al.*, 2001). In addition, the proteins encoded by *R* genes seem to contain domains and motifs that are conserved across different classes of *R* genes. A domain that is particularly common among cloned *R* genes is the NBS domain (Hammond-Kosack and Jones, 1997; Meyers *et al.*, 1999). A few highly conserved motifs within the NBS domain have allowed the development of degenerate oligonucleotide primers able to amplify DNA sequences that are very similar to the NBS–LRR class *R* genes. Subsequently, this polymerase chain reaction (PCR)-based approach has also been used to clone DNA sequences in the kinase or receptor-like kinase class *R* genes (Deng and Gmitter 2003). Several terms are used in the literature to refer to these sequences, including RGCs (resistance gene candidates), RGAs (resistance gene analogues) or RGLs (resistance gene-like sequences); the term RGC is used in the present discussion. The choice of this term is based not only on the strong similarities of these sequences to cloned plant *R* genes and their possession of the typical features of cloned *R* genes, but also on their increased uses in a so-called candidate gene cloning approach and their potential of being functional *R* genes, as confirmed in a few recent cases (Aarts *et al.*, 1998; McDowell *et al.*, 1998; Pflieger *et al.*, 2001).

Though many of the RGCs obtained so far correspond only to part of the full structures of cloned *R* genes, they have found valuable uses in plant breeding and genetics research. Linkage mapping data have confirmed that many RGCs are tightly linked or co-segregate with known disease resistance gene loci that are responsible for resistance to bacterial, fungal or viral pathogens, or parasitic nematodes (Kanazin *et al.*, 1996; Leister *et al.*, 1996; Yu *et al.*, 1996). In this sense, RGCs may provide the ultimate molecular markers for tagging plant disease resistance traits (Michelmore, 1996). Several RGCs, when used with an MBC strategy, have led to isolation of functional resistance genes. Examples include the cloning of lettuce *Dm3* and *Arabidopsis Rpp8* that are responsible for resistance to two different downy mildew pathogens, *Bremia lactucae* and *Pernospora parasitica*, respectively (Aarts *et al.*, 1998; McDowell *et al.*, 1998). Thus, RGCs may serve as valuable resources for using a candidate gene approach to clone plant *R* genes. Because of their close association with *R* genes, RGCs can also provide effective tools to facilitate understanding of some fundamental features of *R* genes, such as *R* gene organization, distribution and evolution (Michelmore, 1996).

Application of a homology-based cloning approach may be of particular significance to citrus and many other horticultural, agricultural crops and forest plants for genetic improvement or engineering of disease resistance. Transposon tagging and MBC, two approaches previously widely used in plant *R* gene cloning, are not readily accomplished in citrus and many other crops, because of their long generation times, large plant sizes or complex genetic behaviours. In addition, each plant genome may contain hundreds of potential *R* gene sequences to specify resistance to different pathogens. Isolating these sequences one by one, as in the case of MBC or transposon tagging, can be time consuming. A homology-based PCR amplification approach may allow access to numerous candidate sequences in a genome within a short period of time, thus making it desirable, perhaps essential, in order to tag and understand the enormous repertories of *R* genes in plant genomes.

Cloning and characterization of citrus RGCs

In citrus, cloning of RGCs initially was attempted as a way to facilitate map-based isolation of *Ctv*. As the previous section in this chapter stated, a high resolution linkage map and molecular markers tightly linked to or co-segregating with the target disease resistance genes are prerequisites for using a map-based gene cloning strategy. The *Ctv* locus had been tagged with a dozen RAPD and several SCAR markers by 1997, but most of these markers were found to be more distant from the resistance gene locus than expected and desired, in subsequent mapping with a large population consisting of 678 backcross progeny. This indicated the need to search for more molecular markers closer to the gene. To meet this need, Deng *et al.* (2000) used two previously reported primers (LM637 and LM638) and four primers (F11, R11, R16 and R18) newly designed from the conserved NBS domain and amplified four major DNA bands from the genomic DNA of USDA 17-47, an intergeneric CTV-resistant hybrid of *C. grandis* and *P. trifoliata*. When these bands were cloned and 39 individual clones were sequenced, they identified 22 RGCs. These sequences were very similar to the NBS–LRR class *R* genes and contained the typical motifs (P-loop, kinase-2 and kinase-3a) of the NBS domain found in this class of *R* genes. The overall amino acid identity between these citrus RGCs and five representative *R* genes (*Arabidopsis* RPS2, RPM1, tobacco N, flax L6 and tomato I2) ranged from 18 to 42%. Using a 70% amino acid identity threshold value, the 22 sequences were grouped into ten classes, RGC1–RGC10. The majority of the citrus RGCs seem to belong to the so-called non-TIR group of NBS–LRR genes, while only two of the classes, RGC1 (clone Pt6) and RGC2 (clone Pt14), fall into the TIR group. Specific primers developed from the divergent regions of 13 representative RGC sequences detected three types of polymorphism between 'Thong Dee' pummelo and USDA 17-40: fragment length difference, restriction site difference, or presence and absence of amplified bands. Three markers (18P33, Pt8a and Pt9a) from the RGCs were found to be closely linked to *Ctv*. Marker Pt8a also segregated in a population for citrus nematode resistance mapping and was located very close to *Tyr1* (Deng *et al.*, 2000; Ling *et al.*, 2000).

This initial success in finding RGCs associated with disease resistance loci in *Poncirus* prompted a more targeted and thorough search for new RGCs more tightly linked to or even co-segregating with *Ctv*. Sixteen combinations of degenerate primers designed from the NBS–LRR class *R* genes were screened on two genomic DNA pools each consisting of eight CTV-resistant and eight CTV-susceptible individuals, respectively, from a segregating backcross population (the R family). One DNA band polymorphic between the two bulks was detected on a polyacrylamide gel. This DNA fragment was highly similar to the NBS–LRR class *R* genes. Two markers derived from this fragment, 11R1-1a (cleaved amplified polymorphic sequence (CAPS) marker) and 11R1-1a5 (restriction fragment length polymorphism (RFLP) marker), co-segregated with *Ctv* in high resolution linkage mapping. Subsequently, they allowed a chromosome landing on BAC clones of the *Ctv* locus, thus greatly expediting the progress in constructing BAC contigs to cover the CTV resistance and susceptibility allelic regions.

Recent efforts in citrus RGC cloning have been focused on the receptor-like kinase class *R* genes that are represented by rice *Xa21*. This class of *R* genes is unique in that they each contain an extracellular LRR domain for signal recognition and a cytoplasmic catalytic kinase domain for signal transduction (Song *et al.*, 1995; Ronald, 1997). Few of this class of RGCs have been amplified or cloned in plants other than rice and its related species, and *Arabidopsis*. Degenerate primers are not readily available for this class of *R* genes, because their LRRs are poorly conserved while their kinase domains are too common in many proteins responsible for diverse functions. Deng and Gmitter (2003) aligned the kinase domains of rice *Xa21* and tomato Pto protein (a kinase class *R* gene for tomato bacterial streak resistance) and found two well-conserved motifs in the kinase subdomains I and VIII. Degenerate primers, kindF1 and kindR1, designed from these motifs, amplified genomic DNA sequences from the intergeneric hybrid USDA 17-47 that are similar to rice *Xa21*. Subsequently, oligonucleotide primers derived from one of the genomic sequences (A2UP and A2LW) amplified *Xa21*-like sequences from citrus BACs. All 53 sequences contained open reading frames (ORFs); their deduced peptide sequences carried the features found in the corresponding region of rice *Xa21* protein and they shared 55–60% amino acid identity and 65–71% similarity. In multiple sequence alignment, the sequences were clustered into five groups, designated as CRK1–5 (citrus receptor-like kinase). Each group was further divided into 2–4 subgroups, according to their sequence similarities. Rice *Xa21* specifies broad-spectrum resistance to more than 30 isolates of *Xanthomonas oryzae* pv *oryzae* (*Xoo*), the causal agent of rice bacterial blight. *Xanthomonas axonopodis* pv *citri* (*Xac*), a pathogen related to *Xoo*, causes the devastating disease citrus canker in citrus. Resistance to citrus canker has been observed in *C. ichangensis* and kumquat (*Fortunella* spp.). It will be an intriguing question to determine whether *Xa21*-like sequences are involved in resistance to citrus canker.

The rice genome may contain more than 700 copies of NBS type sequences, and *Arabidopsis* may have nearly 1% of its total genomic sequences encoding NBS domains. The copy number of NBS–LRR class sequences in citrus genomes has yet to be determined, but it may be expected to be in the hundreds. Capturing some of these sequences, converting them into specific markers and locating them on to genetic linkage maps may provide powerful tools to tag, track and utilize gene loci for disease resistance in *Poncirus*, *Citrus* or *Fortunella*. As a first step to realize this potential of citrus RGCs, Deng *et al.* (2001a) probed several BAC and transformation-competent (TAC) libraries with cloned RGC sequences to identify large-insert clones containing resistance gene-like sequences. Screening of the USDA 17-47 *Bam*HI BAC library with 13 representative NBS sequences from RGC1 to RGC10 yielded 322 positive clones. These clones may correspond to 40–70 genetic loci in the *Citrus* or *Poncirus* genome and contain 80–140 copies of unique NBS sequences. A similar BAC library screening with A2, a prevalent *Xa21*-like sequence identified above, resulted in 79 BAC clones containing receptor-like kinase-encoding capacity. These clones were assembled into 35 contigs, and contain approximately 50–70 copies of *Xa21*-like sequences. These BACs can provide a valuable source in citrus *R* gene tagging, mapping, cloning and characterization. Insert ends of some of these BACs have been sequenced and will be converted into specific PCR markers for further tests of potential association with particular resistance gene loci.

RGCs amplified by PCR with NBS or kinase domain-derived primers usually are 400–650 bp in size and correspond only to part of the full-length structures of the cloned *R* genes. A genuine question is frequently raised regarding whether the amplified RGCs are from structurally and functionally ‘real’ genes. Compared with the rapid increase in the number of cloned RGCs in various plants, progress in this area is lagging remarkably. This may be partly due to the ease of PCR amplification and subsequent cloning, but mostly due to the tremendous amount of work required for (full gene) sequence acquisition, construct development, genetic transformation and expression analysis. So far, such work has been done only in a limited number of cases with a few plants. To answer this critical question, we obtained the upstream and downstream sequences, through primer-based walking, of four NBS-encoding regions in four BACs hybridizing with RGC Pt8, and two kinase-encoding regions in BACs hybridizing with RGC A2. Two of the DNA sequences contain ORFs that can translate into polypeptide sequences without any stop codons. They seem to contain all the structural features of an NBS–LRR class *R* gene. In a BLAST search of the GenBank database, their best hit is *Arabidopsis* RPS2 among cloned *R* genes, showing a 30–35% amino acid identity, 50–55% similarity and an Expect value less than e-110. The other two NBS–LRR class sequences seem to be pseudogenes as they each contain one stop codon in their coding regions. The two receptor-like kinase class RGCs (17o6RLKP and 26m19RLKP) contained all the nine domains of rice *Xa21*, including a signal peptide, extracellular LRRs, transmembrane and kinase domains. Overall, they shared approximately 44% amino acid identity and 51% similarity with rice *Xa21*. Their kinase domains contained all the 12 subdomains and the serine–threonine-specifying motifs of rice *Xa21*, and the 15 invariant amino acids of protein kinases. The LRR domains of 17o6RLKP and 26m19RLKP consist of the same number of imperfect repeats as *Xa21*, and they shared approximately 44% amino acid identity and approximately 51% similarity. Although the functionality of these full-length RGCs remains to be determined, the above data indicate that many RGCs may have complete structures as cloned functional *R* genes.

General procedure and considerations

Homology-based cloning of citrus RGCs involves several steps: identifying conserved domains and motifs in cloned *R* genes, designing appropriate degenerate PCR primers, amplification of genomic or complementary DNAs, cloning of heterogeneous PCR fragments, characterization and selection of plasmid clones, and sequencing and identification of RGCs. Some of these steps such as cloning and sequencing of PCR fragments are similar to standard procedures used in molecular cloning, and therefore discussion will be focused on some of the special conditions used for RGC cloning. When further genetic, structural or functional analyses are intended for cloned RGCs, it may require acquisition of additional DNA sequences, and conducting linkage mapping, gene expression study and/or genetic transformation, most of which are beyond the scope of this chapter.

Conserved motifs and degenerate primers

A number of computer software programs can be used for aligning multiple amino acid sequences to reveal highly conserved motifs in cloned plant *R* genes. CLUSTERX (Thompson *et al.*, 1997) and GCG 'PILEUP' are among the commonly used programs. Successful PCRs require two primers facing each other; thus at least two regions of highly conserved amino acids should be found, and each should extend for at least six amino acids, so primers of at least 17 nucleotides could be derived. The distance between the two chosen motifs needs to be at least 20 amino acids, so that PCR products can be at least 100 bp in size. The two motifs are converted to sense and antisense nucleotide sequences, respectively. Desirable degenerate primers are expected to amplify diverse RGC sequences but few non-specific sequences. Therefore, it is necessary to control the degeneracy of designed primers. Using inosine as an alternative base, and/or synthesizing multiple sets of primers for a given conserved motif are two approaches frequently adopted.

The NBS domain, particularly its P-loop motif, has been frequently targeted in degenerate primer design. This is because the domain appears at a high frequency in the cloned *R* genes and it is most conserved among the genes. In addition, an internal hydrophobic domain (HD) between the NBS and LRR domains of the NBS–LRR class of *R* genes also has been used extensively for amplification of this class of RGCs. In PCRs, these primers are expected to amplify DNA fragments of the size of approximately 500 bp, provided that no introns are located between the DNA sequences encoding the two motifs.

Two short stretches of peptide sequences (FG(K/S)VYKG and GY(A/I)(A/D)PEY) in the kinase subdomains I and VIII were found to be relatively well conserved in the receptor-like kinase class *R* genes. Degenerate primers designed from these regions amplified citrus DNA sequences similar to the receptor-like kinase class *R* genes. Attempts have been made to design degenerate primers for the LRR domain found in the NBS–LRR class and LRR class *R* genes, but the poor conservation of amino acids other than leucine has led to few successes in targeting these classes of RGCs in plant genomes.

PCR amplification

When a degenerate primer is synthesized, each primer actually contains numerous oligonucleotide sequences as dictated by its degeneracy. Because of this nature, it is necessary to increase, compared with standard PCRs, the amount of degenerate primers and DNA templates used in PCRs to amplify RGC sequences. In addition, appropriate annealing temperatures may need to be identified to obtain the expected products.

Cloning, sequencing and sequence analysis

PCR products amplified with degenerate primers are rather heterogeneous, even though they appear as a single band or a few bands on an agarose gel. It is necessary to obtain several dozen or more independent recombinant DNA clones to capture the major part of the sequence diversity initially detected by the degenerate primers. Restriction pattern analysis is then used to identify unique recombinant clones for subsequent sequencing. A BLAST search of the GenBank database seems to be a very convenient and effective way to reveal whether or not obtained sequences share any similarity to cloned *R* genes or the vast number of plant RGCs. In-depth analysis of RGC sequences for their coding capacities, pairwise comparisons, multiple alignment, phylogenetic relationship, etc. can be accomplished using various free or commercial software programs such as DNASISMAX, the GCG SEQWEB software ('GAP', 'PILEUP', etc.) and the CLUSTAL X package.

Reducing non-specific sequences

Many DNA sequences obtained may have no similarity to plant *R* genes. This phenomenon is common in homology-based cloning of plant *R* genes or genes responsible for other functions. It is because of the degenerate nature of primers used, and the tremendous complexity of plant genomes. Several strategies are often used to reduce the number of non-specific sequences in PCR amplification and cloning. Using multiple sets of primers of lower degeneracy, optimizing annealing temperatures for PCR, separation of DNA bands on agarose or polyacrylamide gels, and the choice of DNA bands are various approaches attempting to minimize the problem of non-specificity.

Recent technical developments

Targeted isolation of RGCs for known disease resistance loci

Early studies involved cloning and sequencing of RGCs, followed by designing specific primers and screening them for a potential association with resistance gene loci. Numerous degenerate primers have been designed from various classes of cloned *R* genes in recent years, but it can be rather time consuming to follow this procedure, especially when it is uncertain whether or not any RGCs, or what class of RGCs, might be linked to the target resistance gene loci of interest. In search of new markers for *Ctv*, we developed a so-called BSA-RGC approach, based on the bulked segregant analysis (Michelmore *et al.*, 1991) that has been in use for screening large numbers of random primers in tagging disease resistance loci with RAPD markers. The BSA-RGC approach screens degenerate primers and their amplified products with pairs of resistant and susceptible plant groups before RGCs are cloned, sequenced and converted into molecular markers. It was used in the development of markers 11R1-1a and 11R1-1a5 that co-segregated with *Ctv* and led to chromosome landing on *Ctv*-containing BACs. This approach should be applicable to other disease resistance gene loci, and to allow a large number of primer combinations to be screened in a short period of time and potential RGC sequences linked to the target gene locus to be identified, before engaging with DNA fragment cloning and sequencing.

Improving RGC sequence diversity

One of the goals in RGC cloning is to obtain DNA sequences most similar to cloned *R* genes and from as many different chromosomal locations as possible, i.e. maximum sequence diversity. Frequently it becomes difficult to find new RGCs after dozens of plasmid clones have been sequenced. One recent study indicates that BACs may help overcome this difficulty and allow for capturing more diverse RGC sequences. Deng and Gmitter (2003) compared the diversity of receptor-like kinase class RGC sequences amplified from genomic and BAC clone DNA, respectively. The 29 sequences amplified from genomic DNA with degenerate primers kindF1 and kindR1 fell into two major groups and four subgroups in cluster analysis. The prevalent one among the 29 sequences, A2, was used to screen the USDA 17-47 *Bam*HI library, and positive BACs were used for amplifications with A2-derived specific primers. Surprisingly, the 23 sequences amplified from these BACs fell into five major groups and eight subgroups, i.e. sequences not revealed by amplification of genomic DNA were found among the BAC clone inserts.

Acquisition of full-length gene structures

The acquisition of full-length gene sequences is essential for structural and/or functional analysis of RGCs, and also beneficial for marker development and genetic mapping. Long-range PCR, 5' and 3' RACE (rapid amplification of cDNA ends), and cDNA and genomic library screening have been used in various cases for this purpose. Recent experience indicates that BAC libraries (and clones) are an excellent resource for this endeavour (Deng and Gmitter 2003). Through primer-based walking on BACs, full-length gene structures have been obtained for a number of NBS–LRR class and receptor-like kinase class RGCs. BAC libraries are easy to screen and their clones contain large inserts (50–250 kb). These characteristics facilitate not only recovery of upstream and downstream coding or non-coding regions of specific RGCs, but also the identification and characterization of resistance gene sequence clusters. BACs in some vectors, such as pYLTAC (such clones have been called TACs), may be transferred into *Agrobacterium* cells and used in genetic transformation. Several BAC libraries have been constructed in the last several years, and they cover both the *Poncirus* and *Citrus* genomes with good representation (Deng *et al.*, 2001a).

R Genes and RGCs in the Citrus Genome – Features and Distributions

Clustering of *R* genes and RGCs

Clustering of citrus *R* genes was first observed in genetic mapping of *Ctv* and *Tyr1* (Deng *et al.*, 2000; Ling *et al.*, 2000). Initially, a RAPD marker amplified with random primer OP07 was identified linked to the *Ctv* locus (Gmitter *et al.*, 1996). In a separate effort toward mapping citrus nematode resistance with a separate population, this random primer yielded a DNA fragment linked to *Tyr1* and apparently of the same size as the marker associated with *Ctv*. When the *Ctv*-linked RAPD fragment was cloned and converted into SCAR marker SCO07, its high specificity allowed the confirmation that the two RAPD markers identified in different mapping populations for two resistance traits were the same and that *Ctv* and *Tyr1* are closely linked in coupling phase. This relationship between *Ctv* and *Tyr1* has been confirmed further in subsequent mapping with a number of specific PCR markers derived from RGCs or BAC end sequences. The genetic distance between the two loci was estimated to be approximately 12 cM, using populations of fewer than 100 progeny individuals (Deng *et al.*, 2000; Ling *et al.*, 2000). A more precise estimation is being attempted using a large population (X. Xiang, Q. Zheng, S. Huang, C. Chen, L. W. Duncan, Z. Deng, K.D. Bowman and F.G. Gmitter, JR, unpublished results unpublished). The quantitative inheritance of citrus nematode resistance may hinder pinpointing of *Tyr1* on to a linkage map; consequently, the exact genetic location of *Tyr1* will be not as well defined as that of *Ctv*, which creates difficulties for a precise estimation of the genetic distance.

Physical evidence of *R* gene and RGC clustering has also become available recently from BAC and BAC contig analysis. BAC clones contain genomic DNA inserts in the range of 35–300 kb, therefore they are ideal for examining gene or sequence clustering on a much finer scale compared with genetic mapping. In characterizing BACs identified with NBS–LRR class RGCs, Deng *et al.* (2001a) found that more than 40% of the 29 BAC contain 2–4 copies of NBS–LRR class sequences within each BAC. Deng and Gmitter (2003) analysed 35 BACs identified with receptor-like kinase class RGCs and recognized that approximately 50% of the clones each carry 2–3 copies of *Xa21*-like sequences. The average size of these BACs is approximately 115 kb. Considering the close relationship between RGCs and *R* genes in plants, this clustering feature of RGCs in citrus BACs may serve as a good indication of possible *R* gene distribution in the citrus genome.

Sequencing of overlapping BAC clones at the *Ctv* locus and its immediate vicinity has revealed the existence of five copies of CC–NBS–LRR type complete gene sequences, two copies of pseudogene sequences and nine resistance gene fragments (Yang *et al.*, 2003). Similar gene contents have been predicted from sequencing BACs covering the *Ctv* locus but from a different source (Z. Deng, S. Huang, C. Chen and F.G. Gmitter, Jr, unpublished). These data, therefore, have provided the most direct evidence of *R* gene or RGC clustering around a functional resistance gene locus in citrus.

It was first recognized that the *Ctv–Tyr1* region may contain a major cluster of resistance genes in 2000 (Deng *et al.*, 2000), based on the available data at that time (Fang *et al.*, 1998; Deng *et al.*, 2000; Ling *et al.*, 2000). As more genetic mapping, RGC cloning and BAC sequences accumulate, this region seems to contain more *R* genes or RGCs than previously thought. An up-to-date count of *R* genes and *R* gene-like sequences includes up to 12 copies of NBS–LRR sequences and 3–4 copies of LRR sequences. Though the functions of most of these sequences remain to be determined, the discovery of this cluster may have important implications for disease resistance breeding.

Transposable elements and *R* genes

One surprising feature revealed in BAC clone sequencing was the rich presence of transposable elements around *R* gene sequences. Five *copia*-like and three *gypsy*-like retrotransposons, and two mutator-like transposons have been identified within the 282 kb DNA at the *Ctv* locus (Yang *et al.*, 2003). In addition, six small DNA fragments similar to parts of other transposable elements are present within the 282 kb. The total length of these elements together is greater than 65 kb, i.e. accounting for approximately 23% of the nucleotides. In a sense, *R* gene sequences are surrounded by transposable elements at the *Ctv* locus.

Similar phenomena have been observed in other NBS–LRR sequence-rich BACs (C. Chen and F.G. Gmitter, Jr, unpublished; Z. Deng, S. Huang, C. Chen and F.G. Gmitter, Jr, unpublished). It seems that the NBS–LRR class *R* gene sequences are quite often associated with retrotransposable elements in citrus. The presence of transposable elements with other classes of *R* gene sequences has not yet been characterized in citrus. It has been proposed that transposons might have participated in diversification of the *Xa21* gene in rice for *R* gene evolution (Song *et al.*, 1997). *Arabidopsis* has a small genome and a relatively low portion of transposable elements. In many plants with large genomes, transposable elements often contribute to most of their DNA contents. Compared with these plants, the citrus genome is quite small and has been expected to contain a relatively small percentage of its DNA content. It remains to be determined whether transposable elements contribute to *R* gene evolution in citrus and if they are widely distributed around other *R* genes in citrus.

Future Prospects

As indicated at the outset of the chapter, the science of genomics, driven by advances in the field of human genomics and the rapidly expanding capability of technology, is revolutionizing the entire field of biology and genetics. These developments have already impacted the citrus genetic improvement community worldwide. However, it is expected that the rate of progress toward greater understanding of the genetic control of agriculturally important traits, and the ability to manipulate and modify citrus genomes to improve plant performance, will greatly accelerate. As some of the examples illustrated above reveal, much is possible in achieving quantum improvements both in the ability to make meaningful genetic changes and in the nature of the changes that are possible. It is important to recognize, though, that most of the work described above took more than ten years to accomplish. The long path toward cloning the CTV resistance gene, for example, began in the mid-1990s, and it has not yet been completed. This is not for lack of diligent effort, but more clearly it is related to attempting to achieve the goal with technology that was not as mature as that which exists currently. Since the inception of the *Ctv* cloning project, many new technological developments have been seen. These include the ability to sequence comprehensive expressed sequence tag (EST) libraries and even complete genomes, microarray technology to study global gene expression, bioinformatics capabilities that enable processing volumes of informative data that would have been unimaginable just a decade ago, and high-throughput marker systems for mapping projects that can yield high density maps containing thousands of markers. These new abilities will clearly hasten progress into the future.

There remain some impediments that have plagued citrus breeders over the course of time, and some new impediments as well, that may temper the rate of accelerated progress. Citrus will remain a plant for which the fundamental underlying genetic studies are difficult to conduct. The difficulty with which citrus can be transformed, and the fact that methods to transform several recalcitrant citrus lines have not yet been developed, is a limitation to making the leap from fundamental genomic information and understanding of disease resistance, or any other trait for that matter, to practical deployment of genetically improved citrus plants for the benefit of producers and consumers. Our inability to quickly and accurately phenotype citrus plants, for example CTV resistance, has meant that although we have full sequence information of the resistance genomic region and even expression data, we have yet to confirm the absolute genetic element that distinguishes immune from susceptible plants. Many of the traits of greatest potential commercial value that might be modified are subject to complex genetic and environmental interactions, and there currently are very few reports of cloned QTLs that have contributed to substantial improvements in performance. Finally, the goals for citrus genetic advancement, whether for scion or rootstock cultivars, are almost always expressed in terms of improving multiple characteristics. Nonetheless, outcomes of citrus genomic research are already providing incredible opportunities to achieve genetic goals for disease resistance previously thought to be impossible, from unravelling the basis of genetic resistance to pathogens, to developing efficient tools to select for resistance, to developing highly targeted approaches to modify very specific traits. As the research community continues to explore the potential solutions that already exist within the citrus genomic pool, with currently available and with newer, more effective technologies that are being developed, the impossibilities of today will become tomorrow's realities.

References

Aarts, M.G., te Lintel Hekkert, B., Holub, E.B., Beynon, J.L., Stiekema, W.J. and Pereira, A. (1998) Identification of R-gene homologuous DNA fragments genetically linked to disease resistance loci in *Arabidopsis thaliana. Molecular Plant-Microbe Interactions* 11, 251–258.

Baines, R.C., Bitters, W.P. and Clarke, O.F. (1968) Infectivity of two biotypes of citrus nematode on citrus and on some other hosts. *Journal of Nematology* 1, 150–159.

Bar-Joseph, M., Marcus, R. and Lee R.F. (1989) The continuous challenge of citrus tristeza virus control. *Annual Review of Phytopathology* 27, 291–316.

Bent, A.F., Kundel, B.N., Dahlbeck, D., Brown, K.L., Schmidt, R., Giraudat, J., Leung, J. and Staskawicz, B.J. (1994) RPS2 of *Arabidopsis thaliana*: a leucine-rich repeat class of plant disease resistance genes. *Science* 265, 1856–1860.

Cai, Q., Guy, C.L. and Moore, G.A. (1994) Extension of the linkage map in *Citrus* using random amplified polymorphic DNA (RAPD) markers and RFLP mapping of cold-acclimation-responsive loci. *Theoretical and Applied Genetics* 89, 606–614.

Cameron, J.W., Baines, R.C. and Oscar, F.C. (1954) Resistance of the hybrid seedlings of the trifoliate orange to infestation by the citrus nematode. *Phytopathology* 44, 456–458.

Chen, C. and Gmitter, F.G., Jr (1999) Direct cloning and sequencing of bacterial artificial chromosome (BAC) insert ends based on double digestion. *Plant Molecular Biology Reporter* 17, 231.

Cobb, N.A. (1914) Notes of *Mononchus* and *Tylenchulus*. *Journal of the Washington Academy of Science* 3, 287–288.

Deng, Z. and Gmitter, F.G., Jr (2003) Cloning and characterization of receptor kinase class disease resistance gene candidates in citrus. *Theoretical and Applied Genetics* 108, 53–61.

Deng, Z., Huang, S., Xiao, S. and Gmitter, F.G., Jr (1997) Development and characterization of SCAR markers linked to the citrus tristeza virus resistance gene from *Poncirus trifoliata*. *Genome* 40:697–704.

Deng, Z., Huang, S., Ling, P., Chen, C., Yu, C., Weber, C.A., Moore, G.A. and Gmitter, F.G., Jr (2000) Cloning and characterization of NBS–LRR class resistance gene candidate sequences in citrus. *Theoretical and Applied Genetics* 101, 814–822.

Deng, Z., Tao, Q., Chang, Y.L., Huang, S., Ling, P., Yu, C., Chen, C. and Gmitter, F.G., Jr (2001a) Construction of a bacterial artificial chromosome (BAC) library for citrus and identification of BAC contigs containing resistance gene candidates. *Theoretical and Applied Genetics* 102, 1177–1184.

Deng, Z., Huang, S., Ling, P., Yu, C., Tao, Q., Chen, C., Wendell, M.K., Zhang, H.B. and Gmitter, F.G., Jr (2001b) Fine genetic mapping and BAC contig development for the citrus tristeza virus resistance gene locus in *Poncirus trifoliata* (Raf.). *Molecular Genetics and Genomics* 265, 739–747.

Duncan, L.W. and Cohn, E. (1990) Nematode parasites of citrus. In: Luc, M., Sikora, R.A. and Bridge, J. (eds) *Plant Parasitic Nematodes in Subtropical and Tropical Agriculture.* CABI Publishing, Wallingford, UK, pp. 321–346.

Durham, R.E., Liou, P.C., Gmitter, F.G., Jr and Moore, G.A. (1992) Linkage of restriction fragment length polymorphisms and isozymes in Citrus. *Theoretical and Applied Genetics* 84, 39–48.

Ellis, J., Dodds, P. and Pryor, T. (2000) Structure, function and evolution of plant disease resistance genes. *Current Opinion in Plant Biology* 3, 278–284.

Fang, D.Q., Federici, C.T. and Roose, M.L. (1997) Development of molecular markers linked to a gene controlling fruit acidity in citrus. *Genome* 40, 841–849.

Fang, D.Q., Federici, C.T. and Roose, M.L. (1998) A high-resolution linkage map of the citrus tristeza virus resistance gene region in *Poncirus trifoliata* (L.) Raf. *Genetics* 150, 883–890.

Garnsey, S.M. and Lee, R.F. (1988) Tristeza. In: Whiteside, J.O., Garnsey, S.M. and Timmer, L.W. (eds) *Compendium of Citrus Diseases.* APS Press, St Paul, Minnesota, pp. 48–50.

Garnsey, S.M., Barrett, H.C. and Hutchison, D.J. (1981) Resistance to citrus tristeza virus in citrus hybrids determined by enzyme-linked immunosorbent assay. *Phytopathology* 71, 875.

Garnsey, S.M., Barrett, H.C. and Hutchison, D.J. (1987) Identification of citrus tristeza virus resistance in citrus relatives and its potential applications. *Phytophylactica* 19, 187–191.

Gmitter, F.G., Jr, Xiao, S.Y., Huang, S., Hu, X.L., Garnsey, S.M. and Deng, Z. (1996) A localized linkage map of the citrus tristeza virus resistance gene region. *Theoretical and Applied Genetics* 92, 688–695.

Gmitter, F.G., Jr, Louzada, E.S., Deng, Z.A., Huang, S. and Drew, R.A. (1997) A bacterial artificial chromosome (BAC) library for cloning a citrus tristeza virus resistance gene. *Acta Horticulturae* 461, 355–359.

Hammond-Kosack, K.E. and Jones, J.D.G. (1997) Plant disease resistance genes. *Annual Review of Plant Physiology and Plant Molecular Biology* 48, 575–607.

Jander, G., Norris, S.R., Rounsley, S.D., Bush, D.F., Levin, I.M. and Last, R.L. (2002) *Arabidopsis* map-based cloning in the post-genome era. *Plant Physiology* 129, 440–450.

Jones, D.A., Thomas, C.M., Hammond-Kosack, K.E., Balint-Kurti, P.J. and Jones, J.D.G. (1994) Isolation of the tomato *Cf-9* gene for resistance to *Cladosporium fulvum* by transposon tagging. *Science* 266, 789–793.

Kanazin, V., Marek, L.F. and Shoemaker, R.C. (1996) Resistance gene analogs are conserved and clustered in soybean. *Bulletin of the Fruit Trees Research Station* 93, 11746–11750.

Lehmann, P. (2002) Structure and evolution of plant disease resistance genes. *Journal of Applied Genetics* 43, 403–414.

Leister, D., Ballvora, A., Salamini, F. and Gebhardt, C. (1996) A PCR-based approach for isolating pathogen resistance genes from potato with potential for wide application in plants. *Nature Genetics* 14, 421–429.

Ling, P., Duncan, L.W., Deng, Z., Dunn, D., Hu, X., Huang, S. and Gmitter, F.G., Jr (2000) Inheritance of citrus nematode resistance and its linkage with molecular markers. *Theoretical and Applied Genetics* 100, 1010–1017.

Liou, P.C., Gmitter, F.G., Jr and Moore, G.A. (1996) Characterization of the Citrus genome through analysis of restriction fragment length polymorphisms. *Theoretical and Applied Genetics* 92, 425–435.

Martin, G.B., Brommonschenkel, S.H,, Chunwongse, J., Frary, A., Ganal, M.W., Spivey, R., Wu, T., Earle, E.D. and Tanksley, S.D. (1993) Map-based cloning of a protein kinase gene conferring disease resistance in tomato. *Science* 262, 1432–1436.

McDowell, J.M., Dhandaydham, M., Long, T.A., Aarts, M.G.M., Goff, S., Holub, E.B. and Dangl, J.L. (1998) Intragenic recombination and diversifying selection contribute to the evolution of downy mildew resistance at the *Rpp8* locus of *Arabidopsis*. *Plant Cell* 10, 1861–1874.

Meyers, B.C., Dickerman, A.W., Michelmore, R.W., Sivaramakrishnan, S., Sobral, B.W. and Young, N.D. (1999) Plant disease resistance genes encode members of an ancient and diverse protein family within the nucleotide-binding superfamily. *Plant Journal* 20, 317–332.

Michelmore, R.W. (1996) Flood warning – resistance genes unleashed. *Nature Genetics* 14, 376–378.

Michelmore, R.W., Paran, I. and Kessel,i R.V. (1991) Identification of markers linked to disease resistance by bulked segregant analysis: a rapid method to detect markers in specific genomic regions by using segregating populations. *Bulletin of the Fruit Trees Research Station* 88, 9828–9832.

Pflieger, S., Lefebvre, V. and Causse, M. (2001) The candidate gene approach in plant genetics: a review. *Molecular Breeding* 7, 275–291.

Ronald, P.C. (1997) The molecular basis of disease resistance in rice. *Plant Molecular Biology* 35, 179–196.

Song, W.Y., Wang, G.L., Chen, L.L., Kim, H.S., Pi, L.Y., Holsten, T., Gardner, J., Wang, B., Zhai, W.X., Zhu, L.H., Fauquet, C. and Ronald, P. (1995) A receptor kinase-like protein encoded by the rice disease resistance gene, *Xa21*. *Science* 270, 1804–1806.

Song, W.Y., Pi, L.Y., Wang, G.L., Gardner, J., Holsten, T. and Ronald, P.C. (1997) Evolution of the rice *Xa21* disease resistance gene family. *Plant Cell* 9, 1279–1287.

Staskawicz, B.J., Mudgett, M.B., Dangl, J.L. and Galan, J.E. (2001) Common and contrasting themes of plant and animal diseases. *Science* 292, 2285–2289.

Swingle, W.T. and Reece, P.C. (1967) The botany of citrus and its wild relatives. In: Reuther, W., Webber, H.J. and Batchelor, L.D. (eds) *The Citrus Industry*, Vol. 1. University of California Press, Berkeley, California, pp. 190–430.

Takken, F.L.W. and Joosten, M.H.A.J. (2000) Plant resistance genes: their structure, function and evolution. *European Journal of Plant Pathology* 106, 699–713.

Thompson, J.D., Gibson, T.J., Plewniak, F., Jeanmougin, F. and Higgins, D.G. (1997) The ClustalX windows interface: flexible strategies for multiple sequence alignment aided by quality analysis tools. *Nucleic Acids Research* 24, 4876–4882.

Whitham, S., Dinesh-Kumar, S.P., Choi, D., Hehl, R., Corr, C. and Baker, B. (1994) The product of the tobacco mosaic virus resistance gene N: similarity to Toll and the interleukin-1 receptor. *Cell* 78, 1101–1115.

Xiao, S., Ellwood, S., Calis, O., Patrick, E., Li, T., Coleman, M. and Turner, J.G. (2001) Broad-spectrum mildew resistance in *Arabidopsis thaliana* mediated by RPW8. *Science* 291, 118–120.

Yang, Z.N., Ye, X.R., Molina, J., Roose, M.L. and Mirkov, T.E. (2003) Sequence analysis of a 282-kb region surrounding the citrus tristeza virus resistance gene (*Ctv*) locus in *Poncirus trifoliata* L. Raf. *Plant Physiology* 131, 1–11.

Yoshida, T. (1985) Inheritance of susceptibility to citrus tristeza virus in trifoliate orange. *Bulletin of the Fruit Trees Research Station* 82, 17–26. (In Japanese with English summary.)

Yoshida, T. (1993) Inheritance of immunity to citrus tristeza virus of trifoliate orange in some citrus intergeneric hybrids. *Bulletin of the Fruit Trees Research Station* 90, 33–43. (In Japanese with English summary.)

Young, N.D. (2000) The genetic architecture of resistance. *Current Opinion in Plant Biology* 3, 285–290.

Yu, Y.G., Buss, G.R. and Saghai-Maroof, M.A. (1996) Isolation of a superfamily of candidate disease-resistance genes in soybean based on a conserved nucleotide-binding site. *Proceedings of the National Academy of Sciences of the USA* 93,11751–11756.

14 Genetic Transformation of Citrus for Pathogen Resistance

Vicente J. Febres[1], **R. F. Lee**[2] **and Gloria A. Moore**[1]

[1]University of Florida, Horticultural Sciences Department, Plant Molecular and Cellular Biology Program, PO Box 110690, Gainesville, FL 32611, USA; [2]USDA-ARS National Clonal Germplasm Repository for Citrus and Dates, 1060 Martin Luther King Blvd, Riverside, CA 92507, USA

Introduction

Commercial citrus, wherever it is grown, is plagued by numerous pathogens. For example, Citrus tristeza closterovirus (CTV) is the most economically important virus affecting citrus. Worldwide, it is estimated that more than 50 million trees have been lost to CTV, with economic losses in the order of the hundreds of millions of US dollars (Bar-Joseph *et al.*, 1989). Citrus canker, caused by *Xanthomonas axonopodis* pv. *citri*, severely limits citrus production in Asia, South America and other parts of the world. Canker was re-introduced into Florida for at least the third time in 1995 or earlier, first found near Miami airport. Since this introduction, the disease has moved northward toward major citrus-growing areas, in spite of eradication efforts that have removed approximately three million trees. Also, recently, Brazil has seen the appearance of sudden death, a disease of unknown aetiology, and has also been affected by citrus blight (also of unknown aetiology) and citrus variegated chlorosis (caused by the bacterium *Xylella fastidiosa*) for many years. Leprosis, a viral disease that only recently has been characterized molecularly, is a problem in Brazil, Panama, Venezuela and other countries. In addition, viroids such as cachexia and exocortis, and several fungal diseases can also affect citrus production.

A long-term solution to the problems caused by these pathogens is the production of citrus varieties that are genetically resistant or immune to them. This will not be simple to do using conventional breeding, given the reproductive biology of the genus and the lack of resistance genes in commercially acceptable citrus types. Therefore, one strategy to overcome these limitations is to engineer citrus plants for pathogen resistance. The research to produce CTV-resistant plants using pathogen-derived resistance is the farthest along, and will be discussed in the most detail here, but work we have initiated to produce more generalized disease resistance is also described.

Strategies for Producing Pathogen-resistant Plants

Natural resistance

A potential pathogen has to overcome several barriers in order to infect a plant and become an actual pathogen (Thordal-Christensen, 2003). These barriers can be pre-formed (wax, cell walls, secondary metabolites and antimicrobial enzymes), but are often active responses following the recognition of the pathogen (Thordal-Christensen, 2003). Plants often respond in similar ways to host and non-host pathogens sharing some of the same signal components (Thordal-Christensen, 2003). The outcome of an interaction with a pathogen is thus governed by many factors, including the genotypes of the plant and pathogen as well as a complex exchange of signals between the two (McDowell and Dangl, 2000). Activation of inducible defences is contingent upon recognition of an invasion. Structural molecules from the pathogen such as cell wall components and bacterial flagellin elicit a defence response and are considered 'general elicitors' (Martin *et al.*, 2003). However, a more specific detection system by the plant is through a complex array of constitutively expressed *R* (for resistance) genes (Martin *et al.*, 2003). Individual *R* genes have narrow recognition capabilities and they trigger resistance only when the invading pathogen expresses a corresponding *Avr* (for avirulence) gene. Avr proteins can specifically modify host targets that are detected by the R proteins in resistant plants (Zhu *et al.*, 2004). *R* gene-mediated resistance (also called gene-for-gene resistance) is commonly, although not invariably, associated with rapid necrosis of plant cells at the site of invasion, the hypersensitive response, that stops pathogen infection. Six distinct classes of highly polymorphic but structurally conserved R proteins that mediate resistance against different pathogen taxa have been identified (Lahaye, 2002). The majority of the cloned *R* genes encode putatively cytoplasmic proteins with a nucleotide-binding site (NBS) and a C-terminal leucine-rich repeat (LRR) domain (Lahaye, 2002). NBS–LRR proteins are further divided into two subclasses depending on whether the N-terminal domain is a Toll/interleukin-1 receptor (TIR) or coiled-coil (CC) motif. The LRR backbone is proposed to provide a versatile recognition surface for specific Avr perception, while evidence indicates that the TIR and CC domains mediate downstream signalling (Lahaye, 2002).

Downstream of recognition, which occurs when plants first encounter pathogens, more global resistance mechanisms are induced. The earliest detectable cellular events are ion fluxes across the plasma membrane and a burst of oxygen metabolism that produces reactive oxygen intermediates (ROIs), such as superoxide (O_2^-) and hydrogen peroxide (H_2O_2) (McDowell and Dangl, 2000; Delledonne *et al.*, 2002). Receptor-mediated ion fluxes trigger localized production of nitrous oxide (NO) and ROIs immediately after pathogen recognition. These second messengers synergistically induce cell death, production of salicylic acid (SA), defence gene expression and more ROIs, establishing a putative feedback loop in which the response is amplified. Characterization of these steps in plant defence is an area of very active research.

An important component of this defence system is systemic acquired resistance (SAR) (recently reviewed by Durrant and Dong, 2004). After a hypersensitive response to invading pathogens, plants show elevated accumulation of SA, induced expression of pathogenesis-related (*PR*) genes and SAR to further infection by a broad range of pathogens. Considerable effort has been directed toward identifying the signalling molecules responsible for activating the hypersensitive response and SAR, and there is now compelling evidence that SA plays a crucial role in triggering SAR. Application of SA or its analogues to both tobacco and *Arabidopsis* induces *PR* gene expression and resistance as would a biological agent (Shah *et al.*, 1999).

Transgenic tobacco plants that express the bacterial salicylate hydroxylase (*nahG*) gene cannot accumulate SA or develop SAR, and exhibit heightened susceptibility to pathogen infection (Delaney *et al.*, 1994). Likewise, preventing SA synthesis by specifically inhibiting the activity of phenylalanine ammonia-lyase, the first enzyme in the SA biosynthetic pathway, makes otherwise resistant *Arabidopsis* plants susceptible to *Peronospora parasitica* (Mauch-Mani *et al.*, 1996). Further, at least some *R* genes are upregulated by SA, forming a feedback amplification signal loop (Shirano *et al.*, 2002; Xiao *et al.*, 2003). Two distinct signal transduction pathways that conduct to SAR after pathogen recognition by R proteins have been identified (Fig. 14.1). Two genes from *Arabidopsis*, *EDS1* and *PAD4*, that encode lipase-like proteins and that interact with each other, mediate the downstream signalling of TIR- but not CC-type R protein receptors. Further, *EDS1* and *PAD4* seem to be a converging point of TIR-mediated resistance to bacteria, fungi and viruses (Liu *et al.*, 2002; Whitham *et al.*, 2003). A membrane-associated protein encoded by *NDR1* is required to trigger resistance by many CC R proteins, but not by TIR proteins. Another set of genes also required for signal transduction, *RAR1*, *SGT1* and *PBS3*, overlap the pathways described above, indicating the complexity of the signal pathway. RAR1 is a component of the N-mediated resistance to Tobacco mosaic virus (TMV) and some powdery mildews; however, it is not required by all *R* genes (Liu *et al.*, 2002; Quirino and Bent, 2003). SGT1 and RAR1 physically interact and they have been suggested to be associated with the ubiquitination and subsequent degradation of proteins (Martin *et al.*, 2003; Quirino and Bent, 2003). Further, a molecular chaperone (Hsp90) interacts in *Nicothiana tabacum* with SGT1, RAR1 and the N protein, and is also required for the signal transduction of *R* gene-mediated resistance (Liu *et al.*, 2004). This may indicate that some R proteins are able to assemble into recognition complexes without the help of chaperones, hence explaining the different requirements of *R* genes for SGT1/RAR1 resistance (Liu *et al.*, 2004).

In addition to SA, ethylene (ET) and jasmonic acid (JA) serve as important signals for the induction of various defence responses, including insects, mechanical damage and necrotrophic pathogens; they work with SA to signal some but not all defence responses, and in some cases function independently (Pieterse and van Loon, 1999; Shah *et al.*, 1999). A different form of systemic resistance is the induced systemic resistance (ISR) that requires signal pathways responding to the hormones ET and JA and is independent of SA. In nature, certain non-pathogenic bacteria can trigger this response. Both SAR and ISR pathways are part of a complex signal network with members that seem to act independently but coordinately (Fig. 14.1). Overall, the SA and JA/ET pathways act in opposition to each other although some overlap exists (Glazebrook *et al*, 2003).

One of the key genes involved in both SAR and ISR is the *Arabidopsis thaliana NPR1/NIM1* that functions as a signal modulator (Fig. 14.1) (Cao *et al.*, 1997, 1998; Yu *et al.*, 2001). Upon induction (by pathogen infection and/or SA), *NPR1* expression is elevated and the NPR1 protein is activated, in turn inducing expression of a battery of downstream *PR* genes. The *NPR1* gene encodes a protein containing ankyrin repeat motifs of the sort mediating protein–protein interactions in animals (Cao *et al.*, 1997). *Arabidopsis* mutants lacking a functional *NPR1* gene are unable to express *PR* genes in response to SA or its analogues. In addition, *npr1* mutant plants show increased susceptibility to fungal, bacterial and viral pathogens. The recessive nature of most identified NPR1 mutant alleles strongly suggests that *NPR1* is a positive regulator of the SA signal transduction pathway. In uninduced cells, NPR1 exists as an oligomer formed by intermolecular disulphide bonds. After an oxidative burst during SAR induction and the subsequent accumulation of antioxidants, the cell becomes a more reductive environment and

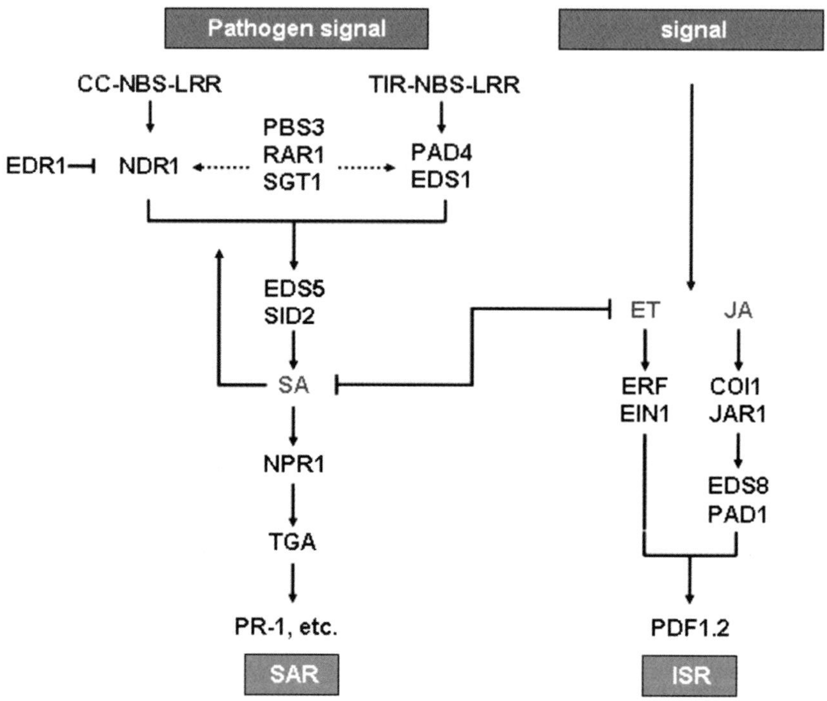

Fig. 14.1. Systemic signalling in *Arabidopsis* disease resistance. Recognition of invading pathogens is mediated by resistance (R) proteins (CC– and TIR–NBS–LRR). Two distinct signal transduction pathways conduct to SAR. *EDS1* and *PAD4,* that encode lipase-like proteins and interact with each other, mediate the downstream signalling of TIR- but not CC-type R protein receptors. NDR1 is required to trigger resistance by many CC R proteins, but not by TIR proteins. RAR1 and SGT1 overlap the pathways described above. After induction, the levels of salicylic acid (SA) increase, the cell becomes a more reductive environment and NPR1 is converted from an oligomeric to a monomeric state. In this reduced state, NPR1 is translocated to the nucleus where it interacts with members of the basic leucine zipper (bZIP) family of transcription factors (TGA) to induce the activity of pathogenesis-related (*PR*) genes. Some non-pathogenic bacteria, necrotrophic pathogens, insects and wounding induce a separate defence pathway that is dependent on jasmonic acid (JA) and ethylene (ET). This pathway ultimately produces the induction of defensins (PDF1.2). SAR = systemic acquired resistance; ISR = systemic induced resistance. Solid arrows indicate a positive effect; dashed arrows indicate a partial requirement; and lines with a bar indicate an inhibitory effect.

NPR1 is converted to a monomeric state by the reduction of intermolecular disulphide bonds (Mou *et al.*, 2003). In this reduced state, NPR1 is translocated to the nucleus where it interacts with members of the basic leucine zipper (bZIP) family of transcription factors (TGA factors) to induce the activity of *PR* genes (Kinkema *et al.*, 2000; Fan and Dong, 2002; Johnson *et al.*, 2003). Interestingly, one TGA factor (TGA1) also has to be reduced before it is able to interact with NPR1 (Despres *et al.*, 2003). Cytoplasmic NPR1 appears to modulate cross-talk between SA and JA defence pathways (Spoel *et al.*, 2003). Constitutive overexpression of NPR1 in *Arabidopsis* did not result in constitutive *PR* gene expression in the absence of pathogens; however, it did lead to enhanced resistance to the bacterium *Pseudomonas syringae* and the oomycete *Peronospora parasitica*, with no obvious detrimental effect on the transgenic plants (Cao *et al.*, 1998; Friedrich *et al.*, 2001). NPR1 homologues have been identified in a variety of economically important plants, including rice, soybean and maize. The ubiquitous existence of NPR1 in different plant species suggests that the findings with NPR1 in *Arabidopsis* are likely to apply to other species. Recently, the *Arabidopsis NPR1* gene was overexpressed in rice plants and the transgenic plants showed enhanced resistance to *Xanthomonas oryzae* pv. *oryzae* (Chern *et al.*, 2001), indicating conserved signal transduction pathways controlling NPR1-mediated resistance among widely diverged plant species (in this case dicots and monocots).

At least some early SA-activated genes (glutathione *S*-transferase, *GST6* and glucosyltransferase, *EIGT*) do not require NPR1 for their induction. This points toward different mechanisms for early and late gene activation despite the fact that they require SA and TGA, and share some promoter elements with late activated genes such as *PR1* (Uquillas *et al.*, 2004).

Protein kinases and signal transduction

One of the early cellular events that occur within minutes of pathogen recognition is the activation of protein kinases. The evidence suggests that they have an essential role in early signal transduction. Some of the rapidly activated kinases have been identified as mitogen-activated protein kinases (MAPKs) (Peck, 2003). MAPKs are activated by general elicitors such as flagellin, chitin and other fungal cell wall components, as well as specific elicitors recognized by the R protein transduction pathway (Ekengren *et al.*, 2003). This is an indication that kinases could be a converging point for host and non-host signal transduction pathways. Further, at least in one system (*Pseudomonas syringae* pv. *tomato* carrying the *AvrPto* or *AvrPtoB* avirulence genes and tomato carrying the *Pto* resistance gene), two MAPKs (*WIPK* and *NTF6*) and two MAPK kinases (*MEK1* and *MEK2*) are essential for the activation of the NPR1-mediated SA pathway (Ekengren *et al.*, 2003). On the other hand, *EDR1* of *Arabidopsis* (a MAPK kinase kinase), is a negative regulator of the SA defence pathway (Frye *et al.*, 2001). Orthologues of *EDR1* are present in several species of dicots and monocots, indicating that it must be part of a conserved pathway in plants (Frye *et al.*, 2001; Tang and Innes, 2002; Kim *et al.*, 2003).

The questions we are asking in our research are: do these same genes and disease signal transduction pathways function in the perennial plant citrus in the same manner as in annual plants? Can overexpression of genes in the pathways be used to provide generalized pathogen resistance? These issues are discussed further below.

Pathogen-derived resistance

In 1985, Sanford and Johnson advocated the use of pathogen-derived genes for generating host resistance. Proof of concept was demonstrated shortly thereafter by Roger Beachy's group, who showed that the expression of a viral coat protein (CP) gene in a transgenic plant could confer resistance to the donor (Powell-Abel *et al.*, 1986). In the intervening years, there have been numerous reports of various virus–host combinations where this strategy has been successful to some degree.

Once this strategy was found to be successful, investigations into the underlying mechanism(s) of resistance began. It was found that there was indeed more than one mechanism operating and that they could be protein mediated or RNA mediated. The best studied example of protein-mediated resistance is that of TMV resistance conferred by expression of the TMV *CP* gene in transgenic tobacco by Beachy's group as referenced above. In this case, it was found that virus resistance depends on the synthesis of transgene-encoded CP, since a non-coding TMV *CP* gene was not effective (Powell *et al.*, 1990). The resistance can be overcome when plants are inoculated with uncapsidated viral RNA rather than virions, leading to the conclusion that resistance is due to interference of the transgenic CP with virion disassembly (Nelson *et al.*, 1997). In grafting experiments, the transgenic CP also inhibited systemic spread of the virus (Wisniewski *et al.*, 1990), which might implicate a second mechanism of protection. However, systemic movement of the TMV virions may also require virion disassembly and assembly (Baulcombe, 1996).

RNA-mediated mechanisms were also documented early on. In 1987, the year following the first description of resistance in CP-transgenic plants, the very first virus-resistant transgenic plants in which resistance was clearly due to RNA-mediated phenomena were reported. In plants expressing a satellite RNA, either Cucumber mosaic virus (Harrison *et al.*, 1987) or Tobacco ringspot virus (Gerlach *et al.*, 1987), resistance was thought to be due to competition for replication between these non-coding RNAs and the viral genomic RNAs. The same mechanism has been invoked more recently when other non-coding viral RNAs, such as the 3′ non-coding region of a viral genome used for transformation (Zaccomer *et al.*, 1993), have also been shown to confer resistance through competition with the viral RNAs.

However, the most important type of RNA-mediated resistance is produced via a quite different mechanism, first denoted as gene silencing, now most often termed RNA interference, or RNAi. This was first shown by Dougherty's group, who demonstrated that resistance to Tobacco etch virus could be conferred by an untranslatable *CP* gene, and also that the strongest RNA-mediated protection was observed in plant lines in which little or no transgene mRNA accumulated. When resistance was observed in plants that initially accumulated transgene mRNA, the plants were first fully infected, after which, recovery, leading to a completely insensitive state, was accompanied by disappearance of the transgene mRNA. They proposed that resistance was conferred by sequence-specific degradation of both the transgene mRNA and the corresponding viral RNA, via a mechanism similar to post-transcriptional gene silencing (PTGS) (Lindbo *et al.*, 1993; Dougherty *et al.*, 1994).

Subsequently, gene silencing, or RNAi, has been the subject of many investigations. Basically, gene silencing is thought to result from the action of a mechanism that surveys the RNAs in a cell and sequence specifically degrades those perceived as unwanted. This phenomenon has been proposed to be a natural plant defence mechanism against viruses and transposons and is also involved in the regulation of the expression levels of certain genes (Waterhouse *et al.*, 2001; Tang *et al.*, 2002; Voinnet, 2002). Evidence of this is that some plant viruses encode proteins that specifically disable this system.

RNA silencing (under different names) has been demonstrated in animals, fungi, plants (Fig. 14.2) and possibly in bacteria (Tchurikov *et al.*, 2000). The mechanism involved in these organisms shares some common features, but the hallmark is the production of small (21–25 nucleotide) RNAs that act as determinants for the RNA degradation (Hamilton and Baulcombe, 1999; Hammond *et al.*, 2000). In plant cells infected by a virus, the RNA first requires the conversion to double-stranded (ds) RNA. This is probably accomplished by the viral replicase, although transgenes with inverted repeats may directly produce dsRNA. The dsRNA is cleaved into the small interfering RNAs (siRNAs) that are the mediators of gene silencing by an RNase III-like enzyme complex (Dicer) that is ATP dependent and contains putative helicase-, RNase III- and dsRNA-binding domains (Fagard and Vaucheret, 2000; Park *et al.*, 2002). Two classes of siRNA are generated in plants by different Dicer enzymes, short siRNA (21 nucleotides) and long siRNA (24 nucleotides) (Hamilton and Baulcombe, 1999; Tang *et al.*, 2002). These siRNAs are double stranded with two-nucleotide 3′ overhangs and hydroxyl termini. It has been proposed that the short siRNAs direct PTGS via mRNA degradation and the long siRNAs trigger systemic

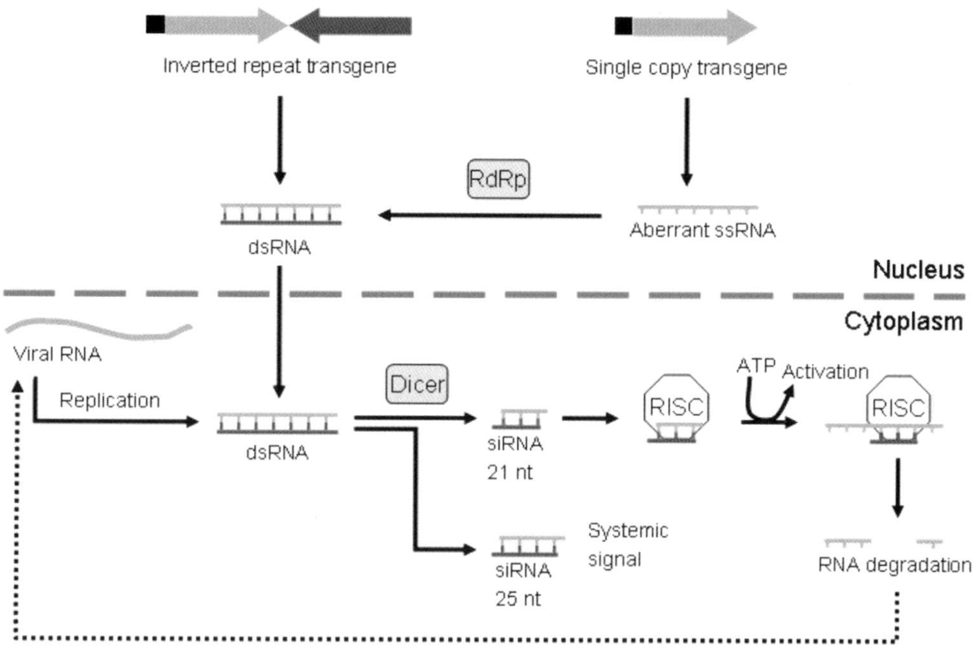

Fig. 14.2. A model for RNA silencing in transgenic plants. An inverted repeat copy of a gene with homology to the target viral sequence forms a double-stranded RNA (dsRNA) molecule upon expression. In the case of single copy transgenes, the mRNA can be perceived as aberrant by an unknown mechanism and is converted to dsRNA by a plant-specific RNA-dependent RNA polymerase (RdRp). The dsRNA is translocated to the cytoplasm where it becomes a target for the Dicer enzyme. Viral replication of many RNA viruses also includes a dsRNA replicative form that can be targeted by Dicer. In plants, two types of Dicer proteins generate short (21 nucleotide) and long (25 nucleotide) micro RNAs (miRNA, also denominated small interfering RNA, siRNA). These siRNAs direct RNA interference in plants. The short ones are associated with local silencing and the long ones with systemic silencing. The short siRNAs associate with a RISC protein and become single stranded in an activation process that requires ATP. The RISC–siRNA complex directs the sequence-specific degradation of homologous RNA. In the diagram, the viral RNA is degraded by the transgene-induced RNA silencing (dotted arrow). It should be noted that the process of degradation can also occur in the nucleus.

silencing and the methylation of homologous DNA (Hamilton *et al.*, 2002). The siRNAs are opened in an ATP-dependent step, leaving them single stranded to be incorporated into the multimeric RNase complex, that is then denominated RISC (RNA interference-specific complex), which is guided by Watson–Crick base paring, ensuring the sequence-specific degradation of the unwanted RNA (Baulcombe, 2002; Wassenegger, 2002). Inducible endogenous plant RNA-dependent RNA polymerases (RdRps) are also involved in RNA silencing. In both *Arabidopsis* and tobacco, virus- and SA-induced RdRps have been identified that are not required for the initiation of RNA silencing but are necessary for its maintenance (Yu *et al.*, 2003).

In plants, constructs designed to produce dsRNA or self-complementary hairpin RNA transcribed from inverted repeats have been shown to be highly efficient inducers of gene silencing. This same strategy can lead to very effective virus resistance or immunity (Waterhouse *et al.*, 1998; Smith *et al.*, 2000; Wang *et al.*, 2000; Tenllado *et al.*, 2004).

Molecular Biology of the Citrus Tristeza Virus

CTV is a member of the genus Closterovirus of the Closteroviridae, and has a positive single-stranded RNA genome of approximately 20 kb. Different strains of the virus can cause diverse disease syndromes that vary from mild, with no visible symptoms; to quick decline and death of scions grafted on sour orange; to stem pitting and fruit size reduction of most species and varieties of citrus (Bar-Joseph *et al.*, 1989). Virions are long flexuous particles encompassed by two CPs, a major CP of 25 kDa (referred to in this work as CP) and a minor CP of 27 kDa (p27) that encapsidates about 5% of the terminal region of the particle (Febres *et al.*, 1996). Including these two genes, the single-stranded, positive-sense genomic RNA encodes 12 open reading frames (ORFs) (Pappu *et al.*, 1994; Karasev *et al.*, 1995). The 5′-proximal ORF 1a encodes a 349 kDa protein containing two papain-like protease domains, plus methyltransferase-like and helicase-like domains; the virus RdRp is encoded by ORF 1b and is expressed by a +1 frameshift (Karasev *et al.*, 1995; Cevik, 2001). The 3′ half of the genome contains ten ORFs that are expressed via 3′-co-terminal subgenomic RNAs. One of the most highly transcribed genes is that for a 20 kDa protein (p20) that accumulates as a cytoplasmatic amorphous inclusion body in infected cells (Gowda *et al.*, 2000) and appears to be a repressor of local and systemic RNA silencing (Lu *et al.*, 2004). The most 3′ gene of CTV, p23, that codes for a 23 kDa protein, is also highly transcribed and has been found to control asymmetric accumulation of CTV RNAs, in addition to being a suppressor of local silencing (Lu *et al.*, 2004). It downregulates negative-stranded RNA accumulation and indirectly increases the expression of 3′ genes (Satyanarayana *et al.*, 2002). The 107 bp 5′- and 271 bp 3′-untranslated regions (UTRs) of the genomic RNA are necessary for virus replication (Mawassi *et al.*, 2000). Several strains of CTV have been fully sequenced, and they reveal a higher degree of nucleotide sequence homology in the 3′ end (>97% identity) than in the 5′ end (<40% identity) of the genomic RNA (Mawassi *et al.*, 1996; Vives *et al.*, 1999; Yang *et al.*, 1999; Albiach-Marti *et al.*, 2000).

In the past, sour orange (*Citrus aurantium* L.) and Mexican lime (*C. aurantifolia* (Christm.) Swing.) have been transformed with the *CP* gene of CTV (Gutierrez *et al.*, 1997; Dominguez *et al.*, 2000; Ghorbel *et al.*, 2000). Only the Mexican lime plants were evaluated for resistance against CTV. Some of the clonal lines showed some resistance (in a proportion of 10–33%), while others showed delay in virus accumulation (Dominguez *et al.*, 2002). However, by the third flush, the resistance was largely overcome. In another published experiment where citrus plants have been transformed with a CTV sequence, Mexican lime transgenic plants expressing the CTV p23 protein showed virus-like symptoms such as vein clearing and stem pitting (Ghorbel *et al.*, 2001).

Experiments Being Carried Out in Our Laboratory

Sources of virus isolates

Florida CTV isolates T66 sub-isolate E, T36 (quick decline-inducing isolate) and T30 (mild isolate) were maintained at the University of Florida's Citrus Research and Education Center, Lake Alfred, Florida. Isolate B249 (stem pitting isolate from Venezuela) was obtained from the Collection of Exotic Citrus Pathogens maintained at the USDA-BARC, Beltsville, Maryland. Isolate DPI 3800 (a mixture of stem pitting-like and T36-like strains from Florida) was acquired from the Florida Department of Plant Industry (DPI), Gainesville, Florida. All isolates were propagated on Mexican lime.

Plasmid vector construction and bacterial strains

Since the process of producing and growing transgenic citrus is labour intensive and rather inefficient, and CTV is a complex virus, our strategy has been to introduce individually a wide variety of CTV sequences to increase the chances of producing tolerant plants. The sequences include the *CP* gene from three different strains with distinct biological characteristics, a non-translatable version of one of the *CP* genes, the RdRp, p27 and p20 genes, and the 3′ end (400 3′-terminal bases, including part of the p23 gene and the 3′ UTR) of the CTV genomic RNA. Additionally, we have also thus far used the *NPR1* gene from *Arabidopsis* in transformation experiments in an attempt to obtain broad-spectrum disease resistance. The construction of these vectors has been described in detail elsewhere (Febres *et al.*, 2003). Four binary vector plasmids were used to clone the sequences for their delivery into grapefruit tissue via *Agrobacterium tumefaciens* transformation. These vectors were pGA482GG, pMON10098, pCAMBIA 2201 and pCAMBIA 2202. A summary of the constructs used for the transformation results presented here is shown in Table 14.1.

Pathogen-derived resistance can be limited to the strain from which the transgene was derived or to closely related strains (Lomonossoff, 1995). These constructs are varied in terms of the genes they contain, the position of the genes in the CTV genome (both 3′ and 5′ regions are represented) and their origin (virus strain). Since the 3′ end of CTV is the most conserved region among different strains of the virus (>97%), we speculated that this region was the most likely to induce resistance to the widest varieties of CTV strains. Other sequences (CP and RdRp) were chosen because they have been widely and

Table 14.1. Description of the plasmids and bacterial strains used for the *Agrobacterium tumefaciens*-mediated transformation of grapefruit.

Construct[a]	Vector	Plant selectable marker[b]	Plant reporter gene[c]	Bacterial selectable marker[d]	*A. tumefaciens* strain
35S-CP T36	pGA448GG	Kan	GUS	Tet	EHA 101
35S-CP T30	pGA448GG	Kan	GUS	Tet	EHA 101
35S-CP B249	pGA448GG	Kan	GUS	Tet	EHA 101
35S-RdRp	pGA448GG	Kan	GUS	Tet	EHA 101
35S-p20	pMON 10098	Kan	GUS	Spc	ABI
35S-p27	pMON 10098	Kan	GUS	Spc	ABI
34S-NT CP	pCAMBIA 2201	Kan	GUS	Cap	Agl1
35S-3END-S	pCAMBIA 2201	Kan	GUS	Cap	Agl1
34S-3END-S	pCAMBIA 2201	Kan	GUS	Cap	Agl1
34S-3END-AS	pCAMBIA 2201	Kan	GUS	Cap	Agl1
34S-NPR1	pCAMBIA 2201	Kan	GUS	Cap	Agl1
34S-NPR1	pCAMBIA 2202	Kan	GFP	Cap	Agl1

[a]35S = CaMV 35S promoter; 34S = FMV 34S promoter; CP T36 = major CP gene from decline-inducing isolate T36; CP T30 = major CP gene from mild isolate T30; CP B249 = major CP gene from stem pitting-inducing isolate B249; RdRp = replicase gene from isolate T36; p20 = repressor of RNA silencing from isolate T36; p27 = minor CP gene from isolate T36; NT CP = non-translatable major CP from isolate T36; 3END-S = 3′ end (400 3′-terminal bases, including part of the p23 gene and the 3′ UTR) in the sense orientation from stem pitting-inducing isolate DPI 3800; 3END-AS = 3′ end from isolate DPI 3800 in the antisense orientation; NPR1 = *A. thaliana* NPR1 gene (*non-expressor of pathogenesis-related proteins 1*).

[b]Kan = kanamycin.

[c]GUS = glucuronidase; GFP = green fluorescent protein.

[d]Tet = tetracycline; Spc = spectinomycin; Cap = chloramphenicol.

successfully used for inducing resistance in other plant–virus systems (Lomonossoff, 1995; Hammond, 1999). The p20 gene was chosen because it is highly expressed and is thought to be important in the virus life cycle, and also it is located in the more conserved 3′ end of the virus. The sequences for the CP T36, NT CP, RdRp, p27 and p20 were cloned by reverse transcription–polymerase chain reaction (RT–PCR) from the Florida CTV isolate T36 and were identical to the publicly available sequence for this isolate (gb U16304). Similarly, the sequence for the CP T30 was identical to the public sequence available (gb AF260651). The 3′ end of CTV was cloned from isolate DPI 3800, a mixture of quick decline and stem pitting strains. The sequences of the 35S-3END-S and 34S-3END-AS constructs were T36-like, whereas the sequence of the 34S-3END-S construct was more similar to stem pitting isolates (data not shown). The *NPR1* gene from *Arabidopsis* was cloned into the pCAMBIA transformation vectors from genomic DNA using specific primers designed based on the sequence publicly available (gb U76707.1).

Plant transformation

Grapefruit (*C. paradisi* Macf. cv Duncan) seeds were extracted from mature fruits and used for *A. tumefaciens*-mediated transformation according to the procedure previously described (Luth and Moore, 1999). Four- to 6-week-old etiolated stem segments were used as explants. An initial β-glucuronidase (GUS) assay was performed on all of the shoots that regenerated on selection medium (some 5–8 weeks after inoculation with *Agrobacterium*) by removing small sections from their basal ends, followed by histochemical GUS staining (Moore *et al.*, 1992). A second GUS assay was performed on leaf segments of rooted plants using a similar staining procedure. Similarly, green fluorescent protein (GFP) was visualized in intact shoots and plants using a fluorescence stereoscope fitted with a 535nm emission filter (Seizz). A standard PCR was performed on the putatively transgenic plants (positive in the first GUS or GFP assay) using gene-specific primers (Febres *et al.*, 2003).

The plants obtained from the transformation experiments resembled the wild type morphologically and displayed normal growth. The results of the plant transformation experiments with the CTV sequences are summarized in Table 14.2. There was variability in the number of explants (stem segments) regenerating shoots (between 4 and 13%, with an average of 9%), the number of GUS-positive shoots (between 6 and 39%, with an average of 14%) and the number of solid (non-chimeric in the histochemical assay) GUS-positive shoots (between 0 and 40%, with an average of 17%). The control plasmid pGA482GG, without any CTV genes, produced an above average number of GUS-positive shoots and the highest number of solid GUS-positive shoots. This could be due to a higher efficiency in the transfer of smaller *Agrobacterium* T-DNA segments or perhaps to a detrimental effect of the CTV genes.

None of the p20-regenerated shoots and only one of the p27-regenerated shoots were PCR positive for the gene of interest (Table 14.3), even though large numbers of stem segments were inoculated with *Agrobacterium* containing these constructs (Table 14.2). This does not seem to be a problem caused by the inefficiency of the pMON10098 vector in comparison with the pGA482GG or pCAMBIA 2201 vectors, since the initial numbers and percentages of GUS-positive shoots produced using the constructs with the p20 and p27 genes was high (19 and 10% of regenerated shoots, respectively, Table 14.2). It is possible that these two genes are toxic or deleterious to the plant and are removed from the genome by rearrangements, or that such transgenic plants die at an early stage. This would be consistent with a role for p20 as a repressor of RNA silencing. It is now accepted that RNA silencing is one of the mechanisms used by the cell to regulate gene expression (Voinnet, 2002). Such a repressor constitutively expressed could alter development in a deleterious way. As for

Table 14.2. Transgenic shoot production from the stem segment explants of grapefruit inoculated with *Agrobacterium tumefaciens* carrying different sequences of CTV.

Construct	No. of segments evaluated	No. of segments (%) producing shoots[a]	Total no. of shoots	No. of GUS+ shoots (%)[b]	No. of solid GUS+ shoots[c]
pGA482GG	5,808	209 (3.6%)	259	102 (39.4%)	41 (40.2%)
35S-CP T36	4,039	513 (12.7%)	687	70 (10.2%)	28 (40.0 %)
35S-CP T30	3,585	398 (11.1%)	536	49 (9.1%)	17 (34.7%)
35S-CP B249	3,504	386 (11.0%)	500	54 (10.8%)	12 (22.2%)
35S-RdRp	3,599	353 (9.8%)	448	42 (9.4%)	9 (21.4%)
35S-p20	8,227	748 (9.7%)	915	173 (18.9%)	14 (8.1%)
35S-p27	38,759	1,747 (4.5%)	2,170	225 (10.4%)	1 (0.4%)
34S-NT CP	28,809	2,287 (7.9%)	3,100	432 (13.9%)	60 (13.9%)
35S-3END-S	27,848	1,789 (6.4%)	2,359	192 (8.1%)	16 (8.3%)
34S-3END-S	15,723	1,783 (11.3%)	2,331	137 (5.9%)	5 (3.6%)
34S-3END-AS	8,171	514 (6.3%)	651	58 (8.9%)	6 (10.3%)

[a]Percentage refers to the number of segments producing shoots versus the number of segments analysed.
[b]Percentage refers to the number of GUS+ shoots versus the total number of shoots.
[c]Percentage refers to the number of solid GUS+ shoots versus the number of GUS+ shoots.

p27, the only known role is that of a structural protein; however, it is not uncommon for viral proteins to have more than one function in the virus life cycle.

After the transgenic plants were established in soil, we performed a second GUS assay or GFP visualization. Not all of the plants that tested positive in the first assay were positive in the second assay (Table 14.3). In most cases, the second assay GUS/GFP-negative plants originated from the chimeric shoots observed in the first assay (data not shown). We were interested in determining whether the second GUS/GFP assay was a good indication for the presence of the transgenes, compared with the first assay. Excluding the results for 35S-p27 and 35S-p20, most of the GUS/GFP-positive plants in the second assay (>75%, and in most cases close to 100%, with an average of 91%) were also PCR positive for the gene of interest (Table 14.3). However, a few of the plants (an average of 12%) that were negative in the second GUS/GFP assay (but positive for GUS/GFP in the first assay) contained the gene of interest (Table 14.3). Even though these percentages are low, since the number of putatively transgenic plants (PCR positive for the gene of interest) is also relatively low, it may be worthwhile to analyse all of the rooted plants by PCR even if they are negative in the second reporter gene assay. Further, these plants may represent a silenced state that could be of more interest for our purpose, at least in the case of the CTV sequences.

A selected number of the putatively transgenic plants were further analysed by Southern blots to corroborate the integration of the genes into the plant genome and to determine their copy number. The results indicated that we obtained stably transformed grapefruit plants with one to several copies of the transgene per genome (data not shown and Febres *et al.*, 2003). Further, western blot analysis of the CP transgenic plants showed the presence of the 25 kDa CP in several lines, demonstrating that the CP from some of these lines was expressed to detectable levels (data not shown).

Virus challenge of the transgenic plants

In nature, CTV is transmitted by aphids (Bar-Joseph *et al.*, 1989). Unfortunately, it is

Table 14.3. PCR analysis of the regenerated reporter gene-positive grapefruit shoots using transgene-specific primers.

Construct	No. of first assay GUS+ shoots	No. of second assay GUS+ shoots	Total	No. (%) of PCR-positive shoots	
				From second GUS+[a]	From second GUS–[b]
35S-CP T36	47	25	29	23 (92%)	6 (27%)
35S-CP T30	27	17	18	17 (100%)	1 (10%)
35S-CP B249	20	16	14	14 (88%)	0 (0%)
35S-RdRp	31	13	18	12 (92%)	6 (33%)
35S-p20	20	4	0	0 (0%)	0 (0%)
35S-p27	31	8	1	1 (13%)	0 (0%)
34S-NT CP	22	8	8	7 (88%)	1 (7%)
35S-3END-S	11	6	5	5 (83%)	0 (0%)
34S-3END-S	17	4	5	3 (75%)	2 (15%)
34S-3END-AS	13	3	4	3 (100%)	1 (10%)
34S-NPR1[c]	45	15	16	14 (99%)	2 (6%)

[a]Percentage refers to the number of PCR-positive shoots versus the number of second assay GUS-*positive* shoots.
[b]Percentage refers to the number of PCR-positive shoots versus the number of second assay GUS-*negative* shoots.
[c]Results from GUS- and GFP-positive plants were combined for this analysis.

very difficult to infect CTV into a citrus plant by mechanical inoculation (Muller and Garnsey, 1984; Bar-Joseph *et al.*, 1989). Therefore, to test whether the transgenic plants we are producing show resistance to CTV, they are being challenged by both aphid and graft inoculations (Fig. 14.3). The results shown here are mostly from graft inoculation with CTV isolate T66-E. Buds from verified transgenic plants (GUS, PCR and Southern blot positive) were grafted on Swingle citrumelo (*Poncirus trifoliata* (L.) Raf. × *C. paradisi* Macf.). Each transgenic plant was budded in five replicates, and one of the transgenic plants was maintained for healthy controls. After the transgenic buds began to grow, the plants were graft challenged with CTV severe isolate T66-E by grafting three infected leaf segments on to the transgenic scion. After 12 and 18 months, the transgenic grapefruit scions were tested by double antibody sandwich indirect enzyme-linked immunosorbent assay (DASI-ELISA) (Garnsey and Cambra, 1991) using the polyclonal antibody UF 1052 for coating and the monoclonal antibody MCA13 as secondary antibody. MCA13 reacts predominantly with the CP of severe isolates of CTV (such as T66) and does not react with T30 (Permar *et al.*, 1990). Each ELISA sample was tested in duplicate. In the case of the aphid inoculations, colonies of *Toxopthera citricida* (brown citrus aphid) were fed for at least 48 h on T66-E-infected Mexican lime plants. Groups of 20 aphids were transferred to the transgenic plants (additional propagations from those used in the grafting experiments) and allowed to feed for several days. Non-transgenic plants were used as controls for the efficiency of virus transmission. Infection by CTV was also tested using ELISA.

Only a small, representative number of all the lines challenged are shown in Table 14.4. Most of the lines did not show resistance to CTV (line 81, for example), testing positive for CTV. A few of the lines, however, showed some of the plants free of CTV (lines 146, 169, 212, 538 and 595). Plant 595 (transformed with the 3END construct in the sense orientation) was the most resistant. Interestingly, a couple of the lines showed plants that seemed to have recov-

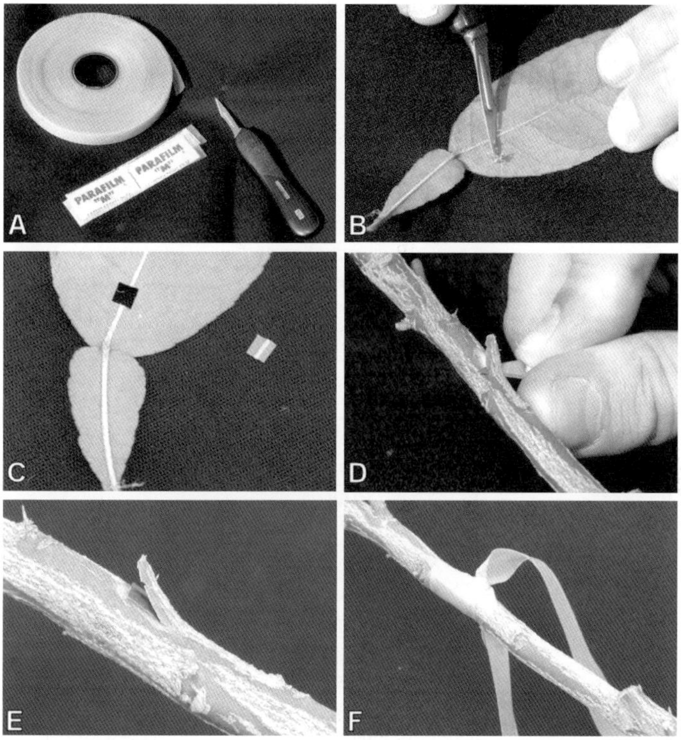

Fig. 14.3. Grafting procedure used to challenge the transgenic plants. (A) The materials required for grafting are a scalpel, blade or sharp knife and grafting tape. If grafting tape is not available, 'Parafilm' can be used instead. However, it tends to 'melt' with time and it becomes more difficult to remove. (B) The main vein of the leaf is scraped off to expose the phloem tissue. (C) A small square or 'chip' of about 3–4 mm wide is cut out from the exposed phloem portion. (D) A cut is made to the stem of the plant to be challenged as shown. Ideally, straight portions of the stem are selected for the graft, away from leaves or other branches, which are removed. (E) The chip is placed between the stem and the flap, with the phloem of the chip toward the phloem of the stem. (F) Grafting tape is wrapped around and used to secure the graft in position. In our experiments, we make three to four grafts onto each transgenic scion. After 4–5 weeks, the tape is removed. By then the graft should have taken. The chip should be green and the flap eventually dies and falls off.

ered from the viral infection (line 212). Overall, 41 lines of transgenic plants have been tested, of which 13 have shown some degree of resistance. These results are similar to those of Dominguez *et al.* (2002) in which only a proportion of the plants showed resistance to CTV.

We initiated the effort to produce CTV-resistant citrus several years ago. At the time, little was known about the mechanism involved in pathogen-mediated disease resistance. During the time we have been producing and challenging these plants, a time-consuming and laborious process, much has been learned about what occurs during pathogen-mediated disease resistance and how to bring it about effectively (Waterhouse, 2001; Rovere *et al.*, 2002; Voinnet, 2002; Wassenegger *et al.*, 2002a, b; Tang *et al.*, 2003). Any of the genes that show promise in the transgenic plants we have produced will be introduced into transgenic plants in constructs that include inverted repeats that result in RNA with a double-stranded 'hairpin' structure (Waterhouse *et al.*, 1998), direct

Table 14.4. Results of the DASI-ELISA of the plants challenged with CTV T66-E.

Construct	Line	Replicate	Treatment[a]	ELISA[b] 12 months	ELISA[b] 18 months
35S-RdRp	81	1	Graft	0.631	1.247
		2	Graft	0.604	1.026
		3	Graft	0.361	1.075
		4	Graft	0.537	1.095
		5	Unchallenged	**0.069**	**0.176**
35S-RdRp	146	1	Graft	0.434	0.797
		2	Graft	1.282	1.149
		3	Graft	**0.091**	0.777
		4	Graft	**0.059**	**0.221**
35S- CP B249	169	1	Graft	0.519	1.004
		2	Graft	0.615	1.013
		3	Graft	0.520	1.041
		4	Graft	**0.069**	**0.215**
35S- CP T30	212	1	Graft	0.991	0.908
		2	Graft	0.576	**0.163**
		3	Graft	0.484	0.938
		4	Graft	0.522	0.921
35S-3END-S	538	1	Graft	0.561	0.951
		2	Graft	0.542	1.041
		3	Graft	0.549	0.998
		4	Graft	**0.064**	**0.179**
35S-3END-S	595	1	Graft	**0.061**	**0.211**
		2	Graft	**0.060**	**0.177**
		3	Graft	**0.072**	**0.194**
		4	Graft	**0.102**	**0.192**
		5	Aphid	**0.061**	**0.159**
	Healthy	1	Unchallenged	**0.066**	**0.098**
		2	Unchallenged	**0.085**	**0.127**
	T66-E	1	Aphid	0.571	0.922
		2	Aphid	**0.063**	0.382
	T66-E	1	Graft	0.744	N/A
		2	Graft	0.810	N/A
	DPI 3800	1	Graft	N/A	1.123

[a]Plants were challenged with CTV by either grafting (Graft) or aphid transmission (Aphid), or remained unchallenged.
[b]Values are averages of OD_{415} from two repetitions estimated 12 or 18 months after inoculation. A positive result is at least twice the average OD of the uninfected controls. Values in bold indicate those samples considered negative. N/A = sample not assayed.

tandem repeats of three or four copies of the gene (Ma and Mitra, 2002) or short (40–60 bp) inverted repeats (Lecomme *et al.*, 2003). These strategies are likely to be more efficient (close to 100% silencing) than the single gene strategy followed to this point.

Experiments on natural disease resistance

Of course, CTV is not the only pathogen that plagues citrus; there are a host of others of bacterial, fungal, viroid and viral origin. As described above, mechanisms of plant disease resistance are being elucidated in annual species such as *Arabidopsis*, and a number of the genes involved have been isolated in recent years. The most promising characteristic of using resistance genes of plant origin such as *NPR1* is that they may lead to broad-spectrum protection against a variety of pathogens. In addition, since they are of plant origin, they should be of less concern to consumers. Our citrus gene isolation and transformation efforts are in their early stages. In the single report where citrus was transformed with a PR protein, a transgenic sweet orange line containing a tomato *PR-5* gene was significantly more resistant to the oomycete *Phytophthora citrophthora* (Fagoaga *et al.*, 2001).

One of our major objectives is to determine whether the genes shown to be important in disease resistance in other species, such as *EDR1*, *EDS1*, *EDS5*, *NDR1*, *NPR1*, *PAD4*, *PBS1*, *PBS3*, *PR1*, *RAR1*, *SGT1* and *RdRp*, are present in citrus. We are using two strategies to do this. One is to BLAST (http://www.ncbi.nlm.nih.gov/BLAST/) the sequence of proteins of known function from other organisms to the *Citrus* expressed sequence tag (EST) database available to identify potentially homologous genes. In recent years, EST sequences from various citrus species and tissues, and under different biological and environmental conditions have been added to GenBank (http://www.ncbi.nlm.nih.gov), allowing easy access to this information for BLAST comparison and identification of genes of interest. As of August 2007, more than 94,000 entries were available for *C. sinensis* (sweet orange) and thousands more for other *Citrus* types (http://www.ncbi.nlm.nih.gov/dbEST/dbEST_summary.html), and this number is rapidly increasing. Several groups around the world, including Japan, California (UC, Riverside's HarVest:Citrus, http://harvest.ucr.edu/) and Florida (USDA), among others, are actively adding more EST sequences to the database.

Alternatively, degenerate primers based on protein sequence alignments can be designed in those cases where BLAST does not identify any homologous sequences. A useful tool for the design of degenerate primers is the CODEHOP program (http://blocks.fhcrc.org/codehop.html; Rose *et al.*, 2003).

One of the genes we have cloned using these strategies is the grapefruit *EDR1* homologue. This *Citrus* gene has the kinase domain and the conserved regions present in EDR1 orthologues but is divergent in paralogues (Fig. 14.4). As mentioned above, this gene is a negative regulator of SAR. A mutated, kinase-deficient version was transformed into *Arabidopsis*, and the regenerated plants showed increased resistance to powdery mildews (Tang and Innes, 2002). We anticipate that a similar strategy could work in citrus.

Because citrus is affected by such a diversity of pathogens, finding individual genetic cures for each one of them is time consuming and inefficient. Ideally broad-spectrum resistance is desired, and the use of the same natural defence mechanisms that are present in plants has the potential of being very effective, based on the limited available evidence. The idea behind using genetic engineering is to produce citrus plants that have a heightened or much earlier defence response than wild types after recognition of a pathogen, stopping it short of infection. Most of the experiments on genetically engineered resistance against pathogens have been developed in annual crops. In such cases, just a delay in the onset of a disease can be enough to guarantee a profitable production. In the case of a perennial crop, such as citrus, this is not enough. In our case, we want to produce plants that are either immune or highly tolerant to pathogen infection over long periods of time.

In summary, some of the defence genes that have been identified in other species and that can provide broad-spectrum resistance appear to be able to act in heterologous systems. However, we are in the process of identifying some of these genes from *Citrus* types as well. We intend to test both heterologous and homologous genes in transgenic plants to determine their potential in

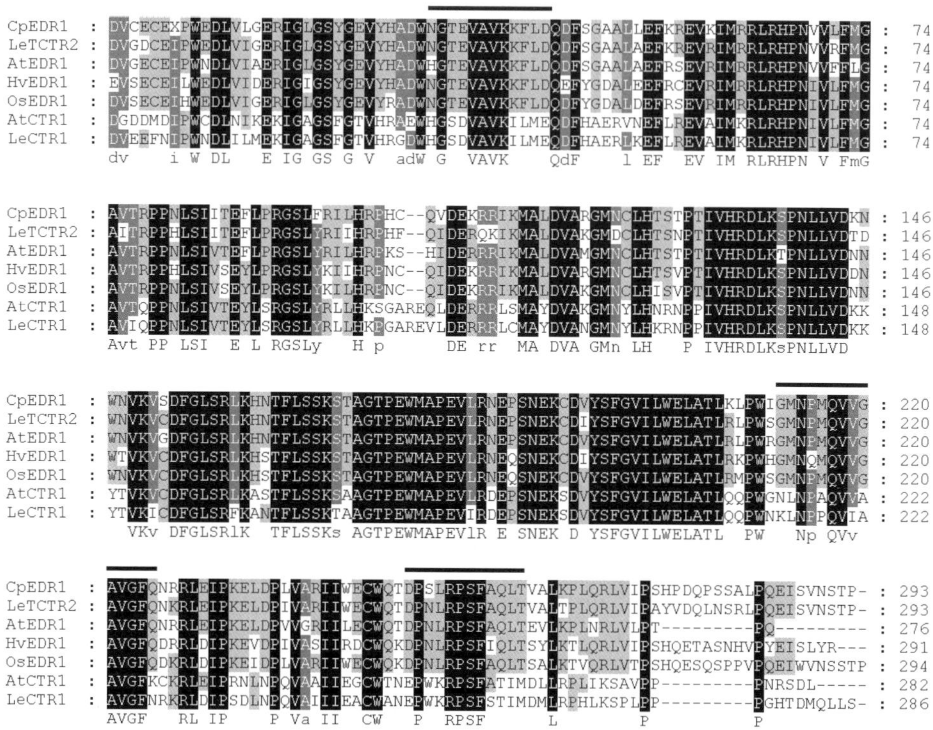

Fig. 14.4. Amino acid sequence alignment of the kinase domain of EDR1-type proteins from several plant species including grapefruit (Cp). EDR1 is a Raf-like mitogen-activated protein kinase kinase (MAPKK) that functions as a negative regulator of disease resistance and SAR. Horizontal bars indicate regions conserved in EDR1 orthologues but divergent in paralogues (CTR1). Cp = *Citrus paradisi*; Le = *Lycopersicon esculentum*; At = *Arabidopsis thaliana*; Hv = *Hordeum vulgare*; Os = *Oryza sativa*.

inducing long-lasting, broad-spectrum disease resistance against the pathogens that plague this important crop.

Acknowledgements

We thank Steve M. Garnsey, Yavuz Yagiz, Suzanne M. Stevens, Karen Champ, Kimberly Niblett, Neil Berger and Reinhart Eckloff for technical assistance; Larry Brown and Susan Halbert for collecting and providing isolate DPI 3800; and David Stark for providing the pMON10098 vector. This research was supported in part by funds from a UF/USDA Specific Cooperative Agreement on Exotic Citrus Diseases, a USDA Special Grant on CTV, and a grant from the Florida Citrus Production Research Advisory Council.

References

Albiach-Marti, M.R., Mawassi, M., Gowda, S., Satyanarayana, T., Hilf, M.E., Shanker, S., Almira, E.C., Vives, M.C., Lopez, C., Guerri, J., Flores, R., Moreno, P., Garnsey, S.M. and Dawson, W.O. (2000) Sequences of citrus tristeza virus separated in time and space are essentially identical. *Journal of Virology* 74, 6856–6865.

Bar-Joseph, M., Marcus, R. and Lee, R.F. (1989) The continuous challenge of citrus tristeza virus control. *Annual Review of Phytopathology* 27, 291–316.

Baulcombe, D.C. (1996) Mechanisms of pathogen-derived resistance to viruses in transgenic plants. *Plant Cell* 8, 1833–1844.

Baulcombe, D. (2002) RNA silencing. *Current Biology* 12, R82–R84.

Cao, H., Glazebrook, J., Clarke, J.D., Volko, S. and Dong, X. (1997) The Arabidopsis NPR1 gene that controls systemic acquired resistance encodes a novel protein containing ankyrin repeats. *Cell* 88, 57–63.

Cao, H., Li, X. and Dong, X.N. (1998) Generation of broad-spectrum disease resistance by overexpression of an essential regulatory gene in systemic acquired resistance. *Proceedings of the National Academy of Sciences of the USA* 95, 6531–6536.

Cevik, B. (2001) Characterization of the RNA-dependent RNA Polymerase Gene of Citrus Tristeza Closterovirus. PhD Thesis. University of Florida, Gainesville, Florida.

Chern, M.S., Fitzgerald, H.A., Yadav, R.C., Canlas, P.E., Dong, X.N. and Ronald, P.C. (2001) Evidence for a disease-resistance pathway in rice similar to the NPR1-mediated signaling pathway in Arabidopsis. *Plant Journal* 27, 101–113.

Delaney, T.P., Uknes, S., Vernooij, B., Friedrich, L., Weymann, K., Negrotto, D., Gaffney, T., Gutrella, M., Kessmann, H., Ward, E. and Ryals, J. (1994) A central role of salicylic acid in plant disease resistance. *Science* 266, 1247–1250.

Delledonne, M., Murgia, I., Ederle, D., Sbicego, P.F., Biondani, A., Polverari, A. and Lamb, C. (2002) Reactive oxygen intermediates modulate nitric oxide signaling in the plant hypersensitive disease-resistance response. *Plant Physiology and Biochemistry* 40, 605–610.

Despres, C., Chubak, C., Rochon, A., Clark, R., Bethune, T., Desveaux, D. and Fobert, P.R. (2003) The Arabidopsis NPR1 disease resistance protein is a novel cofactor that confers redox regulation of DNA binding activity to the basic domain/leucine zipper transcription factor TGA1. *Plant Cell* 15, 2181–2191.

Dominguez, A., Guerri, J., Cambra, M., Navarro, L., Moreno, P. and Peña, L. (2000) Efficient production of transgenic citrus plants expressing the coat protein gene of citrus tristeza virus. *Plant Cell Reports* 19, 427–433.

Dominguez, A., Hermoso de Mendoza, A., Guerri, J., Cambra, M., Navarro, L. and Peña, L. (2002) Pathogen-derived resistance to Citrus tristeza virus (CTV) in transgenic Mexican lime (*Citrus aurantifolia* (Christ.) Swing) plants expressing its p25 coat protein gene. *Molecular Breeding* 10, 1–10.

Dougherty, W.G., Lindbo, J.A., Smith, H.A., Park, T.D., Swaney, S. and Proebsting, W.L. (1994) RNA-mediated virus resistance in transgenic plants: exploitation of a cellular pathway possibly involved in RNA degradation. *Molecular Plant-Microbe Interactions* 7, 544–552.

Durrant, E.E. and Dong, X. (2004) Systemic acquired resistance. *Annual Review of Phytopathology* 42, 185–209.

Ekengren, S.K., Liu, Y., Schiff, M., Dinesh-Kumar, S.P. and Martin, G.B. (2003) Two MAPK cascades, NPR1 and TGA transcription factors play a role in Pto-mediated disease resistance in tomato. *Plant Journal* 36, 905–917.

Fagard, M. and Vaucheret, H. (2000) (Trans)gene silencing in plants: how many mechanisms? *Annual Review of Plant Physiology and Plant Molecular Biology* 51, 167–194.

Fagoaga, C., Roodrigo, I., Conejero, V., Hinarejos, C., Tuset, J.J., Arnau, J., Pina, J.A., Navarro, L. and Pena, L. (2001) Increased tolerance to *Phytophthora citrophthora* in transgenic orange plants constitutively expressing a tomato pathogenesis related protein PR-5. *Molecular Breeding* 7, 175–185.

Fan, W.H. and Dong, X.N. (2002) *In vivo* interaction between NPR1 and transcription factor TGA2 leads to salicylic acid-mediated gene activation in Arabidopsis. *Plant Cell* 14, 1377–1389.

Febres, V.J., Ashoulin, L., Mawassi, M., Frank, A., Bar-Joseph, M., Manjunath, K.L., Lee, R.F. and Niblett, C.L. (1996) The p27 protein is present at one end of citrus tristeza virus particles. *Phytopathology* 86, 1331–1335.

Febres, V.J., Niblett, C.L., Lee, R.F. and Moore, G.A. (2003) Characterization of grapefruit plants (*Citrus paradisi* Macf.) transformed with citrus tristeza closterovirus genes. *Plant Cell Reports* 21, 421–428.

Friedrich, L., Lawton, K., Dietrich, R., Willits, M., Cade, R. and Ryals, J. (2001) NIM1 overexpression in Arabidopsis potentiates plant disease resistance and results in enhanced effectiveness of fungicides. *Molecular Plant-Microbe Interactions* 14, 1114–1124.

Frye, C.A., Tang, D.Z. and Innes, R.W. (2001) Negative regulation of defense responses in plants by a conserved MAPKK kinase. *Proceedings of the National Academy of Sciences of the USA* 98, 373–378.

Garnsey, S.M. and Cambra, M. (1991) Enzyme linked immunosorbent assay for citrus pathogens. In: Roistacher, C.N. (ed.) *Graft-transmissible Diseases of Citrus: Handbook for Detection and Diagnosis.* IOCV, FAO, Rome, pp. 193–216.

Gerlach, W.L., Llewellyn, D. and Haseloff, J. (1987) Construction of a plant disease resistance gene using satellite RNA of Tobacco ringspot virus. *Nature* 328, 802–805.

Ghorbel, R., Dominguez, A., Navarro, L. and Peña, L. (2000) High efficiency genetic transformation of sour orange (*Citrus aurantium*) and production of transgenic trees containing the coat protein gene of citrus tristeza virus. *Tree Physiology* 20, 1183–1189.

Ghorbel, R., López, C., Fagoaga, C., Moreno, P., Navarro, L., Flores, R. and Peña, L. (2001) Transgenic citrus plants expressing the citrus tristeza virus p23 protein exhibit viral-like symptoms. *Molecular Plant Pathology* 2, 27–36.

Glazebrook, J., Chen, W.J., Estes, B., Chang, H.S., Nawrath, C., Metraux, J.P., Zhu, T. and Katagiri, F. (2003) Topology of the network integrating salicylate and jasmonate signal transduction derived from global expression phenotyping. *Plant Journal* 34. 217–228.

Gowda, S., Satyanarayana, T., Davis, C.L., Navas-Castillo, J., Albiach-Marti, M.R., Mawassi, M., Valkov, N., Bar-Joseph, M., Moreno, P. and Dawson, W.O. (2000) The p20 gene product of Citrus tristeza virus accumulates in the amorphous inclusion bodies. *Virology* 274, 246–254.

Gutierrez, M.A., Luth, D. and Moore, G.A. (1997) Factors affecting *Agrobacterium*-mediated transformation in Citrus and production of sour orange (*Citrus aurantium* L.) plants expressing the coat protein gene of citrus tristeza virus. *Plant Cell Reports* 16, 745–753.

Hamilton, A.J. and Baulcombe, D.C. (1999) A species of small antisense RNA in posttranscriptional gene silencing in plants. *Science* 286, 950–952.

Hamilton, A., Voinnet, O., Chappell, L. and Baulcombe, D. (2002) Two classes of short interfering RNA in RNA silencing. *EMBO Journal* 21, 4671–4679.

Hammond, J. (1999) Overview: the many uses and applications of transgenic plants. *Current Topics in Microbiolocial Immunology* 240, 1–19.

Hammond, S.M., Bernstein, E., Beach, D. and Hannon, G.J. (2000) An RNA-directed nuclease mediates post-transcriptional gene silencing in *Drosophila* cells. *Nature* 404, 293–296.

Harrison, B.D., Mayo, M.A. and Baulcombe, D.C. (1987) Virus resistance in transgenic plants that express Cucumber mosaic virus satellite RNA. *Nature* 328, 799–802.

Johnson, C., Boden, E. and Arias, J. (2003) Salicylic acid and NPR1 induce the recruitment of trans-activating TGA factors to a defense gene promoter in Arabidopsis. *Plant Cell* 15, 1846–1858.

Karasev, A.V., Boyko, V.P., Gowda, S., Nikolaeva, O.V., Hilf, M.E., Koonin, E.V., Niblett, C.L., Cline, K., Gumpf, D.J. and Lee, R.F. (1995) Complete sequence of the citrus tristeza virus RNA genome. *Virology* 208, 511–520.

Kim, J.A., Agrawal, G.K., Rakwal, R., Han, K., Kim, K., Yun, C.H., Heo, S.G., Park, S.Y., Lee, Y.H. and Jwa, N.S. (2003) Molecular cloning and mRNA expression analysis of a novel rice (*Oryza sativa* L.) MAPK kinase kinase, OsEDR1, an ortholog of *Arabidopsis* AtEDR1, reveal its role in defense/stress signalling pathways and development. *Biochemical and Biophysical Research Communications* 300, 868–876.

Kinkema, M., Fan, W.H. and Dong, X.N. (2000) Nuclear localization of NPR1 is required for activation of PR gene expression. *Plant Cell* 12, 2339–2350.

Lahaye, T. (2002) The Arabidopsis RRS1-R disease resistance gene – uncovering the plant's nucleus as the new battlefield of plant defense? *Trends in Plant Science* 7, 425–427.

Lecomme, C., Hrubikova, K. and Hein, I. (2003) Enhancement of virus-induced gene silencing through viral-based production of inverted repeats. *Plant Journal* 34, 543–553.

Lindbo, J.A., Silva-Rosales, L., Proebsting, W.M. and Dougherty, W.G. (1993) Induction of a highly specific antiviral state in transgenic plants: implications for regulation of gene expression and virus resistance. *Plant Cell* 5, 1749–1759.

Liu, Y.L., Schiff, M., Marathe, R. and Dinesh-Kumar, S.P. (2002) Tobacco Rar1, EDS1 and NPR1/NIM1 like genes are required for N-mediated resistance to tobacco mosaic virus. *Plant Journal* 30, 415–429.

Liu, Y., Burch-Smith, T., Schiff, M., Feng, S. and Dinesh-Kumar, S.P. (2004) Molecular chaperone Hsp90 associates with resistance protein N and its signaling proteins SGT1 and Rar1 an innate immune response in plants. *Journal of Biological Chemistry* 279, 2101–2108.

Lomonossoff, G.P. (1995) Pathogen-derived resistance to plant-viruses. *Annual Review of Phytopathology* 33, 323–343.

Lu, R., Folimonov, A., Shintaku, M., Li, W.X., Falk, B.W., Dawson, W.O. and Ding, S.W. (2004) Three distinct suppressors of RNA silencing encoded by a 20-kb viral RNA genome. *Proceedings of the National Academy of Sciences of the USA* 101, 15742–15747

Luth, D. and Moore, G. (1999) Transgenic grapefruit plants obtained by *Agrobacterium tumefaciens*-mediated transformation. *Plant Cell Tissue and Organ Culture* 57, 219–222.

Ma, C.L. and Mitra, A. (2002) Intrinsic direct repeats generate consistent post-transcriptional gene silencing in tobacco. *Plant Journal* 31, 37–49.

Martin, G.B., Bogdanove, A.J. and Sessa, G. (2003) Understanding the functions of plant disease resistance proteins. *Annual Review of Plant Biology* 54, 23–61.

Mauch-Mani, B. and Slusarenko, A.J. (1996) Production of salicylic acid precursors is a major function of phenylalanine ammonia-lyase in the resistance of *Arabidopsis* to *Peronospora parasitica. Plant Cell* 8, 203–212.

McDowell, J.M. and Dangl, J.L. (2000) Signal transduction in the plant immune response. *Trends in Biochemical Sciences* 25, 79–82.

Mawassi, M., Mietkiewska, E., Gofman, R., Yang, G. and Bar-Joseph, M. (1996) Unusual sequence relationships between two isolates of citrus tristeza virus. *Journal of General Virology* 77, 2359–2364.

Mawassi, M,, Satyanarayana, T., Albiach-Marti, M.R., Gowda, S., Ayllon, M.A., Robertson, C. and Dawson, W.O. (2000) The fitness of citrus tristeza virus defective RNAs is affected by the lengths of their 5′- and 3′-termini and by the coding capacity. *Virology* 275, 42–56.

Mou, Z., Fan, W.H. and Dong, X.N. (2003) Inducers of plant systemic acquired resistance regulate NPR1 function through redox changes. *Cell* 113, 935–944.

Moore, G.A., Jacono, C.C., Neidigh, J.L., Lawrence, S.D. and Cline, K. (1992) *Agrobacterium*-mediated transformation of citrus stem segments and regeneration of transgenic plants. *Plant Cell Reports* 11, 238–242.

Muller, G.W. and Garnsey, S.M. (1984) Susceptibility of citrus varieties, species, citrus relatives, and non-Rutaceous plants to slash-cut mechanical inoculation with citrus tristeza virus (CTV). In: Garnsey, S.M., Timmer, L.W. and Dodds, J.A. (eds) *Proceedings of the Ninth Conference IOCV*. IOCV, Riverside, California, pp. 33–40.

Nelson, R.E., Powell Abel, P. and Beachy, R.N. (1987) Lesions and virus accumulation in inoculated transgenic tobacco plants expressing the coat protein gene of tobacco mosaic virus. *Virology* 158, 126–132.

Park, W., Li, J.J., Song, R.T., Messing, J. and Chen, X.M. (2002) CARPEL FACTORY, a Dicer homolog, and HEN1, a novel protein, act in microRNA metabolism in *Arabidopsis thaliana. Current Biology* 12, 1484–1495.

Pappu, H.R., Karasev, A.V., Anderson, E.J., Pappu, S.S., Hilf, M.E., Febres, V.J., Eckloff, R.M., McCaffery, M., Boyko, V. and Gowda, S. (1994) Nucleotide sequence and organization of eight 3′ open reading frames of the citrus tristeza closterovirus genome. *Virology* 199, 35–46.

Peck, S. (2003) Early phosphorylation events in biotic stress. *Current Opinion in Plant Biology* 6, 334–338.

Permar, T.A., Garnsey, S.M., Gumpf, D.J. and Lee, R.F. (1990) A monoclonal-antibody that discriminates strains of citrus tristeza virus. *Phytopathology* 80, 224–228.

Pieterse, C.M.J. and van Loon, L.C. (1999) Salicylic acid-independent plant defence pathways. *Trends in Plant Science* 4, 52–58.

Powell, P.A., Sanders, P.R., Tumer, N. and Beachy, R.N. (1990) Protection against tobacco mosaic virus infection in transgenic plants requires accumulation of capsid protein rather than coat protein RNA sequences. *Virology* 175, 124–130.

Powell-Abel, P.A., Nelson, R.S., Barun De, Hoffmann, N., Rogers, S.G., Fraley, R.T. and Beachy, R.N. (1986) Delay of disease development in transgenic plants that express the tobacco mosaic virus coat protein gene. *Science* 232, 738–743.

Quirino, B.F. and Bent, A.F. (2003) Deciphering host resistance and pathogen virulence: the *Arabidopsis/Pseudomonas* interaction as a model. *Molecular Plant Pathology* 4, 517–530.

Rose, T.M., Henikoff, J.G. and Henikoff, S. (2003) CODEHOP (COnsensus-DEgenerate Hybrid Oligonucleotide Primer) PCR primer design. *Nucleic Acids Research* 31, 3763–3766.

Rovere, C.V., del Vasm M. and Hopp, H.E. (2002) RNA-mediated virus resistance. *Current Opinion in Biotechnology* 13, 167–172.

Satyanarayana, T., Gowda, S., Ayllon, M.A., Albiach-Marti, M.R., Rabindran, S. and Dawson, W.O. (2002) The p23 protein of citrus tristeza virus controls asymmetrical RNA accumulation. *Journal of Virology* 76, 473–83.

Sanford, J.C. and Johnston, S.A. (1985) The concept of parasite-derived resistance – deriving resistance genes from the parasites own genome. *Journal of Theoretical Biology* 113, 395–405.

Shah, J., Kachroo, P. and Klessig, D.F. (1999) The Arabidopsis ssi1 mutation restores pathogenesis-related gene expression in npr1 plants and renders defensin gene expression salicylic acid dependent. *Plant Cell* 11, 191–206.

Shirano, Y., Kachroo, P., Shah, J. and Klessig, D.F. (2002) A gain-of-function mutation in an Arabidopsis Toll interleukin-1 receptor-nucleotide binding site-leucine-rich repeat type R gene triggers defense responses and results in enhanced disease resistance. *Plant Cell* 14, 3149–3162.

Smith, N., Singh, S., Wang, M.B., Stoutjesdijk, P., Green, A. and Waterhouse, P.M. (2000) Total silencing by intron-spliced hairpin RNAs. *Nature* 407, 319–320.

Spoel, S.H., Koornneef, A., Claessens, S.M.C., Korzelius, J.P., Van Pelt, J.A., Mueller, M.J., Buchala, A.J., Metraux, J.P., Brown, R., Kazan, K., Van Loon, L.C., Dong, X. and Pieterse, C.M. (2003) NPR1 modulates cross-talk between salicylate- and jasmonate-dependent defense pathways through a novel function in the cytosol. *Plant Cell* 15, 760–770.

Tang, D.Z. and Innes, R.W. (2002) Overexpression of a kinase-deficient form of the EDR1 gene enhances powdery mildew resistance and ethylene-induced senescence in Arabidopsis. *Plant Journal* 32, 975–983.

Tang, G.L., Reinhart, B.J., Bartel, D.P. and Zamore, P.D. (2003) A biochemical framework for RNA silencing in plants. *Genes and Development* 17, 49–63.

Tchurikov, N.A., Chistyakova, L.G., Zavilgelsky, G.B., Manukhov, I.V., Chernov, B.K. and Golova, Y.B. (2000) Gene-specific silencing by expression of parallel complementary RNA in *Escherichia coli*. *Journal of Biological Chemistry* 275, 26523–26529.

Tenllado, F., Llave, C. and Diaz-Ruiz, J.R (2004) RNA interference as a new biotechnological tool for the control of virus diseases in plants. *Virus Research* 102, 85–96.

Thordal-Christensen, H. (2003) Fresh insights into processes of nonhost resistance. *Current Opinion in Plant Biology* 6, 351–357.

Uquillas, C., Letelier, I., Blanco, F., Jordana, X. and Holuigue, L. (2004) NPR1-independent activation of immediate early salicylic acid-responsive genes in *Arabidopsis*. *Molecular Plant-Microbe Interactions* 17, 34–42.

Vives, M.C., Rubio, L., Lopez, C., Navas-Castillo, J., Albiach-Marti, M.R., Dawson, W.O., Guerri, J., Flores, R. and Moreno, P. (1999) The complete genome sequence of the major component of a mild citrus tristeza virus isolate. *Journal of General Virology* 80, 811–816.

Voinnet, O. (2002) RNA silencing: small RNAs as ubiquitous regulators of gene expression. *Current Opinion in Plant Biology* 5, 444–451.

Wang, M.B., Abbott, D.C. and Waterhouse, P.M. (2000) A single copy of a virus-derived transgene encoding hairpin RNA gives immunity to barley yellow dwarf virus. *Molecular Plant Pathology* 1, 347–356.

Wassenegger, M. (2002a) Gene silencing. *International Review of Cytology* 219, 61–113.

Wassenegger, M. (2002b) Gene silencing-based disease resistance. *Transgenic Research* 11, 639–653.

Waterhouse, P.M., Graham, H.W. and Wang, M.B. (1998) Virus resistance and gene silencing in plants can be induced by simultaneous expression of sense and antisense RNA. *Proceedings of the National Academy of Sciences of the USA* 95, 13959–13964.

Waterhouse, P.M., Wang, M.B. and Lough, T. (2001) Gene silencing as an adaptive defence against viruses. *Nature* 411, 834–842.

Whitham, S.A., Quan, S., Chang, H.S., Cooper, B., Estes, B., Zhu, T., Wang, X. and Hou, Y.M. (2003) Diverse RNA viruses elicit the expression of common sets of genes in susceptible *Arabidopsis thaliana* plants. *Plant Journal* 33, 271–283.

Wisniewski, L.A., Powell, P.A., Nelson, R.S. and Beachy, R.N. (1990) Local and systemic movement of tobacco mosaic virus (TMV) in tobacco plants that express the TMV coat protein gene. *Plant Cell* 2, 559–567.

Xiao, S., Brown, S., Patrick, E., Brearley, C. and Turner, J.G. (2003) Enhanced transcription of the Arabidopsis disease resistance genes ROW8.1 and RPW8.2 via salicylic acid-dependent amplification circuit is required for hypersensitive response. *Plant Cell* 15, 33–45.

Yang, Z.N., Mathews, D.M., Dodds, J.A. and Mirkov, T.E. (1999) Molecular characterization of an isolate of citrus tristeza virus that causes severe symptoms in sweet orange. *Virus Genes* 19, 131–142.

Yu, D.Q., Chen, C.H. and Chen, Z.X. (2001) Evidence for an important role of WRKY DNA binding proteins in the regulation of NPR1 gene express ion. *Plant Cell* 13, 1527–1539.

Yu, D., Fan, B., MacFarlane, S.A. and Chen, Z. (2003) Analysis of the involvement of an inducible Arabidopsis RNA-dependent RNA polymerase in antiviraldefense. *Molecular Plant-Microbe Interactions* 16, 206–216.

Zaccomer, B., Cellier, F., Boyer, J.C., Haenni, A.L. and Tepfer, M. (1993) Transgenic plants that express genes including the 3′ untranslated region of the Turnip yellow mosaic virus (TYMV) are partially protected against TYMV infection. *Gene* 136, 87–94.

Zhu, M., Shao, F., Innes, R.W., Dixon, J.E. and Xu, Z. (2004) The crystal structure of *Pseudomonas* avirulence protein AvrPphB: a papain-like fold with a distinct substrate-binding site. *Proceedings of the National Academy of Sciences of the USA* 101, 302–307.

15 Genetic Transformation

Leandro Peña, Magdalena Cervera, Riadh Ghorbel, Antonio Domínguez, Carmen Fagoaga, José Juárez, José Antonio Pina and Luis Navarro

Departamento Protección Vegetal y Biotecnología, Instituto Valenciano de Investigaciones Agrarias (IVIA), Apartado Oficial, 46113-Moncada, Valencia, Spain

Summary

Genetic improvement of citrus through conventional breeding is limited by their genetic and reproductive characteristics. Citrus species have a complex reproductive system, with many cases of cross- and self-incompatibility, apomixis and high heterozygosity, and most of them have very long juvenile periods. In addition, the mode of inheritance of most traits of agricultural importance is unknown. In this context, genetic transformation offers an important alternative for the genetic improvement of citrus. We have developed efficient and reliable transformation systems for many economically important citrus species, including sweet orange, sour orange, lime, lemon, citrange, alemow and mandarin, based on the selection of the appropriate *Agrobacterium* strain for super-transforming citrus, the establishment of the most appropriate infection and co-cultivation conditions and culture media, as well as adequate selection conditions and culture media, the use of source plant material in a good ontological state, the determination of the competent cells for transformation in citrus explants, the use of appropriate marker genes and the rapid production of whole transgenic plants through grafting of regenerating transgenic shoots on to vigorous rootstocks first *in vitro* and later in the greenhouse. Furthermore, we have been able to transform mature material, recovering transgenic plants that flowered and set fruits in 1–2 years after *Agrobacterium* infection. Our research is now oriented to incorporating transgenes of potential agricultural interest into citrus species with the aim of obtaining resistance to citrus tristeza virus, higher tolerance to *Phytophthora*, higher tolerance to salinity and shortening the juvenile period.

Introduction

Importance and distribution of citrus

Citrus is the most important fruit crop in the world, with a production of almost 100 Mt in 2001 (Food and Agriculture Organization, 2001). It is grown in more than 100 countries all over the world, mainly in tropical and subtropical areas

(~40° latitude on each side of the equator) where favourable soil and climatic conditions occur. Major producing countries include Brazil, the USA, China, Spain, Mexico, India, Iran, Italy, Egypt, Argentina, Turkey, Japan, Pakistan, South Africa, Greece, Thailand, Morocco, Israel, Indonesia, Korea and Australia. Citrus fruits are marketed fresh or as processed juice and canned segments.

The general area of origin of citrus is believed to be South-east Asia, including south China, north-eastern India and Burma, though its introduction into cultivation probably started in China. Commercial citrus species and related genera belong to the order Geraniales, family Rutaceae, subfamily Aurantoidea. All rootstocks and varieties used are included in the genus *Citrus*, except for kumquats (*Fortunella* spp.) and trifoliate orange (*Poncirus trifoliata* L. Raf.), the latter used exclusively as a rootstock. Commercial citrus fruits fall into several main groups: sweet oranges (*C. sinensis* (L.) Osb.), mandarins, including satsumas (*C. unshiu* (Mak.) Marc.) and clementines (*C. clementina* Hort. ex Tan.), grapefruits (*C. paradisi* Macf.), pummelos (*C. grandis* (L.) Osb.), lemons (*C. limon* (L.) Burm. f.) and limes (*C. aurantifolia* (Christm.) Swing.). There are other species of relative importance in certain areas, such as sour oranges (*C. aurantium* L.), citrons (*C. medica* L.) and bergamots (*C. bergamia* Risso & Poit.). Some hybrids of commercial interest include citranges (sweet orange × trifoliate orange) and citrumelos (grapefruit × trifoliate orange), used as rootstocks, and tangelos (mandarin × grapefruit), tangors (mandarin × sweet orange) and mandarin hybrids, used as varieties.

Need for genetic improvement

Many different citrus genotypes are commercially grown in a wide diversity of soil and climatic conditions; therefore, trees are subjected to important abiotic and biotic stresses that limit the production and, in some instances, the use of certain rootstocks and varieties. The main abiotic stresses are acid, alkaline and salty soils, flooding and drought, freezing and high temperatures. Citrus trees are also affected by many pests and diseases caused by nematodes, fungi, bacteria, spiroplasmas, phytoplasmas, viruses and viroids. Some diseases are spread throughout the world, such as those produced by the fungus *Phytophthora* sp. or by the citrus tristeza virus (CTV), that preclude the use of certain excellent rootstocks, and severely restrict fruit production and quality of important varieties in some countries. Other diseases are restricted to specific geographic areas, such as those caused by the bacterium *Xylella fastidiosa* in Brazil and *Liberobacter asiaticum* in most countries of South-east Asia. In both cases, these bacteria are currently devastating millions of trees and there are no means for efficient control.

At the same time that the citrus industry is threatened by important biotic and abiotic stresses, the markets in developed countries demand fruit of increasing quality. In this situation, genetic improvement of citrus is a major priority. However, conventional breeding of citrus has important limitations.

Genetic improvement of citrus

Citrus species have a complex reproductive biology. Some important genotypes have total or partial pollen and/or ovule sterility and cannot be used as parents in breeding programmes. There are many cases of cross- and self-incompatibility. Most species are apomictic, which means that adventitious embryos initiate directly from maternal nucellar cells precluding the development of zygotic embryos, and thus the recovery of sexual progeny populations. They have a long juvenile period and most species need at least five years to start flowering in subtropical areas. All these features together with their large plant size, high heterozygosity, lack of basic knowledge about how the most important horticultural traits are inherited, and quantitative inheritance of most characters have greatly impeded genetic improvement of citrus through conventional breeding methods.

Although some breeding programmes for citrus improvement started more than 100 years ago (Soost and Cameron, 1975), nowadays most rootstocks are citrus species without any improvement, and most important varieties have been originated by budsport mutations and chance seedlings. Only a few hybrid varieties are economically relevant in certain local markets. Probably, the most significant results from any improvement programme performed in the world are the hybrids Carrizo and Troyer citranges, originating from a cross made in 1909 that was originally oriented to introduce cold tolerance in edible fruits (Savage and Gardner, 1965). Nowadays, citranges are widely used as rootstocks in countries such as Spain and the USA.

Recently, the development of genetic markers is providing a new potential tool for citrus breeding. Linkage maps have been performed using isozymes, restriction fragment length polymorphism (RFLP), random amplified polymorphic DNA (RAPD), sequence-characterized amplified regions (SCAR), amplified fragment length polymorphism (AFLP), microsatellites (single sequence repeat; SSR) and cleaved amplified polymorphic sequences (CAPS). Although these studies have served to determine the mode of inheritance of these traits and they could be useful for breeding purposes, map-based cloning of the corresponding genes is still a long way off. The only exception could be the CTV resistance gene from *P. trifoliata*. Cloning of this gene is under way in several laboratories (Deng *et al.*, 2001; Yang *et al.*, 2001). Other promising biotechnologies for citrus improvement include somatic hybridization (Grosser *et al.*, 2000) and embryo rescue mainly for obtaining triploid (seedless) varieties (Ollitrault *et al.*, 1998).

Genetic transformation may provide an efficient alternative for citrus improvement, opening the way for the introduction of specific traits into known genotypes without altering their genetic elite background.

Genetic Transformation of Citrus

State of the art

Until recently, citrus species have been recalcitrant to transformation. Kobayashi and Uchimiya (1989) obtained transgenic callus from Trovita sweet orange by polyethylene glycol (PEG) treatment of cell suspensions with a plasmid containing the marker gene *nptII*, but regeneration of transgenic plants from that callus was unsuccessful. Vardi *et al.* (1990) produced transgenic callus from Rough lemon (*C. jambhiri* Lush) by PEG treatment of protoplasts with a plasmid containing the marker genes *cat* and *nptII*, and obtained several stably transgenic embryogenic lines, but only two of them regenerated whole plants. Hidaka *et al.* (1990) produced transformed callus of Washington navel and Trovita oranges by co-cultivation of embryogenic cell suspension lines with *Agrobacterium tumefaciens*, but only one transgenic plantlet of Washington navel was regenerated. Moore *et al.* (1992) produced two transgenic plantlets of Carrizo citrange by co-cultivation of internodal stem segments from *in vitro* grown seedlings with *A. tumefaciens*. More recently, this group reported a slight increase in transformation efficiency using basically the same procedure and regenerated two sour orange, nine lime and nine Carrizo citrange plants with the major coat protein gene of CTV (Gutiérrez *et al.*, 1997). Hidaka and Omura (1993) obtained transformed Ponkan mandarin callus by electroporation of protoplasts, but no plants were regenerated. Yao *et al.* (1996) reported transformation of Page tangelo embryogenic cells using particle bombardment, and produced 15 transgenic embryo lines, but they did not progress further. Kayim *et al.* (1996) bombarded nucellar cells of lemon, but only β-glucuronidase (GUS) expression from those cells was reported.

The first report of efficient and reliable production of citrus transgenic plants was published by Kaneyoshi *et al.* in 1994. They obtained transformation efficiency higher than 25% by co-cultivating etiolated epicotyl segments of trifoliate orange with *A. tumefaciens*. This allowed them efficiently to incorporate the human epidermal growth factor (*hEGF*) (Kobayashi *et al.*, 1996) and the *rolC* gene from *Agrobacterium rhizogenes* (Kaneyoshi and Kobayashi, 1999) into this species. The same procedure with slight modifications has been used by Bond and Roose (1998) to transform Washington navel orange, and by Luth and Moore (2000) to transform grapefruit. Pérez-Molphe and Ochoa-Alejo (1998) have reported efficient production of transgenic lime plants by co-cultivation of internodal stem segments from *in vitro* grown seedlings with *A. rhizogenes*. Yang *et al.* (2000) have used *A. tumefaciens* to transform epicotyl segments of Rio Red grapefruit, and have reported production of transgenic plants containing an untranslatable version of the major coat protein gene from CTV and the *Galanthus nivalis* agglutinin gene. Co-cultivation of epicotyl segments with *A. tumefaciens* has been also used by Gentile *et al.* (1998) and LaMalfa *et al.* (2000) to regenerate transgenic plants of Tarocco orange and Troyer citrange with *rol* genes from *A. rhizogenes* and with the green fluorescent protein gene (*gfp*) from the jellyfish *Aequorea victoria*, respectively, and by Koltunow *et al.* (2000) to produce transgenic limes containing genes for decreased seed set. A different approach was followed by Fleming *et al.* (2000) who used the *gfp* gene as both a selectable and reporter marker to transform Itaborai sweet orange protoplasts with PEG, produce embryogenic callus from them and recover whole transgenic plants through somatic embryogenesis.

The recalcitrance of citrus to genetic transformation is mainly due to several factors: inefficiency of bacterial vectors in the transformation of citrus cells, since citrus species are not natural hosts of *Agrobacterium*; difficulties in regenerating shoots only from the transformed cells at the same time avoiding the recovery of escapes; and difficulties in rooting the transgenic shoots. Our group started to work in genetic transformation of citrus in 1993, and since then we have been able to develop efficient and reliable procedures to produce transgenic plants from Carrizo citrange (Peña *et al.*, 1995a; Cervera *et al.*, 1998a), using epicotyl segments from *in vitro* grown seedlings as source material, and from Pineapple sweet orange (Peña *et al.*, 1995b; Cervera *et al.*, 1998b), lime (Peña *et al.*, 1997; Domínguez *et al.*, 2000), sour orange (Ghorbel *et al.*, 2000), alemow, lemon and Cleopatra mandarin (Ghorbel *et al.*, 2001a), using internodal stem segments from greenhouse-grown seedlings as source material. The use of the appropriate *Agrobacterium* strain super-transforming citrus as vector, the establishment of the appropriate infection and co-cultivation conditions and culture media, as well as adequate selection conditions and culture media, the use of source plant material in a good physiological state, the determination of the cells competent for transformation in citrus explants, the use of appropriate marker genes, and the rapid production of whole transgenic plants through grafting of regenerating transgenic shoots into vigorous rootstocks first *in vitro* and later in the greenhouse have been crucial to enabling the regeneration of transgenic citrus plants at high efficiencies. Furthermore, we have been able to transform mature material, recovering transgenic plants that flowered and set fruits in 1–2 years after *Agrobacterium* infection (Cervera *et al.*, 1998b). We are now incorporating transgenes of potential agricultural interest into citrus species. These procedures and the recent developments of our projects are the subjects of this chapter.

Transformation from seedling explants

Four- to 12-month-old greenhouse-grown (18–27°C) sweet orange, sour orange, lime, alemow, lemon and mandarin seedlings are used as the source of tissue for transformation. Stem pieces (20 cm in length) are stripped of their leaves and thorns, disinfected for 10 min in a 2% (v/v) sodium hypochlorite solution and rinsed three times with sterile distilled water. For Carrizo citrange, stored seeds originating from the same tree stock are peeled, removing both seed coats, disinfected for 10 min in a 0.5% (v/v) sodium hypochlorite solution containing 0.1% (v/v) Tween-20, and rinsed three times with sterile distilled water. Five-week-old germinating seedlings are used as the starting material for genetic transformation. These seedlings are grown in MS salt solution (Murashige and Skoog, 1962) plus 10 g/l agar, pH 5.7, at 26°C in darkness for the first 2 weeks, and under a 16 h photoperiod and illumination of 45 μEm^{-2} s^{-1} for an additional 3 weeks.

Agrobacterium tumefaciens strain EHA105 (Hood *et al.*, 1993) carrying a binary plasmid is used as the vector system for transformation. The T-DNA of the binary plasmid must contain, apart from the gene or genes of interest, a selectable marker gene, such as *nptII*, and a reporter marker gene, such as *uidA*. The binary plasmid is introduced into *Agrobacterium* by electroporation. Bacteria are cultured overnight in an orbital shaker at 28°C and 200 r.p.m. in LB medium containing the appropriate antibiotics for the binary system to grow. Bacterial cells are pelleted at 3500 r.p.m. for 10 min, resuspended and diluted to approximately 4×10^7 cells/ml in liquid inoculation medium, which consists of MS salt solution, 0.2 mg/l thiamine hydrochloride, 1 mg/l pyridoxine hydrochloride, 1 mg/l nicotinic acid and 3% (w/v) sucrose, pH 5.7.

Sweet orange, sour orange, lime, alemow, lemon and mandarin internodal stem segments, and citrange epicotyl segments (~1 cm long) are cut transversely and incubated for 15 min in 10 cm diameter plates containing 15 ml of the bacterial suspension in inoculation medium by gentle shaking. The infected explants are blotted dry on sterile filter paper and placed horizontally on plates with CM medium for a 3 day co-cultivation period. CM medium consists of MS salts, 1 mg/l thiamine hydrochloride, 1 mg/l pyridoxine hydrochloride, 1 mg/l nicotinic acid, 3% (w/v) sucrose, 2 mg/l indole-3-acetic acid, 1 mg/l 2-isopentenyladenine, 2 mg/l 2,4-dichlorophenoxyacetic acid and 0.8% (w/v) agar, pH 5.7. Co-cultivation in a medium rich in auxins provides to the plant cells at the cut ends of the explants an appropriate treatment to shift them to a competent state for transformation, involving dedifferentiation, induction of cell division and callus proliferation (Peña *et al.*, 1997; Cervera *et al.*, 1998a).

After co-cultivation, the explants are blotted dry with sterile filter paper and transferred to SRM medium, which consists of MS salts, 0.2 mg/l thiamine hydrochloride, 1 mg/l pyridoxine hydrochloride, 1 mg/l nicotinic acid, 3% (w/v) sucrose, 1% (w/v) agar, pH 5.7, plus 100 mg/l kanamycin for the selection of transgenic shoots and 250 mg/l vancomycin and 500 mg/l cefotaxime to control bacterial growth. This medium is supplemented with 3 mg/l benzylaminopurine (BAP) for sweet orange and citrange, 1 mg/l BAP for lime, lemon, alemow and mandarin, and 1 mg/l BAP plus 0.3 mg/l naphthaleneacetic acid (NAA) for sour orange. Cultures are maintained in the dark for 4 weeks at 26°C and then are transferred to a 16 h photoperiod, with 45 μEm^{-2} s^{-1} illumination at 26°C. Culture of the explants in the dark favours callus formation and thus the progress of transformation events to regenerate transgenic shoots and, at the same time, avoid regeneration of escape shoots that could be stimulated by the exposure of explants directly to the light (Peña *et al.*, 1997; Cervera *et al.*, 1998a). For sour orange, it can also be speculated that the combination of BAP and NAA in the SRM medium is more favourable than BAP alone in stimulating cell divisions and re-differentiation from the transgenic competent cells to undergo transformation events (Ghorbel *et al.*, 2000).

Small pieces of the shoots emerging from the explants are assayed for histochemical GUS activity (Jefferson *et al.*, 1987) and then apical portions are grafted *in vitro* onto Troyer citrange seedlings (Navarro *et al.*, 1975; Navarro 1992; Peña *et al.*, 1995a, b). Rootstock preparation is as follows: Troyer citrange seeds are peeled, removing both seed coats, disinfected for 10 min in a 0.5% (v/v) sodium hypochlorite solution containing 0.1% (v/v) Tween-20 and rinsed three times with sterile water. The germination medium is MS salts with 10 g/l agar, pH 5.7. Seeds are sown individually in tubes and grown in darkness at 27°C for 2 weeks. Troyer citrange seedlings are decapitated, leaving 1–1.5 cm of the epicotyls. The roots are shortened to 4–6 cm and the cotyledons and their axillary buds are removed. Then, the regenerated shoot apical ends are placed on the top cut surfaces of the decapitated citrange epicotyls, in contact with the vascular ring, or, when larger than 0.4 cm, they are inserted into a lateral incision or in a vertical incision along the length of the epicotyl, starting at the point of decapitation (Navarro *et al.*, 1975; Peña *et al.*, 1995a, b, 1997). Grafted plants are cultured in a liquid medium composed of MS inorganic salts, 100 mg/l m-inositol, 0.2 mg/l thiamine-HCl, 1 mg/l pyridoxine-HCl, 1 mg/l nicotinic acid and 75 g/l sucrose, pH 5.7. The cultures are kept at 25°C, with a 16 h photoperiod and 45 μE m^{-2} s^{-1} illumination. Shoots of only 0.1 cm in length can be used to regenerate transgenic plants following this protocol. A frequency of 100% successful grafts is usually obtained. Scions develop 2–4 expanded leaves 3–4 weeks after grafting. A new grafting of the *in vitro* growing plants on vigorous rootstocks in the greenhouse allows the rapid acclimatization and development of the plants.

Alternatively, for certain purposes, rooting of the transgenic shoots allow rapid analyses of the effect of the transgene of potential interest to be performed even *in vitro* when the new inserted trait could affect not only to the aerial part of the plant but also the roots (i.e. tolerance to salinity, plant size, etc.). Although development of whole plants is slower and less efficient than performing *in vitro* grafting, rooting can be obtained by cutting 0.5–1 cm regenerated shoots from the explants and transferring them to SRM medium supplemented with 0.3 mg/l BAP for 7–10 days, and then to SRM medium supplement with 5 mg/l indolebutyric acid. At least for lime and lemon, roots develop within a 3–6 week period with 90–100% efficiency.

Standard polymerase chain reaction (PCR) techniques are used to detect the presence of the transgene(s) in leaf samples from the regenerated putative transgenic plantlets. Southern blot analyses are performed to confirm the stable integration of the transgene(s), and northern blot, western blot and enzyme-linked immunosorbent assay (ELISA) analyses allow the detection of their expression in the transgenic plants. Defining the transformation efficiency as the frequency of whole transgenic regenerated plants established in a greenhouse per *Agrobacterium*-inoculated explant, efficiencies of more than 40% for citrange, about 20% for sweet orange and lime, between 5 and 10% for alemow and sour orange, and less than 5% for lemon and mandarin are currently obtained in our laboratory.

Transformation of mature plant material

Improvement of woody species through genetic engineering will have limited applications unless tissue from mature plants of elite varieties can be readily transformed. Transformation and regeneration of these plants are usually restricted to juvenile tissues (Cervera *et al.*, 1998b), which exhibit greater regenerative potential and greater competence for transformation and susceptibility to *Agrobacterium* than mature tissues. Regenerated plants from juvenile tissues will have juvenile characters, and several years will be needed before horticultural and commercial traits of the transgenic plants can be evaluated. Development of transformation procedures that could bypass the juvenile stage would greatly reduce the time involved in improving woody trees by genetic engineering.

Citrus species show juvenile phases ranging between 5 and 20 years in subtropical areas, which are marked by vigorous and upright growth habit, a characteristic leaf shape, development of long thorns in the leaf axils and lack of flower initiation. We have chosen Pineapple sweet orange as a model to develop a transformation procedure for mature tissue of a selected cultivar (Fig. 15.1).

Previous experiments in our laboratory had confirmed the limited regenerative potential of explants from aged mature citrus plants. Thus, to increase the regeneration ability of mature explants, buds from adult trees were grafted on vigorous seedlings. Regeneration from stem segments from the first, second and third flushes of newly grafted invigorated mature sweet orange plants was evaluated in comparison with regeneration from stem segments from juvenile plants. The results indicated that explants from the first and second flushes produced similar regeneration frequencies, significantly higher than that of the explants from the third flush. The first flush of the adult plants was selected as the source of tissue for genetic transformation experiments.

Buds collected from Pineapple sweet orange maintained in a screenhouse of the pathogen-free Germplasm Bank Collection

Fig. 15. 1. Procedure for the genetic transformation of adult sweet orange plants. (A) After co-cultivation of internodal stem segments with *Agrobacterium tumefaciens*, the explants are transferred to a regeneration–selection medium which promotes the regeneration of transgenic shoots. (B) Apical portions of shoots are grafted *in vitro* on decapitated citrange seedlings. (C) Several weeks after grafting, developed scions are grafted again on vigorous rootstocks in the greenhouse. (D) After one year, transgenic plants started to flower and set fruit, confirming their mature nature. They showed a normal phenotype and development. (E) Release of genetically modified citrus plants under controlled field conditions at the IVIA.

of the IVIA are grafted on seedlings of *C. volkameriana* Ten & Pasq. in the greenhouse (18–27°C). Then, newly elongated mature shoots are allowed to develop the first flush. Preparation of explants, transformation and regeneration conditions are the same as described above for transformation of juvenile material. Internodal stem segments (1 cm long) in a semi-hardened stage are cut transversely from the stem pieces, inoculated and co-cultivated with *A. tumefaciens*, and transferred to SRM, as described above. Shoot regeneration is observed after 2–5 months in selective medium (Fig. 1A). Whole transgenic plants are obtained from these shoots by *in vitro* grafting (Fig. 1B) and a posterior grafting in the greenhouse (Fig. 1C).

The putative mature transgenic sweet orange plants show the morphology and growth habits of an adult plant, as compared with control mature plants. In fact, whereas juvenile plants show a pronounced thorniness, transgenic mature plants are almost thornless, similar to the mature plants from which the explants are taken for transformation. After 14–18 months in the greenhouse, the transgenic and control plants usually start to flower, confirming their mature nature (Fig. 1D).

These results confirm the maintenance of the ontogenic mature stage of the invigorated mature plants as well as the transgenic plants. Interestingly, transgenic events keep their epigenetic mature state even after a process of dedifferentiation, callus induction and redifferentiation, necessary to shift the cells to a competent stage for transformation. Therefore, we are able to transform and regenerate mature tissues of citrus directly, by-passing the juvenile stage. This process greatly shortens the period of time until flowering and bearing fruit, and decreases the time to achieve horticulturally acceptable characteristics by years. It is the first time that a procedure for genetic transformation and regeneration of mature tissues of woody fruit plants has been developed (Cervera *et al.*, 1998b).

Improvement in the efficiency of generating citrus transgenic plants

Since some citrus genotypes are particularly recalcitrant to transformation, we have investigated the use of proper marker genes and *Agrobacterium* vectors with the aim of increasing the efficiency of generating transgenic citrus plants.

The selection of genetic transformants requires the use of marker genes that function as reporters of gene expression and so permit the recovery of transgenic plants. Since the first demonstration of the *gfp* gene as a vital marker gene in both bacteria and *Caenorhabditis elegans* (Chalfie *et al.*, 1994), it has attracted increasing interest and is considered to have several advantages over other visual marker genes. The fluorescence emission of GFP only requires the excitation of living cells by UV or blue light.

Sour orange is a species among citrus genotypes particularly recalcitrant to transformation. The production of transgenic plants is rare and circumstantial. In many recalcitrant plants, cells competent for transformation are restricted to specific tissues. To develop transformation procedures in these cases, it is important to localize the sites of transgene expression in order to favour the regeneration of whole plants from such competent cells. When *uidA* is used as a reporter gene, the diffusion of the GUS indigo dye reaction, the possibility of the presence of bacteria expressing the *uidA* gene, the possible contact between different transformation events and the destructive character of GUS assays all contribute to the imprecise localization of cells or tissues competent for transformation. GFP provides the possibility to perform *in vivo* monitoring of *Agrobacterium*-inoculated plant tissues and thus allowed localization of competent cells for transformation in dedifferentiated callus from the cambium tissue of citrus explants (Ghorbel *et al.*, 1999). Therefore, treatments favouring the development of such callus tissue are necessary to increase transformation frequencies and to enhance a more efficient recovery of citrus transgenic plants (Ghorbel *et al.*, 2000).

It has been important in many plants to use reporter genes, mainly *uidA*, to test the influence of factors affecting both transformation and the regeneration of transformants and escapes. However, since assays to test reporter gene activity are destructive, such factors cannot be evaluated continuously and simultaneously. Moreover, the destructive character of GUS assays precludes the recovery of the GUS + putative transgenic regenerated shoots. To select transgenic shoots from systems in which escapes and chimeras regenerate at high frequencies, such as in citrus, all the regenerated shoots have to be analysed, which is a very laborious process. We have shown that GFP expression permits a rapid and easy discrimination of transgenic and escape shoots in citrus (Fig. 15.2C). Competition for growth between transformed and non-transformed shoots could be avoided by eliminating the escapes soon after their origin (Ghorbel *et al.*, 1999).

Selectable markers, such as NPTII, have been widely used to produce transgenic plants. However, the presence of these markers may restrict the release of transgenic products due to their adverse influence on consumer perceptions. In fact, the presence and expression of selectable markers coding for antibiotic resistance has been a topic of much discussion for the commercialization of certain transgenic crops in Europe. Recently, a European Union Directive forbid the commercialization of transgenic foods with antibiotic transgenes, effective from 2008. We have used GFP for the early detection and rescue of transgenic buds, circumventing the use of antibiotic marker genes to produce transgenic plants. Green fluorescent shoots can be regenerated from explants inoculated with *Agrobacterium* but cultured in a medium without kanamycin (Ghorbel *et al.*, 1999) (Fig. 15.2C). This opens up the possibility of producing transgenic plants without using selective agents.

Moreover, we have investigated the use of appropriate *Agrobacterium* strains to transform economically important citrus genotypes efficiently. Citrus is not naturally infected by *A. tumefaciens*. In fact, crown

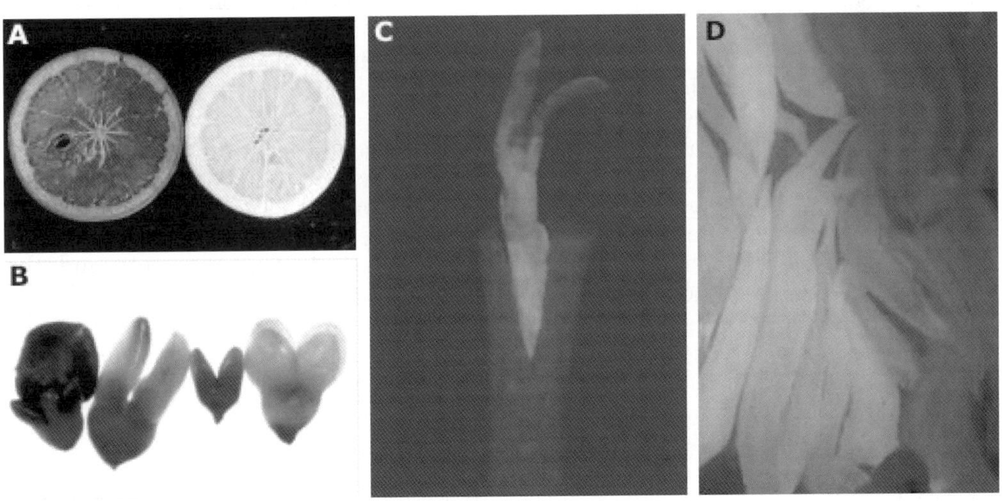

Fig. 15.2. Use of *uidA* and *gfp* as reporter genes in the recovery of transgenic citrus plants. (A) GUS expression of a transgenic sweet orange fruit (left) and a non-transgenic control fruit (right). (B) GUS expression of cultured embryos recovered from seeds of an adult transgenic plant. (C) GFP-positive shoot grafted on a decapitated citrange seedling illuminated with blue light. (D) GFP expression of juice vesicles from a transgenic (left) and a non-transgenic (right) lime fruit.

gall is not a disease affecting citrus. However, we have demonstrated that *A. tumefaciens* strain A281 is not only able to transform citrus efficiently but that it is even super-virulent in many citrus genotypes (Cervera *et al.*, 1998c; R. Ghorbel *et al.*, unpublished results). In addition, we have found a correlation between super-virulence and the super-transformation ability of its disarmed derivative EHA105. The super-virulence of A281 in many plants has been attributed to higher expression of *vir* genes after induction. In fact, we have demonstrated that the *vir* region of its Ti plasmid pTiBo542 provides this ability in citrus (Ghorbel *et al.*, 2001a).

Furthermore, introduction of extra copies of *virG* from pTiBo542 into EHA105 increases stable transformation frequency in citrus. When additional copies of *virG* from pTiBo542 are introduced into C58 (pMP90), stable transformation frequency is drastically increased, approximately 80-fold in lime, suggesting that not only the quantity but also the quality of *virG* from pTiBo542 is a determinant in the super-transformation ability of *Agrobacterium* in citrus. In lemon, introduction of additional copies of *virG* from pTiBo542 in EHA 105 permits the transformation frequency to be doubled. Thus, additional copies of *virG* from pTiBo542 can be used as a tool to increase transformation frequency in those citrus genotypes more recalcitrant to *Agrobacterium*-mediated transformation (Ghorbel *et al.*, 2001a).

Greenhouse and field performance of transgenic plants

In vegetatively propagated and long-lived perennial fruit crops, an important requisite for evaluating the validity of genetic transformation technology in improvement programmes is the stability of the modified genome and transgene integration and expression over long periods of time. We have kept in a screenhouse for a period of 4–5 years a collection of 70 transgenic citrange plants in order to investigate: (i) the origin of morphological variants in the transgenic population; (ii) how factors related to T-DNA integration, the regeneration process and the expression of transgenes may be influenced by transformation conditions; (iii) whether transgenes are stably integrated and expressed over long time periods in citrus plants grown under natural environmental conditions; and (iv) whether correlations between integration patterns and transgene expression could be established in these plants.

A detailed survey of the plants allowed detection of four phenotypic off-type plants and a large variation of transgene integration and expression patterns among the population. Off-type plants were analysed and characterized as nucellar tetraploids, probably originating from tetraploid starting nucellar tissues rather than from somaclonal variation events. Transgene integration and expression analyses revealed that: (i) a significant negative correlation was found between copy number and transgene expression; (ii) rearrangements of the T-DNA inserts did not imply low expression levels; and (iii) stability of integration and expression of transgenes was confirmed for all the transformants grown under natural environmental conditions. These combined features validate transformation as a tool for genetic improvement of citrus (Cervera *et al.*, 2000a).

A release of genetically modified citrus plants under controlled field conditions has also been ongoing since 1997 (Fig. 15.1E). The release site is located at the IVIA, in an experimental field with an extension of 1638 m^2. There are 130 trees, including 16 transgenic plants of Pineapple sweet orange, 16 transgenic plants of lime and 16 transgenic plants of Carrizo citrange (two plants from eight independent transgenic lines for each case). In addition, there are eight non-transgenic control plants from each of the species and an external border of 58 non-transgenic trees of Clemenules clementine. The purpose of the release is to investigate morphological and phenological characteristics of the transgenic trees, expression of the transgenes in leaves, flowers and fruit, stability of the transgenes, transmission of the transgenes to the progeny, and the possibility of transgene dispersal through the pollen to non-transgenic monoembryonic citrus trees (Clemenules clementine). The trial has been approved by the Spanish Ministry of Environment (permit No. B/ES/96/15) and is in accordance with Article 9 of Directive 90/220 of the European Union. This is the first release of transgenic citrus plants to the field. To date, the transgenic plants are morphologically and phenologically normal, as are the non-transgenic controls, and transgenes are stably expressed over different seasons and in different plant tissues and organs (Fig. 15.2A, B and D).

Incorporation of Genes of Potential Interest

The availability of genetic transformation systems for many different citrus genotypes is allowing us to incorporate transgenes of potential agricultural interest into citrus rootstocks and varieties.

Soil salinity significantly limits citrus production in many areas worldwide. Carrizo citrange, considered an excellent citrus rootstock, is very sensitive to salt stress, which restricts its use in salty soils. We have successfully transformed plants of Carrizo citrange with the halotolerance gene *HAL2* (Cervera *et al.*, 2000b), originally isolated from yeast and implicated in salt tolerance mechanisms (Murguía *et al.*, 1995). Plants showing higher transcription levels in northern analyses were chosen to undergo *in vivo* salt stress tolerance assays, by using transgenic plants as rootstocks for a sensitive citrus variety. These tests are currently being performed at our Institute.

Phytophthora citrophthora is the most widespread fungus over all the citrus-growing areas and represents one of the major causes of crop losses. Constitutive overexpression of proteins involved in the plant defence mechanism against disease is one of the strategies proposed to increase plant tolerance to fungal pathogens. p23 is a 23 kDa pathogenesis-related protein induced in tomato (*Lycopersicon esculetum* Mill. cv. Rutgers) plants when these are infected with citrus exocortis viroid, and its antifungal activity has been shown in *in vitro* assays (Rodrigo *et al.*, 1993). We have successfully produced transgenic Pineapple sweet orange plants with a chimeric gene construct comprising the coding region of the *p23* gene. Transgenic lines constitutively expressing the PR protein were challenged with *P. citrophthora* and one of them achieved plant survival rates higher than the control when whole transgenic plants were inoculated with fungal cultures. These results provide evidence for the antifungal activity *in vivo* of the p23 pathogenesis-related protein against *P. citrophthora* and suggest that this may be employed as a strategy aimed at the engineering of *Phytophthora* disease resistance in citrus (Fagoaga *et al.*, 2001).

CTV is the causal agent of the most important virus disease of citrus in the world. Several strategies have been used to engineer plant resistance to viral pathogens. Most are based on the concept of pathogen-derived resistance (Sanford and Johnston, 1985), which proposes that the introduction and expression in plants of viral sequences could interfere with the life cycle of the same or a closely related challenging virus, thus providing resistance to infection. This approach has proved to be applicable to a range of more than 40 virus species in different transgenic hosts. We are introducing coding and non-coding versions of the major coat protein gene of the virus in lime, which is very susceptible to CTV, to evaluate pathogen-derived resistance. More than 40 transgenic lines expressing the coat protein gene of the virus have already been produced (Domínguez *et al.*, 2000). When plants propagated from each transgenic line were graft or aphid inoculated with CTV, two types of response to viral challenge were observed: some lines developed CTV symptoms similar to those of non-transgenic controls, whereas others exhibited protection against the virus (Fig. 15.3A). This protection consisted of a proportion of plants, ranging

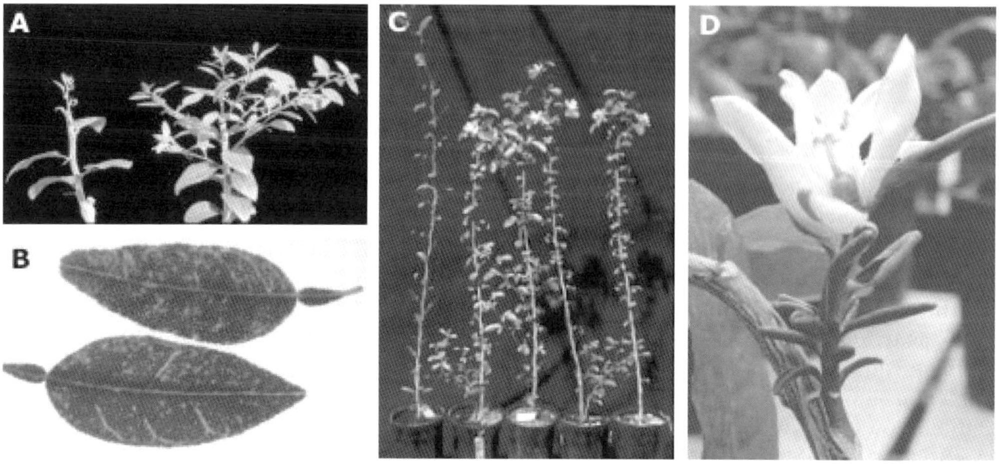

Fig. 15.3. Incorporation of genes of potential interest. (A and B) Engineering citrus resistance to CTV. (A) Non-transgenic control (left) and CP (p25) transgenic (right) lime plants infected with CTV. The transgenic CP plant showed resistance to the virus. (B) Leaves from a non-transgenic control lime plant inoculated with CTV (top) and from a p23-transgenic lime plant (bottom). Expression of p23 in citrus plants causes the appearance of CTV-like symptoms. (C and D) Overexpression of *APETALA1* and *LEAFY* genes from *Arabidopsis* in Carrizo citrange transgenic plants. (C) Transgenic AP1 plants flowering six months after sowing (four plants on the right) compared with a control non-transformed plant (plant on the left). (D) Transgenic LFY plant that flowered after being grafted on a vigorous rootstock in the greenhouse. Transgenic LFY plants generally showed developmental aberrations.

from 10 to 33%, that were resistant to CTV, and the remainder showed a significant delay in virus accumulation and symptom onset (Domínguez *et al.*, 2002). This is the first demonstration of pathogen-derived resistance in transgenic plants against a Closterovirus member in its natural host. The major coat protein gene of CTV has been also inserted into sour orange, and transgenic plants either expressing or not expressing the transgene have been generated (Ghorbel *et al.*, 2000).

The 3'-terminal gene of CTV codes for a 23 kDa protein (p23), which is an RNA-binding protein that contains a motif rich in cysteine and histidine residues. For this reason, a regulatory role in CTV replication or gene expression has been suggested for p23. To explore whether overexpression in transgenic plants of this protein could affect the normal CTV infectious process, we have produced transgenic lime plants carrying the *p23* transgene, or a truncated version thereof. Constitutive expression of p23 induced phenotypic aberrations resembling symptoms caused by CTV in non-transgenic lime plants (Fig. 15.3B), whereas transgenic plants expressing the truncated version of p23 were normal. The onset of CTV-like symptoms in p23-transgenic plants was associated with the expression of p23, and its accumulation level paralleled symptom intensity (Ghorbel *et al.*, 2001b). This demonstrates that p23 is involved in symptom development and that it probably plays a key role in CTV pathogenesis. This finding also delimits a small region of the large CTV genome for future mapping of specific pathogenic determinants.

Results of genetic improvement programmes for citrus are almost non-existent, due, among other reasons, to the extremely long juvenile phases of the trees which prolong the time required to analyse late traits such as fruit features. With the aim of accelerating their flowering time, we have transformed juvenile Carrizo citrange seedlings to constitutively express the *Arabidopsis*

LEAFY or *APETALA1* genes, that are sufficient to promote flower initiation in *Arabidopsis* (Mandel and Yanofsky, 1995; Weigel and Nilsson, 1995). Both types of transgenic citrus plants produced fertile flowers and fruit as early as the first year, notably through a mechanism involving a dramatic shortening of their juvenile phase (Fig. 15.3C and D). This phenomenon had not been observed in *Arabidopsis*. Furthermore, expression of *APETALA1*, being as efficient as *LEAFY* in the initiation of flowers, did not produce any severe developmental abnormality. Both types of transgenic trees flowered again in consecutive years and their flowering response was under environmental control. In addition, sexual and nucellar derived transgenic seedlings had a very short juvenile phase and flowered in their first spring, demonstrating the stability and inheritance of this trait (Peña *et al*., 2001). This opens up the possibility of using independent *APETALA1* transgenic plants as parents in crosses with non-transformed genotypes that would yield 50% of the progeny flowering and setting fruit in 1–2 years, thus providing the opportunity to evaluate fruit features very early and to advance generations rapidly. On the other hand, by retransformation of *APETALA1* transgenic citrus plants, it could be possible to test the effect of the expression of certain transgenes rapidly under flower organ- or fruit-specific promoters as a system to look for biotechnological strategies to develop seedless varieties, modify fruit colour or favour easy-peeling characteristics.

Conclusion and Future Prospects

Just a decade ago, citrus was recalcitrant to genetic transformation. Nowadays, many citrus species are readily transformed, due to the selection of appropriate vectors, the use of vigorous source plant material competent for transformation and regeneration, and the establishment of adequate inoculation, co-cultivation and regeneration–selection culture media and conditions. This has permitted initiation of the introduction of transgenes in citrus with the aim of improving rootstock and variety performance.

Looking ahead, efforts must be made to try to establish genetic transformation systems for other important citrus genotypes, such as mandarins, including clementines and satsumas. Furthermore, desirable traits to incorporate into citrus would include disease and pest resistance, higher tolerance to abiotic stresses, and improved fruit quality and productivity. Although we and others are producing transgenic citrus plants with genes from other plants and non-plant sources for these purposes, the major obstacle in the improvement of citrus through genetic transformation is the lack of adequate useful genes isolated from the citrus genome. Almost no important traits have been characterized in sufficient detail to be incorporated into the citrus genome through gene transfer. Therefore, it is necessary first to identify and clone genes involved in determining traits of interest. Efficient genetic transformation procedures will also be an important tool to perform functional genomic analyses. Only then will it be possible to select efficiently which genes should be inserted to achieve the desired results. In addition, efforts must also focus on understanding in more detail the mechanisms that regulate transgene expression and how the modified genome is affected by the insertion of a foreign gene and its expression, and in investigating tissue-specific or inducible promoters from citrus genes. Moreover, research on the environmental impacts of releasing transgenic citrus plants to the field will also constitute a necessary task in the near future.

Acknowledgements

The authors wish to thank C. Ortega, A. Navarro and M. Romero for their excellent technical assistance, and personnel from the greenhouses who took care of the plants. This research was supported by grants SC97-102 and RTA01-120 from the Instituto Nacional de Investigaciones Agrarias, and IFD97-0822 from CICYT.

References

Bond, J.E. and Roose, M.L. (1998) *Agrobacterium*-mediated transformation of the commercially important citrus cultivar Washington navel orange. *Plant Cell Reports* 18, 229–234.

Cervera, M., Pina, J.A., Juárez, J., Navarro, L. and Peña, L. (1998a) *Agrobacterium*-mediated transformation of citrange: factors affecting transformation and regeneration. *Plant Cell Reports* 16, 271–278.

Cervera, M., Juárez, J., Navarro, A., Pina, J.A., Durán-Vila, N., Navarro, L. and Peña, L. (1998b) Genetic transformation and regeneration of mature tissues of woody fruit plants bypassing the juvenile stage. *Transgenic Research* 7, 51–59.

Cervera, M., López, M.M., Navarro, L. and Peña, L. (1998c) Virulence and supervirulence of *Agrobacterium tumefaciens* in woody fruit plants. *Physiological and Molecular Plant Pathology* 52, 67–78.

Cervera, M., Pina, J.A., Juárez, J., Navarro, L. and Peña, L. (2000a) A broad exploration of a transgenic population of citrus: stability on gene expression and phenotype. *Theoretical and Applied Genetics* 100, 670–677.

Cervera, M., Ortega, C., Navarro, A., Navarro, L. and Peña, L. (2000b) Generation of transgenic citrus plants with the tolerance-to-salinity gene HAL2 from yeast. *Journal of Horticultural Science and Biotechnology* 75, 26–30.

Chalfie, M., Tu, Y., Euskirchen, G., Ward, W.W. and Prasher, D.C. (1994) Green fluorescent protein as a marker for gene expression. *Science* 263, 663–664.

Deng, Z., Huang, S., Ling, P., Yu,. C., Tao, Q., Chen, C., Wendell, M.K., Zhang, H.B. and Gmitter, F.G., Jr (2001) Fine genetic mapping and BAC contig development for the citrus tristeza virus resistance gene locus in *Poncirus trifoliata* (Raf.). *Molecular and General Genomics* 265, 739–747.

Domínguez, A., Guerri, J., Cambra, M., Navarro, L., Moreno, P. and Peña, L. (2000) Efficient production of transgenic citrus plants expressing the coat protein gene of citrus tristeza virus. *Plant Cell Reports* 19, 427–433.

Domínguez, A., Hermoso de Mendoza, A., Guerri, J., Cambra, M., Navarro, L., Moreno, P. and Peña, L. (2002) Pathogen-derived resistance to citrus tristeza virus (CTV) in transgenic Mexican lime (*Citrus aurantifolia* (Christ.) Swing.) plants expressing its *p25* coat protein gene. *Molecular Breeding* 10, 1–10.

Fagoaga, C., Rodrigo, I., Conejero, V., Hinarejos, C., Tuset, J.J., Arnau, J., Pina, J.A., Navarro, L. and Peña, L. (2001) Increased tolerance to *Phytophthora citrophthora* in transgenic orange plants overexpressing a tomato pathogenesis related protein PR-5. *Molecular Breeding* 7, 175–181.

Fleming, G.H., Olivares-Fuster, O., Fatta Del-Bosco, S. and Grosser, J.W. (2000) An alternative method for the genetic transformation of sweet orange. *In Vitro Cellular and Developmental Biology-Plant* 36, 450–455.

Food and Agriculture Organization (2001) http.//apps.fao.org/lim500/nph-wrap.pl

Gentile, A., LaMalfa, S., Deng, Z.N., Domina, F., Nicolosi, E. and Tribulato, E. (1998) Transgenic citrus: first experiences with Rol genes. *Rivisti di Frutticoltura e di Ortofloricoltura* 61, 59–61.

Ghorbel, R., Juárez, J., Navarro, L. and Peña, L. (1999) Green fluorescent protein as a screenable marker to increase the efficiency of generating transgenic woody fruit plants. *Theoretical and Applied Genetics* 99, 350–358.

Ghorbel, R., Domínguez, A., Navarro, L. and Peña, L. (2000) High efficiency genetic transformation of sour orange (*Citrus aurantium* L.) and production of transgenic trees containing the coat protein gene of citrus tristeza virus. *Tree Physiology* 20, 1183–1189.

Ghorbel, R., López, C., Moreno, P., Navarro, L., Flores, R. and Peña, L. (2001a) Transgenic citrus plants expressing the citrus tristeza virus p23 protein exhibit viral-like symptoms. *Molecular Plant Pathology* 2, 27–36.

Ghorbel, R., LaMalfa, S., López, M.M., Petit, A., Navarro, L. and Peña, L. (2001b) Additional copies of *virG* from pTiBo542 provide a super-transformation ability to *Agrobacterium* tumefaciens in citrus. *Physiological and Molecular Plant Pathology* 58, 103–110.

Grosser, J.W., Ollitrault, P. and Olivares-Fuster, O. (2000) Somatic hybridization in Citrus: an effective tool to facilitate variety improvement. *In Vitro Cellular and Developmental Biology-Plant* 36, 439–449.

Gutiérrez, M.A., Luth, D.E. and Moore, G.A. (1997) Factors affecting *Agrobacterium*-mediated transformation in *Citrus* and production of sour orange (*Citrus aurantium* L.) plants expressing the coat protein gene of citrus tristeza virus. *Plant Cell Reports* 16, 745–753.

Hidaka, T. and Omura, M. (1993) Transformation of citrus protoplasts by electroporation. *Journal of the Japanese Society of Horticultural Science* 62, 371–376.

Hidaka, T., Omura, M., Ugaki, M., Tomiyama, M., Kato, A., Ohshima, M. and Motoyoshi, F. (1990) *Agrobacterium*-mediated transformation and regeneration of Citrus spp. from suspension cells. *Japanese Journal of Breeding* 40, 199–207.

Hood, E.E., Gelvin, S.B., Melchers, L.S. and Hoekema, A. (1993) New *Agrobacterium* helper plasmids for gene transfer to plants. *Transgenic Research* 2, 208–218.

Jefferson, R.A., Kavanagh, T.A. and Bevan, M.W. (1987) GUS fusions: β-glucuronidase as a sensitive and versatile gene fusion marker in higher plants. *EMBO Journal* 6, 3901–3907.

Kaneyoshi, J. and Kobayashi, S. (1999) Characteristics of transgenic trifoliate orange (*Poncirus trifoliata* Raf.) possessing the *rolC* gene of *Agrobacterium rhizogenes* Ri plasmid. *Journal of the Japanese Society of Horticultural Science* 68, 734–738.

Kaneyoshi, J., Kobayashi, S., Nakamura, Y., Shigemoto, N. and Doi, Y. (1994) A simple and efficient gene transfer system of trifoliate orange. *Plant Cell Reports* 13, 541–545.

Kayim, M., Koc, N.K. and Tor, M. (1996) Gene transfer into citrus (*Citrus limon* L.) nucellar cells by particle bombardment and expression of GUS activity. *Turkish Journal of Agriculture and Forestry* 20, 349–352.

Kobayashi, S. and Uchimiya, H. (1989) Expression and integration of a foreign gene in orange (*Citrus sinensis* Osb.) protoplasts by direct DNA transfer. *Japanese Journal of Genetics* 64, 91–97.

Kobayashi, S., Nakamura, Y., Kaneyoshi, J., Higo, H. and Higo, K. (1996) Transformation of kiwifruit (*Actinidia chinensis*) and trifoliate orange (*Poncirus trifoliata*) with a synthetic gene encoding the human epidermal growth factor (*hEGF*). *Journal of the Japanese Society of Horticultural Science* 64, 763–769.

Koltunow, A.M., Brennan, P., Protopsaltis, S. and Nito, N. (2000) Regeneration of West Indian limes (*Citrus aurantifolia*) containing genes for decreased seed set. *Acta Horticulturae* 535, 81–91.

LaMalfa, S., Gentile, A., Deng, Z.N. and Domina, F. (2000) Citrus genetic transformation with a vital reporter gene: expression of green fluorescent protein in Troyer citrange. *Italus Hortus* 7, 17–21.

Luth, D. and Moore, G. (2000) Transgenic grapefruit plants obtained by *Agrobacterium tumefaciens*-mediated transformation. *Plant Cell Tissue and Organ Culture* 57, 219–222.

Mandel, M.J. and Yanofsky, M.F. (1995) A gene triggering flower formation in *Arabidopsis*. *Nature* 377, 522–524.

Moore, G.A., Jacono, C.C., Neidigh, J.L., Lawrence, S.D. and Cline, K. (1992) *Agrobacterium*-mediated transformation of citrus stem segments and regeneration of transgenic plants. *Plant Cell Reports* 11, 238–242.

Murashige, T. and Skoog, F. (1962) A revised medium for rapid growth and bioassays with tobacco tissue cultures. *Physiologia Plantarum* 15, 473–479.

Murguía, J.R., Bellés, J.M. and Serrano, R. (1995) A salt-sensitive 3′(2′),5′-bisphosphate nucleotidase involved in sulfate activation. *Science* 267, 232–234.

Navarro, L. (1992) Citrus shoot-tip grafting *in vitro*. In: Bajaj, Y.P.S. (ed.) *Biotechnology in Agriculture and Forestry*, Vol. 18. Springer-Verlag, Berlin, pp. 328–338.

Navarro, L., Roistacher, C.N. and Murashige, T. (1975) Improvement of shoot-tip grafting *in vitro* for virus-free citrus. *Journal of the American Society for Horticultural Science* 100, 471–479.

Ollitrault, P., Dambier, D., Sudahono, Mademba-Sy, F., Vanel, F., Luro, F. and Aubert, B. (1998) Biotechnology for triploid mandarin breeding. *Fruits* 53, 307–317.

Peña, L., Cervera, M., Juárez, J., Ortega, C., Pina, J.A., Durán-Vila, N. and Navarro, L. (1995a) High efficiency *Agrobacterium*-mediated transformation and regeneration of citrus. *Plant Science* 104, 183–191.

Peña, L., Cervera, M., Juárez, J., Navarro, A., Pina, J.A., Durán-Vila, N. and Navarro, L. (1995b) *Agrobacterium*-mediated transformation of sweet orange and regeneration of transgenic plants. *Plant Cell Reports* 14, 616–619.

Peña, L., Cervera, M., Juárez, J., Navarro, A., Pina, J.A. and Navarro, L. (1997) Genetic transformation of lime (*Citrus aurantifolia* Swing.): factors affecting transformation and regeneration. *Plant Cell Reports* 16, 731–737.

Peña, L., Martín-Trillo, M., Juárez, J., Pina, J.A., Navarro, L. and Martínez-Zapater, J.M. (2001) Constitutive expression of Arabidopsis LEAFY and APETALA1 genes in citrus reduces their generation time. *Nature Biotechnology* 19, 263–267.

Pérez-Molphe, E. and Ochoa-Alejo, N. (1998) Regeneration of transgenic plants of Mexican lime from *Agrobacterium rhizogenes*-transformed tissues. *Plant Cell Reports* 17, 591–596.

Rodrigo, I., Vera, P., Tornero, P., Hernández-Yago, J. and Conejero, V. (1993) cDNA cloning of viroid-induced tomato pathogenesis-related protein P23. Characterization as a vacuolar antifungal factor. *Plant Physiology* 102, 939–945.

Sanford, J.C. and Johnston, S.A. (1985) The concept of parasite-derived resistance. Deriving resistance genes from the parasite´s own genome. *Journal of Theoretical Biology* 113, 395–405.

Savage, E.M. and Gardner, F.E. (1965) The Troyer and Carrizo citranges. *California Citrigrapher* 50, 112–116.

Soost, R.K. and Cameron, J.W. (1975) Citrus. In: Janick, J. and Moore, J.N. (eds) *Advances in Fruit Breeding.* Purdue University Press, West Lafayette, Indiana, pp. 507–540.

Vardi, A., Bleichman, S. and Aviv, D. (1990) Genetic transformation of citrus protoplasts and regeneration of transgenic plants. *Plant Science* 69, 199–206.

Weigel, D. and Nilsson, O. (1995) A developmental switch sufficient for flower initiation in diverse plants. *Nature* 377, 495–500.

Yang, Z.N., Ingelbrecht, I.L., Louzada, E., Skaria, M. and Mirkov, T.E. (2000) *Agrobacterium*-mediated transformation of the commercially important grapefruit cultivar Rio Red (*Citrus paradidi* Macf.). *Plant Cell Reports* 19, 1203–1211.

Yang, Z.N., Ye, X.R., Choi, S., Molina, J., Moonan, F., Wing, R.A., Roose, M.L. and Mirkov, T.E. (2001) Construction of a 1.2-Mb contig including the citrus tristeza virus resistance gene locus using a bacterial artificial chromosome library of *Poncirus trifoliata* (L.) Raf. *Genome* 44, 382–393.

Yao, J.L., Wu, J.H., Gleave, A.P. and Morris, B.A.M. (1996) Transformation of citrus embryogenic cells using particle bombardment and production of transgenic embryos. *Plant Science* 113, 175–183.

16 Mutation Breeding

Mikeal L. Roose and Timothy E. Williams

Department of Botany and Plant Sciences, University of California, Riverside, CA 92521, USA

Introduction

Mutation breeding refers to development of new cultivars or germplasm by the deliberate production and selection of new mutations. In some cases, the objective of the breeder is to create a mutation in an existing cultivar that improves a defect such as seediness. After finding and evaluating a selection with the desired trait, the original cultivar containing the mutation would be released. The breeder may also utilize mutation breeding to obtain selections with a desired mutation, and then use this germplasm as a parent in a hybridization-selection programme. This chapter describes the issues that a citrus breeder faces when considering if and how to implement a mutation breeding project. Examples of development of citrus cultivars using mutation breeding are also described. Van Harten (1998) provides a detailed summary of mutation breeding.

The Nature of Mutation

Mutation refers to any heritable change in the DNA, but the breeder is generally interested only in those mutations that alter a phenotype. At the molecular level, mutations can alter DNA by base substitution, insertion, deletion or sequence rearrangement. Any of these types of changes may cause a phenotypic change. It is important to distinguish the 'permanent' change in DNA from types of temporary change that are generally classified as epigenetic. For example, expression of a gene may be silenced by methylation of the DNA. This state may be 'semi-permanent' in that daughter cells derived from the altered cells retain the methylation pattern. This means that it may be possible to propagate trees that retain this character. However, regeneration from tissue culture or meiosis can sometimes restore the non-methylated DNA and corresponding phenotype. It is not clear whether epigenetic changes contribute to the diverse phenotypes seen in citrus.

A further complexity is chimerism (Fig. 16.1). When a new mutation is induced in somatic cells, it will affect cells in one of the three histogen layers: L-I which gives rise to the epidermis and most juice vesicle tissue; L-II which produces subepidermal tissues including the flavedo, nucellus and gametes; and L-III which includes most internal tissues such as the cortex. Generally cells divide to give daughter cells of the same layer, so a mutation

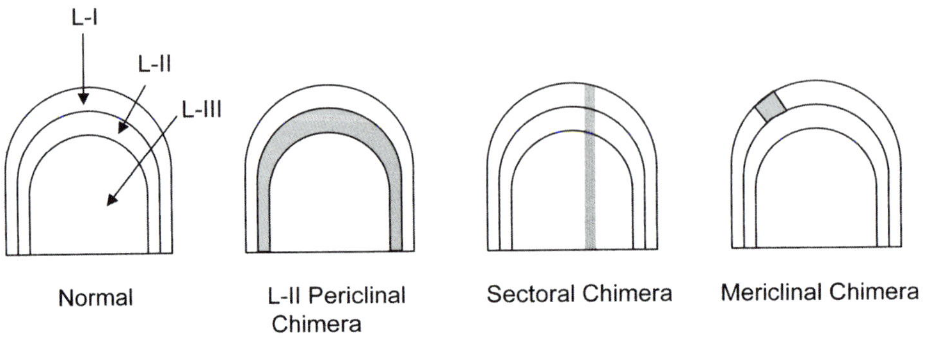

Fig. 16.1. Types of chimeras in a shoot of citrus. Shaded areas represent mutant tissue.

affecting one cell layer can appear stable during vegetative propagation. Such genotypes are called periclinal chimeras. A 'layer substitution' can cause such chimeras to revert to the original type, or become a 'solid' or non-chimeric mutant. Layer substitutions can arise when tissues differentiate from callus such as when pruning or freeze damage leads to regeneration of new branches.

When is Mutation Breeding Appropriate?

Mutation breeding is the strategy of choice when it is essential to retain the characteristics of an existing cultivar, but correct a specific defect in it. The high heterozygosity of most citrus cultivars makes it very difficult to recover progeny with very similar characters from hybridization-selection. The relatively small population sizes that most citrus breeders can produce and evaluate further limit hybridization-selection in many crops. These conditions occur in all important fruit cultivar types except mandarins, so it would seem that mutation breeding should be widely used. This has not been the case because, for most traits, mutation breeding is also limited by the breeder's ability to produce and evaluate very large populations. Mutation breeding procedures (other than genetic engineering) do not allow the breeder to target particular genes or traits. Tissue exposed to mutagenic agents will have random mutations, most of which have no effect or undesirable effects. Therefore, it is necessary to screen large populations to have a reasonable chance of finding the desired mutation. Increasing the dosage of mutagen will increase the frequency of mutations, but also increases mortality. At high dosages, the desired mutation will often be accompanied by undesirable mutations which mean that the mutant line cannot be used directly as a cultivar. The one trait that clearly can be obtained relatively easily with mutation breeding is seedlessness, presumably because the mechanism does not require mutation of a single gene or genes. This is discussed in detail below. Alterations in other traits can also be obtained by mutation breeding, but success is less certain and the effort required is likely to be greater.

The place of mutation breeding in citrus cultivar development will depend greatly on the cost and acceptance of genetic engineering. Genetic engineering promises to achieve the same type of end result as mutation breeding: introduction of a specific trait into an existing cultivar. However, genetic engineering is more precise because it depends not on a chance mutation, but on addition or change in a specific target gene. Induced mutations are not generally different from the natural mutations responsible for variation in nature. Therefore, mutation breeding does not face the same level of controversy as the 'unnatural' changes induced by genetic engineering (Predieri, 2001). Furthermore, approval for release of cultivars produced by genetic engineering is presently very expensive in most countries, while cultivars from mutation breeding are not subject to special regulation.

The stages of a mutation breeding project involve first choosing a target tissue to expose to the mutagenic agent, choosing a mutagen, exposing the tissue, propagating plants from the exposed tissue, screening these plants for the desire mutation and then evaluating selections that carry the mutation, either directly or as parents. Each of these stages is discussed below.

Tissue choice

The breeder can initiate a mutation breeding project with one of several different types of tissues, each with certain advantages and disadvantages (Table 16.1). Choice of tissue may also depend on the target trait and type of mutagen to be used.

The major advantage of treating haploid tissues, generally pollen or microspores, it that chimeras are not produced. The treated pollen is then used to fertilize untreated egg cells, seeds are collected, and seedlings are grown and evaluated. The major disadvantages of this approach are, for fruit traits, the time to evaluation is long because the seedlings will be juvenile, and it may be difficult to distinguish mutants from recombinants in the sexual progeny that result. Furthermore, this approach results in hybrids few of which will resemble the parent cultivar. Pollen irradiation has rarely been used in citrus.

Treatment of seeds is widely practised in seed-propagated crops, but is less common in citrus and other fruit trees. It is easy to treat large numbers of seeds, but identification of mutants is easy only if most embryos are of nucellar origin. It is not clear whether the frequency of chimeras is similar after treatment of seeds and buds. In both cases, the target cells occur in meristems composed of many cells, and chimerism seems fairly likely, but there are few data on rates of chimerism on which to base a conclusion.

Treatment of tissue cultures with mutagens has been used by several researchers. The first step involves establishing a culture capable of regeneration. This can be difficult in some cultivars and is at least tedious if it is desired to induce the mutation in many cultivars. The type of tissue culture influences the likelihood of chimerism and time to evaluation. If protoplasts from an embryogenic callus system are used, then most embryos are thought to regenerate from single cells and therefore chimerism is unlikely, but the regenerated plants will be juvenile. Chimeras are also minimized if plants are regenerated from protoplasts produced before or after mutagen treatment. If stem segments from mature tissue are cultured, then organized meristems are already present and chimeras are as likely as with seeds or budwood. An advantage of tissue culture systems is that

Table 16.1. Advantages and disadvantages of treating various tissue types with mutagens.

Target tissue	Ease of use	Time to evaluation	Chimerism	Identification of mutants
Pollen	Good	Long	None	Difficult
Seed	Good	Long	Low	Varies
Embryogenic callus culture	Fair	Long	Low	Easy
Mature stem segment culture	Fair	Short	High	Easy
Budwood	Good	Short	High	Easy

selection for certain traits can be applied to the culture, increasing the chances of recovery of a desired mutation.

Treatment of buds has been widely used in citrus (e.g. Froneman *et al.*, 1996). A major advantage is that trees propagated from such buds will not display the delayed flowering and thorninesss typical of seedlings. However, the probability of chimerism is high relative to treatment of single-cell systems. Lapins (1983) discusses the organization of primary and axillary buds in detail. A dormant leaf bud contains the primary leaf bud and several axillary bud primordia. These buds are typically induced to grow into shoots that may contain mutant cells. Axillary buds that are larger and more fully differentiated at the time of mutagen treatment are more likely to produce shoots, but these are more likely to be chimeric. The smaller axillary buds have fewer cells and therefore less probability of giving rise to a chimera.

Mutagenic agents

The breeder can choose from a wide range of mutagenic agents. Each agent typically induces mutations by a particular mechanism and therefore results in a specific type or types of DNA alteration. Specific agents may be appropriate for a particular tissue or target trait. Mutagens can be broadly classified as radiation, chemical, transposable element and pathogen.

Radiation is the most widely used type of mutagen for citrus. In part, this is because there has been considerable funding for mutation breeding from national and international atomic energy agencies anxious to characterize the effects of radiation and find useful applications. Although UV irradiation can be used, it generally penetrates tissues less deeply than ionizing radiation. Ionizing radiation includes X-rays and γ-rays, and has been the most widely used and effective type of mutagen for citrus breeding. It can induce a wide range of mutation types including chromosome breaks and rearrangements, and point mutations. It is thought that seedless forms of citrus can be obtained at relatively high frequencies by irradiation because the resulting chromosome breaks frequently lead to rearrangements (inversions and translocations) that cause sterility. This supposition has not been verified experimentally, but many researchers have reported a high frequency (up to 10%) of seedless selections from irradiation (Hearn, 1984, 1986; Vardi *et al.*, 1996)

When using radiation, as well as other mutagens, the breeder must determine the appropriate dose and exposure to apply. The two most commonly used radiation types are X-rays and γ-rays (generally from a ^{60}Co source). Dose is the amount of mutagen applied per unit time, while exposure is the total absorbed radiation. The current unit of absorbed radiation energy is the Gy (Gray), while most of the older literature reports exposure in rads or Roentgens (R, which measures exposure rather than dose). One Gy equals 100 rad. This unit is useful in calculating how long and at what distance to expose tissue to the radiation source. However, it is difficult to predict the effect of a given dose and exposure on a particular tissue. Relatively little is known about the effects of different dose rates for citrus tissues because most breeders have access to a single radiation source which emits at a constant rate. Breeders typically use the LD_{50} as a measure of the biological response of tissue to a particular radiation dose. While LD stands for lethal dose, and therefore the term implies survival, it is also possible to apply it to vegetative growth, i.e. a 50% reduction in vegetative growth. Sensitivity to ionizing radiation typically varies with the physiological condition of the tissue, particularly water content, and the cultivar, so breeders generally identify an appropriate dose empirically. In the breeding programme at the University of California, Riverside, we use a dose rate of 4–8 Gy/min (50 Gy/min would be ideal if available) and exposure of 30–50 Gy with budwood to induce mutations to seedlessness. Hearn (1986) also reported good results from irradiating grapefruit budwood with 30–50 Gy from a ^{137}Cs source. Vardi *et al.* (1996) found about 10–13% of low-seeded mutants in two mandarin cultivars after exposures of 27.5–65 Gy, but another cultivar had only about 8.5% after exposures above 40 Gy.

Chemical mutagens have been used less commonly in citrus than radiation. Different chemical mutagens induce different spectra of mutations. A very wide range of chemical mutagens are now available, but few of these have been evaluated in citrus. If vegetative tissues such as buds or meristems are to be treated, it is generally considered better to dissect away the outer layer of leaves that cover the bud or meristem to improve penetration of the chemical. Tissue cultures can be treated with mutagens rather conveniently, as summarized by Predieri (2001), but their use for citrus mutation breeding has not been reported.

Transposable elements (TEs) are natural genetic elements with the ability to move from one genome location to another, either with or without duplication of the element. A variety of different classes have been identified (Feschotte *et al.*, 2002). TEs can be used to induce mutations, a strategy known as transposon tagging (Maes *et al.*, 1999) because a gene mutated by a transposon insertion can be cloned using the transposable element as a 'tag sequence'. No TEs in citrus have been shown to be active, but there is indirect evidence of such activity since some clementine cultivars derived by mutation have differences in DNA markers detected using primers from retrotransposon elements (Breto *et al.*, 2001). TEs can be transferred from one plant species to another, but do not always work in unrelated species. If a TE inserts into a gene or regulatory region, it can induce a mutation that alters expression of that gene. At present, use of TEs as an experimental mutagenesis system for citrus has not been reported.

Targeted mutation, in which mutations are induced in a specific gene or DNA sequence, is not yet efficient in plants. If this technique were successfully developed, it would certainly reinvigorate mutation breeding research because it would open up an alternative path to apply genomics tools to citrus, one which is likely to be less controversial than transformation.

Propagation of mutagen-treated materials

When the plant material has been treated with the mutagen, the next step is to propagate trees from the potentially mutant cells. The precise method depends on the tissue that was treated, since this determines appropriate propagation mechanisms and influences the probability of chimeras. If trees or buds are treated with radiation, then DuPlooy *et al.* (1993) suggest that the mv1 generation should be budded on to rootstocks and the mv2 generation buds from these trees should be budded on to a new set of rootstock and the resulting trees planted in the field for evaluation. This is expected to reduce the frequency of chimeras, but detailed results were not reported. In the UCR programme, to induce mutations to low seed content, we irradiate budwood, propagate trees from this and evaluate these mv1 trees. About 50% of irradiated buds result in trees that are planted in the field for evaluation. Most buds that grow are the primary bud, but secondary buds also push in some cases. Of these, perhaps 1–5% are low seeded, the percentage varying greatly with cultivar. We have rarely observed chimeras in this material in that all fruit from a tree is as seedy as that of the original parent, or all fruit have greatly reduced seed number. However, this does not test for periclinal chimeras. Selections are then re-propagated using standard budding methods to produce about 60–100 trees for trials in several locations. It is expected that chimeras may be revealed at this stage. A rapid method to test for chimerism would be a valuable addition.

Screening

Screening refers to initial evaluation of the mutant population. The method chosen will depend on the target trait. If trees are not expected to be juvenile then they can be screened fairly quickly and a high planting density can be used with little difficulty. If trees result from mutagen treatment of seed, pollen or tissue cultures so that juvenility is expected, then they are unlikely to fruit until 5–7 years old. Training trees to a long central leader (Vardi and Spiegel-Roy, 1988) and certain cultural practices can speed fruiting, but may require more labour than is available to some programmes. Trees expressing juvenility must either be planted at wider spacing or pruned more severely to maintain plantings. In any case, when the trait is expressed in a reasonable percentage of the population, the breeder can begin screening. Each tree and sector of the tree should be evaluated for the trait, and for other visible mutations. If sectoral chimeras occur, these may be detectable as branches with different leaf colour, internode distance or other visible differences that correlate with fruit character effects. Evaluation of fruit for 2–3 years is advisable to reduce the number of selections that do not consistently express the mutation. An example of a mutation breeding programme is shown in Fig. 16.2.

Evaluation

Field evaluation of selections from mutation breeding is similar to that for other cultivars, but a few specific considerations are relevant. Cultivars from mutation breeding are less likely to be genetically stable and to be hybrids, and therefore particular attention should be paid to evaluating uniformity within and between trees. Trees should be subject to normal cultural practices that might influence stability. Pruning in particular may lead to layer substitutions which cause reversion of chimeras, so branches that regrow after pruning should be carefully studied. On the other hand, if the cultivar treated with mutagen has already been characterized for performance attributes, it should be possible to evaluate the selections for somewhat fewer years than would be required for a hybrid cultivar that may have many novel attributes and/or defects.

Examples

We present two examples that illustrate mutation breeding in citrus. The 'Star Ruby' grapefruit resulted from a mutation breeding programme in Texas (Hensz, 1971, 1977). Development of 'Star Ruby' was initiated in 1959 by irradiation of seed of 'Hudson' grapefruit, a red-fleshed seedy cultivar with thermal neutrons. The exact dosage of thermal neutrons used was not reported. Seeds were planted in 1959. In 1960, buds from the seedlings were grafted on to sour orange rootstock and the resulting trees were planted in the field in the winter of 1961–62. Trees first set fruit in 1966. Budwood was released in March 1970, at which time only the original tree had produced fruit. Thus no replicated yield testing was done prior to release. 'Star Ruby' has more intense flesh colour than 'Hudson' as well as few (1–2) seeds per fruit. The deeper red rind and flesh colour of 'Star Ruby' is not attributed to mutation because all nucellar seedlings of 'Hudson' have a colour similar to that of 'Star Ruby' (Hensz, 1981). Evidently the 'Hudson' is chimeric, carrying genes for deep red colour in the L-II layer which are expressed in the non-chimeric nucellar seedlings. Despite its deep colour and few seeds, 'Star Ruby' shows a number of horticultural problems including poor production in some areas, low tree vigour and increased susceptibility to some herbicides (Gmitter, 1995). It is interesting that most of these problems are reported in the 1971 paper, and thus were apparent in the original tree. Possibly some of these traits are due to deleterious mutations induced by irradiation. 'Star Ruby' does not appear to be chimeric – its characteristics have remained quite stable.

The 'Mor' mandarin is a low-seeded selection of 'Murcott' produced by budwood irradiation (Vardi *et al.*, 1993). About 400 buds of 'Murcott' were irradiated with

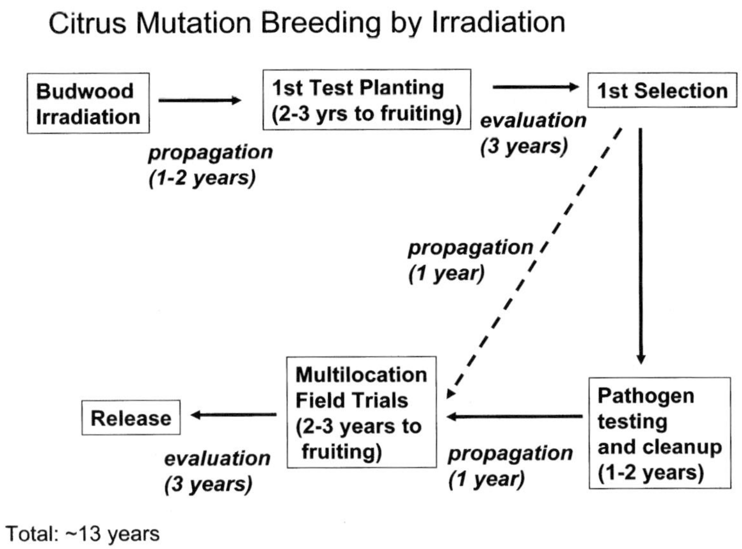

Fig. 16.2. Mutation breeding programme at the University of California, Riverside.

35 Gy of γ-rays from a ^{60}Co source. Individual buds were grafted on sour orange rootstocks. Buds from these trees were regrafted about 6–9 months later and the trees were field planted in 1987. Fruits were observed in 1989 and 1990. One tree, named selection 13/24, was distinct from 'Murcott' in having 5–7 seeds per fruit whereas 'Murcott' had about 20–25 seeds per fruit. This tree had a more compact growth (one-third to one-half the size of 'Murcott') and flowers with low pollen fertility. Productivity is reported to be similar to that of 'Murcott'. Details on field trial evaluation of this variety have not been reported.

References

Breto, M.P., Ruiz, C., Pina, J.A. and Asins, M.J. (2001) The diversification of *Citrus clementina* Hort. ex Tan., a vegetatively propagated crop species. *Molecular Phylogenetics and Evolution* 21, 285–293.

DuPlooy, C.P., Froneman, S., Human, C.F. and Koekemoer, P.J.J. (1993) Mutation breeding strategy for Citrus in South Africa. *Citrus Journal* 3, 26–28.

Feschotte, C., Jiang, N. and Wessler, S.R. (2002) Plant transposable elements: where genetics meets genomics. *Nature Reviews: Genetics* 3, 329–341.

Froneman, I.J., Breedt, H.J., Koekemoer, P.J.J. and Van Rensburg, P.J.J. (1996) Producing seedless citrus cultivars with gamma irradiation. *Proceedings of the International Society of Citriculture* 1, 159–163.

Gmitter, F.G., Jr (1995) Origin, evolution, and breeding of the grapefruit. *Plant Breeding Reviews* 13, 345–363.

Hearn, C.J. (1984) Development of seedless orange and grapefruit cultivars through seed irradiation. *Journal of the American Society for Horticultural Science* 109, 270–273.

Hearn, C.J. (1986) Development of seedless grapefruit cultivars through budwood irradiation. *Journal of the American Society for Horticultural Science* 111, 304–306.

Hensz, R.A. (1971) 'Star Ruby', a new deep-red-fleshed grapefruit variety with distinct tree characteristics. *Journal of the Rio Grande Valley Horticultural Society* 25, 54–58.

Hensz, R.D. (1977) Mutation breeding and the development of the 'Star Ruby' grapefruit. *Proceedings of the International Society of Citriculture* 2, 582–585.

Hensz, R.A. (1981) Bud mutations in citrus cultivars in Texas. In: *Proceedings of the International Society of Citriculture* pp. 89–91.

Lapins, K.O. (1983) Mutation breeding. In: Janick J. and Moore, J.N. (eds) *Methods in Fruit Breeding.* Purdue University, West Lafayette, Indiana, pp 74–99.

Maes, T., De Keukeleire, P. and Gerats, T. (1999) Plant tagnology. *Trends in Plant Science* 4, 90–96.

Predieri, S. (2001) Mutation induction and tissue culture in improving fruits. *Plant Cell Tissue and Organ Culture* 64, 185–210.

Van Harten, A.M. (1998) *Mutation Breeding: Theory and Practical Applications.* Cambridge University Press, Cambridge.

Vardi, A. and Spiegel-Roy, P. (1988) A new approach to selection for seedlessness. In: Goren, R. and Mendel, K. (eds) *Proceeding of the Sixth International Citrus Congress, Middle East,* Vol. 1. Balaban, Rehovot, Israel, pp. 131–134.

Vardi, A., Spiegel-Roy, P. and Elchanati, A. (1993) Mandarin tree named Mor. US Patent PP8,378.

Vardi, A., Elchanati, A., Frydman-Shani, A. and Neumann, H. (1996) Strategies and considerations in mandarin improvement programmes. *Proceedings of the International Society of Citriculture* 1, 109–112.

17 Shoot-tip Grafting *In vitro*: Impact in the Citrus Industry and Research Applications

L. Navarro and J. Juárez

Centro de Protección Vegetal y Biotechnología, Instituto Valenciano de Investigaciones Agrarias (IVIA), Apartado Oficial, 46113-Moncada, Valencia, Spain

Introduction

Citrus graft-transmissible diseases produced by viruses, viroids, some bacteria, spiroplasmas and phytoplasmas produce very important economic losses in most citrus-growing areas. In general, they cause decline, loss of vigour and short commercial life of trees, low yields and poor fruit quality, and they restrict the use of some rootstocks. Thus, they have the potential to become primary limiting factors of production. Only preventive measures are useful for control of graft-transmissible pathogens, such as use of tolerant or resistant germplasm, exclusion of potential diseases from the citrus area, and establishing orchards using pathogen-free, high quality nursery trees.

Frequently pathogen-free plants of many cultivars were not available and it was necessary to recover healthy plants from infected ones. The procedure used in the past to recover pathogen-free citrus plants was the selection of nucellar seedlings of polyembryonic or apomictic cultivars (Weathers and Calavan, 1959; Roistacher, 1977). Nucellar embryony is effective because most citrus pathogens are not transmitted through the process of embryogenesis, and because nucellar plants are produced by asexual embryogenesis *in vivo*, and thus have the same genotype as the mother plants that produced the seeds. The limitation of the procedure is that nucellar plants are juvenile, and consequently they are excessively vigorous, thorny and late in bearing, and they have to be grown for many years until these characters disappear and are acceptable for commercial propagation (Roistacher, 1977). In addition, seed transmission of psorosis-like pathogens and citrus leaf blotch virus has been reported (Guerri *et al.*, 2004). Recovery of pathogen-free nucellar plants of monoembryonic cultivars by nucellus culture *in vitro* has been reported (Rangan *et al.*, 1968; Juárez *et al.*, 1976), but most of the plants are not true to type (Navarro *et al.*, 1985).

Thermotherapy has been used to recover pathogen-free citrus plants without juvenile characters. However, this technique is not effective for the elimination of pathogens which replicate well under warm conditions, such as citrus exocortis, cachexia and other citrus viroids, citrus stubborn or citrus leaf blotch virus (dweet mottle) (Roistacher 1977).

In this situation, a method to recover citrus plants free of all graft-transmissible pathogens and without juvenile characters was required to produce healthy trees for commercial plantings. The initial attempts to accomplish this objective were made by shoot-tip culture *in vitro*, a technique widely used to recover healthy plants of many plant species, particularly herbaceous ones. However, all attempts failed (Murashige *et al.*, 1972), and still today it has not been possible to regenerate citrus plants from small shoot-tips cultured *in vitro* (L. Navarro and J. Juárez, unpublished results). Murashige *et al.* (1972) were able to recover a few citrus plants by grafting shoot-tips from diseased plants on young rootstock seedlings growing *in vitro*. Some of these plants were free of the exocortis viroid and did not have juvenile characters. This technique, named shoot-tip grafting *in vitro* (STG), was studied in detail by Navarro *et al.* (1975), who developed a routine procedure that allowed a 30–50% incidence of successful grafts that were transplanted to soil with over 95% survival. The resulting plants did not have juvenile characters, and most of them were free of graft-transmissible pathogens.

Several reviews of the STG technique have been published previously (Navarro 1981, 1988, 1992; Jonard, 1986). In this chapter, the different steps of this technique are described in detail, and the elimination of pathogens and its research and industry applications are reviewed.

Technique of Shoot-tip Grafting *In Vitro*

The standard procedure described by Navarro *et al.* (1975) is routinely used in most laboratories. It includes the following steps: rootstock preparation; scion preparation; grafting; culture *in vitro* of grafted plants; and transfer to soil.

Rootstock preparation

Seedlings recovered by seed germination *in vitro* are used as rootstocks. Seeds are peeled, removing both seed coats (Fig. 17.1a), surface sterilized and sown in 25 × 150 mm culture tubes containing 25 ml of the plant cell culture salt solution of Murashige and Skoog (1962), solidified with 1% Bacto agar. Cultures are incubated at 27°C in continuous darkness for 2 weeks (Fig. 17.1b).

The age of the rootstock has an important influence on grafting success (Navarro *et al.*, 1975). The highest rate of successful grafts using Troyer citrange (*Poncirus trifoliata* (L.) Raft. × *Citrus sinensis* (L.) Osb) as rootstock was obtained with 2-week-old seedlings. Most shoot-tips grafted on younger seedlings (1 week old) were covered with precocious callus formation by the rootstock, whereas most shoot-tips grafted on older rootstock (3 and 4 weeks old) dried, turned brown and died. These results suggest that grafting success depends on the degree of tissue differentiation of the rootstock, which is affected by age.

Age may be only indicative of seedling rate of development, which is influenced by embryo size, germination temperature, seed batch and storage period. Stem height and diameter are parameters more appropriate to determine the optimal stage of seedling for grafting. We routinely use Troyer citrange seedlings 3–5 cm tall with a diameter of 1.6–1.8 mm at the point of grafting. This size is reached in 12–16 days at 27°C.

Seedlings can be stored at 4°C for at least 2 weeks without any detrimental effect on grafting success. This allows an easy selection of uniform seedlings for experimental purposes.

Troyer citrange is the most commonly used rootstock for STG. It has the advantage that its trifoliate leaves serve as a morphological marker for identification of adventitious shoots produced by the rootstock. However, any rootstock graft compatible with the shoot-tip scion variety could be used for STG. At least the following rootstocks have been successfully used for STG:

Fig. 17.1.

Fig. 17.1.
Technique of shoot-tip grafting *in vitro*. (a) Troyer citrange seeds intact (top) and without integuments (bottom); (b) freshly sown seed (left) and seedling after 2 weeks (right); (c) 2-week-old seedling (left) and rootstock prepared for grafting (right), with epicotyl decapitated, cotyledons and root tip eliminated, and an inverted-T incision made at the top of the decapitated epicotyl; (d) young flushes produced by plants defoliated and grown in a growth chamber at 32°C; (e) flushes produced by a budstick cultured *in vitro* at 32°C; (f) vegetative young flush (left) and shoot prepared for surface disinfection (right); (g) histological section of a sweet orange vegetative flush; (h) freshly excised sweet orange shoot-tip (0.14 mm in height) on a razor blade sliver attached to a surgical handle; (i) detail of a freshly made graft of the shoot-tip inside the incision of the rootstock; (j) freshly made graft in a test tube; (k) histological section of a sweet orange shoot-tip 5 days after grafting on a Troyer citrange seedling; (l) sweet orange on Troyer citrange 3 weeks after grafting; (m) sweet orange on Troyer citrange 6 weeks after grafting; (n) Clausellina mandarin on Troyer citrange 2 years after micrografting.

Fig. 17.4.

Fig. 17.4.
Different types of shoot-tip grafting *in vitro* for propagation of elite genotypes.

Fig. 17.5.

Fig. 17.5.
Abnormal embryos produced after protoplast fusion experiments (a, b and c) and a grafted plant *in vitro* (d) that produced a viable plant.

Troyer and Carrizo citranges, Rough lemon (*C. jambhiri* Lush.), *P. trifoliata*, Alemow (*C. macrophylla* West), Etrog citron (*C. medica* L.), sour orange (*C. aurantium* L.), Rangpur lime (*C. limonia* Osb.), Volkamer lemon (*C. volkameriana* Pasq.), sweet orange (*C. sinensis* (L.) Osb.) and Cleopatra mandarin (*C. reshni* Hort. ex. Tan.). Differences in grafting success have been observed among these rootstocks, but detailed comparative studies have not been done. In our laboratory, we routinely use Troyer citrange as rootstock for STG.

The rootstock seedling is removed from the test tube under aseptic conditions and it is decapitated, leaving about 1.5 cm of the epicotyl. The root is cut to a length of 4–6 cm and the cotyledons and their axillary buds removed (Fig. 17.1c). Grafting can be done by several methods, but it is recommended to place the shoot-tip in an inverted-T incision (Navarro *et al.* 1975) (Fig. 17.1i). This is made by a 1 mm long vertical incision, starting at the point of decapitation, and a 1–2 mm wide horizontal cut. The cuts are done through the cortex, and the flaps of the incision are slightly lifted to expose the cortex.

Scion preparation

Shoot-tips can be excised from growing vegetative flushes of field or greenhouse plants, or from budwood cultured *in vitro*. Shoot-tips can also be excised from dormant buds or shoots from buds cultured *in vitro*, but these are not recommended sources for routine work, since the incidence of successful grafts and pathogen elimination was reduced in comparison with other sources (Navarro *et al.*, 1975, 1976).

Growing vegetative flushes from the different origins are the recommended source of shoot-tips for STG. Field trees are the most direct source of shoot-tips, but have the disadvantage of the seasonal flushing and the lower rate of elimination of some pathogens. Greenhouse plants have the advantage that flushes can be induced when necessary, avoiding the seasonal dependency of field trees. In addition, they can be subjected to a warm pre-treatment to improve the efficiency of elimination of some pathogens (Navarro *et al.*, 1980b). Infected citrus plants growing in pots in the greenhouse could be completely defoliated by hand and placed in a warm greenhouse, or in a growth chamber. As a routine procedure in our laboratory, the infected cultivars are propagated and grown in containers in a greenhouse at 18–25°C. They are defoliated by hand when needed, placed in a growth chamber at constant 32°C and exposed to 350 $\mu E\ m^{-2}\ s^{-1}$ illumination 16 h daily. After 8–12 days, new flushes are produced (Fig. 17.1d) and used as source of shoot-tips for STG.

Budwood cultured *in vitro* at constant 32°C and exposed 16 h daily to 80 $\mu E\ m^{-2}\ s^{-1}$ illumination in a culture medium containing the plant cell culture salt solution of Murashige and Skoog (1962) solidified with 1.2% Bacto agar is also an excellent source of shoot-tips for STG. New flushes are produced in 8–16 days (Fig. 17.1e) and used for isolation of shoot-tips for STG. This source is routinely used in our laboratory for plant introduction through the Citrus Quarantine Station (see below) (Navarro *et al.*, 1984) and is being increasingly used for local genotypes, since it is faster than propagation of plants in the greenhouse, and it gives a better grafting success (Navarro *et al.*, 2002).

Grafting

Flushes 3 cm long or shorter are collected to avoid the abscission stage. They are stripped of larger leaves, cut to about 1 cm long (Fig. 17.1f) and surface sterilized by immersion in a mild 0.25% sodium hypochlorite solution. Larger leaves of shoots are removed with the aid of dissecting instruments under a microscope, and shoot-tips composed of the apical meristem and three leaf primordia, and measuring 0.1–0.2 mm in length, are then excised with a razor blade sliver attached to a surgical handle (Fig. 17.1g and h). This shoot-tip does not have a vascular connection with the rest of the plant (Fig. 17.1g), and this explains why most pathogens are eliminated by shoot-tip grafting. The shoot-tip is placed inside the incision of the rootstock with its cut surface in contact with the cortex exposed by the horizontal cut of the incision (Fig. 17.1i) made at the top of the decapitated epicotyl (Navarro *et al.*, 1975).

The frequency of successful grafts increases with shoot-tip size, but the incidence of recovery of healthy plants decreases (Navarro *et al.*, 1976). Consequently, it is necessary to choose a size that will give a realistic degree of grafting success with the highest possible number of pathogen-free plants, since indexing is more expensive and time consuming than STG. In our laboratory, we routinely use shoot-tips composed of the apical meristem and three leaf primordia. The size, measured from the cut end to the tip of the largest leaf primordium, varies from 0.1 to 0.2 mm, depending on the citrus species.

Culture *in vitro* of grafted plants

Micrografted plants are cultured in a liquid nutrient medium composed of the plant cell culture salt solution of Murashige and Skoog (1962), modified White's vitamins and 75 g/l sucrose (Navarro *et al.*, 1975). The medium is distributed into 25 × 150 mm test tubes in 25 ml aliquots. A folded paper platform, perforated at its centre for insertion of the root portion of the rootstock, is placed in the nutrient solution (Fig. 17.1j). The cultures are kept at constant 27°C and exposed 16 h daily to 40–50 $\mu E\ m^{-2}\ s^{-1}$ illumination.

Several experiments have been done using auxins and cytokinins in the culture medium or as applications at the grafting point to increase the incidence of grafting success (Navarro *et al.*, 1975; Edriss and Burger, 1984; Starrantino *et al.*, 1986). The treatments gave either negative results or some small increase in grafting success, not confirmed in other laboratories. In practice, no growth regulators are used for STG.

Sucrose concentration in the nutrient medium of grafted plants has an important influence on grafting success and growth of plants (Navarro *et al.*, 1975). The highest rate of successful grafts was obtained with 75 g/l of sucrose. There was also a progressive increase in the number and size of new leaves arising from shoot-tips of successful grafts and in the production of new lateral roots by the rootstock with increasing sucrose concentration up to 75 and 100 g/l.

Histological studies showed that three days after grafting there was already some callus development between the rootstock and the shoot-tip, and five days after grafting the callus was completely developed at the graft union (Fig. 17.1k). Initiation of vascular differentiation was observed seven days after grafting, and there was a complete vascular connection between the shoot-tip and the rootstock 11 days after grafting (Li Xiancai, 1997).

Growing small leaves from the shoot-tip coming out from the incision can be observed about 3–4 weeks after grafting (Fig. 17.1l), and 4–6 weeks after grafting the successful grafts already have 2–4 expanded leaves and they can be transplanted to soil (Fig. 17.1m).

In the Citrus Variety Improvement Programme in Spain (CVIPS), carried out in our laboratory, this standard procedure has been used with many genotypes of at least 43 *Citrus* species, and with *Poncirus* and *Fortunella* species. The overall percentage of grafting success has been 38%, but differences were found between species and shoot-tip sources (Table 17.1). Comparative studies using the same genotype have not been done, but shoot-tips excised from budwood cultured *in vitro* gave a slightly higher incidence of grafting success than shoot-tips from plants growing in a growth chamber, albeit that this is species dependent. In the case of satsumas the grafting success has been much higher with shoot-tips isolated from budwood cultured *in vitro* (Table 17.1). Overall, more than 40% graft-

Table 17.1. Average grafting success (%) of shoot-tip grafting *in vitro* carried out in the Citrus Variety Improvement Program in Spain between 1985 and 2003.

Group of genotypes	Shoot-tips excised from plants growing in growth chambers	Shoot-tips excised from budwood cultured *in vitro*
All genotypes	33.6	43.5
Sweet oranges	40	40
Clementines	46	48.2
Satsumas	18.1	39.8
Other mandarins	39.3	45.8
Lemons	19.4	21.2
Grapefruit	–	58.3

ing success was usually obtained with most commercially grown species using budwood cultured *in vitro* as the source of shoot-tips, with the exception of lemons that gave only about 20% success.

Transplanting to soil

Scions of successful grafts should have at least two expanded leaves before being transplanted to soil. This stage is usually reached 4–6 weeks after grafting (Fig. 17.1n). Micrografted plants are transferred to pots containing a steam-sterilized artificial soil mix suitable to grow citrus. Pots are enclosed in polyethylene bags that are closed with rubber bands and placed in a shaded area of a temperature-controlled greenhouse set at 18–25°C. After 8–10 days, the bags are opened, and after another 8–10 days, the bags are removed and the plants are grown under regular greenhouse conditions (Fig. 17.1l). Alternatively, the epicotyl of the micrografted plants can be regrafted on vigorous seedlings growing in the greenhouse (De Lange, 1978). With both procedures, we usually obtain over 95% survival, but growth is faster with the latter method.

Plants recovered by STG do not have juvenile characters, if the shoot-tips are excised from adult plants. They usually flower and set fruits within two years from grafting (Fig. 17.1l). Several thousands plants have been obtained by STG in different laboratories, and all available data indicate that they are true to type.

Applications of STG

Control of graft-transmissible pathogens

To control diseases caused by graft-transmissible pathogens, it is necessary to use healthy and high quality trees in the new plantings. The production of these trees requires the establishment of three different but related programmes: sanitation, quarantine and certification (Navarro, 1993). Sanitation programmes have the objective to recover healthy plants from local varieties, and quarantine programmes have the objective of importing foreign varieties avoiding introduction of new pests and diseases. STG plays a key role in these two programmes that produce healthy plants for the certification programmes, which have the objective to guarantee that the sanitary status of the initial material is maintained during the process of commercial propagation at the nurseries. Figure 17.2 shows a diagrammatic representation of these programmes.

Sanitation programmes

The most important application of STG is the recovery of healthy plants. It has been effective to recover plants free from all pathogens assayed, including the causal agents of the following diseases: cachexia, canker, concave gum, cristacortis, dweet mottle, exocortis, huanglonbing (ex greening), impietratura, infectious variegation-crinkly leaf, leaf blotch, psorosis A and B,

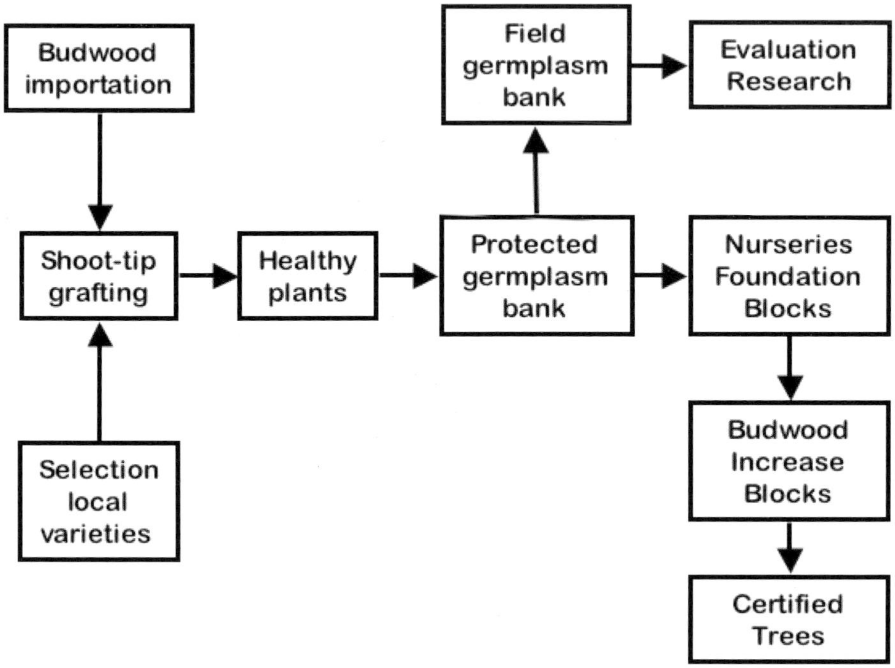

Fig. 17.2. Diagrammatic representation of the citrus sanitation, quarantine and certification programmes.

ringspot, rough lemon incompatibility, seedling yellows-tristeza, stubborn, tatter leaf, tristeza and yellow vein (Navarro *et al.*, 1975, 1976, 1980a, b, 1981, 1988, 1991, 2002; Roistacher *et al.*, 1976; Roistacher and Kitto, 1977; Arroyo, 1984; Koizumi 1984; Su and Chu, 1984). These include diseases produced by viruses, viroids, mycoplasma, phloem-restricted bacteria and graft-transmissible agents of unknown nature.

The most important factors that influence elimination of citrus pathogens by STG are the pathogen itself, the size of the shoot-tip and the growing temperature of the shoot-tip source plants. Some pathogens are very easy to eliminate, such as citrus viroids (cachexia, exocortis), infectious variegation-crinkly leaf, stubborn, huanglonbing (ex greening), tristeza and vein enation, and almost 100% of the plants recovered by STG are free of these diseases. Seedling yellows-tristeza and yellow vein are easy to eliminate by STG, and around 80% of micrografted plants are free of these diseases. Finally, concave gum, cristacortis, dweet mottle, leaf blotch, impietratura, psorosis, ringspot and tatter leaf are difficult to eliminate, and usually fewer than 25% of the micrografted plants are free of these diseases. These data were obtained when shoot-tips were collected from field trees or plants growing in a greenhouse at 18–25°C.

Shoot-tip size has an important influence on the incidence of healthy plants recovered by STG. It was shown that increasing shoot-tip size resulted in higher incidence of successful grafts, but with an important decrease of the incidence of healthy plants (Navarro *et al.*, 1976). The relatively low frequency of healthy plants obtained in some laboratories is probably due to the use of larger shoot-tips. As mentioned above, a shoot-tip composed of the apical meristem and three leaf primordia measuring 0.1–0.2 mm gives a realistic frequency of successful grafts and healthy plants, and is recommended for routine work.

Fig. 17.6.

Fig. 17.6.
Regeneration of plants from irradiated shoot-tips. (a) Irradiated shoot-tips of Nules clementine in a Petri dish. (b) Growing shoot from an irradiated shoot-tip after micrografting. (c) Original Nules clementine (left) and selected irradiated Nulessin with reduced fertility (Photo: Dr M.J. Asins).

Fig. 17.7.

Fig. 17.7.
Regeneration of haploid Nules clementine plants. (a and b) Abnormal proliferation of tissues of embryos from aborted seeds produced after *in situ* parthenogenesis. (c) Developing graft. (d) Flowering haploid Nules clementine. (e) Flowers from diploid (left) and haploid (right) Nules clementine.

Fig. 17.8.

Fig. 17.8.
Regeneration of clementine plants from somaclonal variation experiments. (a) Internode cultured *In vitro* from adult Nules clementine. (b) Micrografted plant recovered from the small buds produced by the internodes. (c) Field evaluation of potential somaclonal variants.

Fig. 17.9.

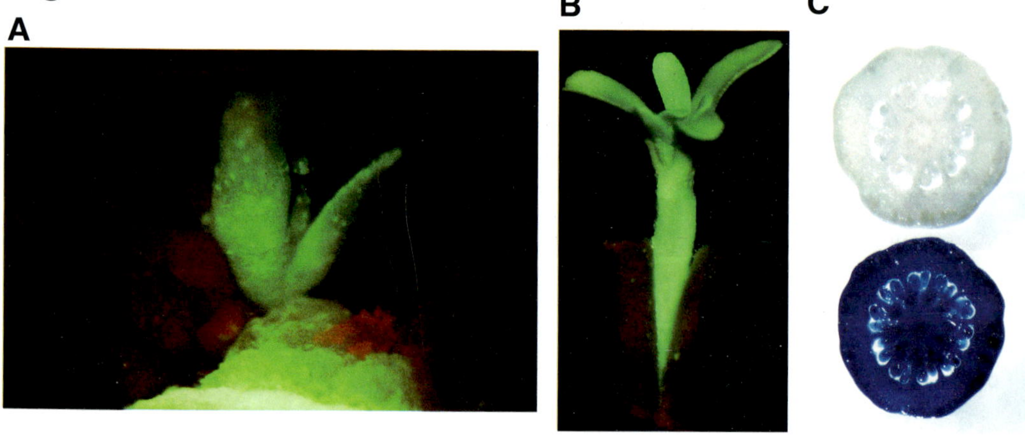

Fig. 17.9.
Regeneration of transgenic plants, (a) Transgenic shoot expressing *gfp*. (b) Transgenic shoot grafted on an untransformed Troyer citrange rootstock. (c) Pineapple sweet orange fruits from adult control (top) and transformed plants (bottom).

The incidence of recovery of plants free from diseases difficult to eliminate increased by growing the shoot-tip source plants under warm conditions (Navarro *et al.*, 1980b; Koizumi, 1984). As mentioned above, in the routine procedure in the CVIPS, shoot-tips are isolated from plants growing in a growth chamber at constant 32°C or from budwood cultured *in vitro* at 32°C. With this procedure, more than 90% of micrografted plants are usually free of pathogens, including those difficult to eliminate (Navarro *et al.*, 1988, 2002). The only exception is citrus leaf blotch virus, that is eliminated in fewer than 50% of micrografted plants.

STG is the best available technique to recover pathogen-free citrus plants for sanitation programmes, since it is possible to recover plants free of pathogens that cannot be eliminated by thermotherapy, the resulting plants do not have juvenile characters, as occurs with plants recovered by nucellar embryony, and all recovered plants are true to type. STG is being used in citrus sanitation programmes in all the major citrus-growing countries.

In Spain, there is a very extensive citrus sanitation programme based on STG (Navarro, 1976; Navarro *et al.*, 1980a, 1981, 1988, 2002). About 120 million healthy certified nursery trees, originally recovered by this technique, have already been planted in the field. These healthy trees have 15–20% more production of fruit of higher quality that the original infected trees. Diseases produced by graft-transmissible diseases, that were the main limitation of the Spanish citrus industry in the past, do not now produce damage in the new plantings.

Quarantine procedures

Movement of citrus species and varieties between different citrus areas for commercial and scientific purposes is often desirable. However, uncontrolled importation of budwood carries the risk of introducing new pests and pathogens that in some instances may be devastating or may cause very important economic damage. This risk may be minimized by controlled introduction through quarantine stations, that have the objective of importing foreign varieties while avoiding the introduction of new pests and diseases that may be carried in the original material. In citrus there are two different quarantine procedures that can be safely used for importation of plant material (Frison and Taher, 1991)

The classic method consists of propagating the imported budwood in quarantine greenhouses located far away from citrus-growing areas. Then the newly propagated plants can be indexed or submitted directly to STG followed by indexing. This procedure requires the availability of facilities and trained personnel on citrus pests, diseases and cultural practices in areas not directly involved with citrus research. It is used in some countries with a long tradition in quarantine that have central facilities and personnel for importation of plant material of several crops, but it is very expensive and difficult to establish only for citrus in most countries.

An alternative citrus tissue culture procedure was developed for safe introduction of citrus genotypes (Navarro *et al.*, 1984, 1991) that has been proved to be very efficient in excluding citrus pests and diseases (Fig. 17.3). Budsticks that are received from another country are thoroughly cleaned and surface sterilized and then cultured *in vitro* at 32°C in a growth chamber as described above to induce the sprouting of lateral buds and formation of flushes (Fig. 17.1e) from which shoot-tips are isolated and micrografted *in vitro*. The only material really imported is a small shoot-tip that usually is free of pests and pathogens. This process includes many different controls (Fig. 17.3) and minimizes the possibility of escape of harmful pathogens, and it allows rapid processing of new entries.

This tissue culture method has several advantages over the traditional quarantine method. Pests and diseases that might be in the original material are eliminated at the early stages of introduction, thus shortening the quarantine period. With tissue cul-

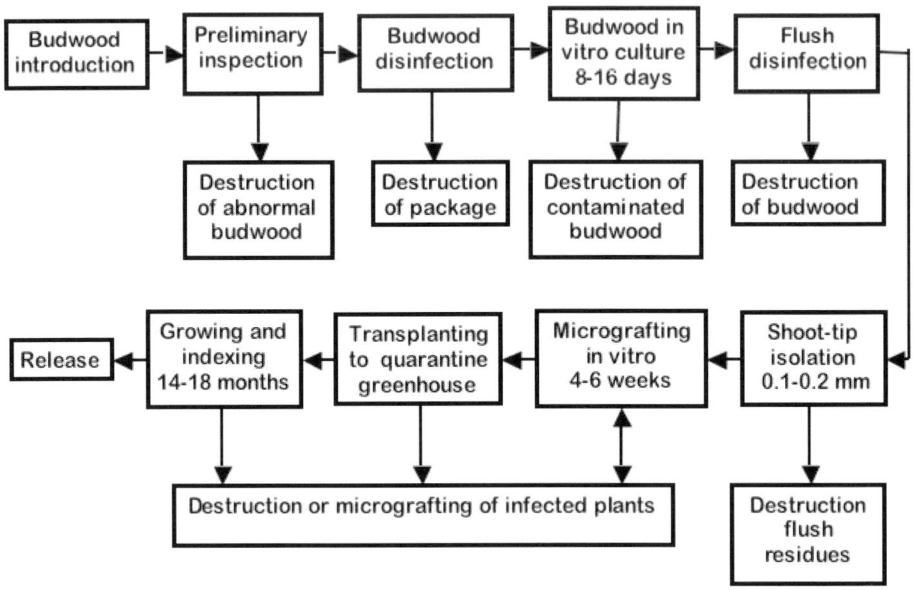

Fig. 17.3. Diagrammatic representation of the procedure of *in vitro* quarantine of citrus.

ture, the test tubes serve as a substitute for the greenhouses located in isolated areas, and thus the quarantine station may be located at citrus research stations. At many of these stations STG is being used for sanitation of local cultivars, and the required facilities and personnel are usually available. Consequently the tissue culture procedure can be easily established in many countries for safe import of citrus vegetative material. This method is recommended for exchange of citrus germplasm (Frison and Taher, 1991) and has been legally accepted in several areas, such as the European Union. Due to these advantages, the tissue culture method is recommended for the safe importation of citrus budwood. In Spain, it has been used successfully to import over 200 varieties from different citrus areas.

Research applications

In addition to research with graft-transmissible pathogens, STG is becoming a very useful technique for production, propagation and regeneration of elite genotypes in several areas of research. *In vitro* grafting for this purpose may be done using larger shoots (at least up to 1 cm) using different types of incisions (Fig. 17.4) with close to 100% grafting success. Some research applications of STG are described below.

Regeneration of somatic hybrids

In protoplast fusion experiments, abnormal embryos are very often produced. These include multiple fasciated cotyledons, embryos that only produce shoots, germinating embryos without a good vascular connection between shoot and root, and abnormal shoot proliferation, among others (Fig. 17.5) (Olivares-Fuster, 1988). These embryos do not produce plants that can be established in the greenhouse, thus reducing the efficiency of recovery of somatic hybrids and losing potentially valuable genotypes. In our laboratory, we routinely graft *in vitro* shoots produced by these embryos to recover plants that are established in the greenhouse with high efficiency (Fig. 17.5).

Regeneration of plants from irradiated shoots

Irradiation is used in citrus improvement programmes in attempts to reduce the number of seeds produced by high quality genotypes, particularly with mandarins. However, in many cases, unstable chimeras are produced that revert to the original variety after some cycles of propagation. In assays to produce seedless clementines, shoot-tips were isolated, placed facing upwards in a Petri dish with agar media and irradiated. The assumption was that cells of the exposed meristem would be mutated, producing stable plants. After irradiation, whole plants were recovered by STG, transplanted to soil and evaluated (Fig. 17.6). A relatively high number of apparently stable mutants have been produced and a new variety Nulessín selected and released (Asins *et al.*, 2002).

Regeneration of haploid plants

Haploid plants have interesting applications in citrus genetics and genomics. They can be recovered by *in situ* parthenogenesis after pollination with irradiated pollen, which induce aborted seeds that in some cases contain haploid embryos, that after *in vitro* culture produce a few plants very difficult to establish in the greenhouse (Ollitrault *et al.*, 1996). We have used this approach in our laboratory, and haploid embryos produced abnormal clusters of proliferating tissues that did not allow the regeneration of plants. However, grafting *in vitro* some of these shoot-like tissues allowed us to regenerate several Nules clementine plants that were successfully established in the greenhouse, where they are even flowering (Fig. 17.7).

Production of stable tetraploid plants of monoembryonic genotypes

Tetraploid plants of monoembryonic genotypes are very important for use as female parents in triploid mandarin breeding programmes. However, this type of genotype is not available. They could be recovered by colchicine treatment of budwood, but usually only unstable chimeras without any value for breeding are produced. We have attempted the recovery of tetraploid plants by treatment of shoot-tips with colchicine solution and regeneration of plants by STG or by adding a drop of colchicine solution on a shoot-tip two weeks after grafting *in vitro* (Juárez *et al.*, 2004). The objective was to induce chromosome duplication in the exposed cells of the meristem. We were able to recover stable tetraploid plants of Nules, Fina and Marisol clementines and Moncada mandarin that are now the most widely used female parents in our triploid breeding programme (Navarro *et al.*, 2003).

Regeneration of plants from somaclonal variation experiments of adult material

Somaclonal variation is an interesting approach for citrus improvement that is being used for this purpose in some laboratories (Grosser *et al.*, 2003). Clementines are a good candidate for somaclonal variation studies, since they are genetically unstable and frequently produce budsports in the field. We have attempted to recover somaclonal variants through the process of adventitious organogenesis from adult internode tissues, but these only produce very tiny buds that cannot be rooted to produce plants (Fig. 17.8). However, grafting these buds *in vitro* allowed us to recover more than 450 plants that are under field evaluation (Fig. 17.8).

Regeneration of transgenic plants

One of the major limitations of citrus transformation was the difficulty in recovery of plants from transgenic shoots (Peña *et al.*, 2003; see also Chapter 15). With many genotypes, the rooting efficiency of transgenic shoots is very low and rooting of shoots from adult material is almost impossible. The use of STG to regenerate plants from transformed shoots or buds has become a routine procedure in most laboratories, allowing an important increase of the efficiency of the genetic transformation protocols (Fig. 17.9).

Conclusions

The technique of STG is having a very important impact in the citrus industry worldwide. Some graft-transmissible diseases are disappearing from commercial plantings in several countries, there is a diminution of risks of introducing exotic pests and diseases with exchange of genotypes, and there are reduced limitations of rootstock use as a consequence of variety infection. In addition, there is an increase of production and fruit quality as a consequence of control of graft-transmissible diseases.

Furthermore, STG is increasingly being used as a tool in different research areas to regenerate elite genotypes or produce plants that cannot be recovered by other means. This includes genetic transformation, somatic hybridization, recovery of haploid and tetraploid plants, and somaclonal variation.

References

Asins, M.J., Juárez, J., Pina, J.A., Puchades, J., Carbonell, E.A. and Navarro, L. (2002) Nulessín, una nueva clementina. *Levante Agrícola* 359, 36–40.

Arroyo, L.E. (1984) Elimination of citrus ringspot virus by shoot-tip grafting. In: Garnsey, S.M., Timmer, L.W. and Dodds, J.A. (eds) *Proceedings of the Ninth Conference of the International Organization of Citrus Virologists*. IOCV, Riverside, California, pp. l77–179.

De Lange, J.H. (1978) Shoot-tip grafting – a modified procedure. *Citrus Subtropical Fruit Journal* 539, 13–15.

Edriss, M.H. and Burger, D.W. (1984) Micrografting shoot-tip culture of citrus on three trifoliate rootstocks. *Science Horticulturae* 23, 255–259.

Frison, E. and Taher, M.M. (eds) (1991) *FAO/IBPGR Technical Guidelines for the Safe Movement of Citrus Germplasm*. Food and Agriculture Organization of the United Nations/International Board for Plant Genetic Resources, Rome.

Grosser, J.W., Chandler, J.L. and Gmitter, F.G., Jr (2003) Development of improved sweet oranges via somaclonal variation. *Proceedings of the International Society of Citriculture* 1, 42–45.

Guerri, J., Pina, J.A, Vives, M.C., Navarro, L. and Moreno, P. (2004) Seed transmission of *Citrus leaf blotch virus*: implications in quarantine and certification programs. *Plant Disease* 88, 906.

Jonard, R. (1986) Micrografting and its application to tree improvernent. In: Bajaj, Y.P.S. (ed.) *Biotechnology in Agriculture and Forestry, Vol 1: Trees I*. Springer, Berlin, pp. 31–48.

Juárez, J., Aleza, P., Olivares-Fuster, O. and Navarro, L. (2004) Recovery of tetraploid clementine plants (*citrus. clementine* hort. ex Tan.) by in vitro colchicine treatment of shoot tips. *Proceedings of the International Society of Citriculture 1*, 151–154.

Juárez, J., Navarro, L. and Guardiola, J.L. (1976) Obtention de plants nucellaires de divers cultivars de clémentiniers au moyen de la culture de nucelle *in vitro*. *Fruits* 31, 751–762.

Koizumi, M. (1984) Elimination of tatter leaf-citrange stunt virus from satsuma mandarin by shoot-tip grafting following pre-heat treatment. In: Garnsey, S.M., Timmer, L.W. and Dodds, J.A. (eds) *Proceedings of the Ninth Conference of the International Organization Citrus Virologists*. IOCV, Riverside, California, pp. 229–233.

Li Xiancai (1997) Cultivo de Tejidos *In Vitro* de Especies Relativas de los Cítricos. PhD thesis, Universidad Politécnica de Valencia, Valencia, Spain.

Murashige, T. and Skoog, F. (1962) A revised medium for rapid growth and bioassays with tobacco tissue cultures. *Physiologia Plantarum* 15, 473–497.

Murashige, T., Bitters, W.P., Rangan, T.S., Nauer, E.M., Roistacher, C.N. and Holliday, B.P. (1972) A technique of shoot apex grafting and its utilization towards recovering virus-free citrus clones. *HortScience* 7, 118–119.

Navarro, L. (1976) The citrus variety improvement program in Spain. In: Calavan, E.C. (ed.) *Proceedings of the Seventh Conference of the International Organization of Citrus Virologists*. IOCV, Riverside, California, pp. 198–202.

Navarro, L. (1981) Citrus shoot-tip grafting *in vitro* (STG) and its applications: a review. *Proceedings of the International Society of Citriculture* 1, 452–456.

Navarro, L. (1988) Application of shoot-tip grafting *in vitro* to woody species. *Acta Horticulturae* 227, 43–55.

Navarro, L. (1992) Citrus shoot-tip grafting *in vitro*. In: Bajaj, J.P.S. (ed.) *Biotechnology in Agriculture and Forestry, Vol. 18. High-tech and Micropropagation*. Springer-Verlag, Berlin, pp. 327–338.

Navarro, L. (1993) Citrus sanitation, quarantine and certification programs. In: Moreno, P., da Graca, J.V. and Timmer, L.W. (eds) *Proceedings of the 12th Conference of the International Organization of Citrus Virologists*. IOCV, Riverside, California, pp. 383–391.

Navarro, L. and Juárez, J (1977) Elimination of citrus pathogens in propagative budwood. II. *In vitro* propagation. *Proceedings of the International Society of Citriculture* 3, 973–987.

Navarro, L., Roistacher, C.N. and Murashige, T. (1975) Improvement of shoot-tip grafting *in vitro* for virus-free citrus. *Journal of the American Society for Horticultural Sciences* 100, 471–479.

Navarro, L., Roistacher, C.N. and Murashige, T. (1976) Effect of size and source of shoot-tips on psorosis-A and exocortis content of navel orange plants obtained by shoot-tip grafting *in vitro*. In: Calavan, E.C. (ed.) *Proceedings of the Seventh Conference of the International Organization of Citrus Virologists*. IOCV, Riverside, California, pp. 194–197.

Navarro, L., Ballester, J.F., Juárez, J., Pina, J.A., Arregui, J.M., Bono, R., Fernandez de Cordova, L. and Ortega, C. (1980a) The citrus variety improvement program in Spain (CVIPS) after four years. In: Calavan, E.C., Garnsey, S.M. and Timmer, L.W. (eds) *Proceedings of the Eighth Conference of the International Organization of Citrus Virologists*. IOCV, Riverside, California, pp. 289–294.

Navarro, L., Juárez, J., Ballester, J.F. and Pina, J.A. (1980b) Elimination of some citrus pathogens producing psorosis-like leaf symptoms, by shoot-tip grafting *in vitro*. In: Calavan, E.C., Garnsey, S.M. and Timmer, L.W. (eds) *Proceedings of the Eighth Conference of the International Organization of Citrus Virologists*. IOCV, Riverside, California, pp. 162–166.

Navarro, L., Ballester, J.F., Juárez, J., Pina, J.A., Arregui, J.M. and Bono, R. (1981) Development of a program for disease-free citrus budwood in Spain. *Proceedings of the International Society of Citriculture* 1, 452–456.

Navarro, L., Juárez, J., Pina, J.A., Ballester, J.F. (1984) The citrus quarantine station in Spain. In: Garnsey, S.M., Timmer, L.W. and Dodds, J.A. (eds) *Proceedings of the Ninth Conference of the International Organization of Citrus Virologists*. IOCV, Riverside, California, pp. 365–370.

Navarro, L., Ortiz, J.M. and Juárez, J. (1985) Aberrant citrus plants obtained by somatic embryogenesis *in vitro*. *HortScience* 20, 214–215.

Navarro, L., Juárez, J., Pina, J.A., Ballester, J.F. and Arregui, J.M. (1988) The citrus variety improvement program in Spain after 11 years. In: Timmer, L.W., Garnsey, S.M. and Navarro, L. (eds) *Proceedings of the Tenth Conference of the International Organization of Citrus Virologists*. IOCV, Riverside, California, pp. 400–406.

Navarro, L., Civerolo, E.L., Juárez, J. and Garnsey, S.M. (1991) Improving therapy methods for citrus germplasm exchange. In: Brlansky, R.H., Lee, R.F. and Timmer L.W. (eds) *Proceedings of the 11th Conference of the International Organization of Citrus Virologists*. IOCV, Riverside, California, pp. 400–408.

Navarro, L., Pina, J.A., Juárez, J., Ballester-Olmos, J.F., Arregui, J.M., Ortega, C., Navarro, A., Duran-Vila, N., Guerri, J., Moreno, P., Cambra, M. and Zaragoza. S. (2002) The citrus variety improvement program in Spain in the period 1975–2001. In: Duran-Vila, N., Milne, R.G. and da Graça, J.V. (eds) *Proceedings of the 15th Conference of the International Organization of Citrus Virologists*. IOCV, Riverside, California, pp. 306–316.

Navarro, L., Juárez, J., Aleza, P. and Pina, J.A. (2003) Recovery of triploid seedless mandarin hybrids from 2n × 2n and 2n × 4n crosses by embryo rescue and flow cytometry. In: Vasil, I.K. (ed.) *Plant Biotechnology 2002 and Beyond*. Kluwer Academic Publishers, Dordrecht, pp. 541–544.

Olivares-Fuster, O. (1988) Fusión de Protoplastos de Cítricos. PhD thesis, Universidad Politécnica de Valencia, Valencia, Spain.

Ollitrault, P., Allent, V. and Luro, F. (1996) Production of haploid plants and embryogenic calli of clementine (*Citrus reticulata* Blanco) after *in situ* parthenogenesis induced by irradiated pollen. *Proceedings of the International Society of Citriculture* 2, 913–917.

Peña, L., Cervera, M., Ghorbel, R., Dominguez, A., Fagoaga, C., Juárez, J., Pina, J.A. and Navarro, L. (2003) Transgenic citrus. In: Singh, R.P. and Jaiwal, P.K. (eds) *Plant Genetic Engineering Vol 3, Improvement of Commercial Plants*. Sci-Tech Publishing LLC, Houston, Texas, pp. 261–282.

Rangan, T.S., Murashige, T. and Bitters, W.P. (1968) *In vitro* initiation of nucellar embryos in monoembryonic citrus. *HortScience* 3, 226–278.

Roistacher, C.N. (1977) Elimination of citrus pathogens in propagative budwood. I. Budwood selection, indexing and thermotherapy. *Proceedings of the International Society of Citriculture* 3, 965–972.

Roistacher, C.N. and Kitto, S.L. (1977) Elimination of additional citrus viruses by shoot-tip grafting *in vitro*. *Plant Disease Reports* 61, 594–596.

Roistacher, C.N., Navarro, L. and Murashige, T. (1976) Recovery of citrus selections free of several viruses, exocortis viroid and spiroplasma citri by shoot-tip grafting *in vitro*. In: Calavan, E.C. (ed.) *Proceedings of the Seventh Conference of the International Organization of Citrus Virologists*. IOCV, Riverside, California, pp. 186–193.

Starrantino, A., Guo Zhi-yong and Caruso, A. (1986) Influenza di alcuni fitoregolatori sul l'attecchimento dei microinnesti degli agrumi. *Rivista Ortoflorofrutticoltura Italiana* 70, 117–126.

Su, H.J. and Chu J.Y. (1984) Modified technique of citrus shoot-tip grafting and rapid propagation method to obtain citrus budwood free of citrus viruses and likubin organism. *Proceedings of the International Society of Citriculture* 1, 332–334.

Weathers, L.G. and Calavan, E.C. (1959) Nucellar embryony as a means of freeing citrus clones of viruses. In: Wallace, J.M. (ed.) *Citrus Virus Diseases*. University of California, Division of Agricultural Sciences, Berkeley, pp. 197–200.

Index